Post-Shrinkage Strategies in Statistical and Machine Learning for High Dimensional Data

This book presents some post-estimation and predictions strategies for the host of useful statistical models with applications in data science. It combines statistical learning and machine learning techniques in a unique and optimal way. It is well-known that machine learning methods are subject to many issues relating to bias, and consequently the mean squared error and prediction error may explode. For this reason, we suggest shrinkage strategies to control the bias by combining a submodel selected by a penalized method with a model with many features. Further, the suggested shrinkage methodology can be successfully implemented for high-dimensional data analysis. Many researchers in statistics and medical sciences work with big data. They need to analyze this data through statistical modeling. Estimating the model parameters accurately is an important part of the data analysis. This book may be a repository for developing improve estimation strategies for statisticians. This book will help researchers and practitioners for their teaching and advanced research and is an excellent textbook for advanced undergraduate and graduate courses involving shrinkage, statistical, and machine learning.

- The book succinctly reveals the bias inherited in machine learning method and successfully provides tools, tricks, and tips to deal with the bias issue.
- Expertly sheds light on the fundamental reasoning for model selection and post-estimation using shrinkage and related strategies.
- This presentation is fundamental because shrinkage and other methods appropriate for model selection and estimation problems, and there is a growing interest in this area to fill the gap between competitive strategies.
- Application of these strategies to real-life data set from many walks of life.
- Analytical results are fully corroborated by numerical work, and numerous worked examples are included in each chapter with numerous graphs for data visualization.
- The presentation and style of the book clearly makes it accessible to a broad audience. It offers rich, concise expositions of each strategy and clearly describes how to use each estimation strategy for the problem at hand.
- This book emphasizes that statistics/statisticians can play a dominant role in solving Big Data problems and will put them on the precipice of scientific discovery.
- The book contributes novel methodologies for HDDA and will open a door for continued research in this hot area.
- The practical impact of the proposed work stems from wide applications. The developed computational packages will aid in analyzing a broad range of applications in many walks of life.

Post-Shrinkage Strategies in Statistical and Machine Learning for High Dimensional Data

Syed Ejaz Ahmed
Feryaal Ahmed
Bahadir Yüzbaşi

CRC Press
Taylor & Francis Group
Boca Raton London New York

CRC Press is an imprint of the
Taylor & Francis Group, an **informa** business
A CHAPMAN & HALL BOOK

Designed cover image: © Askhat Gilyakhov

First edition published 2023
by CRC Press
6000 Broken Sound Parkway NW, Suite 300, Boca Raton, FL 33487-2742

and by CRC Press
4 Park Square, Milton Park, Abingdon, Oxon, OX14 4RN

CRC Press is an imprint of Taylor & Francis Group, LLC

© 2023 Syed Ejaz Ahmed, Feryaal Ahmed and Bahadir Yüzbaşi

Reasonable efforts have been made to publish reliable data and information, but the author and publisher cannot assume responsibility for the validity of all materials or the consequences of their use. The authors and publishers have attempted to trace the copyright holders of all material reproduced in this publication and apologize to copyright holders if permission to publish in this form has not been obtained. If any copyright material has not been acknowledged, please write and let us know so we may rectify in any future reprint.

Except as permitted under U.S. Copyright Law, no part of this book may be reprinted, reproduced, transmitted, or utilized in any form by any electronic, mechanical, or other means, now known or hereafter invented, including photocopying, microfilming, and recording, or in any information storage or retrieval system, without written permission from the publishers.

For permission to photocopy or use material electronically from this work, access www.copyright.com or contact the Copyright Clearance Center, Inc. (CCC), 222 Rosewood Drive, Danvers, MA 01923, 978-750-8400. For works that are not available on CCC please contact mpkbookspermissions@tandf.co.uk

Trademark notice: Product or corporate names may be trademarks or registered trademarks and are used only for identification and explanation without intent to infringe.

ISBN: 978-0-367-76344-2 (hbk)
ISBN: 978-0-367-77205-5 (pbk)
ISBN: 978-1-003-17025-9 (ebk)

DOI: 10.1201/9781003170259

Typeset in CMR10
by KnowledgeWorks Global Ltd.

Publisher's note: This book has been prepared from camera-ready copy provided by the authors.

Dedicated in loving memory to Don Fraser and Kjell Doksum.

Contents

Preface

The discipline of statistical science is ever changing and evolving from the investigation of classical finite-dimensional data to high-dimensional data analysis. We are commonly encountering data sets containing huge numbers of predictors where in some cases the number of predictors exceeds the number of sample observations. Many modern scientific investigations require the analysis of enormous, complex, high-dimensional data far beyond the classical statistical methodologies developed decades ago. For example, data from genomic, proteomic, spatial-temporal, social network, and many other disciplines fall into this category. Modeling and making statistical sense of high-dimensional data is a challenging problem. A range of different models with increasing complexity can be considered, and a model that is optimal in some sense needs to be selected from a set of candidate models. Simultaneous variable selection and model parameter estimation play a central role in such investigations. There is a massive literature on variable selection and penalized regression methods that are currently available. A plethora of interesting and useful developments have been recently published in scientific and statistical journals. This area of research continues to grow in the foreseeable future.

The application of regression models for high-dimensional data analysis is challenging and rewarding task. Regularization/penalization methods have attracted much attention in this arena. Penalized regression is a technique for mitigating the difficulties that arise from collinearity and high dimensionality. This approach inherently incurs an estimation bias while reducing the variance of the estimator. A tuning parameter is needed to adjust the penalization effects so that a balance between model parsimony and goodness-of-fit can be achieved. Different forms of penalty functions have been studied intensively over the last three decades. However, development in this area is still in its infancy. For example, methods may require the assumption of sparsity in the model where most coefficients are exactly zero and the nonzero coefficients are big enough to be separated from the zero ones. There are situations where noise cannot easily be separated from the signal, especially in the presence of weak signals. Furthermore, penalty estimators are not efficient when the number of variables is extremely large compared to the sample size. To mitigate these problems, I suggested the shrinkage strategy, which combines a model containing strong signals with a model with weak signals. One of the goals of this book is to improve the understanding of high-dimensional modeling from an integrative perspective and to bridge the gap among statisticians, computer scientists, applied mathematicians and others in understanding each other's tools. This book highlights and expands the breadth of the existing methods in high-dimensional data analysis and their potential to advance both statistical learning and machine learning for future research in the theory of shrinkage strategies.

This book is intended to provide Stein-type shrinkage strategies in a variety of regression modeling problems. Since the inception of the shrinkage strategy there has been much progress in developing improved estimation methods, both in terms of theoretical developments and their applications in solving real-life problems. LASSO and related penalty-type estimation have become popular in problems related to variable selection and predictive modeling. The book focuses on the shrinkage strategy and provides a unified approach for estimation and prediction when many weak signals in the regression model are under consideration. The shrinkage method considered in this book relies on prior information of inactive predictors when estimating the coefficients of active predictors. Conversely, the penalty methods identify inactive variables by producing zero solutions for their associated regression coefficients. Different kinds of shrinkage estimators have been proposed in situations where the number of predictors dominates the sample size. In this book we emphasize the applications of the shrinkage strategy and in each chapter a different regression model is considered. In each chapter, low and high-dimensional shrinkage estimation strategies are proposed to improve the prediction performance based only on a predefined subset. The asymptotic property of the shrinkage estimator is developed, and its relative performance is critically assessed with respect to the full model and submodel estimators using a quadratic loss function. The results in the chapter show both analytically and numerically that the high-dimensional shrinkage estimator performs better than the full model estimator and, in many instances it performs better than the penalty estimators. The work in the book indicates that if the number of inactive predictors is correctly specified, the shrinkage method would be expected to perform better than the penalty method. If the number of inactive predictors is incorrectly specified, the penalty methods would be expected to do better than the shrinkage strategy.

In this book, selected penalty techniques have been compared with the full model, submodel, and shrinkage estimation in some regression models. Further, one chapter is dedicated to machine learning methods. Several real data examples have been presented along with Monte Carlo simulations to appraise the performance of the estimators in real settings. It showcases the applications and methodological developments in both low and high-dimensional cases dealing with interesting and challenging problems concerning the analysis of complex, high-dimensional data with a focus on model selection, post-estimation, and prediction in a host of useful regression model. The chapters contained in this book deal with submodel selection and post-shrinkage estimation for an array of interesting models.

In summary, several directions for statistical inference in high-dimensional statistics are highlighted in this book. This book conveys some of the surprises, puzzles, and success stories in big and high-dimensional data analysis.

We anticipate that the chapters published in this book will represent a meaningful contribution to the development of new ideas in big data analysis and will provide interesting applications.

The book is suitable as a reference book for a graduate course in modern regression analysis and data analytics. The selection of topics and the coverage will be equally useful for the researchers and practitioners in host of fields. This book is organized into ten chapters. The chapters are standalone, so that anyone interested in a particular topic or area of application may read that specific chapter. Those new to this area may read the first four chapters and then skip to the topic of their interest. A brief outline of the contents is available in chapter 1.

Acknowledgments

I would like to express my appreciation to my several Ph.D. students and collaborator for their interest and support in preparing this book.

More specifically, I would like to express my thanks to my former Ph.D. students, Drs. S. Hossain, Eugene Opoku, Orawan Reangsephet, and Janjira Piladaeng for their valuable contributions in the preparation of the manuscript. Further, I want to thank my former Ph.D. students Drs. Andrei Volodin, Enayat Raheem, Kashif Ali, Nighat Zahra, and Hira Nadeem for the interesting discussions on the topic. I would also like to thank one of my current Ph.D. students Ersin Yilmaz who was always available and eager to help me on this project. Further, I owe thanks to my former colleague Dr. Abdul Hussein for his support and help over the years!

Feryaal would like to express her deepest gratitude to her mom, Ghazala, and brother, Jazib, for always keeping her spirits and motivation during this project. She also extends her sincere thanks to her husband Ali, for his support and encouragement over the past year.

Dr. Yüzbaşi gives his heartfelt thanks to his wife Zühal, his daughter Beril, and his son Buğra for their continued love, support, and patience. I would like to express my gratitude to Prof. Muhammed Fatih Talu (İnönü University, Turkey) for letting make use of his server to run the intensive computations for the book.

I will take this opportunity to extend my gratitude to my colleagues and collaborators, specifically to Drs. Yi Li, Shuangge Ma, Xiaoli Gao, Yang Feng, Jiwei Zhao, Mohamed Amezziane, Supranee Lisawadi, Farouk Nathoo, Serge Provost, Abbas Khalili, and Dursun Aydin for thoughtful research discussions and collaborations.

This book would not have been possible without the assistance of everybody at the incredible CRC team, particularly Curtis and David. My special thank goes to David for his encouragements and support during the preparation of this book; he is a man with an infinite amount of patience, who never gives up!

S. Ejaz Ahmed November 2022 - Canada

Author/editor biographies

Dr. S. Ejaz Ahmed is a Professor of Statistics and Dean of the Faculty of Math and Science at Brock University, Canada. Previously, he was Professor and Head of the Mathematics and Statistics Department at the University of Windsor, Canada, and the University of Regina, Canada as well as Assistant Professor at the University of Western Ontario, Canada. He holds adjunct professorship positions at many Canadian and International universities. He has supervised more than 20 Ph.D. students and organized several international workshops and conferences around the globe. He is a Fellow of the American Statistical Association and held prestigious ASEAN Chair Professorship position. His areas of expertise include big data analysis, statistical learning, and shrinkage estimation strategy. Having authored several books, he edited and co-edited several volumes and special issues of scientific journals. He is Technometrics Review Editor for past ten years. Further, he is the Editor and Associate Editor of many statistical journals. Overall, he published more than 200 articles in scientific journals and reviewed more than 100 books. Having been among the Board of Directors of the Statistical Society of Canada, he was also Chairman of its Education Committee. Moreover, he was Vice President of Communications for The International Society for Business and Industrial Statistics (ISBIS) as well as a member of the "Discovery Grants Evaluation Group" and the "Grant Selection Committee" of the Natural Sciences and Engineering Research Council of Canada.

Feryaal Ahmed is a Management Science Ph.D. candidate at Ivey Business School, Western University. Her research interests are in data analytics, machine learning, and revenue management, specifically in modeling pricing strategies for service industries that offer ancillary items.

Bahadir Yüzbaşi is an Associate Professor at İnönü University. He received his Doctorate from İnönü University in 2014 under the co-supervision of Professor Ahmed. He has been working on big data and statistical machine learning techniques with theory and applications, as well as professionally coding his studies in R and publishing them on CRAN. He has written a number of articles and chapters for books that have been published by well-known publishers.

List of Figures

List of Tables

Contributors

S. Ejaz Ahmed
Brock Univsersity
St. Catharines, Canada

Feryaal Ahmed
Ivey Business School, Western University
London, ON, Canada

Bahadir Yüzbaşi
İnönü University
Malatya, Turkey

Abbreviations

ADB	Asymptotic Distributional Bias
ADR	Asymptotic Distributional Risk
AIC	Akaike Information Criterion
ALASSO	Adaptive LASSO
ANN	Artifcial Neural Network
BGM	Gradient Boosting Machine
BIC	Bayesian Information Criterion
BSS	Best Subset Selection
CN	Condition Number
CNI	Condition Number Index
CTFR	Tuning-Free Regression Method
ENET	Elastic Net
FM	Full Model
FN	False Negatives
FP	False Positive
GLM	Generalized Linear Model
GLS	Generalized Least Squares Estimator
HDD	High-Dimensional Data
HDDA	High-Dimensional Data Analysis
IPT	Improved Pretest Estimator
kNN	k-Nearest Neighbours
LAD	Least Absolute Deviation
LASSO	Least Absolute Shrinkage and Selection Operator
LDA	Linear Discriminant Analysis
LMM	Linear Mixed Effects Model
LS	Linear Shrinkage
LSE	Least Squares Estimation
MAPE	Mean Absolute Prediction Error
MARS	Multivariate Adaptive Regression Spline
MCP	Minimax Concave Penalty
MLE	Maximum Likelihood Estimator
MLR	Multiple Linear Regression
MSE	Mean Squared Error
MSPE	Mean Square Prediction Error
NB	Negative Binomial
NN	Neural Network
OF	Overfitted
PCA	Principle Component Analysis
PE	Prediction Error
PLM	Partially Linear Regression Model
PS	Positive Part of the Shrinkage
PT	Pretest Estimator
QADB	Quadratic Asymptotic Distributional Bias
RF	Random Forest
RFM	Ridge Full Model
RMAPE	Relative Mean Absolute Prediction Error
RMSE	Relative Mean Squared Error
RMSPE	Relative Mean Square Prediction Error
RPE	Relative Prediction Error
RTMSPE	Relative Trimmed Mean Squared Prediction Error
S	Shrinkage or Stein-Type
SCAD	Smoothly Clipped Absolute Deviation
SE	Dtandard Error
SM	Submodel
SPT	Shrinkage Pretest Estimation
SVM	Support Vector Machine
TCGA	The Cancer Genome Atlas
TMSPE	Trimmed Mean Squared Prediction Error
TN	True Negatives
TP	True Positive
UF	Underfitted
VIF	Variance Inflation Factor

1

Introduction

There are a host of buzzwords in today's data-centric world. We encounter data in all walks of life, and for analytically and objectively minded people, data is crucial to their goals. However, making sense of the data and extracting meaningful information from it may not be an easy task. Contaminated data has increasingly emerged from different fields including signal processing, eCommerce, financial economics, and genomic studies. The rapid growth in the size and scope of data sets in a host of disciplines has created a need for innovative statistical strategies to understand such data. A variety of statistical and computational tools are needed to reveal the story contained in the data. Although the buzzword Big Data is nebulously defined, its problems are real, and statisticians play a vital role in this data-centric world. Complex big data analysis is a very challenging but rewarding research area as data sets include a larger number of features, data contamination, unstructured patterns, and so on. We are living in an era with an abundance of data stemming from diverse fields such as spectroscopy, gene array, functional magnetic resonance imaging, engineering, high-energy physics, financial markets, text retrieval, and social sciences. In these cases, comprehending data is a daunting task. The question is how to extract useful and important messages from big data sets. Biomedical studies are providing abundant survival data with high through predictors. In finance and marketing, big data sets are usually available because most market participants' activities are now online. Most business models are now data-driven with a large number of predictors.

Big data can take many forms. One of which is the existence of many predictors for a relatively small number of observations, defined as high-dimensional data (HDD). Some examples of HDD that have prompted demand are gene expression arrays, social network modeling, clinical, and phenotypic data. Undoubtedly, overcoming the challenges of HDD is key to successful research in a host of fields. Clearly, there is an increasing demand for efficient prediction strategies for analyzing HDD. Shrinkage analysis has been one of my main research fields for many years. Previously, the focus was to shrink a full estimator in the direction of an estimator under a subset model. However, in a high-dimensional setting there is no unique solution for a full estimator. Thus, it becomes an interesting but very challenging problem to study shrinkage analysis in a high-dimensional setting. Most of the existing methods for dealing with HDD begin with selection of a submodel for further investigation. Penalized methods are unstable and biased unless very stringent conditions are imposed. This research proposal in HDD focuses on post-selection strategies to combat some of the issues inherited in penalized methods. The overarching objective is to provide answers to the question: what are the tools and tricks, pitfalls, applications, challenges, and opportunities in HDD analysis?

This book provides a framework for high-dimensional shrinkage analysis when both strong signals and weak signals co-exist. It emphasizes that statisticians can play a dominant role in solving HDD problems moving statisticians from the cellar of scientific discovery to the penthouse. The chapters provide opportunities for training student researchers at all levels. The training will be three-fold: methodological, coding/computational, and analysis of data from real-life examples. More public and private sectors are now acknowledging the importance of statistical tools and its critical role in analyzing Big Data. There are

DOI: 10.1201/9781003170259-1

millions of jobs available globally for Big Data analysts. This book will train individuals for these lucrative positions. In this book, we consider the estimation problem of regression parameters when the model is sparse, when there are many potential predictors in the model that may not have any influence on the response of interest. Some of the predictors may have a strong influence (strong signals), and some may have a weak-to-moderate influence (weak-moderate signals) on the response of interest. It is also possible that there may be extraneous predictors in the model. Consider a clinical example: if one is concerned with a treatment effect or the effect of biomarkers, extraneous nuisance variables may be experimental effects when several labs are involved or the age and sex of patients. The analysis will be more meaningful if "nuisance variables" can be deleted from the model. More importantly, we should not automatically remove all the predictors with weak/moderate signals from the model. This may result in selecting a biased model. A logical way to deal with this is to apply a pretest strategy that tests whether the coefficients with weak-moderate effects are zero and then estimates the model parameters that include coefficients that were rejected by the test. Alternatively, the Stein-rule estimation strategy can be applied where the estimated regression coefficient vector is shrunk in the direction of the candidate subspace. This "soft threshold" modification of the pretest method has been shown to be efficient in various frameworks. Ahmed (1997a) and Ahmed and Yüzbaşı (2016), among others have investigated the properties of shrinkage and pretest methodologies for a host of models.

Due to the trade-off between model prediction and model complexity, model selection is an extremely important and challenging problem in high-dimensional data analysis (HDDA). Over the past two decades, many penalized regularization approaches have been developed to perform variable selection and estimation simultaneously as seen in Ahmed (2014). These techniques that deal with HDDA generally rely on various L_1 penalty regularizes. However, these penalized methods may force the relatively large number of weaker coefficients toward zeros and are subject to a larger selection bias in the presence of a significant number of weak/moderate signals. This leads to the consideration of two models: (1) M_1 that includes all predictors with strong signals and possible variables with weak and moderate signals and (2) M_2 that includes the predictors with only strong signals. In other words, in an effort to achieve meaningful estimation and selection properties, most penalized strategies make the following assumptions:

- Most of the regression coefficients are zeros except for a few ones
- All non-zero β_j's are larger than noise level, $c\sigma\sqrt{(2/n)\log(p)}$ with $c \geq 1/2$

Over the years, many penalized regularization approaches have been developed to do variable selection and estimation simultaneously. Among them, the least absolute shrinkage and selection operator (LASSO) is commonly used Tibshirani (1996). It is a useful estimation technique in part due to its convexity and computational efficiency. The LASSO approach is based on an ℓ_1 penalty for the regularization of regression parameters. Zou (2006) provides a comprehensive summary of the consistency properties of the LASSO approach. Related penalized likelihood methods have been extensively studied in the literature, see for example Tran (2011); Huang et al. (2008); Kim et al. (2008); Wang and Leng (2007); Yuan and Lin (2006); Leng et al. (2006). The penalized likelihood methods have a close connection to Bayesian procedures. Thus, the LASSO estimate corresponds to a Bayes method that puts a Laplacian (double-exponential) prior on the regression coefficients Park and Casella (2008); Greenlaw et al. (2017).

It is possible that some weak-moderate signals can be also forced out from the model by an aggressive variable selection method. Further, it is possible that the method at hand may not be able separate weak signals from sparse signals; we refer to Zhang and Zhang (2014) and others. Interestingly, Hansen (2016) demonstrated using simulation studies that post-selection least squares estimate can do better than penalty estimators under such conditions;

we refer to Belloni and Chernozhukov (2013) and Liu and Yu (2013). Therefore, there is one less aggressive variable selection strategy with a larger tuning parameter value that may yield a model with more predictors, which may include strong and some weak signals. In other words, it retains predictors of strong and weak-moderate signals alike. However, it is not certain that the weak signals are truly weak or sparse. Conversely, other aggressive penalized methods and/or with an optimal tuning parameter value yields a model with a few predictors of strong influence. Thus, the predictors with weak-moderate influence should be subject to further scrutiny to improve the prediction error. An appealing way to deal with regression parameter uncertainty is to use a pretest strategy, one that tests whether the coefficients of the variables with weak/moderate signals are zero and estimates model parameters that include the rejected coefficients.

In this book, we consider both low and high-dimensional regression sparse models. To begin, let us consider the following regression model:

$$\boldsymbol{Y} = \boldsymbol{X}\boldsymbol{\beta} + \boldsymbol{\varepsilon}, \tag{1.1}$$

where $\boldsymbol{Y} = (y_1, \ldots, y_n)^\top$ is a vector of responses, $\boldsymbol{X} = (\boldsymbol{x}_1, \ldots, \boldsymbol{x}_n)^\top$ is a $n \times p$ fixed design matrix, where $\boldsymbol{x}_i = (x_{i1}, \ldots, x_{ip})^\top$, $\boldsymbol{\beta} = (\beta_1, \ldots, \beta_p)^\top$ is an unknown vector of parameters, $\boldsymbol{\varepsilon} = (\varepsilon_1, \ldots, \varepsilon_n)^\top$ is the vector of unobservable random errors, and the superscript $(^\top)$ denotes the transpose of a vector or matrix. We do not make any distributional assumptions about the errors except that $\boldsymbol{\varepsilon}$ has a cumulative distribution function $F(\cdot)$ with $E(\boldsymbol{\varepsilon}) = \boldsymbol{0}$ and $E(\boldsymbol{\varepsilon}\boldsymbol{\varepsilon}^\top) = \sigma^2 \boldsymbol{I}_n$, where $\sigma^2 < \infty$.

However, when $p > n$, the inverse of the Gram matrix $(\boldsymbol{X}^\top \boldsymbol{X})^{-1}$ does not exist, meaning there will be infinitely many solutions for the least squares minimization. As a matter of fact, when $p \leq n$ and p are close to n the LSE estimates are not stable due to high standard deviations of estimators. In sparse high-dimensional regression modeling, it is assumed that only some of the predictors are significant for prediction purposes. Thus, the true model has only a relatively small number of non-zero predictors.

Generally, the least squares estimation method does not yield zero estimates for many of the regression parameters; we refer to Hastie et al. (2009).

The penalized least square regression methods are recommended when $p > n$ for the model parameters estimation in (1.1). The key idea in penalized regression methods to obtain the estimates of the parameter by minimizing objection function $L_{\rho,\lambda}$ of the form

$$L_{\rho,\lambda}(\boldsymbol{\beta}) = (\boldsymbol{Y} - \boldsymbol{X}\boldsymbol{\beta})^\top (\boldsymbol{Y} - \boldsymbol{X}\boldsymbol{\beta}) + \lambda \rho(\boldsymbol{\beta}). \tag{1.2}$$

The first component of the function is the sum of the squared error loss, and the second term is a penalty function ρ with λ as a tuning parameter to control the trade-off between the two components. The penalty function is usually chosen as a norm on $\mathbb{R}^p$,

$$\rho_q(\boldsymbol{\beta}) = \sum_{j=1}^{p} |\beta_j|^q, \; q > 0, \tag{1.3}$$

this class of estimator is called the bridge estimators proposed by Frank and Friedman (1993). For $q = 2$, the ridge regression (Hoerl and Kennard (1970) Frank and Friedman (1993)) minimizes the residual sum of squares subject to an l_2-penalty:

$$\widehat{\boldsymbol{\beta}}^{\text{Ridge}} = \arg\min_{\boldsymbol{\beta}} \left\{ \sum_{i=1}^{n} (y_i - \sum_{j=1}^{p} x_{ij}\beta_j)^2 + \lambda \sum_{j=1}^{p} \beta_j^2 \right\}. \tag{1.4}$$

Clearly, the ridge estimator is a continuous shrinkage process and has a better prediction performance than LSE through the bias-variance trade-off. However, it does not set the LSE estimates to zero and fails to yield a sparse model. In the case of l_q-penalty with $q \leq 1$, some coefficients are set exactly to zero and the optimization problem for (1.2) becomes a convex optimization problem.

There are several other penalized methods with more sophisticated penalty functions that not only shrink all the coefficients toward zero but also set some of them exactly to zero. As a result, this class of estimators usually produce biased estimates for the parameters due to shrinkage but still have some advantages such as producing more interpretable submodels and reducing estimate variance.

The following methods perform parameter estimation and model selection simultaneously: *least absolute shrinkage and selection operator* (LASSO) Tibshirani (1996), *smoothly clipped absolute deviation* (SCAD) Fan and Li (2001), *the adaptive LASSO* (ALASSO) Zou (2006), and the *minimax concave penalty* (MCP) method Zhang (2010). Generally speaking, the LASSO and its relatives have an edge over ridge and bridge estimates in terms of variable selection performance; we refer to Tibshirani (1996) and Fu (1998). More importantly, penalized techniques can be used when $p > n$. However, most penalized methods make assumptions on both the true model and the design matrix, see Hastie et al. (2009) and Bühlmann and Van De Geer (2011). Chatterjee (2015) suggested a *tuning-free regression method* (CTFR). We refer to Ahmed and Yüzbaşı (2016) for more insights on these methods. Now, we provide a brief and gentle introduction to some penalized methods.

1.1 Least Absolute Shrinkage and Selection Operator

The LASSO estimates are defined by

$$\widehat{\boldsymbol{\beta}}_n^{\text{LASSO}} = \arg\min_{\boldsymbol{\beta}} \left\{ \sum_{i=1}^{n} (y_i - \sum_{j=1}^{p} x_{ij}\beta_j)^2 + \lambda \sum_{j=1}^{p} |\beta_j| \right\}. \tag{1.5}$$

To gain some insight, let us consider the case when $n = p$ and the design matrix $\boldsymbol{X} = \boldsymbol{I}_n$ is the identity matrix. In this case, the LSE solution minimizes

$$\sum_{i=1}^{p} (y_i - \beta_i)^2$$

with

$$\widehat{\beta}_i^{\text{LSE}} = y_i \quad \text{for all} \quad 1 \leq i \leq p.$$

The ridge regression solution minimizes

$$\sum_{i=1}^{p} (y_i - \beta_i)^2 + \lambda \sum_{i=1}^{p} \beta_i^2$$

with

$$\widehat{\beta}_i^{\text{Ridge}} = y_i/(1+\lambda) \quad \text{for all} \quad 1 \leq i \leq p.$$

Similarly, the LASSO solution minimizes

$$\sum_{i=1}^{p} (y_i - \beta_i)^2 + \lambda \sum_{i=1}^{p} |\beta_i|$$

with

$$\widehat{\beta}_i^{\text{LASSO}} = \begin{cases} y_i - \lambda/2 & \text{if} \quad y_i > \lambda/2; \\ y_i + \lambda/2 & \text{if} \quad y_i < -\lambda/2; \\ 0 & \text{if} \quad |y_i| \le \lambda/2. \end{cases}.$$

The shrinkage applied by ridge and LASSO affects the estimated parameters in a different way. In the LASSO solution, the least square coefficients with absolute values less than $\lambda/2$ are set exactly equal to zero, and other least squares coefficients are shrunken toward zero by a constant amount $\lambda/2$. Hence, sufficiently small coefficients are all estimated as zero. On the other hand, the ridge regression shrinks each LSE toward zero by multiplying each one by a constant proportional to $1/\lambda$. For a more general design matrix, we refer to Hastie et al. (2009) and Hastie et al. (2015) for more on this topic.

Meinshausen and Bühlmann (2006) showed that if the penalty parameter λ is tuned to obtain optimal prediction, then the consistent variable selection can not hold: the LASSO solution includes many noise variables besides the true signals. Leng et al. (2006) reveal this story by considering a model with an orthogonal design.

The LASSO is an l_1 penalized least squares regression method that can fit the observation data well while also seeking a sparse solution. It is known, however, that LASSO may not be the optimal method if a group of columns in a measurement matrix are highly correlated. To overcome this limitation of LASSO, Zou and Hastie (2005) proposed the elastic net (ENET), which is created by linearly combining a l_1 penalty term and a l_2 penalty term.

1.2 Elastic Net

The ENET was proposed by Zou and Hastie (2005) to overcome the limitations of the LASSO and ridge methods.

$$\widehat{\boldsymbol{\beta}}^{\text{ENET}} = \arg\min_{\boldsymbol{\beta}} \left\{ \sum_{i=1}^{n} (y_i - \sum_{j=1}^{p} x_{ij}\beta_j)^2 + \lambda_1 \sum_{j=1}^{p} |\beta_j| + \lambda_2 \sum_{j=1}^{p} \beta_j^2 \right\},$$

where λ_2 is the ridge penalty parameter, penalizing the sum of the squared regression coefficients and λ_1 is the LASSO penalty, penalizing the sum of the absolute values of the regression coefficients, respectively.

1.3 Adaptive LASSO

Zou (2006) introduced the ALASSO by modifying the LASSO penalty by using adaptive weights on the l_1-penalty with the regression coefficients. It has been shown theoretically that the ALASSO estimator is able to identify the true model consistently, and the resulting estimator has the oracle property.

The ALASSO of $\boldsymbol{\beta}$ is obtained by

$$\widehat{\boldsymbol{\beta}}^{\text{ALASSO}} = \arg\min_{\boldsymbol{\beta}} \left\{ \sum_{i=1}^{n} (y_i - \sum_{j=1}^{p} x_{ij}\beta_j)^2 + \lambda \sum_{j=1}^{p} \widehat{w}_j |\beta_j| \right\}, \tag{1.6}$$

where the weight function is

$$\widehat{w}_j = \frac{1}{|\widehat{\beta}_j^*|^\gamma}; \quad \gamma > 0,$$

and $\widehat{\beta}_j^*$ is a root-n-consistent estimator of β. The minimization procedure for the ALASSO solution does not induce any computational difficulty and can be solved very efficiently; for more details see Section 3.5 in Zou (2006). Zou (2006) proved that if $\lambda_n/\sqrt{n} \to 0$ and $\lambda_n n^{(\gamma-1)/2} \to \infty$, then the ALASSO has variable selection consistency with probability one as n tends to infinity and

$$\sqrt{n}(\widehat{\boldsymbol{\beta}}_n^{\text{ALASSO}} - \boldsymbol{\beta}) \to_d N(\mathbf{0}, \sigma^2 \boldsymbol{C}_{11}^{-1})$$

where $\boldsymbol{C}_{11}^{-1}$ is the submatrix of $\boldsymbol{C}$ which corresponds to the non-zero entries of $\boldsymbol{\beta}$.

1.4 Smoothly Clipped Absolute Deviation

Although the LASSO method does both shrinkage and variable selection by setting many coefficients identically to zero, it does not possess oracle properties, Zou (2006) then proposed SCAD. This method not only retains the good features of both subset selection and ridge regression but also produces sparse solutions. The estimates are obtained as:

$$\widehat{\boldsymbol{\beta}}^{\text{SCAD}} = \arg\min_{\boldsymbol{\beta}} \left\{ \sum_{i=1}^{n} (y_i - \sum_{j=1}^{p} x_{ij}\beta_j)^2 + \lambda \sum_{j=1}^{p} p_{\alpha,\lambda}|\beta_j| \right\},$$

where $p_{\alpha,\lambda}(\cdot)$ is the smoothly clipped absolute deviation penalty. The SCAD penalty is a symmetric and a quadratic spline on $[0, \infty)$ with knots at λ and $\alpha\lambda$ whose first order derivative is given by

$$p_{\alpha,\lambda}(x) = \lambda \left\{ I(|x| \le \lambda) + \frac{(\alpha\lambda - |x|)_+}{(\alpha - 1)\lambda} I(|x| > \lambda) \right\}, \quad x \ge 0, \tag{1.7}$$

where $\lambda > 0$ and $\alpha > 2$ are the tuning parameters. For $\alpha = \infty$, the expression (1.7) is equivalent to the l_1-penalty.

1.5 Minimax Concave Penalty

Zhang (2007) suggested

$$\widehat{\boldsymbol{\beta}}_n^{\text{MCP}} = \arg\min_{\boldsymbol{\beta}} \left\{ \sum_{i=1}^{n} \left(y_i - \sum_{j=1}^{p} x_{ij}\beta_j \right)^2 + \sum_{j=1}^{p} \rho(|\beta_j|; \lambda) \right\}, \tag{1.8}$$

where $\rho(\cdot; \lambda)$ is the MCP penalty given by

$$\rho(t; \lambda) = \lambda \int_0^t (1 - \frac{x}{\gamma\lambda})^+ \, dx$$

where $\gamma > 0$ and λ are the regularization and penalty parameters, respectively. The MCP has the threshold value $\gamma\lambda$. The penalty is a quadratic function for values less than the threshold and is constant for values greater than it. The regularization parameter $\gamma > 0$ controls the convexity and therefore the bias of the estimators. This choice enables one to remove almost all the bias of the estimators and to obtain consistent variable selection under less restricted assumptions than those required by LASSO. The MCP solution path converges to the LASSO path as $\gamma \to \infty$. Zhang (2010) proves that the estimator possesses selection consistency at the universal penalty level $\lambda = \sigma\sqrt{2/n \log p}$ under the sparse Riesz condition on the design matrix $\boldsymbol{X}$; we refer to Zhang (2007) and Zhang (2010).

Additional assumptions made regarding the designed covariates include the adaptive irrepresentable condition and the restricted eigenvalue conditions. We refer to Zhao and Yu (2006), Huang et al. (2008), and Bickel et al. (2009) for some insights.

1.6 High-Dimensional Weak-Sparse Regression Model

Now, let us consider a high-dimensional regression model that includes strong, weak, and sparse signals. Again, the model is

$$\boldsymbol{Y} = \boldsymbol{X}\boldsymbol{\beta} + \boldsymbol{\varepsilon}, \quad p > n,$$

where $\boldsymbol{\varepsilon}$'s are random errors distributed to be independent and identically distributed.

We partition the design matrix such that $\boldsymbol{X} = (\boldsymbol{X}_1|\boldsymbol{X}_2|\boldsymbol{X}_3)$, where $\boldsymbol{X}_1$ is $n \times p_1$, $\boldsymbol{X}_2$ is $n \times p_2$, and $\boldsymbol{X}_3$ is $n \times p_3$ sub-matrix of predictors. We make the usual assumption that $p_1 + p_2 < n$ and $p_3 > n$, where p_1 is the dimension of strong signals, p_2 is for weak signals, and p_3 is associated with no signals.

The model can be rewritten as

$$\boldsymbol{Y} = \boldsymbol{X}_1\boldsymbol{\beta}_1 + \boldsymbol{X}_2\boldsymbol{\beta}_2 + \boldsymbol{X}_3\boldsymbol{\beta}_3 + \boldsymbol{\varepsilon}, \quad p = p_1 + p_2 + p_3 > n. \tag{1.9}$$

Thus, the predictors with no signals can be discarded by existing variable selection methods since we assume that the model is sparse.

For the models with weak signals, we use variable selection method which keeps both strong and weak-moderate signals as follows:

$$\boldsymbol{Y} = \boldsymbol{X}_1\boldsymbol{\beta}_1 + \boldsymbol{X}_2\boldsymbol{\beta}_2 + \boldsymbol{\varepsilon}, \quad p_1 + p_2 < n. \tag{1.10}$$

Generally, a variable selection method with a larger tuning parameter value may eliminate the sparse signals and retain predictors with strong and weak signals in the resulting model. For brevity, we characterize such models as an over-fitted model or the full model (FM).

For models with strong signals, suppose an aggressive variable selection method with the optimal tuning parameter value may only keep predictors with strong signals and removes all other predictors; we may call it an under-fitted model or submodel (SM).

$$\boldsymbol{Y} = \boldsymbol{X}_1\boldsymbol{\beta}_1 + \boldsymbol{\varepsilon}, \quad p_1 < n. \tag{1.11}$$

We would like to remark here that some weak signals can also be combined with strong signals. We are primarily interested in estimating $\boldsymbol{\beta}_1$ when weak signals may or may not be significant. In other words, $\boldsymbol{\beta}_2$ may be a null vector, but we are not certain that this is the case. We propose pretest and shrinkage strategies for estimating $\boldsymbol{\beta}_1$ when a model is sparse and $\boldsymbol{\beta}_2$ may be a null vector. It is natural to combine estimates of the over-fitted model with the estimates of an under-fitted model to improve the performance of an under-fitted model.

1.7 Estimation Strategies

A logical way to deal with the issue of weak coefficients is to apply a pretest strategy that tests whether the coefficients with weak effects are zero, then estimate parameters in the model that include coefficients that are rejected by the test. Thus, providing a post-pretest estimator (PE) by performing a test on the weak coefficients. The strategy is defined as follows:

1.7.1 Pretest Estimation Strategy

The pretest estimator (PT) of $\boldsymbol{\beta}_1$ defined as follows:

$$\widehat{\boldsymbol{\beta}}_1^{\mathrm{PT}} = \widehat{\boldsymbol{\beta}}_1^{\mathrm{UF}} \mathrm{I}\left(W_n < \chi^2_{p_2,\alpha}\right) + \widehat{\boldsymbol{\beta}}_1^{\mathrm{OF}} \mathrm{I}\left(W_n \geq \chi^2_{p_2,\alpha}\right), \tag{1.12}$$

or equivalently,

$$\widehat{\boldsymbol{\beta}}_1^{\mathrm{PT}} = \widehat{\boldsymbol{\beta}}_1^{\mathrm{OF}} - \left(\widehat{\boldsymbol{\beta}}_1^{\mathrm{OF}} - \widehat{\boldsymbol{\beta}}_1^{\mathrm{UF}}\right) \mathrm{I}\left(W_n < \chi^2_{p_2,\alpha}\right), \tag{1.13}$$

where the weight function W_n is defined by

$$W_n = \frac{n}{\widehat{\sigma}^2}(\widehat{\boldsymbol{\beta}}_2^{\mathrm{LSE}})^\top (\boldsymbol{X}_2^\top \boldsymbol{M}_1 \boldsymbol{X}_2)\widehat{\boldsymbol{\beta}}_2^{\mathrm{LSE}}, \tag{1.14}$$

$\boldsymbol{M}_1 = \boldsymbol{I}_n - \boldsymbol{X}_1\left(\boldsymbol{X}_1^\top \boldsymbol{X}_1\right)^{-1}\boldsymbol{X}_1^\top$, $\widehat{\boldsymbol{\beta}}_2^{\mathrm{LSE}} = \left(\boldsymbol{X}_2^\top \boldsymbol{M}_1 \boldsymbol{X}_2\right)^{-1}\boldsymbol{X}_2^\top \boldsymbol{M}_1 \boldsymbol{Y}$ and $\widehat{\sigma}^2 = \frac{1}{n-1}(\boldsymbol{Y} - \boldsymbol{X}_1\widehat{\boldsymbol{\beta}}_1^{\mathrm{UF}})^\top(\boldsymbol{Y} - \boldsymbol{X}_1\widehat{\boldsymbol{\beta}}_1^{\mathrm{UF}})$.

1.7.2 Shrinkage Estimation Strategy

In the spirit of Ahmed (2014), the Stein-type shrinkage estimator of $\boldsymbol{\beta}_1$ is defined by combining the over-fitted model estimator $\widehat{\boldsymbol{\beta}}_1^{\mathrm{OF}}$ with the under-fitted $\widehat{\boldsymbol{\beta}}_1^{\mathrm{UF}}$ as follows

$$\widehat{\boldsymbol{\beta}}_1^{\mathrm{S}} = \widehat{\boldsymbol{\beta}}_1^{\mathrm{UF}} + \left(\widehat{\boldsymbol{\beta}}_1^{\mathrm{OF}} - \widehat{\boldsymbol{\beta}}_1^{\mathrm{UF}}\right)\left(1 - (p_2 - 2)W_n^{-1}\right), \; p_2 \geq 3. \tag{1.15}$$

This soft threshold modification of the pretest method has been shown to be efficient in various frameworks; we refer to Ahmed (1997a, 2001); Ahmed et al. (2007); Ahmed and Nicol (2012); Ahmed et al. (2012); Hossain et al. (2015); Yüzbaşı and Ahmed (2015); Ahmed et al. (2006); SE (1999); Hossain and Ahmed (2012); Ahmed and Raheem (2012); Hossain et al. (2009); Yılmaz et al. (2022); Piladaeng et al. (2022); Yüzbaşı et al. (2022); Opoku et al. (2021); Lisawadi et al. (2021); Reangsephet et al. (2021); Fang et al. (2021); Yüzbaşı et al. (2020); Zareamoghaddam et al. (2021); Phukongtong et al. (2020); Reangsephet et al. (2020); Yüzbaşı and Ahmed (2020); Ahmed et al. (2016); Al-Momani et al. (2017); Ahmed and Yüzbaşı (2017); Yüzbaşı and Ahmed (2016); Hossain et al. (2016); Fallahpour and Ahmed (2014); Hossain and Ahmed (2014); Gao and Ahmed (2014); Ahmed et al. (2015); Ahmed and Fallahpour (2012).

In an effort to avoid the over-shrinking problem inherited by $\widehat{\boldsymbol{\beta}}_1^{\mathrm{S}}$ we suggest using the positive part of the shrinkage estimator of $\boldsymbol{\beta}_1$ defined by

$$\widehat{\boldsymbol{\beta}}_1^{\mathrm{S+}} = \widehat{\boldsymbol{\beta}}_1^{\mathrm{UF}} + \left(\widehat{\boldsymbol{\beta}}_1^{\mathrm{OF}} - \widehat{\boldsymbol{\beta}}_1^{\mathrm{UF}}\right)\left(1 - (p_2 - 2)W_n^{-1}\right)^+, \tag{1.16}$$

where $z^+ = \max(0, z)$.

We refer to Ahmed (2014) for historical background on the pretest and shrinkage strategies. In this book we concentrate on shrinkage strategy and compare it with penalty, full model, and submodel estimators.

1.8 Asymptotic Properties of Non-Penalty Estimators

In each chapter, the asymptotic distributional bias (ADB), quadratic asymptotic distributional bias (QADB), and asymptotic distributional risk (ADR) of the full model, submodel, and shrinkage estimators is derived to assess the relative performance of the listed estimators. Under fixed alternatives, the distribution of various shrinkage estimators is equivalent to the benchmark estimator. Therefore, for a large-sample situation, there is not much to investigate. For $\boldsymbol{\beta}_2$ the pivot is taken as $\mathbf{0}$, we consider a shrinkage neighborhood of $\mathbf{0}$ and take the sequence of local alternatives $\left\{K_{(n)}\right\}$ given by

$$K_{(n)} : \boldsymbol{\beta}_2 = \boldsymbol{\beta}_{2(n)} = \frac{\boldsymbol{\xi}}{\sqrt{n}}, \quad \boldsymbol{\xi} = (\xi_{p^*-p_2+1}, \ldots, \xi_{p^*})^T \in \mathbb{R}^{p_2} \tag{1.17}$$

1.8.1 Bias of Estimators

The ADB of the estimator $\boldsymbol{\beta}_1^*$ is defined as

$$ADB(\boldsymbol{\beta}_1^*) = \lim_{n\to\infty} E\left[\sqrt{n}(\boldsymbol{\beta}_1^* - \boldsymbol{\beta}_1)\right] \tag{1.18}$$

The bias expressions for all estimators are not in scalar form, hence we use QADB. The QADB for an estimator $\boldsymbol{\beta}_1^*$ has form

$$QADB(\boldsymbol{\beta}_1^*) = [ADB(\boldsymbol{\beta}_1^*)]^\top C_{11.2}[ADB(\boldsymbol{\beta}_1^*)]. \tag{1.19}$$

1.8.2 Risk of Estimators

Associated with the quadratic error loss of the form

$$n\left(\widehat{\boldsymbol{\beta}}_1^* - \boldsymbol{\beta}_1\right)^\top W\left(\widehat{\boldsymbol{\beta}}_1^* - \boldsymbol{\beta}_1\right),$$

for a positive definite matrix W, the ADR of $\widehat{\boldsymbol{\beta}}_1^*$ is defined by

$$ADR(\boldsymbol{\beta}_1^*; \boldsymbol{\beta}_1) = \mathrm{tr}(W\Gamma) \tag{1.20}$$

where

$$\Gamma = \int\int\cdots\int \boldsymbol{y}\boldsymbol{y}^\top dG(\boldsymbol{y}) = \lim_{n\to\infty} E\left[n(\widehat{\boldsymbol{\beta}}_1^* - \boldsymbol{\beta}_1)(\widehat{\boldsymbol{\beta}}_1^* - \boldsymbol{\beta}_1)^\top\right]$$

is the dispersion matrix obtained from

$$G(\boldsymbol{y}) = \lim_{n\to\infty} E\left[\sqrt{n}(\boldsymbol{\beta}_1^* - \boldsymbol{\beta}_1)\right].$$

For $W = I$, we get squared error loss functions.

For practical purposes and comparing non-penalty estimators with penalty estimators, we conduct extensive numerical studies under various natural settings to appraise the relative performance of the estimators in each chapter. The relative performance of estimators is evaluated by using the relative mean squared error (RMSE) criterion. The RMSE of an estimator $\boldsymbol{\beta}_1^*$ with respect to $\widehat{\boldsymbol{\beta}}_1^{\mathrm{OF}}$ is defined as follows

$$\mathrm{RMSE}\left(\boldsymbol{\beta}_1^*\right) = \frac{\mathrm{MSE}\left(\widehat{\boldsymbol{\beta}}_1^{\mathrm{OF}}\right)}{\mathrm{MSE}\left(\boldsymbol{\beta}_1^*\right)} \tag{1.21}$$

where $\boldsymbol{\beta}_1^*$ is one of the listed estimators. Finally, we apply the penalty and non-penalty estimation strategies to some data sets and calculate the prediction error of the listed estimators.

1.9 Organization of the Book

This book is divided into ten chapters. In Chapter 2, we present a gentle introduction to machine learning strategies. We highlight both supervised and unsupervised learning in the context of statistical learning. R codes are provided, and a case study based on genomic data is included for a smooth learning of the chapter. In Chapter 3, we consider the classical multiple regression model and extend the idea of shrinkage strategies alongside the penalty estimation. We demonstrate that the suggested shrinkage strategy is superior to the classical estimation strategy and performs better than penalty procedures in practical settings. Asymptotic bias and risk expressions of the estimators have been derived, and the performance of shrinkage estimators is compared with the classical estimators and penalty through Monte Carlo simulation experiments. The listed strategies are applied to prostrate cancer data, and the relative prediction error is computed and compared. Further, the relative performance of listed estimators are compared with machine learning strategies. R-codes for simulation and data examples are given to provide access to practitioners. In Chapter 4, we extend the shrinkage strategies to a high-dimensional multiple regression model. Basically, we integrate an over-fitted model and an under-fitted model in an optimal way. Several penalty estimators such as LASSO, ALASSO, and SCAD estimators have been discussed. Monte Carlo simulation studies are used to compare the performance of shrinkage and penalty estimators.

We discuss shrinkage and penalty estimation in partially linear models in Chapter 5. In the low-dimensional case, the risk properties of the non-penalty estimators are studied using asymptotic distributional risk and Monte Carlo simulation studies. Two real data examples are given to illustrate the applications of estimators. We assess the relative performance of the estimators in a high-dimensional case via an extensive simulation study including the models with weak signals. A high-dimensional data example is analyzed using the suggested procedure, and R-codes are given.

In Chapter 6, we consider shrinkage and penalized estimation strategies in the generalized regression model. More specifically, we consider the estimation and prediction problems in logistic regression and negative binomial regression models. The analytic solution for the classical estimator, submodel estimator, and shrinkage estimators are showcased. The numerical analysis by virtue of simulation and real data examples is implemented. In the high-dimensional case, we appraise the performance of shrinkage, penalty, and maximum likelihood estimators with a real data example and through Monte Carlo simulation experiments.

Chapter 7 focuses on estimation and prediction problems in the linear mixed model. For this model, we use the ridge estimator as a base estimator for full model estimation to deal with the multicollinearity issue. Using a variable selection method, we then obtain a submodel for both the low and high-dimensional cases. Finally, we combine the two models to construct the shrinkage estimators. The theoretical properties of the estimators are investigated in the low-dimensional case. A numerical analysis including a data example is considered. In the high-dimensional case, we provide important features of ridge and other penalty estimators and are strongly supported through simulation and data examples. The chapter also offers the R-codes for computational purposes. In Chapter 8, we consider applications of shrinkage and penalty methodologies to nonlinear regression models. We develop estimation and prediction strategies when the model's sparsity assumption may or may not hold. We consider both low- and high-dimensional regimes using a nonlinear regression model. We consider full model, submodel, shrinkage, and penalty estimations. The mean squared error criterion is used to assess the characteristics of the estimators. For low-dimensional cases, we provide some asymptotic properties of non-penalty estimators. We also conduct a simulation study to provide the relative performance of penalty and non-penalty estimators. A real high-dimensional data example and R-codes are given.

Chapter 9 offers shrinkage estimation strategies in multiple regression models containing some outlying observations. Thus, we consider a robust estimator and the least absolute deviation estimator for estimating the regression parameters. We used this estimator to build a shrinkage strategy. Their asymptotic properties are given in a low-dimensional case. A Monte Carlo simulation study is conducted to numerically appraise the relative performance of the listed estimators for both low and high-dimensional data situations. The suggested strategies are applied to some real data sets to demonstrate the usefulness of the suggested procedures. Finally in Chapter 10, we outline the full and submodel estimators based on a Liu regression and build the shrinkage estimators using the two estimators. The large-sample properties of the estimators are given alongside the properties of the estimators. The results of a Monte Carlo simulation experiment are presented, and a numerical comparison with some penalized estimators is also showcased. For illustrative purposes, a real data example and R-codes are also available. In this chapter, we only considered the low-dimensional case. The research on high-dimensional data is under consideration and will be shared in a separate communication and may be added to the book later.

2

Introduction to Machine Learning

The popularity and prevalence of *open science* has benefited the academic health of many industries. Open science data, by its name, is the open or availability of data to researchers. Researchers can further advance their fields without the cumbersome task of having to collect similar data that has previously been collected. For instance, open science data is particularly lucrative in medicine since clinical data is complex and unwieldy to collect. Accessible databases have become a saving grace for many researchers. As technology advances, the value of data becomes especially valuable due to our ability to translate data into imperative information. Buzzwords such as, big data, artificial intelligence, supervised/unsupervised learning, etc., have saturated industries as the need to analyze large datasets has become imperative to businesses, healthcare institutions, governments, and most areas of research. With so many available techniques, it is easy to get lost in the jargon and technicalities of each methodology.

This chapter is to gently ease the reader into the most popular statistical and machine learning techniques. Data analysis, or commonly referred to just analytics, has now become colloquially synonymous with value creation. There are three popular stages of analytics that can provide a researcher with valuable information: descriptive, predictive, and prescriptive analysis. Descriptive analysis, namely, describes the data available, as it allows one to discover trends in the data and observe statistics. But it only scratches the surface of analytics, as it does not offer direct solutions or improvement to ones research questions. Predictive and prescriptive analysis is the bread and butter of value creation. The ability to predict and optimize a solution to ones research question is a powerful tool in finding out how this world operates. Researchers across the globe have been developing all sorts of predictive and prescriptive algorithms to make analytics accessible to the everyday person. As computers continue to grow more sophisticated, so does the accessibility of these algorithms. These algorithms that perform prediction are what we call machine learning. Machines that learn patterns in the data and output information and/or predictions.

2.1 What is Learning?

The term learning in computer science is referred to as a branch of artificial intelligence the design and development of algorithms that allows computers to evolve based on empirical data. The type of algorithm is what dictates the success of the machine learning system. Recently, machine learning has also been referred to as statistical learning because these algorithms have foundations in statistics. There are two main classes of algorithms, supervised and unsupervised learning. Models that people have most likely encountered are usually supervised learning, for example, prediction via logistic or multinomial regression. The differentiating factor between these two classes, is that supervised learning is concerned with data that has labels and the researcher has an idea of what they are looking to predict.

DOI: 10.1201/9781003170259-2

In contrast, unsupervised learning is used when the researcher does not necessarily know what the data entails. In current times, people are overloaded with excessive data and simply figuring out what we are looking at can be a daunting task. Unsupervised learning techniques aids in exploring data, where no assumptions can be made. For example, survey data or medical data can be multidimensional and difficult to interpret; unsupervised algorithms can provide guidance on which variables are pertinent by providing representative variable identification. Unsupervised learning looks at how the data is grouped naturally based on where the data points exist in its multidimensional space. Unsupervised learning used with supervised learning can be a very powerful tool. One can input variables found via unsupervised methods into supervised prediction models, creating a stronger prediction model if such variables are found significant.

To provide a clear guide, this chapter will investigate popular classification and regression machine learning techniques. Classification methods range from the simplest of models to black-box learning. Regression models will build up from the basics and grow in complexity. Logistic regression, multivariate adaptive regression spline (MARS), k-nearest neighbours (kNN), neural nets, support vector machine, random forest, and gradient boosting machine will be discussed in this chapter.

2.2 Unsupervised Learning: Principle Component Analysis and k-Means Clustering

As mentioned, unsupervised learning can be advantaged for the researcher overwhelmed with multidimensional data. We are always looking for the most parsimonious model, unsupervised learning techniques such as principal component analysis and k-nearest neighbours can aid in dimensionality reduction. For example, if the data has 50+ variables, these techniques can suggest which variables are too similar to others or negligible. Sometimes, we can make out which variables can be correlated with one another just from knowing the data collection process. If we remove these seemingly redundant variables, we impose bias into the modeling. Unsupervised techniques aims to avoid these biases by looking at the data for what it is.

2.2.1 Principle Component Analysis (PCA)

Principal component analysis was first developed by Pearson (1901) and later named by Hotelling (1933) as a statistical technique to reduce a large dataset into a smaller one. Principal Component Analysis' goal is to summarize a high-dimensional data matrix by a few principal components. For example, if our data has p variables, PCA reduces them to $\overline{p}$ which is defined as the linear combinations of the full variables, namely, the principal components. The reduction is done so by preserving the distances between the data points. PCA creates new variables and projects the data onto these variables. PCA measures data in terms of these principal components rather than on a normal x-y axis. Figure 2.1 is a simplified demonstration of how dimensionality reduction occurs.

In other words, PCA is essentially an optimization problem. The solution is the collection of the principal components that are representative of the directions of the data. The principal components can be then plotted, and the linear combinations of the original variables can even be visualized in 2-Dimensions.

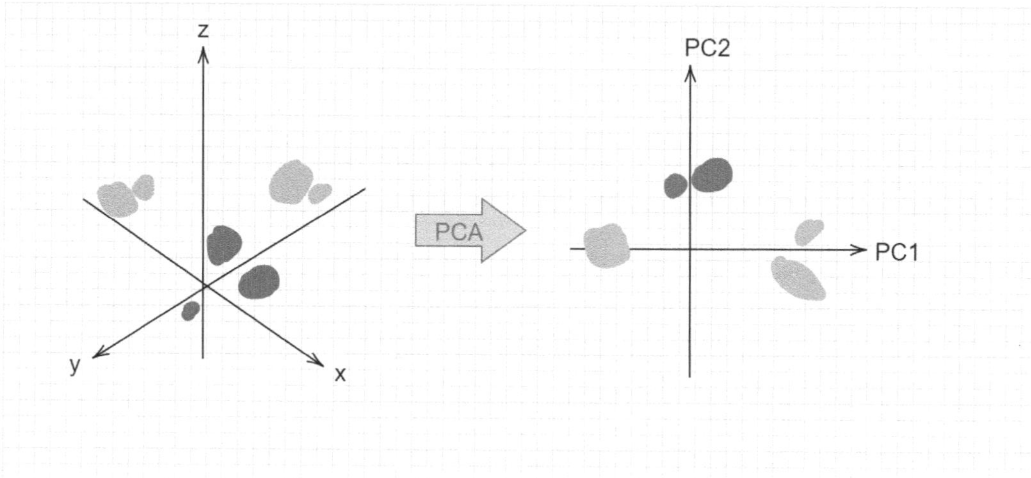

FIGURE 2.1: A 3D Plot with Data is Projected onto a 2D Plot with New Axes from the Principle Components.

The PCA algorithm can be built two different ways, using either maximum variance or minimum error. In maximum variance, the objective function is to find the orthogonal projection of the data onto a lower dimensional linear space to maximize the variance of the projected data. In the minimum error algorithm, the objective function is to minimize the "projection cost" known commonly as the mean squared error between the data and their projection. Both methods reduce dimensionality following four stages:

1. Center the data: Subtract the mean from all the data such that the data is centered around zero
2. Calculate the covariance matrix
3. Calculate the eigenvalues and eigenvectors using eigenvalue decomposition: transforms the coordinates such that the covariance between the new axes is zero.
4. Dimensionality reduction: eigenvectors with only the largest eigenvalues are the principle components now in a reduced space.

R has extensive documentation on the math behind PCA along with built-in functions to perform along with various libraries that can be installed for the visualizations. The general process of how a PCA algorithm can be built is given below, using either the maximum variance or minimum error methods.

ALGORITHM I: MAXIMUM VARIANCE METHOD

1. Given centered data $X = \{x_1, \ldots, x_m\}$ and m points in R^d all dimensionality d
2. Transform vectors into a lower dimensionality space $X^\top = \{x_1^\top, \ldots, x_m^\top\}$ of size k such that $k << d$ by defining the directions of the space with a unit vector $u_1 \in R^d$
3. Project the centered data onto the new space
4. Define the variance of the projected data:
   $max \frac{1}{2N} \sum_{n=1}^{N} \left(u_1^T x_n - u_1^T \bar{x_n}\right)^2 = u_1^T S u_1$

where the sample mean and covariance matrix are $\bar{x} = \frac{1}{N}\sum_{n=1}^{N} x_n$, $S = \frac{1}{N}\sum_{n=1} (x_n - \bar{x})(x_n - \bar{x})^T$

5. Solve for u_1 thus maximizing the variance of the projection in the residual space

PCA ALGORITHM II: MINIMUM ERROR METHOD

1. Given centered data $X = \{x_1, \ldots, x_m\}$ and m points in R^d all dimensionality d
2. Transform vectors into a lower dimensionality space $X^\top = \{x_1^\top, \ldots, x_m^\top\}$ of size k such that $k << d$ by representing the vectors as a linear combination of basis vectors
3. Minimize the following reconstruction error expression $min \frac{1}{N}\sum_{n=1}^{N} ||(u_1^T x_n - u_1^T \bar{x_n}||)^2$ which simplifies using the covariance matrix and solve for the principle components

2.2.2 k-Means Clustering

K-means clustering's goal is to segments the data into different groups, where the data in each group have similar properties. Developed and named by MacQueen (1967), this technique is quite common in data exploration in order to visualize, where the data is grouped naturally. This technique starts by partitioning the data into k distinct clusters, a tuning parameter decided by the researcher. The algorithm then finds the best k centroids (means) of the data and assigns each data point to the closest centroid via Euclidean distance. Since k is often first chosen by the researcher, k-Means clustering is sometimes referred to as "semi-supervised learning" since the choice of initial clusters can impose some bias in the modeling. The steps in the algorithm are as follows: start by assigning random clusters to each data point, compute the average of each cluster (the centroid), reassign the data points such that each data point is assigned to the cluster wit the closes centroid, recompute the closest center, and repeat until the assignments stop changing (Figure 2.2).

K-means clustering is also an optimization problem, where it aims to minimize the within cluster variation. Within cluster variation can be defined by multiple ways but Euclidean

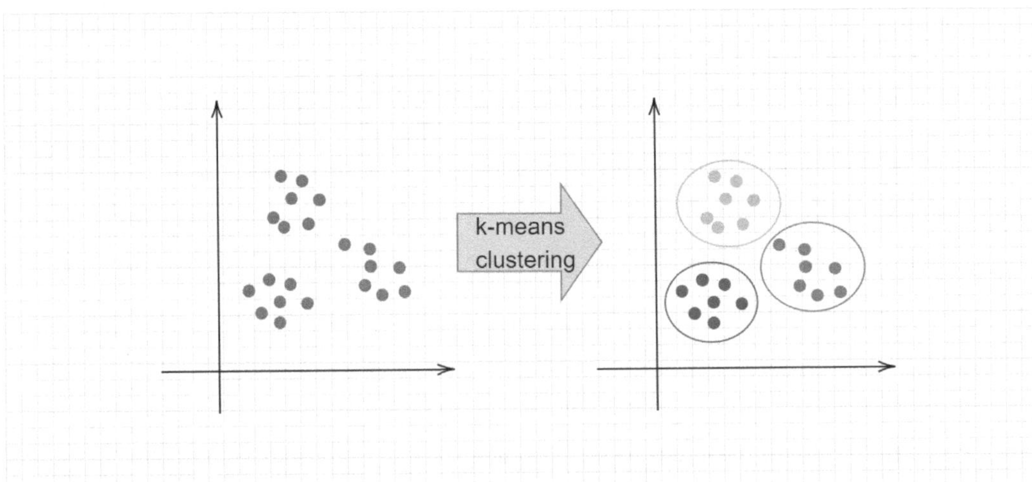

FIGURE 2.2: A Collection of Data can be Categories into 3 Groups via k-Means Clustering using Their Proximity within the Group and Their Separation from Other Groups.

distance is the most popular. More specifically, the algorithm aims to minimize the within cluster variation as the sum of all pairwise Euclidean distances, divided by the number of data points within the cluster.

$$\min \sum_{n=1}^{k} W(C_n)$$

Where $C_n, i \in 1, .., k$ are the number clusters and $W(C_n)$ is the within cluster variation of C_n given by the Euclidean distance:

$$W(C_n) = 1/|C_n| \sum_{i,i\prime \in C_n} \sum_{j=1}^{p} (x_{ij} - x_{i\prime j})^2$$

These steps are quite intuitive, and the researcher is left with the task to interpret what these clusters might mean when the data is multidimensional. For example, clustering techniques are often used in consumer behavior data exploration projects to identify the demographics that explain a certain type of customer (i.e. supermarket collects data and finds cluster of people who purchase baby related products, suggesting the consumer are expecting a child and the supermarket can advertise baby product deals). There are other clustering techniques, but k-means clustering is simple to understand and effective for data exploration. R has documentation on k-means clustering as well as their built-in functions to perform the algorithm and visualize the clusters.

2.2.3 Extension: Unsupervised Text Analysis

Data is vastly accessible, some might even say that we have too much than we can handle. A lot of that data is text data. Posts on social media, clinical reports, published papers, etc. are all pieces of data that can be used for understanding human speech, word association, predictive text, and many more applications. With so much written data, unsupervised learning can be used understand text. For instance, one could webscrape and collect text data on any particular subject. Unsupervised learning can find the natural word associations that we may not pick up at a first glance. Which words appear the most when describing some product? Which comments on a YouTube video are associated with certain age groups? Can we find certain words that doctors jot down in their notes when observing particular types of patients? These questions could be answered using text mining techniques. Unsupervised learning techniques like k-means clustering can help build new vocabularies and understand sentiments without the bias of how one interprets certain words or phrases. Word clouds are the simplest form of unsupervised text analysis (Figure 2.3), they visualize the most frequent words in a corpus of text with size. It allows a researcher to see at a glance the words or phrases that occur the most under a certain topic.

A recent study by Chojnicka and Wawer (2020) perform a computational analysis of linguistics and use it as a tool to aid in language assessments among individuals with autism spectrum disorder. Individuals with autism spectrum disorder often have difficulty with social conversations. Neurotypical individuals are more adept to social cues and certain inflections in our voices to indicate different emotions or feelings. The authors performed natural language processing techniques to help aid in the assessment with those with autism spectrum disorder since their sentiments of neurodivergent individuals differ from those who are neurotypical. Unsupervised learning of text helps us garner an understanding of people who speak differently from the general populous even within the same language. Unsupervised learning helps us discover patterns that are not obvious and provide an excellent starting point to developing sophisticated models.

FIGURE 2.3: Wordcloud Generated from Wikipedia Text on Analytics.

2.3 Supervised Learning

The unsupervised learning methods are great for data exploration and dimensionality reduction without imposing any bias from the researcher. These methods are also important tools for creating new variables (i.e. principal components, or clustered variables), also known as feature engineering in computer science terminology. However, they provide no predictive power. Supervised techniques provide the researcher with predictive ability in order to create actionable solutions for many industries. So much of our current reality relies on supervised learning techniques. Every time we are composing a text message, suggested words or spellings pop up, a product of predictive machine learning. Any time we need to decide if it is too warm for a jacket by consulting the weather forecast, they are all based on a predictive model built on years of data. Learning from our mistakes is part of human nature. So researchers have built several models to give predictions so we can be better prepared and avoid such mistakes. But which model do we choose?

Supervised learning models differ in many ways. Some have impeccable prediction capability but are difficult to interpret, some take more computational power, others are quick but lack the prediction confidence of the same caliber as others. The choice in methodology is up to the researcher and what they value the most. Here we will discuss several models: logistic regression with different penalty functions, multivariate adaptive regression spline (MARS), k-nearest neighbours (kNN), neural nets, support vector machine, random forest, and gradient boosting machine.

2.3.1 Logistic Regression

Logistic regression is a classical statistical prediction model used widely throughout the data science community for its interpretability and predictive prowess. Introduced in many undergraduate statistics courses, logistic regression is common for binary classification. In graduate classes, regularization techniques are often introduced with logistic regression. Regularization techniques are important for dimensionality reduction of the data via some sort of penalty function. The results from the regularization will dictate how many variables are used.

LASSO is an L_1 regularization technique that shrinks the parameter estimates of insignificant variables to zero. By doing so, LASSO also performs variable selection. The remaining explanatory variables, post-correlation and chi-square tests, can be used in the final, most parsimonious model. The LASSO model is implemented such that the free parameter λ that minimises out of sample error is found via grid-search cross validation.

In the case of Ridge regression (L2), the effect of the penalty term is to shrink the coefficients that contribute most to the error. Here, we then compare the beta estimators and manually review which are close to zero. Ridge allows one to inspect the features that were removed in case one that is known to be valuable is assumed to be not. The same implementation for LASSO is done for Ridge to find the optimal λ and misclassification error. However, since Ridge does not perform variable selection, the model tends to overfit.

To mediate underfitting with LASSO and overfitting with Ridge, elastic-net regularization combines the two free parameters and controls the mix between LASSO and Ridge. These two parameters optimal value can be found via cross-validation methods as well. Discussed in this book are more advanced extensions to regularization that controls for bias. These regularization techniques are well documented, and the math is easy to follow at the graduate student level. Application in R is also easy to implement as the software has built in functions for all three regularization techniques via the glmnet package.

2.3.2 Multivariate Adaptive Regression Splines (MARS)

MARS is similar to traditional regression techniques; however, this machine learning algorithm is able to capture non-linearities and interactions. First proposed by Friedman (1991), MARS approach to modeling effectively uncovers important data patterns and relationships that are difficult for other regression methods to reveal. For example, there are cases where the response variable has abrupt changes at certain values of the explanatory variables. These abrupt changes are difficult to account for in common regressions. MARS is a type of regression that uses derived inputs, commonly known as basis functions, that are hinge functions. These hinge functions produce a reflected pair, namely linear spline functions. These reflected pairs do not impose or assume any particular class of relationship (e.g. linear, logistic, etc.) between the predictor variable and the dependent variable of interest. In a sense, the method is based on the divide and conquer strategy. The input space is partitioned by the spline functions into regions with each a regression equation. After fitting, the algorithm computes the predicted value if a numerical value is sought after; or scores, and assigns each case to the class with the highest score predicted for classification. This type of application yields heuristic classifications that work well in practice, but it is not based on a statistical model for deriving classification probabilities and can be prone to over fitting.

MARS ALGORITHM

1. Suppose $x_1, \ldots, x_p$ are n observations of p inputs, given as $\chi = (x_{i,j})_{nxp}$
2. The algorithm reads the data and creates a set of basis functions: $S = \{(x_j - c)_+, (c - x_j)_+ : c\epsilon\{x_{1j}, \ldots, x_{nj}\}, j=1,..,p$
3. Range of predictor values is partition into several groups by knots given by c, the outputs of basis functions are added together and given by hinge functions with the form of:

   $(x-c)_+ = x - c, x \geq c \; (c-x)_+ = c - x, x < c$

   Each knot is assigned a pair of basis functions:(a) The first basis function describes the relationship between the independent variable and the response. (b) The second is the opposite of the first.
4. MARS estimates a least-squares models with its hinge functions as the independent variables. The regression model is given as such:

   $f(x) = \beta_0 + \sum_K^{k=1} \beta_k h_k(x) \; + \; \varepsilon$

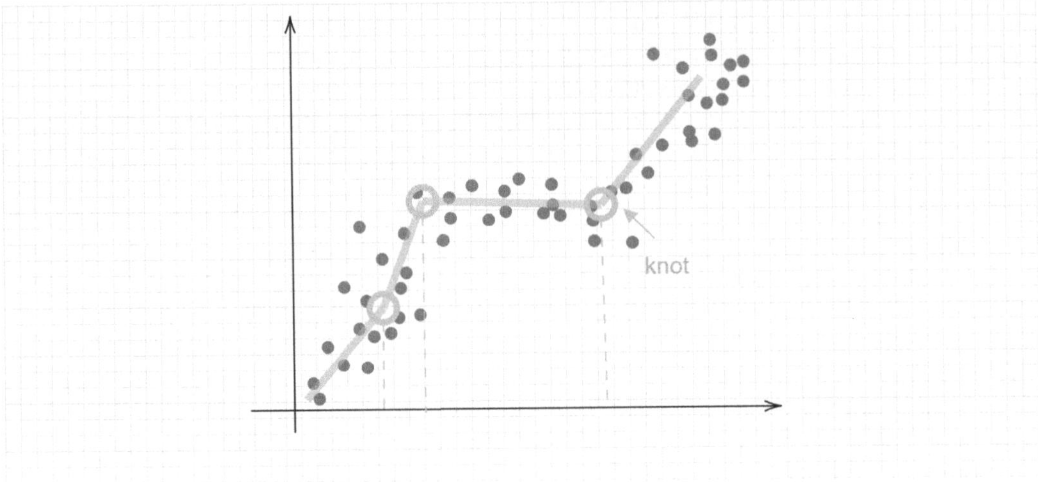

FIGURE 2.4: The Data is Split into Sections at the Knots where There are a Pair of Basis Functions. The Algorithm Fits a Regression Line to the Data Depending on where the Data is Sectioned off by the Knots.

Where $h_k(x)\epsilon S$ are hinge functions in S and the beta coefficients are estimated by least squares.

5. The algorithm begins a forward model building method from the left and starts with the basis functions, then calculates the products of each pair of the hinge function. The pair that decreases the residual error is added to the model.
6. To fit, backwards stepwise selection is used to find the best model either using AIC or GCV depending on computational efficiency and stability.

Figure 2.4 shows a simplified version of how MARS works.

2.3.3 k Nearest Neighbours (kNN)

K nearest neighbours stores all available cases from the data and classifies the new cases based on a similarity measure (e.g. a distance function). kNN has been used in statistical estimation since the 1970s as a non-parametric technique. k is the number of nearest neighbours, and k is the core deciding factor. Given a set of data and a chosen value for k, it considers all points and the new point that is meant to be predicted in an n-dimensional space. The algorithm finds the k closes point to the first predictor and then either takes a weighted average of its neighbours or classifies the point by majority vote of its neighbours (Figure 2.5).

For finding the closest, most similar points it finds the distance between points using measures such as Euclidean distance, Hamming distance, Manhattan distance, and Minkowski distance. A caveat to this method is that majority voting classification occurs when the distribution is skewed, and a more frequent class is known to dominate the prediction of the new example. Another caveat is the choice of k, the researcher must input the parameter and use cross validation techniques to decide the optimal number of neighbours.

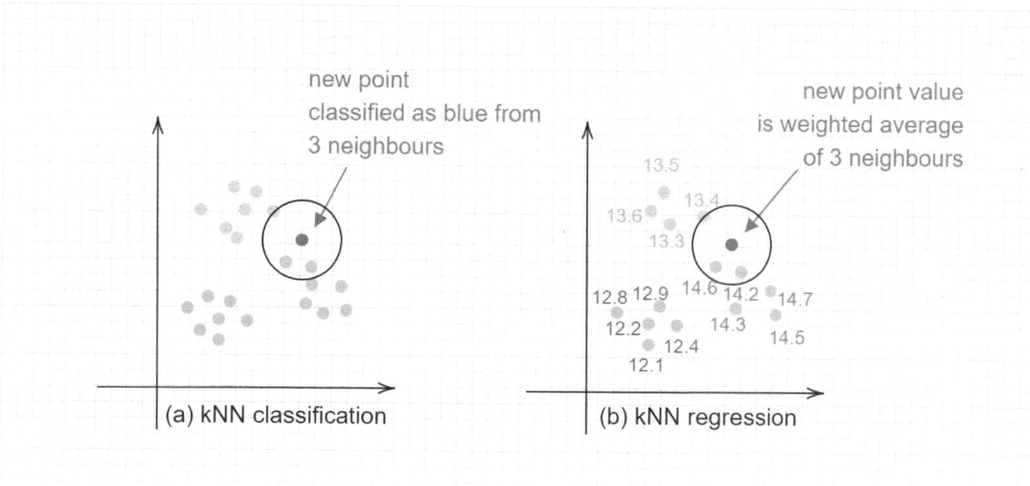

FIGURE 2.5: k Nearest Neighbours Visualization of Classification (a) and Regression (b) when k = 3.

kNN ALGORITHM (CLASSIFICATION)

1. Read data
2. Measure distance from unclassified data to all other data that is already classified using Euclidean distance
   $d\left(x, x^{\top}\right)=\sqrt{\left(x_{1}-x_{1}^{\top}\right)^{2}+\ldots+\left(x_{n}-x_{n}^{\top}\right)^{2}}$
3. Receives k smallest distances
4. Checks which classes have the shortest distance and calculates the probability of each class that appears using the following and sorts the data
   $P(y=j \mid X=x)=\frac{1}{k} \sum_{i \epsilon A} I\left(y^{\{y(i)=j\}}\right)$
5. Returns the highest probable class from the rows and returns the predicted class

kNN ALGORITHM (REGRESSION)

1. Read data
2. Measure distance between new point and each training point in the data using Euclidean distance
   $d\left(x, x^{\top}\right)=\sqrt{\left(x_{1}-x_{1}^{\top}\right)^{2}+\ldots+\left(x_{n}-x_{n}^{\top}\right)^{2}}$
3. Closest k data points are selected
4. Average of chosen data points is final prediction

kNN is a lazy learning method. In computer science, lazy learning is a learning method that does not go beyond the training data and does not attempt to make any generalizations. Due to this lazy learning method, kNN need not require training and can learn non-linear decision boundaries. The logistic regression predicts the probability of the binary classifier while kNN predicts the label as is. Since they are fundamentally different, the way to compare their performance as classification methods is to use them in practice.

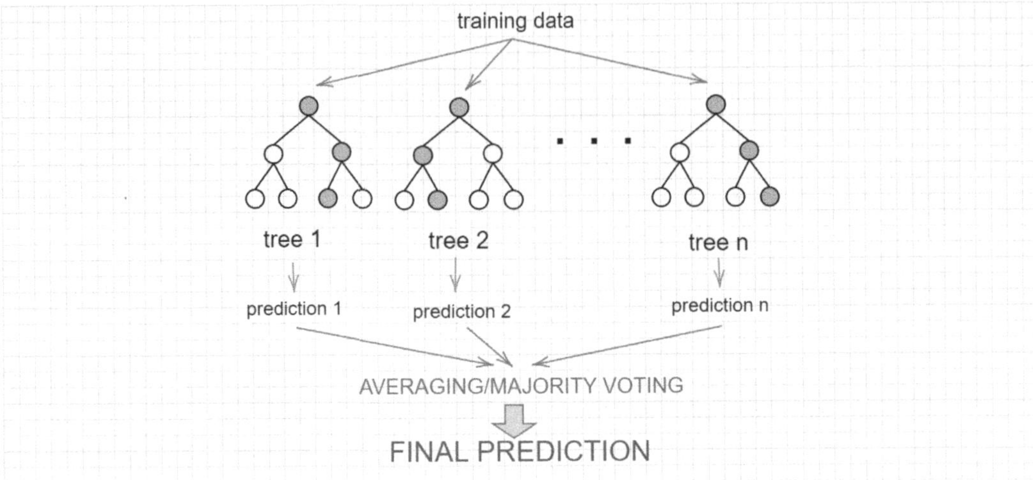

FIGURE 2.6: Random Forest Schematic for Prediction.

2.3.4 Random Forest

Random Forest is an extension of decision trees, a collection of them. Random Forest is also known as a black-box prediction based method. The term "black-box" is often used in the artificial intelligence realm to describe modeling that is defined simply by its inputs and output. How it produces an output is "opaque/black", the inner workings of the algorithm are unknown to many and the algorithm learns in a manner that is difficult for users and researchers to interpret. The human brain is often thought of as a black box, we receive inputs and our brain has a vast amount of data that has learned patterns and behaviors that it often produces a reasonable response, but we do not quite know how exactly it works. We know the fundamentals but the details, including errors, can get lost in the process.

As aforementioned, random forests are built from an ensemble of decisions trees, fundamentals we understand. It combines the predictions from several of these trees in parallel and predicts the average value or the highest ranked class (Figure 2.6). Random forests are likely to increase the accuracy of the model while maintaining the same benefits as a decision tree, where it is robust to outliers. Random forest also handles categorical data well. However, it is difficult to visualize and interpret and offers no statistical information. Unlike a logistic regression where we can obtain information about the explanatory variables and their relationship with the response variable, we have no such luck with random forests.

Random forest regression is more straightforward than classification. Random data points are chosen from the training set, and a decision tree is associated with this set of points. The number of trees is chosen and the first step is repeated for each tree. For the new data point, each tree predicts a value and the average of all predictions is assign to the new data point. Random forest regression performs well for many non-linear problems, however they tend to overfit and finding the optimal number of trees can be difficult.

Random forest classification is slightly more complicated since it uses a ranking algorithm, the details are given below.

RANDOM FOREST ALGORITHM: CLASSIFICATION

1. Suppose training set $S = \{(\boldsymbol{x}_1, y_1), \ldots, (\boldsymbol{x}_n, y_n)\}$ randomly drawn from a probability distribution of $(\boldsymbol{x}_i, y_i) \sim (\boldsymbol{X}, Y)$

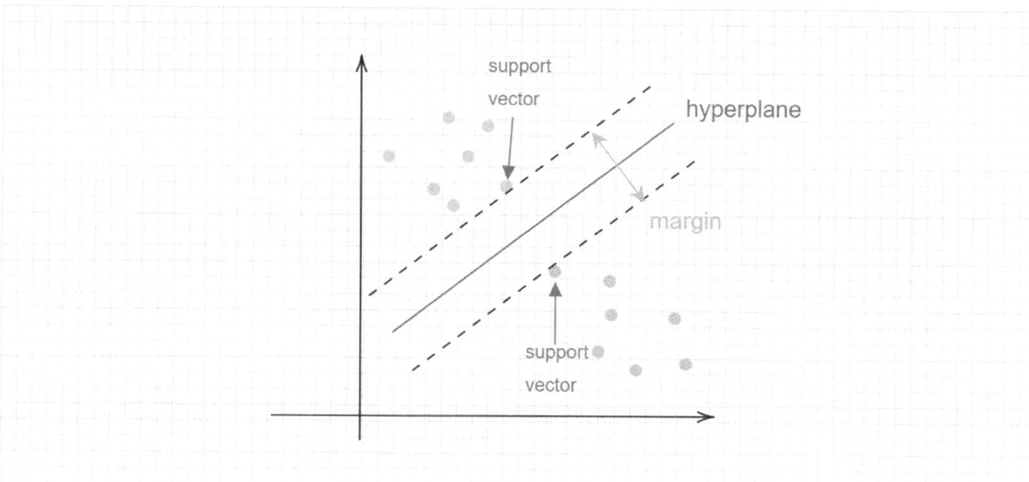

FIGURE 2.7: Support Vector Machine Example Boundary Between Two Classes.

2. Suppose there is a set of classifiers $T = \{t_1(\boldsymbol{x}), \ldots, t_M(\boldsymbol{x})\}$, assume each $t_m(\boldsymbol{x})$ is a decision tree, thus the set T is a random forest

3. Suppose parameters of the decision tree classifiers $t_m(\boldsymbol{x})$ are $\beta_m = (\beta_{m1}, \ldots, \beta_{mp})$ that define the structure of the tree, i.e. which variables are split into which node

4. Decision tree m leads to a classifier of $t_m(\boldsymbol{x}) = t(\boldsymbol{x}|\beta_m)$

5. Variables that appear in nodes of the mth tree are chosen randomly from a model variable vector β

6. Random forest is a classifier based on family of classifiers $t_m(\boldsymbol{x}) = t(\boldsymbol{x}|\beta_1), \ldots, t_m(\boldsymbol{x}) = t(\boldsymbol{x}|\beta_M)$

7. The final classification combines classifiers $\{t_m(\boldsymbol{x})\}$ and each tree casts a vote for the most popular class at input $\boldsymbol{x}$

8. Class with the most votes is chosen to be the predictor value i.e. given $S = \{(x_i, y_i)\}_{i=1}^{n}$, family of $t_m(\boldsymbol{x})$ classifiers are trained, and each classifier $t_m(\boldsymbol{x}) \equiv t(\boldsymbol{x}|\beta_m)$ is a predictor of n, where $y = pm1 =$ outcome associated with $\boldsymbol{x}$

2.3.5 Support Vector Machine (SVM)

Support vector machine was derived for the classification with two classes in the early 1960s. When the two classes are separable the maximal separating decision boundary is a hyperplane that can be found using quadratic programming (Figure 2.7). SVM aims to maximize the margins between the closest support vectors while logistic regression uses the posterior class probability. SVM uses kernel tricks that transforms the data into rich features space, so that complex problems can be dealt with in the same linear fashion in a lifted hyperspace. There are different kernels that can be implemented in R, linear, radial, and polynomial, to name a few. The use of these kernals to separate data can be difficult to interpret and thus also places SVM in the category of black-box learning methods, making it tough to interpret or gain any insight on the classifications.

SUPPORT VECTOR MACHINE ALGORITHM: CLASSIFICATION

1. Suppose there is an optimal hyperplane that separates the data with maximum margin
2. Suppose the point u is unknown and needs to be classified on each side of the hyperplane
3. Suppose a vector **w** is perpendicular to the optimal hyperplane such that the vector **u** is a point vector to u
4. Suppose vectors $\boldsymbol{xa}$ and $\boldsymbol{xb}$ are vectors for points classified as a and b respectively
5. The dot product of $\boldsymbol{w}$ and $\boldsymbol{u}$ determine if the point u is classified as a or b

   $w \cdot \boldsymbol{x_b} + b \geq 1$ and $w \cdot \boldsymbol{x_a} + b \leq 1$
6. Suppose $y = 1, ifina$ and $-1, ifinb$, then $y(w\dot{x}_{a|b} + b) - 1 \geq 0$, this is our constraint
7. The width of the margin is given by the dot product of unit vector in the direction of $\boldsymbol{w}$ and $\boldsymbol{xb} - \boldsymbol{xa}$

Support vector machines are excellent classifiers for when the groups are clearly separated, and advancements in algorithms have fine tuned them to situations, where the separation is not as clear. SVM has been a pioneer for image classifiers. Consider any search engine, and you are interested in looking at images of cats. The search engine must scrape through vasts amounts of images to curate only images of cats. Support vector machines are used by these search engines to analyze a collection of pixels that constitute cats and a collection of pixels that constitute dogs, and finds the best boundary between them to classify a new image as either a cat or a dog. Sophisticated SVM's have been developed to classify breast tissue images as benign or cancerous.The applications of SVM in image classification can be as serious as medical imaging or as seemingly innocuous like object detection on website security checks to make sure we are not a robot!

2.3.6 Linear Discriminant Analysis (LDA)

Linear Discriminant Analysis is quite an old technique that was originally developed in 1936 by Fisher (1936) that was formulated to classify data into one of two classes. Further adapted to be a multi-classifier by Rao in the 60s Rao (1969), LDA still has kept it's lustre as a machine learning technique. Like Principle Component Anlaysis, it performs dimensionality reduction by projecting features in higher dimension space to lower dimensions. In addition to finding the component axes, LDA is interested in finding the anxes that maximizes the separation between multiple classes.

LDA models the distribution of predictors separately in each of the response classes. It then, maximizes the component axes and then employs Bayes theorem to estimate the probability. For example, Fishers LDA (2 classes LDA) estimates probability using a classification device (such as naive Bayes) that uses conditional and marginal information. LDA predicts two normal density functions, one for each class, and creates a linear boundary where they intersect. This is an older method; it unfortunately is sensitive to outliers and requires a lot of assumptions from the researcher. Without normally distributed data LDA is not ideal for classification, however for dimensionality reduction it is robust to the distribution of the data.

LDA can be boiled down to 5 steps that are very similar to PCA. (1) Compute the dimensional mean vectors, (2) Compute the scatter matrices, (3) solve the eigenvalue problem, (4) select the largest eigenvectors values and (5) project the data onto the new subspace.

LINEAR DISCRIMINANT ANALYSIS ALGORITHM: CLASSIFICATION

1. Consider $K \in 1, .., K$ classes and to classify input$X$

2. The prior probabilities are calculated using the training data via Bayes Theorem that assumes a normal distribution

   $P\left(Y = i|X = a\right) = \frac{\frac{1}{\sqrt{2\pi}\sigma}e^{-\frac{1}{2}\left(\frac{X-\mu_i}{\sigma}\right)^2} P(Y=i)}{P(X=a)}$

3. Test of variances homogeneity is performed to determine if linear or quadratic discriminant analysis is needed

4. The parameters of the likelihoods are estimated

5. Discriminant functions are calculated to classify the new data into the known populations

   $\delta\left(X\right) = \frac{\mu_i X}{\sigma^2} - \frac{\mu_i^2}{2\sigma^2} + log\left(P\left(y = i\right)\right)$

6. Cross validation is used to estimate misclassification probabilities

7. Predictions are made for the new data observations

2.3.7 Artificial Neural Network (ANN)

These last two algorithms are the more difficult machine learning techniques to understand because they learn from the training set to give a prediction. The neural network is designed after our own brains neural pathways to recognize patterns. Psychologist Rosenblatt (1958) developed an artificial neural network called perceptron in order to model how our brains visually process and recognize objects. An artificial neural network is a system, and this system is a structure which receives an input, processes the data and provides an output. Input is presented to the neural network, the required target response is set at the output, and every time the NN learns, an error is obtained. The error information is fed back to the system and it makes many adjustments to their parameters in a systematic order which is commonly known as the learning rule. It is the repeated until the desired output is accepted with the lowest error.

The structure of an artificial neural network is where one can lose the reader. The structure is built of individual units called neuron (2.8. Each neuron has inputs, which are the features or explanatory variables that are fed into the model. A weight is assigned to these features, and a transfer function sums these weighted inputs into one output value. To make sure the output is not simply a linear combination of the features, an activation function introduces non-linearity to capture the patterns in the data. To control for the value produced by the activation function a bias in introduced. Multiple neurons stacked in parallel create a layer.

The input layer is the data we provide from external sources, and it is the only layer visible to the researcher. The input layer is fed into one or more hidden layers. These hidden layers constitute "deep learning", they are the layers that do all the dirty work, all of the calculations we are unaware of. The more hidden layers, the deeper the learning is. The output layer is fed information from the hidden layers and provides a final prediction based on the model's training. The architecture described passes information in one direction, from input to output and no loops are in between hidden layers. This is known as a feed-forward neural network 2.9 , one of the most common ANN's used today.

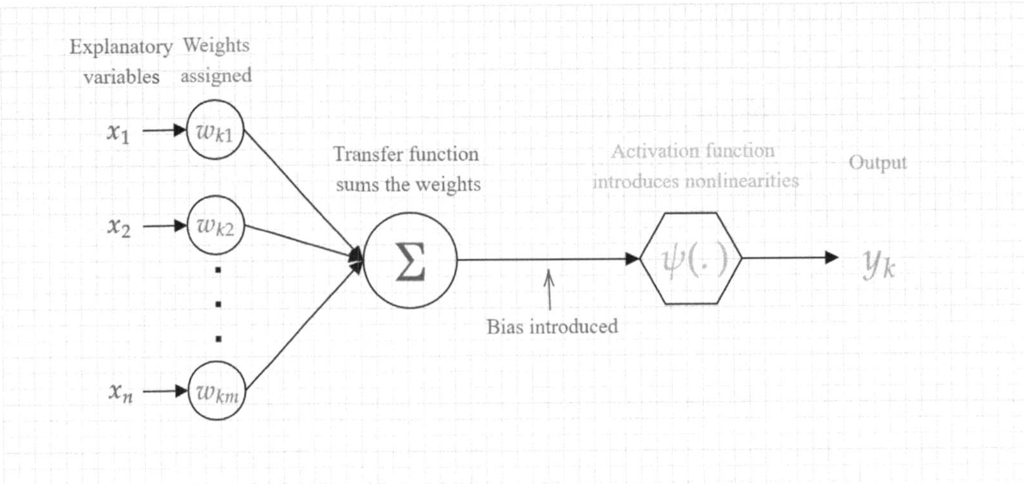

FIGURE 2.8: Architecture of a Neuron

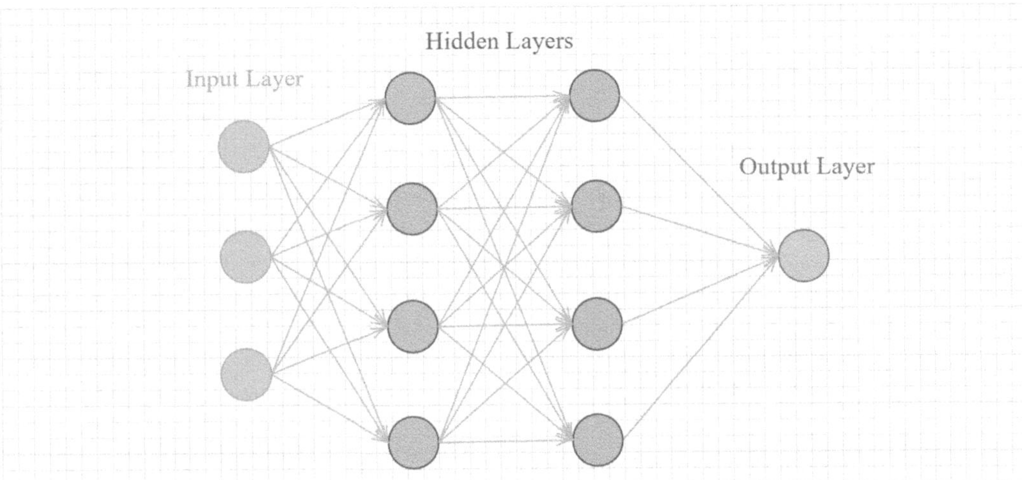

FIGURE 2.9: Architecture of a Feed Forward Neural network

ARTIFICAL NEURAL NETWORK ALGORITHM

1. Define input and target layers
2. Normalize the data
3. Separate data into training, testing, and validation set
4. Initialize number of hidden neurons, retrain and change if needed
5. Weights and biases are selected by random
6. evaluate performance (MSE, misclassfication)
7. If error goal not reach adjust weights and biases again until error goal reached
8. Obtain best network structure and parameters
9. Predict

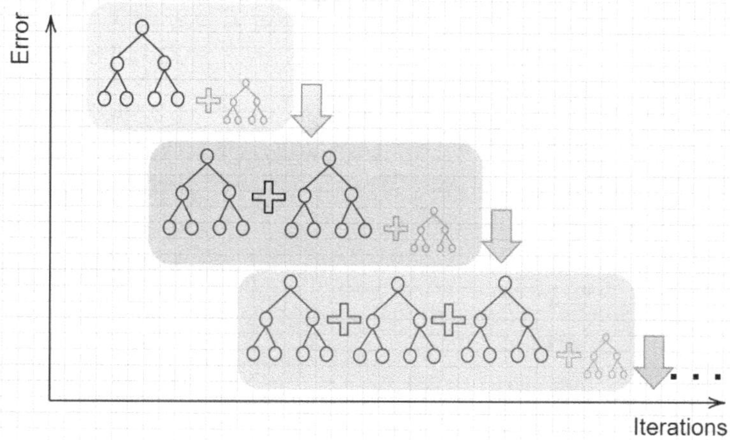

FIGURE 2.10: Gradient Boosting Machine: Error Minimized with Each Iteration of Adding More Trees.

Artificial neural networks are a superset, they can include other regressions and classifiers that can generate more complex decision boundaries. As another black-box technique, they are difficult to interpret but they provide better model accuracy. High accuracy comes at a cost of overfitting. ANN is best used for cases when historical data is likely to occur again in the same fashion. It is machine learning technique with lots of memory, but when fed data who's samples often vary from the population, it tends to overfit.

2.3.8 Gradient Boosting Machine (GBM)

Statisticians and computer scientists wanted to know if a weak learning model could be improved upon. Boosting algorithms began to develop in the late 1990s, where predictions were given by combining several statistical learning techniques. It seems intuitive, with all these statistical learning models available with faster computers, combining them all together are sure to provide us with better accuracy. And the talented data scientists of the world have done so! Sequentially adding the models, the next model is built to rectify the errors presented in the previous one (Figure 2.10).

Several frameworks lead up to the development of Gradient boosting machines where the objective was to minimize the loss by adding inaccurate models using gradient-descent technique. In other words, GBM repetitively leverages the patterns in residuals and strengthens a model with weak predictions and improves it. Once the residuals do not have any pattern that could be modeled, we stop! Algorithmically, we are minimizing our loss function, such that test loss reach its minima. GBM is a series of combinations of additive decision models, estimated iteratively resulting in a stronger learning model. Usually the gradient boosting machines is used in decision models or logistic regressions. There is a performance limit when dealing with high-dimensional data. The more complex the model gets in deep learning, the more they are prone to overfitting, back in the black box with little interpretability.

GRADIENT BOOSTING MACHINE ALGORITHM

1. Calculate the average value of the target variable to act as a baseline for predictions.

2. Calculate the residual values using the mean of the target variable

TABLE 2.1: Machine Learning Technique Packages in R.

Learning technique	R package	Function
PCA	Built in base package factoextra (for plotting)	prcomp() fviz_pca_biplot()
K-means clustering	Built-in base/stats package	kmeans()
Logistic Regression	Built-in base	glm(, family=binomial)
KNN	class	knn(,k=n)
LDA	MASS caret	lda() qda() for quadratic DA
Random Forest	randomForest	randomForest()
SVM	e1071	svm(~,kernel=linear)
ANN	neuralnet	neuralnet()
GBM	gbm xgboost	gbm() xgb.cv()

3. Construct a decision tree with the goal of predicting the residuals first
4. Predict the target label using all the trees within the ensemble. To prevent overfitting as learning rate is multiplied by the residuals to lean toward adding more trees
5. Compute new residuals
6. Repeat 3-5 until number of iterations equal the number of features
7. Final prediction is calculated by using the mean in addition to all the residuals predicted by the trees in the ensemble multiplied by the learning rate

2.4 Implementation in R

R has been a saving grace for many data scientists and statisticians alike. As research progresses, so have the computing packages. Table 2.1 is a list of R packages available that can perform the aforementioned machine learning techniques.

2.5 Case Study: Genomics

Using the topics introduced in addition to the popular machine learning applications we will investigate which ClinVar human genetic variants features or combinations thereof will help researchers predict classification conflicts with the most suitable model. This is a binary classification problem, and with so many options out there for prediction, we will explore some of the hot topic machine learning techniques that sound more like buzzwords these days and evaluate their predictive power. The motivation comes from open science, as accessible

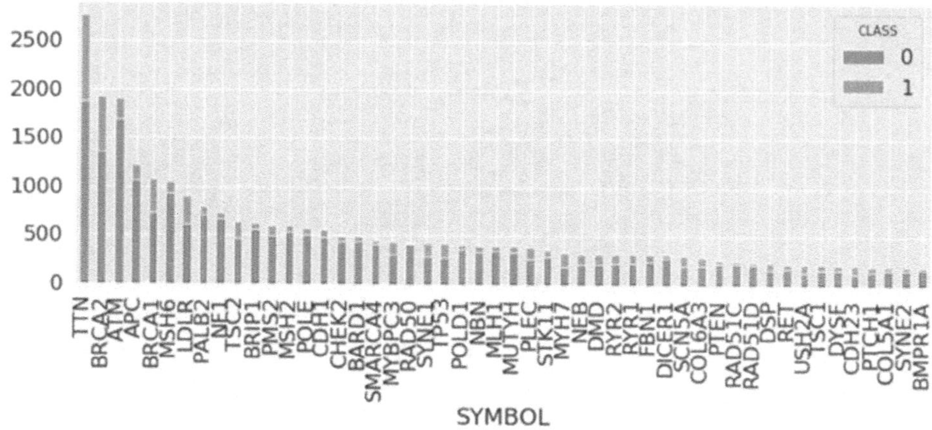

FIGURE 2.11: Frequency Bar Plot of Class Distribution Amongst Genes.

databases like ClinVar have become public, they are a saving grace for many researchers. The problem at hand is that there is often clinical uncertainty. The level of confidence in the accuracy of claims of clinical significance are reliant on the supporting evidence and the origin of reports. The data available to us has $N = 65118$ observations and $p = 46$ predictors and the binary target variable is variant classification conflict of CLASS=1 or concordant CLASS=0. We will explore machine learning and find which genetic variants features or combinations thereof will help predict classification conflicts with the most suitable model.

2.5.1 Data Exploration

Since the data has 46 attributes, comprised of continuous, categorical, and text features, it creates complexity and the task visualization and exploratory analysis is formidable. Let's see how the target variable is distributed by gene to give ourselves an idea of what the data looks like.

There is an obvious class imbalance problem to which the data will be balanced by under-sampling and oversampling such that the minority class, classification conflict, is over-sampled with replacement and majority class , classification concordant, is under-sampled without replacement.

With 40+ features that have several categorical variables that need to be accounted for with indicator variables in each of the models, it can be very computationally taxing. Those who understand their own data or possibly collected it would be able to perform some dimensionality reduction by inspection or employ other methods. However, dimensionality reduction without knowing how your data behaves can limit your analysis! Especially in this case where genomics data is so complex it is difficult to tell by inspection which predictors are important. This is where some unsupervised learning can help find the natural tendencies in your data! For example: One of the features that is difficult to analyze and draw conclusions from are the text manually entered by the laboratory that report notes on diseases with respect to classification conflicts. Here is a chance to implement some unsupervised text mining. These laboratory notes come from professional doctors and often time variables that contain strings are often dropped. BUT open-ended descriptions by patients or the lab describing symptoms might yield useful clues for the medical diagnosis that go unnoticed.

Text mining finds similarities between words or how they may relate to variables and turn text into numbers or meaningful indices. The word cloud generated below shows the most

FIGURE 2.12: Wordcloud Generated from Laboratory Reports Showing Frequency of Most Common Words.

TABLE 2.2: Frequency Ranking for Words with Highest Occurrences.

Rank	Word
1	Hereditary
2	CancerPredisposing
3	Cardiomyopathy
4	Dystrophy
5	Recessive

common clinical words describing diagnoses that are associated with classification conflicts as it reports the top words from the laboratory reports for variant classification conflicts. These top words could potentially aid in prediction of whether or not a classification conflict could occur. Using the top five frequent words in conflicts, a new binary variable is created called POTENTIAL shown in Table 2.2. The POTENTIAL variable is 1 if the top words are included in the clinical report for an individual and 0 otherwise. This feature is engineered by filtering each observations laboratory report note string by the top words found. This is how auto-complete on ones phone functions based on the frequency of words associated with another it returns the most frequent association! This is a simple unsupervised algorithm that can be helpful in constructing models.

The next step is to prepare the data for modeling, this part is done already where the student must check for multicollinearity and any missing values and how to deal with them. Note: chi-square and correlation tests can be regarded as unsupervised learning techniques since they make no assumptions and simply report back data patterns.

2.5.2 Modeling

Employing LASSO (L_1 regularization) shrinks the parameter estimates of insignificant variables to zero, thus performing variable selection. We then use cross-validation for the free parameter lambda that minimized the out of sample error found via grid search using 5 folds

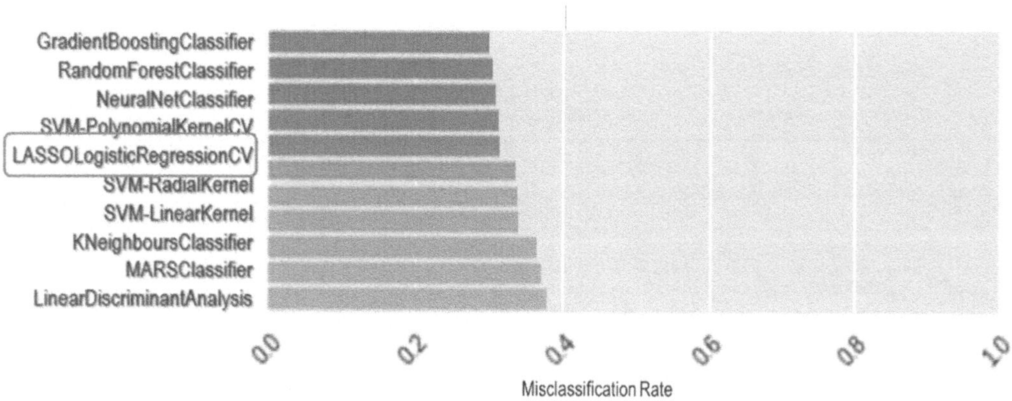

Model	Interpretability	Computational Expense
Gradient Boosting Machine	POOR	HIGH
Random Forest	POOR	HIGH
Neural Net	POOR	HIGH
Support Vector Machine	MODERATE	MODERATE
LASSO-Logistic	HIGH	LOW
K Nearest Neighbours	HIGH	HIGH
Multi Adaptive Regression Splines	MODERATE	LOW
Linear Discriminant Analysis	MODERATE	LOW

FIGURE 2.13: Modeling Results Based on Misclassification Rate

for computational ease. The most parsimonious model obtained from LASSO returned 13 significant predictors to predict variant classification conflict and are used for each machine learning model to compare model accuracy and overall performance. These results are based on the assumption that LASSO has provided us with the best candidate submodel to employ any of these machine learning methods! Using a training dataset into each of the aforementioned models using the R packages available, the misclassification rate is calculated for each. The results below show us that the logistic classifier competes with the sophisticated machine learning techniques!

Method	Brief Description + PRO\CON	Method & Optimal Parameters	R package	Computational Expense
MARS	Similar to traditional regression but captures nonlinearities and interactions. MARS approach to modeling effectively uncovers important data patterns and relationships that are difficult for other regression methods to reveal + EASY TO UNDERSTAND	CV: 10-fold, repeated 7 times	earth, caret	Low
KNN	It need not require training and can learn non-linear decision boundaries. The logistic regression predicts the probability of the binary classifier while KNN predicts the label as is. Since they are fundamentally different, the way to compare their performance as classification methods is to use in practice.	Optimal k=21 (see appendix for graph)	FNN, class	Moderate
Random Forest	Likely increase the accuracy of our model while maintaining the same benefits as a decision tree, where it is robust to outliers. Handles categorical data well (much of this dataset is categorical). Difficult to visualize and interpret.	Optimal trees n=55 vis grid search (see appendix for graph)	random Forest	High
SVM	SVM aims to maximize the margin between the closest support vectors while logistic regression uses the posterior class probability. SVM uses kernel tricks that transforms data into rich features space, so that complex problems can be dealt with in the same "linear" fashion in a lifted hyperspace.	1. Linear kernel (see if data can be linearly separated) 2. Radial kernel 3. Polynomial kernel Found polynomial to be optimal SVM model and then iterated to find best cost value = 70	e1071, wavethresh	Moderate
LDA	Fisher's LDA (2 class LDA): estimates probability using classificatory device (such as naive Bayes) uses conditional and marginal information. Predicts two normal density functions (one for each class) that creates a linear boundary where they intersect: older method, sensitive to outliers, requires lots of assumptions	CV: 10-fold, repeated 7 times	MASS	Low
Neural Net	Neural Networks: A superset that includes Logistic regression and other classifiers that can generate more complex decision boundaries. Difficult to interpret but better model accuracy	Optimal: hidden layers = 1, nodes = 9, learning rate = 0.6 (See Appendix)	neuralnet	High
GBM	Series of combinations of additive models, estimated iteratively resulting in a stronger learning model. Usually the gradient boosting method is used of decision tree models or logistic. Prone to overfitting, performance limit when dealing with high dimensional data	5-fold CV with logistic GBM	xgboost	High

FIGURE 2.14: Model Methodology Breakdown

3

Post-Shrinkage Strategies in Sparse Regression Models

3.1 Introduction

The world is multi-faceted and complex, and multiple linear regression is a simple tool that statisticians can use to help us understand some of the complexities. Most of the questions that arise out of research, industry, and our daily lives are often answered by the phrase, "it depends." Our decisions depend on the variables that contribute to making them. The multiple linear regression model is an extension of the linear regression model to incorporate more than one predictor. This is a statistical technique that allows researchers to infer relationships between predictor variables and the variable of interest, the response variable. Since it depends on the predictors, the response variable is also called the dependent variable.

Multiple regression models are very flexible and can take many forms, depending on how the predictors are entered into the model. It allows us to include a mixture of continuous and categorical predictors, as well as any interaction terms. Interaction terms allow us to simultaneously assess the combined effect of parameters on the response variable. These interaction terms can be between continuous variables, categorical variables, or a mixture of the two. For example, the price of an automobile made by a motor company depends on a variety of characteristics such as car model, body type,engine size, interior style, leather seats, adaptive cruise control, and number of cameras, among others. But there can be information hidden between the combined effect of some of these variables, and interaction terms can aid in extracting that information.

The ultimate objective of a multiple regression analysis is to develop a model that will most accurately predict the response mean (conditional mean of the response variable) as a function of a set of predictor variables. For example, if we wish to develop a regression model to predict the retail price of a new car, one of the primary issues is to determine which predictor to include in the model and later which one to leave out. Adding all predictors to the model can result in a cumbersome model that would be difficult to interpret. Complex may mean accurate, but it does not mean comprehensible. As a researcher, finding the most parsimonious model is the goal, to be able to produce great results but to also explain how to achieve them.

On the other hand, if a model includes only a few predictors, it oversimplifies the problem and may provide substantially different predictions than the initial model including all the predictors. As statisticians, we strive to find the best model that can handle such a delicate balance.

Therefore, estimation, prediction, and variable selection are important features for implementing models and doing data analysis. As it is beneficial to keep the number of predictors as small as possible for interpretation purposes, a submodelbased on a few predictors selected from a larger model (full model) may be considerably biased. Many models have been developed for large data sets, but mostly for data where the number of predictors does

DOI: 10.1201/9781003170259-3

not exceed the number of observations. The classical estimation methods can be used only when the number of observations (sample size, n) in the data set exceeds the number of predictors (p) in the model. There are also situations where the number of observations is very large, and such data sets are classified as big data. Finally, there are situations when both n and p are large.

These days, there are many data sets where the number of predictors is greater than the number of observations. Genomic data falls into this category, where there are a large number of genes but minimal observations for gene expression. This is known as high-dimensional regression analysis in the reviewed literature. For this reason, we suggest a post-shrinkage strategy to combine the full model (the model including all the predictors) with a selected submodel in an adaptive way.

To provide some background, let's return to the $p \leq n$ setting. The maximum likelihood and least squares methods are the most widely used techniques for estimating regression parameters for a given model, either the full model or the submodel. However, in this chapter we focus on integrating full model and submodel estimation using classical shrinkage strategies.

There is a considerable amount of research on improving the maximum likelihood estimators. For example, there has recently been great attention on applying James-Stein shrinkage ideas to parameter estimation in regression models Ahmed (1994, 1997a, 1998); Ahmed and Basu (2000); Ahmed (2001, 2014); Ahmed et al. (2016); Yüzbaşı et al. (2017a); Liang and Song (2009); An et al. (2009); Yüzbaşı et al. (2022). It was inspired by Stein's result that if the parameter dimension is three or larger, the estimators can be improved by incorporating auxiliary information into the estimation procedure, and one may obtain relatively more efficient estimators than when the auxiliary information is ignored.

Statistical methods for developing efficient estimators can be classified into three choices to select a reduced model or submodel from the full model that contains manypredictors. Generally speaking, practitioners prefer to work with models with a reasonable number of predictors. Thus, after building an appropriate full model, possibly with many available predicting variables, one can select a candidate submodel with a small number of influential predictors. This can be achieved usingclassical variable selection strategies, using either the stepwise or subset selection procedures. People also often use the Akaike information criterion (AIC), the Bayesian information criterion (BIC), or some other penalized model selection methods.

Specifying the statistical model is, as always, a critical component in estimation, prediction, and inference. One typically studies the consequences of some forms of model misspecification. A common type of model misspecification is the inclusion of unnecessary predictors in the full model or the omission of necessary variables in the submodel. A delicate balance that researchers are always trying to find.

The validity of eliminating statistical uncertainty through the specification of aparticular parametric formulation depends on information that is generally not available. Theaim of this communication is to analyze some of the issues involved in the estimation of a model that may be over-parameterized or under-parameterized. For example, in the data analyzed by Engle et al. (1986), the electricity demand may be affected by weather, price, income, strikes,and other factors. If we have reason to suspect that a strike has no effect on electricity demand, we may want to decrease the influence of, or delete, this variable. Recently, Cui et al. (2005)developed an estimator of the error variance that can borrow information across genes usingthe James–Stein shrinkage concept. For linear models, Tibshirani (1996) proposed the "least absolute shrinkage and selection operator" (LASSO) method to shrink some coefficients andto set others to zero, and hence tries to retain the good features of both subset selection andridge regression. A penalty on the sum of the absolute ordinary least squares coefficients isintroduced to achieve both continuous shrinkage and automatic variable deletion. The idea ofusing an absolute penalty was used by Chen and Donoho (1994) and Chen et al. (2001) to shrink and delete basic coefficients. Ahmed (2014) advocated using the shrinkage strategy, which combines the full and submodel estimators and improves the performance of the maximum likelihood estimator with respect to the mean squared error (MSE).

The methodologies for variable selection are specific to the statistical method used to estimate the model. It is important to note the difference between classical variable selection and penalized methods. For example, stepwise regression with AIC or BIC criteria or penalty methods selects predictors in the linear regression model. However, the modern penalized likelihood methods are used for simultaneous variable selection and parameter estimation, and they are extremely useful for high-dimensional regression models when $n < p$. These methods deal with ill-defined regression problems in classical frameworks. As mentioned earlier, one of the penalty methods is LASSO Tibshirani (1996), which shrinks some or many less important coefficients to zero. There are other penalized methods such as the smoothly clipped absolute deviation (SCAD) by Fan and Li (2001), ENET by Zou and Hastie (2005), and ALASSO by Zou (2006). The SCAD estimator has many important properties, including continuity, sparsity, and unbiasedness. It also has the Oracle property when the dimension of predictors is fixed or diverges more slowly than the sample size. The ALASSO penalty is a modified version of the LASSO penalty that allows for different amounts of shrinkage for different regression coefficients. It has been shown theoretically that the ALASSO estimator is able to identify the true model consistently, and the resulting estimator is as efficient as Oracle.

The purpose of this chapter is to concentrate on Ahmed (2014) shrinkage strategy for estimating regression parameters, which results in efficient regression parameters and response mean prediction. In an effort to formulate the shrinkage estimators, suppose that the full model has a large number of predictors. In practice, a given variable selection method can be used to get the subset of predictors to keep in the submodel.

To establish the theoretical properties of the estimators, let us partition the regression parameters vector of the full model as follows: $\boldsymbol{\beta} = (\boldsymbol{\beta}_1^\top, \boldsymbol{\beta}_2^\top)^\top$, where ("T") denotes the transpose of a vector or matrix, $\boldsymbol{\beta}_1$ be the parameter vector for the important predictors, to be retained in the model, and $\boldsymbol{\beta}_2$ may be considered as the nuisance parameter vector, maybe removed from the model, assuming that these predictors provide no improvement in prediction. With this motivation, the submodel is then defined simply the submodel subject to the constraint $\boldsymbol{\beta}_2 = \mathbf{0}$, essentially this is so-called sparsity condition. However, the shrinkage strategy does not discard this information but incorporates and retains it to improve the estimation of the submodel. By design, the post-shrinkage estimators are obtained by shrinking the full model estimators toward the submodel estimators, with the shrinkage direction may be defined by the restriction on $\boldsymbol{\beta}_2$.

The model and some estimation strategies, including least squares estimation, maximum likelihood estimation, full and submodel estimation, and shrinkage estimation, are introduced in Section 3.2. In Section 3.3, the asymptotic properties are given. In Section 3.4, we compare the pairwise risk performance of the proposed estimators. Section 3.5 contains a Monte Carlo simulation evaluation to quantify the relative performance of the estimators listed. The real data example is considered in Section 3.6. The R codes are available in Section 3.7. Our findings are summarized in Section 3.8.

3.2 Estimation Strategies

For a smooth reading of this chapter, we review some of the estimation strategies for the parameters of the multiple regression models. Further, we show how to obtain the full model, submodel, and shrinkage estimators.

3.2.1 Least Squares Estimation Strategies

Consider the sparse linear regression model

$$\boldsymbol{y} = \boldsymbol{X}\boldsymbol{\beta} + \boldsymbol{\varepsilon}, \tag{3.1}$$

where

$$\boldsymbol{X} = \begin{bmatrix} 1 & \boldsymbol{x}_{11} & \boldsymbol{x}_{12} & \cdots & \boldsymbol{x}_{1p} \\ 1 & \boldsymbol{x}_{21} & \boldsymbol{x}_{22} & \cdots & \boldsymbol{x}_{2p} \\ \vdots & \vdots & \vdots & \vdots & \vdots \\ 1 & \boldsymbol{x}_{n1} & \boldsymbol{x}_{n2} & \cdots & \boldsymbol{x}_{np} \end{bmatrix}_{n\times(p+1)},$$

where $\boldsymbol{y} = (y_1, y_2, \cdots, y_n)^\top$ is a $n\times 1$ vector and $\boldsymbol{\beta} = (\beta_0, \beta_1, \beta_2, \cdots, \beta_p)^\top$ is a $p\times 1$ vector of parameter. The error vector $\boldsymbol{\varepsilon} = (\varepsilon_1, \varepsilon_2, \cdots, \varepsilon_n)^\top$ is independent and identically distributed random variables with $\mathrm{E}(\boldsymbol{\varepsilon}) = \mathbf{0}$ and variance $\mathrm{Var}(\boldsymbol{\varepsilon}) = \boldsymbol{I}\sigma^2$. The regression parameters are estimated using the least squares principle. Note that it is not necessary to assume that the error term has a normal distribution in order to find the least squares estimator (LSE) of regression parameters. However, under the normality assumption, the LSEs are exactly the same as the maximum likelihood estimators (MLEs).

The matrix form of the regression model (3.1) allows us to discuss and present many properties of the model more conveniently and efficiently. The least squares estimate $\widehat{\boldsymbol{\beta}}^{\mathrm{FM}}$ of $\boldsymbol{\beta}$ is chosen to minimize the residual sum of squares function

$$\widehat{\boldsymbol{\beta}}^{\mathrm{FM}} = \underset{\boldsymbol{\beta}}{\operatorname{argmin}}(\boldsymbol{y} - \boldsymbol{X}\boldsymbol{\beta})^\top(\boldsymbol{y} - \boldsymbol{X}\boldsymbol{\beta}). \tag{3.2}$$

By taking partial derivative in the right side with respect to each component of $\boldsymbol{\beta}$ and set to $\mathbf{0}$ to obtain the normal equation $\boldsymbol{X}^\top\boldsymbol{X}\widehat{\boldsymbol{\beta}}^{\mathrm{FM}} = \boldsymbol{X}^\top\boldsymbol{y}$. The OLS estimates $\widehat{\boldsymbol{\beta}}^{\mathrm{FM}}$ is given by the following formula

$$\widehat{\boldsymbol{\beta}}^{\mathrm{FM}} = (\boldsymbol{X}^\top\boldsymbol{X})^{-1}\boldsymbol{X}^\top\boldsymbol{y}, \tag{3.3}$$

provided that the inverse $(\boldsymbol{X}^\top\boldsymbol{X})^{-1}$ exists.

3.2.2 Maximum Likelihood Estimator

For maximum likelihood method, we assume that $\boldsymbol{\varepsilon}$ is normally distributed. The $\boldsymbol{y}|\boldsymbol{X} \sim \mathrm{N}(\boldsymbol{X}\boldsymbol{\beta}, \boldsymbol{I}\sigma^2)$. The likelihood function $L(\boldsymbol{\beta}, \sigma^2|\boldsymbol{y})$ is the joint probability density function of $f(\boldsymbol{y}|\boldsymbol{\beta}, \sigma^2)$:

$$\begin{aligned}
L(\boldsymbol{\beta}, \sigma^2|\boldsymbol{y}) &= (2\pi\sigma^2)^{-n/2} exp\left[-\frac{(\boldsymbol{y} - \boldsymbol{X}\boldsymbol{\beta})^\top(\boldsymbol{y} - \boldsymbol{X}\boldsymbol{\beta})}{2\sigma^2}\right] \\
l(\boldsymbol{\beta}, \sigma^2|\boldsymbol{y}) &= log L(\boldsymbol{\beta}, \sigma^2|\boldsymbol{y}) = -\frac{n}{2}log(2\pi\sigma^2) - \frac{(\boldsymbol{y} - \boldsymbol{X}\boldsymbol{\beta})^\top(\boldsymbol{y} - \boldsymbol{X}\boldsymbol{\beta})}{2\sigma^2} \\
&= -\frac{n}{2}log(2\pi\sigma^2) - \frac{(\boldsymbol{y} - \boldsymbol{X}\boldsymbol{\beta})^\top(\boldsymbol{y} - \boldsymbol{X}\boldsymbol{\beta})}{2\sigma^2} \\
\frac{\partial l}{\partial \boldsymbol{\beta}} &= 0 - \frac{1}{2\sigma^2}\frac{\partial}{\partial \boldsymbol{\beta}}[(\boldsymbol{y} - \boldsymbol{X}\boldsymbol{\beta})^\top(\boldsymbol{y} - \boldsymbol{X}\boldsymbol{\beta})] = \mathbf{0}.
\end{aligned}$$

Now

$$\begin{aligned}
& -\frac{1}{2\sigma^2}\frac{\partial}{\partial\boldsymbol{\beta}}[(\boldsymbol{y}-\boldsymbol{X\beta})^\top(\boldsymbol{y}-\boldsymbol{X\beta})] = \mathbf{0} \\
\Rightarrow\ & \frac{\partial}{\partial\boldsymbol{\beta}}[(\boldsymbol{y}-\boldsymbol{X\beta})^\top(\boldsymbol{y}-\boldsymbol{X\beta})] = \mathbf{0} \\
\Rightarrow\ & \frac{\partial}{\partial\boldsymbol{\beta}}[(\boldsymbol{y}^\top-\boldsymbol{\beta}^\top\boldsymbol{X}^\top)(\boldsymbol{y}-\boldsymbol{X\beta})] = \mathbf{0}, \quad \textit{since} \quad (\boldsymbol{X\beta})^\top = \boldsymbol{\beta}^\top\boldsymbol{X}^\top \\
\Rightarrow\ & \frac{\partial}{\partial\boldsymbol{\beta}}[(\boldsymbol{y}^\top\boldsymbol{y}-\boldsymbol{y}^\top\boldsymbol{X\beta}-\boldsymbol{\beta}^\top\boldsymbol{X}^\top\boldsymbol{y}+\boldsymbol{\beta}^\top\boldsymbol{X}^\top\boldsymbol{X\beta}] = 0 \\
\Rightarrow\ & -\boldsymbol{X}^\top\boldsymbol{y}-\boldsymbol{X}^\top\boldsymbol{y}+[(\boldsymbol{X}^\top\boldsymbol{X})+(\boldsymbol{X}^\top\boldsymbol{X})^\top]\boldsymbol{\beta} = \mathbf{0} - - - (**) \\
\Rightarrow\ & -2\boldsymbol{X}^\top\boldsymbol{y}+[2(\boldsymbol{X}^\top\boldsymbol{X})]\boldsymbol{\beta} = \mathbf{0}, \quad \textit{since} \quad (\boldsymbol{X}^\top\boldsymbol{X})^\top = (\boldsymbol{X}^\top\boldsymbol{X}) \\
\Rightarrow\ & (\boldsymbol{X}^\top\boldsymbol{X})\boldsymbol{\beta} = \boldsymbol{X}^\top\boldsymbol{y} \\
\Rightarrow\ & (\boldsymbol{X}^\top\boldsymbol{X})^{-1}(\boldsymbol{X}^\top\boldsymbol{X})\boldsymbol{\beta} = (\boldsymbol{X}^\top\boldsymbol{X})^{-1}\boldsymbol{X}^\top\boldsymbol{y} \\
\Rightarrow\ & \widehat{\boldsymbol{\beta}}^{\mathrm{FM}} = (\boldsymbol{X}^\top\boldsymbol{X})^{-1}\boldsymbol{X}^\top\boldsymbol{y}
\end{aligned}$$

The estimator $\widehat{\boldsymbol{\beta}}^{\mathrm{FM}}$ is called least squares estimator or maximum likelihood estimator of $\boldsymbol{\beta}$. Here $\boldsymbol{y}^\top\boldsymbol{X\beta}$ and $\boldsymbol{\beta}^\top\boldsymbol{X}^\top\boldsymbol{y}$ are scalars, so

$$\begin{aligned}
& \frac{\partial(\boldsymbol{X}^\top\boldsymbol{\beta})}{\partial\boldsymbol{\beta}} = \frac{\partial(\boldsymbol{\beta}^\top\boldsymbol{X})}{\partial\boldsymbol{\beta}} = \boldsymbol{X} \\
& \frac{\partial(\boldsymbol{\beta}^\top(\boldsymbol{X}^\top\boldsymbol{X})\boldsymbol{\beta})}{\partial\boldsymbol{\beta}} \\
=\ & \frac{\partial}{\partial\boldsymbol{\beta}}(\boldsymbol{\beta})^\top(\boldsymbol{X}^\top\boldsymbol{X})\boldsymbol{\beta} + \frac{\partial}{\partial\boldsymbol{\beta}}(\boldsymbol{\beta})[(\boldsymbol{X}^\top\boldsymbol{X})^\top\boldsymbol{\beta}] \\
=\ & (\boldsymbol{X}^\top\boldsymbol{X})\boldsymbol{\beta} + [(\boldsymbol{X}^\top\boldsymbol{X})^\top\boldsymbol{\beta}] \\
=\ & [(\boldsymbol{X}^\top\boldsymbol{X}) + (\boldsymbol{X}^\top\boldsymbol{X})^\top]\boldsymbol{\beta} \\
=\ & 2\boldsymbol{X}^\top\boldsymbol{X\beta}, \quad \textit{since} \quad \boldsymbol{X}^\top\boldsymbol{X} \quad \textit{is symmetric.}
\end{aligned}$$

Further,

$$\begin{aligned}
\frac{\partial l}{\partial\sigma^2} &= \frac{\partial}{\partial\sigma^2}\left[-\frac{n}{2}log(2\pi\sigma^2)\right] - \frac{\partial}{\partial\sigma^2}\left[\frac{(\boldsymbol{y}-\boldsymbol{X\beta})^\top(\boldsymbol{y}-\boldsymbol{X\beta})}{2\sigma^2}\right] \\
&= -\frac{n}{2\sigma^2} + \frac{(\boldsymbol{y}-\boldsymbol{X\beta})^\top(\boldsymbol{y}-\boldsymbol{X\beta})}{2\sigma^4} = 0, \\
\text{Hence} \quad & \widehat{\sigma}^2 = \frac{(\boldsymbol{y}-\boldsymbol{X}\widehat{\boldsymbol{\beta}}^{\mathrm{FM}})^\top(\boldsymbol{y}-\boldsymbol{X}\widehat{\boldsymbol{\beta}}^{\mathrm{FM}})}{n}
\end{aligned}$$

3.2.3 Full Model and Submodel Estimators

We consider experiments where the vector of coefficients $\boldsymbol{\beta}$ in model (3.1) can be partitioned as $(\boldsymbol{\beta}_1^\top, \boldsymbol{\beta}_2^\top)^\top$, where $\boldsymbol{\beta}_1$ is the coefficient vector of active predictors and $\boldsymbol{\beta}_2$ is a vector for "nuisance" effects. In this situation, inferences about $\boldsymbol{\beta}_1$ may benefit from moving the MLE for the full model in the direction of MLE without the nuisance variables or from dropping the nuisance variables if there is evidence that they do not provide useful information for the response.

We let $\boldsymbol{X} = (\boldsymbol{X}_1, \boldsymbol{X}_2)$, where $\boldsymbol{X}_1$ is an $n \times p_1$ submatrix containing the regressors of interest, that is, active covariate and $\boldsymbol{X}_2$ is an $n \times p_2$ submatrix that may or may not be active for the response. Accordingly, let $\boldsymbol{\beta}_1$ and $\boldsymbol{\beta}_2$ have dimensions p_1 and p_2, respectively with $p_1 + p_2 = p$. The model (3.1) can be written as

$$\boldsymbol{y} = \boldsymbol{X}_1\boldsymbol{\beta}_1 + \boldsymbol{X}_2\boldsymbol{\beta}_2 + \boldsymbol{\varepsilon}, \tag{3.4}$$

We are essentially interested in the estimation of $\boldsymbol{\beta}_1$ when it is suspected that $\boldsymbol{\beta}_2$ is close to $\mathbf{0}$. The log-likelihood for model (3.4) can be written as

$$\begin{aligned}
l(\boldsymbol{\beta}, \sigma^2|\boldsymbol{y}) &= -\frac{n}{2}log(2\pi\sigma^2) - \frac{(\boldsymbol{y} - \boldsymbol{X}_1\boldsymbol{\beta}_1 - \boldsymbol{X}_2\boldsymbol{\beta}_2)^\top(\boldsymbol{y} - \boldsymbol{X}_1\boldsymbol{\beta}_1 - \boldsymbol{X}_2\boldsymbol{\beta}_2)}{2\sigma^2} \\
&= -\frac{n}{2}log(2\pi\sigma^2) - \frac{1}{2\sigma^2}\left[\boldsymbol{y}^\top\boldsymbol{y} - \boldsymbol{\beta}_1^\top\boldsymbol{X}_1^\top\boldsymbol{y} - \boldsymbol{\beta}_2^\top\boldsymbol{X}_2^\top\boldsymbol{y} - \boldsymbol{y}^\top\boldsymbol{X}_1\boldsymbol{\beta}_1\right. \\
&+ \boldsymbol{\beta}_1^\top\boldsymbol{X}_1^\top\boldsymbol{X}_1\boldsymbol{\beta}_1 + \boldsymbol{\beta}_2^\top\boldsymbol{X}_2^\top\boldsymbol{X}_1\boldsymbol{\beta}_1 \\
&- \left.\boldsymbol{y}^\top\boldsymbol{X}_2\boldsymbol{\beta}_2 + \boldsymbol{\beta}_1^\top\boldsymbol{X}_1^\top\boldsymbol{X}_2\boldsymbol{\beta}_2 + \boldsymbol{\beta}_2^\top\boldsymbol{X}_2^\top\boldsymbol{X}_2\boldsymbol{\beta}_2\right].
\end{aligned} \tag{3.5}$$

The full model estimator $\widehat{\boldsymbol{\beta}}_1^{\text{FM}}$ of $\boldsymbol{\beta}_1$ can be obtained by maximizing the likelihood function (3.5). By taking the derivatives of log-likelihood (3.5) with respect to $\boldsymbol{\beta}_1$ and $\boldsymbol{\beta}_2$ and set to zero, we obtained

$$\widehat{\boldsymbol{\beta}}_1^{\text{FM}} = (\boldsymbol{X}_1^\top\boldsymbol{S}\boldsymbol{X}_1)^{-1}\boldsymbol{X}_1^\top\boldsymbol{S}\boldsymbol{y}, \tag{3.6}$$

where $\boldsymbol{S} = \boldsymbol{I} - \boldsymbol{X}_2(\boldsymbol{X}_2^\top\boldsymbol{X}_2)^{-1}\boldsymbol{X}_2^\top$. Further,

$$\begin{aligned}
l(\boldsymbol{\beta}, \sigma^2|\boldsymbol{y}) &= -\frac{n}{2}log(2\pi\sigma^2) - \frac{1}{2\sigma^2}\left[\boldsymbol{y}^\top\boldsymbol{y} - \boldsymbol{\beta}_1^\top\boldsymbol{X}_1^\top\boldsymbol{y} - \boldsymbol{\beta}_2^\top\boldsymbol{X}_2^\top\boldsymbol{y} - \boldsymbol{y}^\top\boldsymbol{X}_1\boldsymbol{\beta}_1\right. \\
&+ \boldsymbol{\beta}_1^\top\boldsymbol{X}_1^\top\boldsymbol{X}_1\boldsymbol{\beta}_1 + \boldsymbol{\beta}_2^\top\boldsymbol{X}_2^\top\boldsymbol{X}_1\boldsymbol{\beta}_1 \\
&- \boldsymbol{y}^\top\boldsymbol{X}_2\boldsymbol{\beta}_2 + \boldsymbol{\beta}_1^\top\boldsymbol{X}_1^\top\boldsymbol{X}_2\boldsymbol{\beta}_2 + \boldsymbol{\beta}_2^\top\boldsymbol{X}_2^\top\boldsymbol{X}_2\boldsymbol{\beta}_2).\\
&\frac{\partial l}{\partial \boldsymbol{\beta}_1} = 0 - \frac{1}{2\sigma^2}(0 - 2\boldsymbol{X}_1^\top\boldsymbol{y} + 2\boldsymbol{X}_1^\top\boldsymbol{X}_1\boldsymbol{\beta}_1 + 2\boldsymbol{X}_1^\top\boldsymbol{X}_2\boldsymbol{\beta}_2) \\
\text{Set} \quad &\boldsymbol{X}_1^\top\boldsymbol{X}_1\boldsymbol{\beta}_1 + \boldsymbol{X}_1^\top\boldsymbol{X}_2\boldsymbol{\beta}_2 = \boldsymbol{X}_1^\top\boldsymbol{y} = \mathbf{0} \\
&\frac{\partial l}{\partial \boldsymbol{\beta}_2} = 0 - \frac{1}{2\sigma^2}(0 - 2\boldsymbol{X}_2^\top\boldsymbol{y} + 2\boldsymbol{X}_2^\top\boldsymbol{X}_1\boldsymbol{\beta}_1 + 2\boldsymbol{X}_2^\top\boldsymbol{X}_2\boldsymbol{\beta}_2) \\
\text{Set} \quad &\boldsymbol{X}_2^\top\boldsymbol{X}_1\boldsymbol{\beta}_1 + \boldsymbol{X}_2^\top\boldsymbol{X}_2\boldsymbol{\beta}_2 = \boldsymbol{X}_2^\top\boldsymbol{y} = \mathbf{0} \\
&\frac{\partial l}{\partial \sigma^2} = -\frac{n}{2\sigma^2} + \frac{(\boldsymbol{y} - \boldsymbol{X}_1\boldsymbol{\beta}_1 - \boldsymbol{X}_2\boldsymbol{\beta}_2)^\top(\boldsymbol{y} - \boldsymbol{X}_1\boldsymbol{\beta}_1 - \boldsymbol{X}_2\boldsymbol{\beta}_2)}{2\sigma^4} \\
&\frac{\partial^2 l}{\partial \boldsymbol{\beta}_1 \partial \boldsymbol{\beta}_1^\top} = -\frac{1}{\sigma^2}(\boldsymbol{X}_1^\top\boldsymbol{X}_1) \\
&\frac{\partial^2 l}{\partial \boldsymbol{\beta}_1 \partial \boldsymbol{\beta}_2} = -\frac{1}{\sigma^2}(\boldsymbol{X}_2^\top\boldsymbol{X}_1) \\
&\frac{\partial^2 l}{\partial \boldsymbol{\beta}_2 \partial \boldsymbol{\beta}_1} = -\frac{1}{\sigma^2}(\boldsymbol{X}_1^\top\boldsymbol{X}_2) \\
&\frac{\partial^2 l}{\partial \boldsymbol{\beta}_2 \partial \boldsymbol{\beta}_2^\top} = -\frac{1}{\sigma^2}(\boldsymbol{X}_2^\top\boldsymbol{X}_2) \\
&\frac{\partial^2 l}{\partial \boldsymbol{\beta}_1 \partial \sigma^2} = \frac{1}{\sigma^4}(\boldsymbol{X}_1^\top\boldsymbol{X}_1\boldsymbol{\beta}_1 + \boldsymbol{X}_1^\top\boldsymbol{X}_2\boldsymbol{\beta}_2 - \boldsymbol{X}_1^\top\boldsymbol{y}) \\
&\frac{\partial^2 l}{\partial \boldsymbol{\beta}_2 \partial \sigma^2} = \frac{1}{\sigma^4}(\boldsymbol{X}_2^\top\boldsymbol{X}_1\boldsymbol{\beta}_1 + \boldsymbol{X}_2^\top\boldsymbol{X}_2\boldsymbol{\beta}_2 - \boldsymbol{X}_2^\top\boldsymbol{y}) \\
&\frac{\partial^2 l}{\partial \sigma^4} = -\frac{n}{2\sigma^4} - \frac{(\boldsymbol{y} - \boldsymbol{X}_1\boldsymbol{\beta}_1 - \boldsymbol{X}_2\boldsymbol{\beta}_2)^\top(\boldsymbol{y} - \boldsymbol{X}_1\boldsymbol{\beta}_1 - \boldsymbol{X}_2\boldsymbol{\beta}_2)}{2\sigma^6}
\end{aligned}$$

The Hessian matrix is

$$\boldsymbol{H}(\boldsymbol{\beta},\sigma^2)=\begin{bmatrix}\frac{\partial^2 l}{\partial\boldsymbol{\beta}_1\partial\boldsymbol{\beta}_1^\top} & \frac{\partial^2 l}{\partial\boldsymbol{\beta}_1\partial\boldsymbol{\beta}_2} & \frac{\partial^2 l}{\partial\boldsymbol{\beta}_1\partial\sigma^2}\\ \frac{\partial^2 l}{\partial\boldsymbol{\beta}_2\partial\boldsymbol{\beta}_1} & \frac{\partial^2 l}{\partial\boldsymbol{\beta}_2\partial\boldsymbol{\beta}_2^\top} & \frac{\partial^2 l}{\partial\boldsymbol{\beta}_2\partial\sigma^2}\\ \frac{\partial^2 l}{\partial\sigma^2\partial\boldsymbol{\beta}_1} & \frac{\partial^2 l}{\partial\sigma^2\partial\boldsymbol{\beta}_2} & \frac{\partial^2 l}{\partial\sigma^4}\end{bmatrix}.$$

The information matrix is

$$\begin{aligned}\boldsymbol{I}(\boldsymbol{\beta},\sigma^2) &= -E(\boldsymbol{H}(\boldsymbol{\beta},\sigma^2))\\ &= \begin{bmatrix}-E\left(\frac{\partial^2 l}{\partial\boldsymbol{\beta}_1\partial\boldsymbol{\beta}_1^\top}\right) & -E\left(\frac{\partial^2 l}{\partial\boldsymbol{\beta}_1\partial\boldsymbol{\beta}_2}\right) & -E\left(\frac{\partial^2 l}{\partial\boldsymbol{\beta}_1\partial\sigma^2}\right)\\ -E\left(\frac{\partial^2 l}{\partial\boldsymbol{\beta}_2\partial\boldsymbol{\beta}_1}\right) & -E\left(\frac{\partial^2 l}{\partial\boldsymbol{\beta}_2\partial\boldsymbol{\beta}_2^\top}\right) & -E\left(\frac{\partial^2 l}{\partial\boldsymbol{\beta}_2\partial\sigma^2}\right)\\ -E\left(\frac{\partial^2 l}{\partial\sigma^2\partial\boldsymbol{\beta}_1}\right) & -E\left(\frac{\partial^2 l}{\partial\sigma^2\partial\boldsymbol{\beta}_2}\right) & -E\left(\frac{\partial^2 l}{\partial\sigma^4}\right)\end{bmatrix}\end{aligned}$$

which is equal to

$$\boldsymbol{I}(\boldsymbol{\beta},\sigma^2)=\begin{bmatrix}\frac{\boldsymbol{X}_1^\top\boldsymbol{X}_1}{\sigma^2} & \frac{\boldsymbol{X}_2^\top\boldsymbol{X}_1}{\sigma^2} & \boldsymbol{0}\\ \frac{\boldsymbol{X}_1^\top\boldsymbol{X}_2}{\sigma^2} & \frac{\boldsymbol{X}_2^\top\boldsymbol{X}_2}{\sigma^2} & \boldsymbol{0}\\ \boldsymbol{0} & \boldsymbol{0} & \frac{n}{2\sigma^4}\end{bmatrix}=\begin{bmatrix}\mathcal{I}(\boldsymbol{\beta}) & \boldsymbol{0}\\ \boldsymbol{0} & \frac{n}{2\sigma^4}\end{bmatrix}.$$

where

$$\mathcal{I}(\boldsymbol{\beta})=\begin{bmatrix}\frac{\boldsymbol{X}_1^\top\boldsymbol{X}_1}{\sigma^2} & \frac{\boldsymbol{X}_2^\top\boldsymbol{X}_1}{\sigma^2}\\ \frac{\boldsymbol{X}_1^\top\boldsymbol{X}_2}{\sigma^2} & \frac{\boldsymbol{X}_2^\top\boldsymbol{X}_2}{\sigma^2}\end{bmatrix}.$$

Solving $\frac{\partial l}{\partial\boldsymbol{\beta}_1}=\boldsymbol{0}$ and $\frac{\partial l}{\partial\boldsymbol{\beta}_2}=\boldsymbol{0}$ we have

$$\begin{aligned}\boldsymbol{X}_1^\top\boldsymbol{X}_1\boldsymbol{\beta}_1 &= \boldsymbol{X}_1^\top\boldsymbol{y}-\boldsymbol{X}_1^\top\boldsymbol{X}_2\boldsymbol{\beta}_2\\ &= \boldsymbol{X}_1^\top\boldsymbol{y}-\boldsymbol{X}_1^\top\boldsymbol{X}_2((\boldsymbol{X}_2^\top\boldsymbol{X}_2)^{-1}\boldsymbol{X}_2^\top\boldsymbol{y}-(\boldsymbol{X}_2^\top\boldsymbol{X}_2)^{-1}\boldsymbol{X}_2^\top\boldsymbol{X}_1\boldsymbol{\beta}_1)\\ &= \boldsymbol{X}_1^\top\boldsymbol{y}-\boldsymbol{X}_1^\top\boldsymbol{X}_2(\boldsymbol{X}_2^\top\boldsymbol{X}_2)^{-1}\boldsymbol{X}_2^\top\boldsymbol{y}\\ &- \boldsymbol{X}_1^\top\boldsymbol{X}_2(\boldsymbol{X}_2^\top\boldsymbol{X}_2)^{-1}\boldsymbol{X}_2^\top\boldsymbol{X}_1\boldsymbol{\beta}_1\\ &= \boldsymbol{X}_1^\top\boldsymbol{S}\boldsymbol{y}+\boldsymbol{X}_1^\top(\boldsymbol{I}-\boldsymbol{S})\boldsymbol{X}_1\boldsymbol{\beta}_1\\ &\Rightarrow \boldsymbol{X}_1^\top\boldsymbol{X}_1\boldsymbol{\beta}_1-\boldsymbol{X}_1^\top(\boldsymbol{I}-\boldsymbol{S})\boldsymbol{X}_1\boldsymbol{\beta}_1=\boldsymbol{X}_1^\top\boldsymbol{S}\boldsymbol{y}\\ &\Rightarrow \boldsymbol{X}_1^\top(\boldsymbol{I}-\boldsymbol{I}+\boldsymbol{S})\boldsymbol{X}_1\boldsymbol{\beta}_1=\boldsymbol{X}_1^\top\boldsymbol{S}\boldsymbol{y}\\ &\Rightarrow \boldsymbol{X}_1^\top\boldsymbol{S}\boldsymbol{X}_1\boldsymbol{\beta}_1=\boldsymbol{X}_1^\top\boldsymbol{S}\boldsymbol{y}\\ &\Rightarrow \widehat{\boldsymbol{\beta}}_1^{\mathrm{FM}}=(\boldsymbol{X}_1^\top\boldsymbol{S}\boldsymbol{X}_1)^{-1}\boldsymbol{X}_1^\top\boldsymbol{S}\boldsymbol{y}\end{aligned}$$

We call $\widehat{\boldsymbol{\beta}}_1^{\mathrm{FM}}$ the full model estimator of $\boldsymbol{\beta}_1$. Suppose the assumption of sparsity is correct then we can drop $\boldsymbol{X}_2$ from the model (3.4). Then, we obtain a submodel as follows:

$$\boldsymbol{y}=\boldsymbol{X}_1\boldsymbol{\beta}_1+\boldsymbol{\varepsilon}. \tag{3.7}$$

Finally, under the sparsity condition, $\boldsymbol{\beta}_2=\boldsymbol{0}$, the submodel estimator (SM) $\widehat{\boldsymbol{\beta}}_1^{\mathrm{SM}}$ of $\boldsymbol{\beta}_1$ is obtained by maximizing the log-likelihood (3.7) with respect to $\boldsymbol{\beta}_1$ and this has the form

$$\widehat{\boldsymbol{\beta}}_1^{\mathrm{SM}}=(\boldsymbol{X}_1^\top\boldsymbol{X}_1)^{-1}\boldsymbol{X}_1^\top\boldsymbol{y}.$$

In real-world data applications, this situation may arise when there is over-modeling and the researcher wishes to cut down the irrelevant parts of the model (3.4). This can be achieved by using one of the available variable selection methods. The main goal of this chapter, however, is to develop an efficient estimation for the regression parameter $\boldsymbol{\beta}_1$ by combining full and submodel estimators.

3.2.4 Shrinkage Strategies

The shrinkage or Stein-type regression estimator $\widehat{\boldsymbol{\beta}}_1^S$ of $\boldsymbol{\beta}_1$ is defined by

$$\widehat{\boldsymbol{\beta}}_1^{\mathrm{S}} = \widehat{\boldsymbol{\beta}}_1^{\mathrm{SM}} + \left(\widehat{\boldsymbol{\beta}}_1^{\mathrm{FM}} - \widehat{\boldsymbol{\beta}}_1^{\mathrm{SM}}\right)\left(1-(p_2-2)\mathscr{T}_n^{-1}\right), \, p_2 \geq 3.$$

where $\mathscr{T}_n$ is defined as follows:

$$\mathscr{T}_n = \frac{n}{\widehat{\sigma}^2}\left(\widehat{\boldsymbol{\beta}}_2^{\mathrm{LSE}}\right)\boldsymbol{X}_2^\top \boldsymbol{M}_1 \boldsymbol{X}_2 \left(\widehat{\boldsymbol{\beta}}_2^{\mathrm{LSE}}\right),$$

where $\widehat{\boldsymbol{\beta}}_2^{\mathrm{LSE}} = (\boldsymbol{X}_2^\top \boldsymbol{S}_1 \boldsymbol{X}_2)^{-1}\boldsymbol{X}_2^\top \boldsymbol{S}_1 \boldsymbol{y}$ and $\boldsymbol{S}_1 = \boldsymbol{I} - \boldsymbol{X}_1(\boldsymbol{X}_1^\top \boldsymbol{X}_1)^{-1}\boldsymbol{X}_1^\top$. $\widehat{\sigma}^2$ is a consistent estimator of σ^2.

The positive part of the shrinkage regression estimator $\widehat{\boldsymbol{\beta}}_1^{PS}$ of $\boldsymbol{\beta}_1$ defined by

$$\widehat{\boldsymbol{\beta}}_1^{\mathrm{PS}} = \widehat{\boldsymbol{\beta}}_1^{\mathrm{SM}} + \left(\widehat{\boldsymbol{\beta}}_1^{\mathrm{FM}} - \widehat{\boldsymbol{\beta}}_1^{\mathrm{SM}}\right)\left(1-(p_2-2)\mathscr{T}_n^{-1}\right)^+,$$

where $z^+ = max(0, z)$.

3.3 Asymptotic Analysis

We investigate the asymptotic properties of the estimators under the following sequence:

$$K_{(n)} : \boldsymbol{\beta}_2 = \frac{\boldsymbol{\delta}}{\sqrt{n}}, \tag{3.8}$$

where $\boldsymbol{\delta} = (\delta_1, \cdots, \delta_{p_2})^\top \in \mathbb{R}^{p_2}$ is a real fixed vector.

We derive the asymptotic joint normality of the full model and submodel estimators under the above sequence. Let $\boldsymbol{\beta} = (\boldsymbol{\beta}_1^\top, \boldsymbol{\beta}_2^\top)^\top$, with $\boldsymbol{\beta}_1$ and $\boldsymbol{\beta}_2$ being of orders $p_1 \times 1$ and $p_2 \times 1$, respectively. Correspondingly, the information matrix $\boldsymbol{\mathcal{I}}(\boldsymbol{\beta})$ is partitioned as

$$\boldsymbol{\mathcal{I}}(\boldsymbol{\beta}) = \begin{bmatrix} \boldsymbol{\mathcal{I}}_{11} & \boldsymbol{\mathcal{I}}_{12} \\ \boldsymbol{\mathcal{I}}_{21} & \boldsymbol{\mathcal{I}}_{22} \end{bmatrix}, \tag{3.9}$$

and $\boldsymbol{\Sigma} = \boldsymbol{\mathcal{I}}(\boldsymbol{\beta})^{-1}$ is the covariance matrix of $\widehat{\boldsymbol{\beta}}^{\mathrm{FM}}$ which can partitioned as

$$\boldsymbol{\Sigma} = \begin{bmatrix} \boldsymbol{\Sigma}_{11} & \boldsymbol{\Sigma}_{12} \\ \boldsymbol{\Sigma}_{21} & \boldsymbol{\Sigma}_{22} \end{bmatrix} \tag{3.10}$$

Theorem 3.1 *Under (3.8) and the assumed regularity conditions, we have*

$$\begin{bmatrix} \sqrt{n}(\widehat{\boldsymbol{\beta}}_1^{\mathrm{FM}} - \boldsymbol{\beta}_1) \\ \sqrt{n}(\widehat{\boldsymbol{\beta}}_1^{\mathrm{SM}} - \boldsymbol{\beta}_1) \\ \sqrt{n}(\widehat{\boldsymbol{\beta}}_1^{\mathrm{FM}} - \widehat{\boldsymbol{\beta}}_1^{\mathrm{SM}}) \end{bmatrix} \xrightarrow{\mathcal{L}} N\left(\begin{bmatrix} \mathbf{0} \\ \gamma \\ -\gamma \end{bmatrix}, \begin{bmatrix} \boldsymbol{\Gamma}_{11} & \boldsymbol{\Gamma}_{12} & \boldsymbol{\Gamma}_{13} \\ \boldsymbol{\Gamma}_{21} & \boldsymbol{\Gamma}_{22} & \boldsymbol{\Gamma}_{23} \\ \boldsymbol{\Gamma}_{31} & \boldsymbol{\Gamma}_{32} & \boldsymbol{\Gamma}_{33} \end{bmatrix}\right),$$

where $\gamma = \boldsymbol{\Sigma}_{11}^{-1}\boldsymbol{\Sigma}_{12}\boldsymbol{\delta}$, $\boldsymbol{\Sigma}_{11.2} = \boldsymbol{\Sigma}_{11} - \boldsymbol{\Sigma}_{12}\boldsymbol{\Sigma}_{22}^{-1}\boldsymbol{\Sigma}_{21}$, $\boldsymbol{\Sigma}_{22.1} = \boldsymbol{\Sigma}_{22} - \boldsymbol{\Sigma}_{21}\boldsymbol{\Sigma}_{11}^{-1}\boldsymbol{\Sigma}_{12}$, $\boldsymbol{\Gamma}_{11} = \boldsymbol{\Sigma}_{11.2}^{-1}$, $\boldsymbol{\Gamma}_{12} = \boldsymbol{\Sigma}_{11.2}^{-1} - \boldsymbol{\Sigma}_{11}^{-1}\boldsymbol{\Sigma}_{12}\boldsymbol{\Sigma}_{22.1}^{-1}\boldsymbol{\Sigma}_{21}\boldsymbol{\Sigma}_{11}^{-1}$, $\boldsymbol{\Gamma}_{13} = \boldsymbol{\Sigma}_{11}^{-1}\boldsymbol{\Sigma}_{12}\boldsymbol{\Sigma}_{22.1}^{-1}\boldsymbol{\Sigma}_{21}\boldsymbol{\Sigma}_{11}^{-1}$, $\boldsymbol{\Gamma}_{21} = \boldsymbol{\Gamma}_{12}^\top$, $\boldsymbol{\Gamma}_{22} = \boldsymbol{\Gamma}_{12}$, $\boldsymbol{\Gamma}_{23} = \boldsymbol{\Gamma}_{32} = \mathbf{0}$ $\boldsymbol{\Gamma}_{31} = \boldsymbol{\Gamma}_{13}^\top$, *and* $\boldsymbol{\Gamma}_{33} = \boldsymbol{\Gamma}_{12}$.

Proof *Let* $\boldsymbol{\xi}_1 = \sqrt{n}(\widehat{\boldsymbol{\beta}}_1^{\text{FM}} - \boldsymbol{\beta}_1)$, $\boldsymbol{\xi}_2 = \sqrt{n}(\widehat{\boldsymbol{\beta}}_1^{\text{SM}} - \boldsymbol{\beta}_1)$, *and* $\boldsymbol{\xi}_3 = \sqrt{n}(\widehat{\boldsymbol{\beta}}_1^{\text{FM}} - \widehat{\boldsymbol{\beta}}_1^{\text{SM}})$.

$$
\begin{aligned}
E(\boldsymbol{\xi}_1) &= E(\sqrt{n}(\widehat{\boldsymbol{\beta}}_1^{\text{FM}} - \boldsymbol{\beta}_1)) = \mathbf{0}.\\
E(\boldsymbol{\xi}_2) &= E(\sqrt{n}(\widehat{\boldsymbol{\beta}}_1^{\text{SM}} - \boldsymbol{\beta}_1))E(\sqrt{n}(\widehat{\boldsymbol{\beta}}_1^{\text{FM}} - \boldsymbol{\beta}_1 + \boldsymbol{\Sigma}_{11}^{-1}\boldsymbol{\Sigma}_{12}\widehat{\boldsymbol{\beta}}_2^{\text{LSE}}) = \boldsymbol{\gamma}\\
E(\boldsymbol{\xi}_3) &= E(\sqrt{n}(\widehat{\boldsymbol{\beta}}_1^{\text{FM}} - \widehat{\boldsymbol{\beta}}_1^{\text{SM}}))\\
&= E(\sqrt{n}(\widehat{\boldsymbol{\beta}}_1^{\text{FM}} - \boldsymbol{\beta}_1) - (\widehat{\boldsymbol{\beta}}_1^{\text{SM}} - \boldsymbol{\beta}_1)) = -\boldsymbol{\gamma}\\
Var(\boldsymbol{\xi}_1) &= \boldsymbol{\Sigma}_{11.2}^{-1} = \boldsymbol{\Gamma}_{11}.\\
Var(\boldsymbol{\xi}_2) &= Var(\sqrt{n}(\widehat{\boldsymbol{\beta}}_1^{\text{SM}} - \boldsymbol{\beta}_1))\\
&= Var(\sqrt{n}\widehat{\boldsymbol{\beta}}_1^{\text{FM}} - \boldsymbol{\beta}_1) + \boldsymbol{\Sigma}_{11}^{-1}\boldsymbol{\Sigma}_{12}\, Var(\sqrt{n}\widehat{\boldsymbol{\beta}}_2^{\text{LSE}})\boldsymbol{\Sigma}_{21}\boldsymbol{\Sigma}_{11}^{-1}\\
&+ 2Cov(\sqrt{n}(\widehat{\boldsymbol{\beta}}_1^{\text{FM}} - \boldsymbol{\beta}_1), \boldsymbol{\Sigma}_{11}^{-1}\boldsymbol{\Sigma}_{12}\sqrt{n}\widehat{\boldsymbol{\beta}}_2^{\text{LSE}})\\
&= \boldsymbol{\Sigma}_{11.2}^{-1} + \boldsymbol{\Sigma}_{11}^{-1}\boldsymbol{\Sigma}_{12}\boldsymbol{\Sigma}_{22.1}^{-1}\boldsymbol{\Sigma}_{21}\boldsymbol{\Sigma}_{11}^{-1}\\
&+ 2Cov(\sqrt{n}(\widehat{\boldsymbol{\beta}}_1^{\text{FM}} - \boldsymbol{\beta}_1), \sqrt{n}\widehat{\boldsymbol{\beta}}_2^{\text{LSE}})\boldsymbol{\Sigma}_{21}\boldsymbol{\Sigma}_{11}^{-1}\\
&= \boldsymbol{\Sigma}_{11.2}^{-1} + \boldsymbol{\Sigma}_{11}^{-1}\boldsymbol{\Sigma}_{12}\boldsymbol{\Sigma}_{22.1}^{-1}\boldsymbol{\Sigma}_{21}\boldsymbol{\Sigma}_{11}^{-1} - 2\boldsymbol{\Sigma}_{11}^{-1}\boldsymbol{\Sigma}_{12}\boldsymbol{\Sigma}_{22.1}^{-1}\boldsymbol{\Sigma}_{21}\boldsymbol{\Sigma}_{11}^{-1}\\
&= \boldsymbol{\Sigma}_{11.2}^{-1} - \boldsymbol{\Sigma}_{11}^{-1}\boldsymbol{\Sigma}_{12}\boldsymbol{\Sigma}_{22.1}^{-1}\boldsymbol{\Sigma}_{21}\boldsymbol{\Sigma}_{11}^{-1} = \boldsymbol{\Gamma}_{12} = \boldsymbol{\Gamma}_{22}.\\
Var(\boldsymbol{\xi}_3) &= Var(\sqrt{n}(\widehat{\boldsymbol{\beta}}_1^{\text{FM}} - \widehat{\boldsymbol{\beta}}_1^{\text{SM}}))\\
&= Var(\sqrt{n}(\widehat{\boldsymbol{\beta}}_1^{\text{FM}} - \boldsymbol{\beta}_1) - (\widehat{\boldsymbol{\beta}}_1^{\text{SM}} - \boldsymbol{\beta}_1)) = Var(\boldsymbol{\xi}_1 - \boldsymbol{\xi}_2)\\
&= Var(\boldsymbol{\xi}_1) + Var(\boldsymbol{\xi}_2)\\
&- 2Cov(\boldsymbol{\xi}_1, \boldsymbol{\xi}_2)\\
&= \boldsymbol{\Sigma}_{11.2}^{-1} + \boldsymbol{\Sigma}_{11.2}^{-1} - \boldsymbol{\Sigma}_{11}^{-1}\boldsymbol{\Sigma}_{12}\boldsymbol{\Sigma}_{22.1}^{-1}\boldsymbol{\Sigma}_{21}\boldsymbol{\Sigma}_{11}^{-1}\\
&- 2\boldsymbol{\Sigma}_{11.2}^{-1} + 2\boldsymbol{\Sigma}_{11}^{-1}\boldsymbol{\Sigma}_{12}\boldsymbol{\Sigma}_{22.1}^{-1}\boldsymbol{\Sigma}_{21}\boldsymbol{\Sigma}_{11}^{-1}\\
&= \boldsymbol{\Sigma}_{11}^{-1}\boldsymbol{\Sigma}_{12}\boldsymbol{\Sigma}_{22.1}^{-1}\boldsymbol{\Sigma}_{21}\boldsymbol{\Sigma}_{11}^{-1} = \boldsymbol{\Gamma}_{33}\\
Cov(\boldsymbol{\xi}_1, \boldsymbol{\xi}_3) &= Cov(\sqrt{n}(\widehat{\boldsymbol{\beta}}_1^{\text{FM}} - \boldsymbol{\beta}_1), \sqrt{n}(\widehat{\boldsymbol{\beta}}_1^{\text{FM}} - \widehat{\boldsymbol{\beta}}_1^{\text{SM}}))\\
&= Cov(\sqrt{n}(\widehat{\boldsymbol{\beta}}_1^{\text{FM}} - \boldsymbol{\beta}_1), \sqrt{n}((\widehat{\boldsymbol{\beta}}_1^{\text{FM}} - \boldsymbol{\beta}_1) - (\widehat{\boldsymbol{\beta}}_1^{\text{SM}} - \boldsymbol{\beta}_1)))\\
&= Var(\sqrt{n}(\widehat{\boldsymbol{\beta}}_1^{\text{FM}} - \boldsymbol{\beta}_1)) - Cov(\sqrt{n}(\widehat{\boldsymbol{\beta}}_1^{\text{FM}} - \boldsymbol{\beta}_1), \sqrt{n}(\widehat{\boldsymbol{\beta}}_1^{\text{SM}} - \boldsymbol{\beta}_1))\\
&= \boldsymbol{\Sigma}_{11.2}^{-1} - Cov(\sqrt{n}(\widehat{\boldsymbol{\beta}}_1^{\text{FM}} - \boldsymbol{\beta}_1), \boldsymbol{\Sigma}_{11}^{-1}\boldsymbol{\Sigma}_{12}\sqrt{n}\widehat{\boldsymbol{\beta}}_2^{\text{LSE}})\\
&= \boldsymbol{\Sigma}_{11.2}^{-1} - \boldsymbol{\Sigma}_{11.2}^{-1} + Cov(\sqrt{n}(\widehat{\boldsymbol{\beta}}_1^{\text{FM}} - \boldsymbol{\beta}_1), \sqrt{n}\widehat{\boldsymbol{\beta}}_2^{\text{LSE}})\boldsymbol{\Sigma}_{21}\boldsymbol{\Sigma}_{11}^{-1}\\
&= \boldsymbol{\Sigma}_{11}^{-1}\boldsymbol{\Sigma}_{12}\boldsymbol{\Sigma}_{22.1}^{-1}\boldsymbol{\Sigma}_{21}\boldsymbol{\Sigma}_{11}^{-1} = \boldsymbol{\Gamma}_{13}.\\
Cov(\boldsymbol{\xi}_1, \boldsymbol{\xi}_2) &= Cov(\sqrt{n}(\widehat{\boldsymbol{\beta}}_1^{\text{FM}} - \boldsymbol{\beta}_1), \sqrt{n}(\widehat{\boldsymbol{\beta}}_1^{\text{SM}} - \boldsymbol{\beta}_1))\\
&= \boldsymbol{\Sigma}_{11.2}^{-1} - \boldsymbol{\Sigma}_{11}^{-1}\boldsymbol{\Sigma}_{12}\boldsymbol{\Sigma}_{22.1}^{-1}\boldsymbol{\Sigma}_{11}^{-1}\boldsymbol{\Sigma}_{21} = \boldsymbol{\Gamma}_{12}\\
Cov(\boldsymbol{\xi}_3, \boldsymbol{\xi}_2) &= Cov(\sqrt{n}(\widehat{\boldsymbol{\beta}}_1^{\text{FM}} - \widehat{\boldsymbol{\beta}}_1^{\text{SM}}), \sqrt{n}(\widehat{\boldsymbol{\beta}}_1^{\text{SM}} - \boldsymbol{\beta}_1))\\
&= Cov(\sqrt{n}((\widehat{\boldsymbol{\beta}}_1^{\text{FM}} - \boldsymbol{\beta}_1) - (\widehat{\boldsymbol{\beta}}_1^{\text{SM}} - \boldsymbol{\beta}_1)), \sqrt{n}(\widehat{\boldsymbol{\beta}}_1^{\text{SM}} - \boldsymbol{\beta}_1))\\
&= Cov(\sqrt{n}((\widehat{\boldsymbol{\beta}}_1^{\text{FM}} - \boldsymbol{\beta}_1), \sqrt{n}(\widehat{\boldsymbol{\beta}}_1^{\text{SM}} - \boldsymbol{\beta}_1)) - Var(\sqrt{n}(\widehat{\boldsymbol{\beta}}_1^{\text{SM}} - \boldsymbol{\beta}_1))\\
&= \boldsymbol{\Sigma}_{11.2}^{-1} - \boldsymbol{\Sigma}_{11}^{-1}\boldsymbol{\Sigma}_{12}\boldsymbol{\Sigma}_{22.1}^{-1}\boldsymbol{\Sigma}_{21}\boldsymbol{\Sigma}_{11}^{-1} - \boldsymbol{\Sigma}_{11.2}^{-1}\\
&+ \boldsymbol{\Sigma}_{11}^{-1}\boldsymbol{\Sigma}_{12}\boldsymbol{\Sigma}_{22.1}^{-1}\boldsymbol{\Sigma}_{11}^{-1}\boldsymbol{\Sigma}_{21} = \mathbf{0} = \boldsymbol{\Gamma}_{23}.
\end{aligned}
$$

We will get to the main results with the help of the following lemma.

Lemma 3.2 *Let* $\boldsymbol{Z}$ *be a* $p_2 \times 1$ *vector that follows a normal distribution with mean* $\boldsymbol{\mu}_z$ *vector and covariance matrix* $\boldsymbol{\Sigma}_{p_2}$ *as* $\boldsymbol{Z} \sim N_{p_2}(\boldsymbol{\mu}_z, \boldsymbol{\Sigma}_{p_2})$. *Then, for a measurable function* *of ϕ, we have*

$$\begin{aligned} E(\boldsymbol{Z}\phi(\boldsymbol{Z}^\top\boldsymbol{Z})) &= \boldsymbol{\mu}_z E(\phi(\chi^2_{p_2+2}(\Delta))) \\ E(\boldsymbol{Z}\boldsymbol{Z}^\top\phi(\boldsymbol{Z}^\top\boldsymbol{Z})) &= \boldsymbol{\Sigma}_{p_2} E(\phi(\chi^2_{p_2+2}(\Delta))) + \boldsymbol{\mu}_z\boldsymbol{\mu}_z^\top E(\phi(\chi^2_{p_2+4}(\Delta))), \end{aligned}$$

where $\Delta = \boldsymbol{\mu}_z^\top \boldsymbol{\Sigma}_{p_2}^{-1}\boldsymbol{\mu}_z$.

The proof can be found in Judge and Bock (1978).

3.3.1 Asymptotic Distributional Bias

Consider a sequence of parameter values $\boldsymbol{\beta}_1$ and a sequence of estimators $\widehat{\boldsymbol{\beta}}_1^*$. Assume that $\sqrt{n}(\widehat{\boldsymbol{\beta}}_1^* - \boldsymbol{\beta}_1)$ converges in distribution as $n \to \infty$ to some random variable Z with distribution $\tilde{G}$. Then the asymptotic distributional bias (ADB) of $\widehat{\boldsymbol{\beta}}_1^*$ is defined by

$$\mathrm{ADB}(\widehat{\boldsymbol{\beta}}_1^*) = \mathrm{E}(\boldsymbol{Z}) = \int \boldsymbol{z}d\tilde{G}(\boldsymbol{z}).$$

Let Ξ_1 be a $\chi^2_{p_2+2}(\Delta)$ random variable and Ξ_2 be a $\chi^2_{p_2+4}(\Delta)$ random variable. The distribution function of a non-central χ^2 variable with non-centrality parameter Δ and degrees of freedom g is denoted by

$$\Psi_g(z, \Delta) = \Pr(\chi^2_g(\Delta) \le z).$$

Finally, let $\boldsymbol{\gamma} = \boldsymbol{\Sigma}_{11}^{-1}\boldsymbol{\Sigma}_{12}\boldsymbol{\delta}$. We present the respective expressions for the asymptotic distributional biases of the estimators in the following Theorem.

Theorem 3.3 *If the conditions of Theorem 1 hold, then*

$$\begin{aligned} &ADB(\widehat{\boldsymbol{\beta}}_1^{\mathrm{FM}}) = \mathbf{0} \\ &ADB(\widehat{\boldsymbol{\beta}}_1^{\mathrm{SM}}) = \boldsymbol{\Sigma}_{11}^{-1}\boldsymbol{\Sigma}_{12}\boldsymbol{\delta} \\ &ADB(\widehat{\boldsymbol{\beta}}_1^{S}) = -\nu E(\Xi_1^{-1})\boldsymbol{\Sigma}_{11}^{-1}\boldsymbol{\Sigma}_{12}\boldsymbol{\delta}, \quad \nu = p_2 - 2 \\ &ADB(\widehat{\boldsymbol{\beta}}_1^{PS}) = ADB(\widehat{\boldsymbol{\beta}}_1^{S}) + \left(\Psi_{\nu+4}(\nu, \Delta) - \nu E(\Xi_1^{-1} I(\Xi_1 < \nu))\right)\boldsymbol{\Sigma}_{11}^{-1}\boldsymbol{\Sigma}_{12}\boldsymbol{\delta}. \end{aligned}$$

Proof *Clearly,* $ADB(\widehat{\boldsymbol{\beta}}_1^{\mathrm{FM}}) = 0$ *and the ADB of the submodel:*

$$\begin{aligned} ADB(\widehat{\boldsymbol{\beta}}_1^{\mathrm{FM}}) &= E\left[\lim_{n\to\infty}\sqrt{n}(\widehat{\boldsymbol{\beta}}_1^{\mathrm{FM}} - \boldsymbol{\beta}_1)\right] = \mathbf{0}. \\ ADB(\widehat{\boldsymbol{\beta}}_1^{\mathrm{SM}}) &= E\left[\lim_{n\to\infty}\sqrt{n}(\widehat{\boldsymbol{\beta}}_1^{\mathrm{SM}} - \boldsymbol{\beta}_1)\right] \\ &= E\left[\lim_{n\to\infty}\sqrt{n}(\widehat{\boldsymbol{\beta}}_1^{\mathrm{FM}} - \boldsymbol{\beta}_1)\right] + \boldsymbol{\Sigma}_{11}^{-1}\boldsymbol{\Sigma}_{12}\sqrt{n}\boldsymbol{\delta}/\sqrt{n} = \boldsymbol{\Sigma}_{11}^{-1}\boldsymbol{\Sigma}_{12}\boldsymbol{\delta}. \end{aligned}$$

Now, the ADB of the shrinkage estimator can be obtained in the following steps:

$$\begin{aligned} ADB(\widehat{\boldsymbol{\beta}}_1^{S}) &= E\left[\lim_{n\to\infty}\sqrt{n}(\widehat{\boldsymbol{\beta}}_1^{S} - \boldsymbol{\beta}_1)\right] \\ &= E\left[\lim_{n\to\infty}\sqrt{n}\left((\widehat{\boldsymbol{\beta}}_1^{\mathrm{FM}} - \boldsymbol{\beta}_1) - (\widehat{\boldsymbol{\beta}}_1^{\mathrm{FM}} - \widehat{\boldsymbol{\beta}}_1^{\mathrm{SM}})\right)\nu\widehat{\Lambda}^{-1}\right] \\ &= \mathbf{0} - E\left[\lim_{n\to\infty}\left(\sqrt{n}(\widehat{\boldsymbol{\beta}}_1^{\mathrm{FM}} - \boldsymbol{\beta}_1) - \sqrt{n}(\widehat{\boldsymbol{\beta}}_1^{\mathrm{SM}} - \boldsymbol{\beta}_1)\right)\nu\widehat{\Lambda}^{-1}\right] \\ &= -\nu E\left[\boldsymbol{\xi}_2\widehat{\Lambda}^{-1}\right] = -\nu\boldsymbol{\Sigma}_{11}^{-1}\boldsymbol{\Sigma}_{12}\boldsymbol{\delta}E(\Xi_1^{-1}). \end{aligned}$$

Finally, the ADB of $\widehat{\boldsymbol{\beta}}_1^{PS}$ is

$$\begin{aligned}
ADB(\widehat{\boldsymbol{\beta}}_1^{PS}) &= E\left[\lim_{n\to\infty}\sqrt{n}(\widehat{\boldsymbol{\beta}}_1^{PS}-\boldsymbol{\beta}_1)\right] \\
&= E\Big[\lim_{n\to\infty}\sqrt{n}(\widehat{\boldsymbol{\beta}}_1^{S}-\boldsymbol{\beta}_1) \\
&- \lim_{n\to\infty}\sqrt{n}(\widehat{\boldsymbol{\beta}}_1^{\mathrm{FM}}-\widehat{\boldsymbol{\beta}}_1^{\mathrm{SM}})(1-\nu\widehat{\Lambda}^{-1})I(\widehat{\Lambda}<\nu)\Big] \\
&= ADB(\widehat{\boldsymbol{\beta}}_1^{S}) - E\left[\boldsymbol{\xi}_2(1-\nu\widehat{\Lambda}^{-1})I(\widehat{\Lambda}<\nu)\right] \\
&= ADB(\widehat{\boldsymbol{\beta}}_1^{S}) + \boldsymbol{\Sigma}_{11}^{-1}\boldsymbol{\Sigma}_{12}\boldsymbol{\delta}E\left[(1-\nu/\Xi_1)I(\Xi<\nu)\right] \\
&= ADB(\widehat{\boldsymbol{\beta}}_1^{S}) + \boldsymbol{\Sigma}_{11}^{-1}\boldsymbol{\Sigma}_{12}\boldsymbol{\delta}\left[\Psi_{\nu+4}(\nu,\Delta)-\nu E\left(\Xi_1^{-1}I(\Xi_1<\nu)\right)\right].
\end{aligned}$$

The bias expressions for all the estimators are not in the scalar form. We therefore take recourse by converting them into the quadratic forms. Thus, we define the quadratic asymptotic distributional bias (QADB) of an estimator $\widehat{\boldsymbol{\beta}}_1^*$ of $\boldsymbol{\beta}_1$ by

$$QADB(\widehat{\boldsymbol{\beta}}_1^*) = ADB(\widehat{\boldsymbol{\beta}}_1^*)^\top\boldsymbol{\Sigma}_{11.2}^{-1}ADB(\widehat{\boldsymbol{\beta}}_1^*).$$

Theorem 3.4 *Assume the condition of Theorem 2 holds, the QADB of the estimators are*

$$\begin{aligned}
QADB(\widehat{\boldsymbol{\beta}}_1^{\mathrm{FM}}) &= \mathbf{0} \\
QADB(\widehat{\boldsymbol{\beta}}_1^{\mathrm{SM}}) &= b \\
QADB(\widehat{\boldsymbol{\beta}}_1^{S}) &= b\nu^2(E(\Xi_1^{-1}))^2 \\
QADB(\widehat{\boldsymbol{\beta}}_1^{PS}) &= b\left(\Psi_{\nu+4}(\nu,\Delta)-\nu E(\Xi_1^{-1}I(\Xi_1<\nu))-\nu E(\Xi_1^{-1})\right)^2,
\end{aligned}$$

where $b = \boldsymbol{\delta}^\top\boldsymbol{\Sigma}_{11}^{-1}\boldsymbol{\Sigma}_{12}\boldsymbol{\Sigma}_{11.2}^{-1}\boldsymbol{\Sigma}_{21}$

The above results establish the following results.

- As per design, the only full model estimator is asymptotically unbiased for the regression parameters vector.
- The QADB of the submodel estimator is an unbounded function of b or in $\boldsymbol{\delta}$. This is the main problem with the estimators based on any submodel regardless which penalized or any other methods are used to select a submodel, unless $\boldsymbol{\delta}$ is a null vector. In other words, the selected submodel is a correct one; that is, the sparsity condition is justified. This is a serious problem for estimators based on any submodel. The bias will not go away simply by increasing the sample size. Making a clear statement that a submodel estimator should not be used in its own right. However, it can be combined with an unbiased estimator to control the magnitude of the bias, giving rise to a shrinkage strategy.
- As expected, the quadratic bias of $\widehat{\boldsymbol{\beta}}_1^{S}$, and $\widehat{\boldsymbol{\beta}}_1^{PS}$ are bounded in b, since $\mathrm{E}(\Xi_1^{-1})$ is a decreasing function of Δ, the bias function of $\widehat{\boldsymbol{\beta}}_1^{S}$ starts from the origin at $b=0$, increases to a point, and then decreases toward 0. The characteristics of $\widehat{\boldsymbol{\beta}}_1^{PS}$ are similar to those of $\widehat{\boldsymbol{\beta}}_1^{S}$. However, the bias curve of $\widehat{\boldsymbol{\beta}}_1^{PS}$ is below or equal to the bias curve of $\widehat{\boldsymbol{\beta}}_1^{S}$ for all the values of b. We may conclude that a positive shrinkage estimator is less biased than its counterpart, a usual shrinkage estimator. The shrinkage strategies yield estimators with bounded bias, unlike the submodel estimator.

Since the bias is a part of the mean squared error or risk (for a quadratic loss function) and the control of the risk would control both the bias and variance, we shall only focus on the risk comparison going forward. First, we introduced the notion of the asymptotic distributional risk (ADR).

3.3.2 Asymptotic Distributional Risk

Let us first define a quadratic loss function

$$\mathcal{L}(\widehat{\boldsymbol{\beta}}_1^*; \boldsymbol{W}) = \left(\sqrt{n}(\widehat{\boldsymbol{\beta}}_1^* - \boldsymbol{\beta}_1)\right)^\top \boldsymbol{W} \left(\sqrt{n}(\widehat{\boldsymbol{\beta}}_1^* - \boldsymbol{\beta}_1)\right),$$

where $\boldsymbol{W}$ is a suitable positive semi-definite weight matrix (typically, $\boldsymbol{W} = \boldsymbol{I}_{p_1 \times p_1}$, which is the usual quadratic loss). Using a general $\boldsymbol{W}$ gives a loss function that weights different $\boldsymbol{\beta}_1$'s differently. If $\sqrt{n}(\widehat{\boldsymbol{\beta}}_1^* - \boldsymbol{\beta}_1)$ converges in distribution to some random variable Z with distribution $\tilde{G}$, then the ADR of $\widehat{\boldsymbol{\beta}}_1^*$ is defined by

$$\text{ADR}(\widehat{\boldsymbol{\beta}}_1^*; \boldsymbol{W}) = \int \cdots \int \boldsymbol{z}^\top \boldsymbol{W} \boldsymbol{z} d\tilde{G}(\boldsymbol{z}) = \text{tr}\left(\boldsymbol{W}\boldsymbol{\Sigma}^*(\widehat{\boldsymbol{\beta}}_1^*)\right), \tag{3.11}$$

where $\boldsymbol{\Sigma}^*(\widehat{\boldsymbol{\beta}}_1^*) = \int \cdots \int \boldsymbol{z}\boldsymbol{z}^\top d\tilde{G}(\boldsymbol{z})$ is the dispersion matrix for the distribution $\tilde{G}(\boldsymbol{z})$.

Theorem 3.5 *Under the sequence $K_{(n)}$ in (3.8) and the assumed regularity conditions,*

$$\begin{aligned}
ADR(\widehat{\boldsymbol{\beta}}_1^{\text{FM}}; \boldsymbol{W}) &= tr(\boldsymbol{W}\boldsymbol{\Sigma}_{11.2}^{-1}) \\
ADR(\widehat{\boldsymbol{\beta}}_1^{\text{SM}}; \boldsymbol{W}) &= tr(\boldsymbol{W}\boldsymbol{\Gamma}_{12}) + \boldsymbol{\gamma}^\top \boldsymbol{W}\boldsymbol{\gamma}, \quad \textit{where } \boldsymbol{\gamma} = \boldsymbol{\Sigma}_{11}^{-1}\boldsymbol{\Sigma}_{12}\boldsymbol{\delta} \\
ADR\left(\widehat{\boldsymbol{\beta}}_1^{S}; \boldsymbol{W}\right) &= ADR\left(\widehat{\boldsymbol{\beta}}_1^{\text{FM}}; \boldsymbol{W}\right) + \left(\nu^2 E\left(\Xi_1^{-2}\right) - 2\nu E\left(\Xi_1^{-1}\right)\right) tr\left(\boldsymbol{W}\boldsymbol{\Gamma}_{13}\right) \\
&+ \left(\nu^2 E\left(\Xi_2^{-2}\right) + 2\nu E\left(\Xi_1^{-1}\right) - 2\nu E\left(\Xi_2^{-1}\right)\right) tr\left(\boldsymbol{\gamma}^\top \boldsymbol{W}\boldsymbol{\gamma}\right) \\
ADR\left(\widehat{\boldsymbol{\beta}}_1^{PS}; \boldsymbol{W}\right) &= ADR\left(\widehat{\boldsymbol{\beta}}_1^{S}; \boldsymbol{W}\right) - E\left((1 - \nu\Xi_1^{-1})^2 I\left(\Xi_1 < \nu\right)\right) tr\left(\boldsymbol{W}\boldsymbol{\Gamma}_{13}\right) \\
&+ \left(2\Psi_{\nu+4}(\nu, \Delta) - 2\nu E\left(\Xi_1^{-1} I\left(\Xi_1 < \nu\right)\right)\right. \\
&\left. - E\left(\left(1 - \nu\Xi_2^{-1}\right)^2 I\left(\Xi_2 < \nu\right)\right)\right) tr\left(\boldsymbol{\gamma}^\top \boldsymbol{W}\boldsymbol{\gamma}\right).
\end{aligned}$$

Proof *To show that this theorem is true, we must first figure out the asymptotic covariance matrices for the four estimators. The covariance matrix $\boldsymbol{\Sigma}^*(\widehat{\boldsymbol{\beta}}_1^*)$ of any estimator $\widehat{\boldsymbol{\beta}}_1^*$ is defined as:*

$$\boldsymbol{\Sigma}^*(\widehat{\boldsymbol{\beta}}_1^*) = E\left(\lim_{n\to\infty} n(\widehat{\boldsymbol{\beta}}_1^* - \boldsymbol{\beta}_1)(\widehat{\boldsymbol{\beta}}_1^* - \boldsymbol{\beta}_1)^\top\right).$$

First, we derive the covariance matrices of the full model and submodel estimators as follows:

$$\begin{aligned}
\boldsymbol{\Sigma}^*(\widehat{\boldsymbol{\beta}}_1^{\text{FM}}) &= E\left(\lim_{n\to\infty} \sqrt{n}(\widehat{\boldsymbol{\beta}}_1^{\text{FM}} - \boldsymbol{\beta}_1)\sqrt{n}(\widehat{\boldsymbol{\beta}}_1^{\text{FM}} - \boldsymbol{\beta}_1)^\top\right) \\
&= E(\boldsymbol{\xi}_1\boldsymbol{\xi}_1^\top) = Var(\boldsymbol{\xi}_1) + E(\boldsymbol{\xi}_1)E(\boldsymbol{\xi}_1^\top) \\
&= Var(\boldsymbol{\xi}_1) = \boldsymbol{\Sigma}_{11.2}^{-1}. \\
\boldsymbol{\Sigma}^*(\widehat{\boldsymbol{\beta}}_1^{\text{SM}}) &= E\left(\lim_{n\to\infty} \sqrt{n}(\widehat{\boldsymbol{\beta}}_1^{\text{SM}} - \boldsymbol{\beta}_1)\sqrt{n}(\widehat{\boldsymbol{\beta}}_1^{\text{SM}} - \boldsymbol{\beta}_1)^\top\right) \\
&= E(\boldsymbol{\xi}_2\boldsymbol{\xi}_2^\top) = Var(\boldsymbol{\xi}_2) + E(\boldsymbol{\xi}_2)E(\boldsymbol{\xi}_2^\top) \\
&= \boldsymbol{\Gamma}_{12} + \boldsymbol{\gamma}\boldsymbol{\gamma}^\top.
\end{aligned}$$

Next, we derive the covariance matrices of the shrinkage and positive shrinkage estimators:

$$\begin{aligned}
\boldsymbol{\Sigma}^*(\widehat{\boldsymbol{\beta}}_1^S) &= E\left(\lim_{n\to\infty}\sqrt{n}(\widehat{\boldsymbol{\beta}}_1^S-\boldsymbol{\beta}_1)\sqrt{n}(\widehat{\boldsymbol{\beta}}_1^S-\boldsymbol{\beta}_1)^\top\right)\\
&= E\Bigg(\lim_{n\to\infty}\sqrt{n}\left(\widehat{\boldsymbol{\beta}}_1^{\text{FM}}-\boldsymbol{\beta}_1-\nu\widehat{\Lambda}^{-1}(\widehat{\boldsymbol{\beta}}_1^{\text{FM}}-\widehat{\boldsymbol{\beta}}_1^{\text{SM}})\right)\\
&\qquad \sqrt{n}\left(\widehat{\boldsymbol{\beta}}_1^{\text{FM}}-\boldsymbol{\beta}_1-\nu\widehat{\Lambda}^{-1}(\widehat{\boldsymbol{\beta}}_1^{\text{FM}}-\widehat{\boldsymbol{\beta}}_1^{\text{SM}})\right)^\top\Bigg)\\
&= E(\boldsymbol{\xi}_1\boldsymbol{\xi}_1^\top)+\nu^2E\left(\boldsymbol{\xi}_3\boldsymbol{\xi}_3^\top\lim_{n\to\infty}\widehat{\Lambda}^{-2}\right)-2\nu E\left(\boldsymbol{\xi}_3\boldsymbol{\xi}_1^\top\lim_{n\to\infty}\widehat{\Lambda}^{-1}\right)\\
&= \boldsymbol{\Sigma}_{11.2}^{-1}+\boldsymbol{\Gamma}_{13}E\left(\Xi_1^{-2}\right)+\boldsymbol{\gamma}\boldsymbol{\gamma}^\top E\left(\Xi_2^{-2}\right)-2\nu E\left(\boldsymbol{\xi}_3\boldsymbol{\xi}_1^\top\lim_{n\to\infty}\widehat{\Lambda}^{-1}\right).
\end{aligned}$$

Using the Lemma 3.2, we can simplify the last term as

$$\begin{aligned}
&E(\boldsymbol{\xi}_3\boldsymbol{\xi}_1^\top\lim_{n\to\infty}\widehat{\Lambda}^{-1}) = E\left(E(\boldsymbol{\xi}_3\boldsymbol{\xi}_1^\top\lim_{n\to\infty}\widehat{\Lambda}^{-1}|\boldsymbol{\xi}_3)\right)\\
&= E\left(\boldsymbol{\xi}_3E(\boldsymbol{\xi}_1^\top|\boldsymbol{\xi}_3)\lim_{n\to\infty}\widehat{\Lambda}^{-1}\right)+E\left(\boldsymbol{\xi}_3\left(\boldsymbol{\xi}_3^\top-E(\boldsymbol{\xi}_3)\right)^\top\lim_{n\to\infty}\widehat{\Lambda}^{-1}\right)\\
&= E\left(\boldsymbol{\xi}_3\boldsymbol{\xi}_3^\top\lim_{n\to\infty}\widehat{\Lambda}^{-1}\right)-E\left(\boldsymbol{\xi}_3\lim_{n\to\infty}\widehat{\Lambda}^{-1}\right)E\left(\boldsymbol{\xi}_3\right)^\top\\
&= \boldsymbol{\Gamma}_{13}E\left(\chi_{p_2+2}^{-2}(\Delta)\right)+\boldsymbol{\gamma}\boldsymbol{\gamma}^\top E\left(\chi_{p_2+4}^{-2}(\Delta)\right)-\boldsymbol{\gamma}\boldsymbol{\gamma}^\top E\left(\chi_{p_2+2}^{-2}(\Delta)\right)\\
&= \boldsymbol{\Gamma}_{13}E\left(\Xi_1^{-1}\right)+\boldsymbol{\gamma}\boldsymbol{\gamma}^\top E\left(\Xi_2^{-1}\right)-\boldsymbol{\gamma}\boldsymbol{\gamma}^\top E\left(\Xi_1^{-1}\right).
\end{aligned}$$

Hence

$$\begin{aligned}
\boldsymbol{\Sigma}^*(\widehat{\boldsymbol{\beta}}_1^S) &= \boldsymbol{\Sigma}_{11.2}^{-1}+\nu^2\left(\boldsymbol{\Gamma}_{13}E(\Xi_1^{-2})+\boldsymbol{\gamma}\boldsymbol{\gamma}^\top E(\Xi_2^{-2})\right)\\
&\quad -2\nu\left(\boldsymbol{\Gamma}_{13}E(\Xi_1^{-1})+\boldsymbol{\gamma}\boldsymbol{\gamma}^\top E(\Xi_2^{-1})-\boldsymbol{\gamma}\boldsymbol{\gamma}^\top E(\Xi_1^{-1})\right)\\
&= \boldsymbol{\Sigma}_{11.2}^{-1}+\left(\nu^2E\left(\Xi_1^{-2}\right)-2\nu E(\Xi_1^{-1})\right)\boldsymbol{\Gamma}_{13}\\
&+ \left(\nu^2E(\Xi_2^{-2})+2\nu E(\Xi_1^{-1})-2\nu E(\Xi_2^{-1})\right)\boldsymbol{\gamma}\boldsymbol{\gamma}^\top.
\end{aligned}$$

Let $g_{n+l}(\Delta)=\left(1-\nu\widehat{\Lambda}^{-1}\right)^l I\left(\widehat{\Lambda}<\nu\right)$, *where* $l=1,2$

$$\begin{aligned}
\boldsymbol{\Sigma}^*(\widehat{\boldsymbol{\beta}}_1^{PS}) &= E\left(\lim_{n\to\infty}\sqrt{n}(\widehat{\boldsymbol{\beta}}_1^{PS}-\boldsymbol{\beta}_1)\sqrt{n}(\widehat{\boldsymbol{\beta}}_1^{PS}-\boldsymbol{\beta}_1)^\top\right),\\
&= E\left(\lim_{n\to\infty}\sqrt{n}(\widehat{\boldsymbol{\beta}}_1^S-\boldsymbol{\beta}_1)\sqrt{n}(\widehat{\boldsymbol{\beta}}_1^S-\boldsymbol{\beta}_1)^\top\right)\\
&\quad +E\left(\lim_{n\to\infty}g_{n+2}(\Delta)\sqrt{n}(\widehat{\boldsymbol{\beta}}_1^{\text{FM}}-\widehat{\boldsymbol{\beta}}_1^{\text{SM}})\sqrt{n}(\widehat{\boldsymbol{\beta}}_1^{\text{FM}}-\widehat{\boldsymbol{\beta}}_1^{\text{SM}})^\top\right)\\
&\quad -2E\left(\lim_{n\to\infty}g_{n+1}(\Delta)\sqrt{n}(\widehat{\boldsymbol{\beta}}_1^{\text{FM}}-\widehat{\boldsymbol{\beta}}_1^{\text{SM}})\sqrt{n}(\widehat{\boldsymbol{\beta}}_1^S-\boldsymbol{\beta}_1)^\top\right)\\
&= \boldsymbol{\Sigma}^*(\widehat{\boldsymbol{\beta}}_1^S)+E\left(\lim_{n\to\infty}g_{n+2}(\Delta)\boldsymbol{\xi}_3\boldsymbol{\xi}_3^\top\right)\\
&- 2E\left(\lim_{n\to\infty}g_{n+1}(\Delta)\boldsymbol{\xi}_3\left(\boldsymbol{\xi}_2^\top+\left(1-\nu\widehat{\Lambda}^{-1}\right)\boldsymbol{\xi}_3^\top\right)\right),\\
&= \boldsymbol{\Sigma}^*(\widehat{\boldsymbol{\beta}}_1^S)-E\left(\lim_{n\to\infty}g_{n+2}(\Delta)\boldsymbol{\xi}_3\boldsymbol{\xi}_3^\top\right)-2E\left(\lim_{n\to\infty}g_{n+1}(\Delta)\boldsymbol{\xi}_3\boldsymbol{\xi}_2^\top\right).
\end{aligned}$$

Using the Lemma 3.2, we can simplify the second term as:

$$\begin{aligned}
&-E\left(\lim_{n\to\infty}g_{n+2}(\Delta)\boldsymbol{\xi}_2\boldsymbol{\xi}_2^\top\right) = -E\left(\lim_{n\to\infty}\left(1-\nu\widehat{\Lambda}^{-1}\right)^2 I\left(\widehat{\Lambda}<\nu\right)\boldsymbol{\xi}_3\boldsymbol{\xi}_3^\top\right)\\
&= -\boldsymbol{\Gamma}_{13}E\left(I\left(\Xi_1<\nu\right)\left(1-\nu\Xi_1^{-1}\right)^2\right)-\boldsymbol{\gamma}\boldsymbol{\gamma}^\top E\left(I\left(\Xi_2<\nu\right)\left(1-\nu\Xi_2^{-1}\right)^2\right).
\end{aligned}$$

Using the Lemma 3.2, we can simplify the third term as:

$$
\begin{aligned}
&-2E\Big(\lim_{n\to\infty} g_{n+1}(\Delta)\boldsymbol{\xi}_3\boldsymbol{\xi}_2^\top\Big) = -2E\Big(\lim_{n\to\infty} \boldsymbol{\xi}_3 E\left(g_{n+1}(\Delta)\boldsymbol{\xi}_2^\top|\boldsymbol{\xi}_3\right)\Big)\\
=\ & -2E\Big(\lim_{n\to\infty} \boldsymbol{\xi}_3\left(E\left(\boldsymbol{\xi}_3^\top\right) + cov\left(\boldsymbol{\xi}_3,\boldsymbol{\xi}_2\right)\left(\boldsymbol{\xi}_3 - E\left(\boldsymbol{\xi}_3\right)\right)\right) g_{n+1}(\Delta)\Big)\\
=\ & -2E\Big(\lim_{n\to\infty} \boldsymbol{\xi}_3 E\left(\boldsymbol{\xi}_2^\top\right) g_{n+1}(\Delta) + \mathbf{0}\Big)\\
=\ & -2E\Big(\lim_{n\to\infty} \boldsymbol{\xi}_3 I\left(\widehat{\Lambda} < \nu\right) - \nu\widehat{\Lambda}^{-1}\boldsymbol{\xi}_3 I\left(\widehat{\Lambda} < \nu\right)\Big) E\left(\boldsymbol{\xi}_2^\top\right)\\
=\ & 2\Psi_{\nu+4}(\nu,\Delta)\boldsymbol{\gamma}\boldsymbol{\gamma}^\top - 2\nu E\left(\Xi_1^{-1} I\left(\Xi_1 < \nu\right)\right)\boldsymbol{\gamma}\boldsymbol{\gamma}^\top\\
=\ & \left(2\Psi_{\nu+4}(\nu,\Delta) - 2\nu E\left(\Xi_1^{-1} I\left(\Xi_1 < \nu\right)\right)\right)\boldsymbol{\gamma}\boldsymbol{\gamma}^\top.
\end{aligned}
$$

Finally,

$$
\begin{aligned}
\boldsymbol{\Sigma}^*\left(\widehat{\boldsymbol{\beta}}_1^{PS}\right) &= \boldsymbol{\Sigma}^*\left(\widehat{\boldsymbol{\beta}}_1^{S}\right) - E\Big(\left(1-\nu\Xi_1^{-1}\right)^2 I\left(\Xi_1 < \nu\right)\Big)\boldsymbol{\Gamma}_{13}\\
&+ \left(2\Psi_{\nu+4}(\nu,\Delta) - 2\nu E\left(\Xi_1^{-1} I\left(\Xi_1 < \nu\right)\right)\right.\\
&- \left. E\Big(\left(1-\nu\Xi_2^{-1}\right)^2 I\left(\Xi_2 < \nu\right)\Big)\right)\boldsymbol{\gamma}\boldsymbol{\gamma}^\top.
\end{aligned}
$$

The risk expressions in Theorem 3.5 are readily obtained from (3.11), which completes the proof.

3.4 Relative Risk Assessment

In this section, we compare the pairwise risk performance of the proposed estimators $\widehat{\boldsymbol{\beta}}_1^{\mathrm{SM}}$, $\widehat{\boldsymbol{\beta}}_1^{S}$, and $\widehat{\boldsymbol{\beta}}_1^{PS}$ with respect to full model estimator, $\widehat{\boldsymbol{\beta}}_1^{\mathrm{FM}}$. Noting that, risk expressions reveal that if $\boldsymbol{\Sigma}_{12} = \mathbf{0}$ then $\boldsymbol{\Sigma}_{11.2} = \boldsymbol{\Sigma}_{11}$ and the respective risk of all the estimators are asymptotically equivalent, and reduced to the risk of $\widehat{\boldsymbol{\beta}}_1^{\mathrm{FM}}$. In the sequel, we assume that $\boldsymbol{\Sigma}_{12} \neq \mathbf{0}$, and $\boldsymbol{W} = \boldsymbol{\Sigma}_{11.2}$ and the remaining discussions follows.

For $\boldsymbol{W} = \boldsymbol{\Sigma}_{11.2}$, the risk of $\widehat{\boldsymbol{\beta}}_1^{\mathrm{FM}}$ is p_1. Noting that risk function of $\widehat{\boldsymbol{\beta}}_1^{\mathrm{FM}}$ is independent of parameter Δ, that is, independent of sparsity assumption. However, the risk function of all other estimators involves the parameter Δ (function of $\boldsymbol{\delta}$). In a sense, the parameter $\Delta > 0$ can be viewed as the sparsity parameter. Thus, it makes sense to assess the relative properties of the estimators in terms of Δ.

Under the sparsity assumption, $\Delta = 0$, that is, when $\boldsymbol{\delta} = \mathbf{0}$, the submodel estimator $\widehat{\boldsymbol{\beta}}_1^{\mathrm{SM}}$ is the best choice and it will perform better than $\widehat{\boldsymbol{\beta}}_1^{\mathrm{FM}}$ and shrinkage estimators. However, when $||\boldsymbol{\delta}||$ moves away from zero, the ADR of $\widehat{\boldsymbol{\beta}}_1^{\mathrm{SM}}$ monotonically increases and goes to ∞. This clearly indicates that the performance of $\widehat{\boldsymbol{\beta}}_1^{\mathrm{SM}}$ depends on the validity of of the sparsity assumption, that is, $\boldsymbol{\beta}_2 = \mathbf{0}$. This is an extremely undesirable characteristic of the estimators based on the selected submodel for practical purposes and is frequently ignored by practitioners and most researchers alike.

Interestingly, it can be seen that the respective risk functions of shrinkage estimators are bounded function of Δ, unlike submodel estimator, and outperform $\widehat{\boldsymbol{\beta}}_1^{\mathrm{FM}}$ in the entire parameter space induced by Δ.

Now, we provide a detailed pairwise comparison of the listed estimators.

3.4.1 Risk Comparison of $\hat{\beta}_1^{\text{FM}}$ and $\hat{\beta}_1^{\text{SM}}$

The difference of between the risks of $\widehat{\boldsymbol{\beta}}_1^{\text{FM}}$ and $\widehat{\boldsymbol{\beta}}_1^{\text{SM}}$ is:

$$
\begin{aligned}
& \text{ADR}(\widehat{\boldsymbol{\beta}}_1^{\text{SM}}; \boldsymbol{W}) - \text{ADR}(\widehat{\boldsymbol{\beta}}_1^{\text{FM}}; \boldsymbol{W}) \\
= \; & \text{tr}(\boldsymbol{W}\boldsymbol{\Gamma}_{12}) + \boldsymbol{\gamma}^\top \boldsymbol{W}\boldsymbol{\gamma} \\
= \; & \text{tr}(\boldsymbol{W}\boldsymbol{\Sigma}_{11.2}^{-1}) - \text{tr}(\boldsymbol{W}\boldsymbol{\Sigma}_{11}^{-1}\boldsymbol{\Sigma}_{12}\boldsymbol{\Sigma}_{22.1}^{-1}\boldsymbol{\Sigma}_{21}\boldsymbol{\Sigma}_{11}^{-1}) + \boldsymbol{\delta}^\top\boldsymbol{\Sigma}_{21}\boldsymbol{\Sigma}_{11}^{-1}\boldsymbol{W}\boldsymbol{\Sigma}_{11}^{-1}\boldsymbol{\Sigma}_{12}\boldsymbol{\delta} \\
- \; & \text{tr}(\boldsymbol{W}\boldsymbol{\Sigma}_{11.2}^{-1}) \\
= \; & \boldsymbol{\delta}^\top\boldsymbol{\Sigma}_{21}\boldsymbol{\Sigma}_{11}^{-1}\boldsymbol{\Sigma}_{11.2}\boldsymbol{\Sigma}_{11}^{-1}\boldsymbol{\Sigma}_{12}\boldsymbol{\delta} - \text{tr}(\boldsymbol{\Sigma}_{11.2}\boldsymbol{\Sigma}_{11}^{-1}\boldsymbol{\Sigma}_{12}\boldsymbol{\Sigma}_{22.1}^{-1}\boldsymbol{\Sigma}_{21}\boldsymbol{\Sigma}_{11}^{-1}). \qquad (3.12)
\end{aligned}
$$

The second term of equation (3.12) can be written as

$$
\begin{aligned}
& \text{tr}(\boldsymbol{\Sigma}_{11.2}\boldsymbol{\Sigma}_{11}^{-1}\boldsymbol{\Sigma}_{12}\boldsymbol{\Sigma}_{22.1}^{-1}\boldsymbol{\Sigma}_{21}\boldsymbol{\Sigma}_{11}^{-1}) \\
= \; & \text{tr}((\boldsymbol{\Sigma}_{11} - \boldsymbol{\Sigma}_{12}\boldsymbol{\Sigma}_{22}^{-1}\boldsymbol{\Sigma}_{21})\boldsymbol{\Sigma}_{11}^{-1}\boldsymbol{\Sigma}_{12}\left(\boldsymbol{\Sigma}_{22} - \boldsymbol{\Sigma}_{21}\boldsymbol{\Sigma}_{11}^{-1}\boldsymbol{\Sigma}_{12}\right)^{-1}\boldsymbol{\Sigma}_{21}\boldsymbol{\Sigma}_{11}^{-1}) \\
= \; & \text{tr}((\boldsymbol{\Sigma}_{11} - \boldsymbol{\Sigma}_{12}\boldsymbol{\Sigma}_{22}^{-1}\boldsymbol{\Sigma}_{21})\boldsymbol{\Sigma}_{11}^{-1}\boldsymbol{\Sigma}_{12} \\
\times \; & (\boldsymbol{\Sigma}_{22}^{-1} + \boldsymbol{\Sigma}_{22}^{-1}\boldsymbol{\Sigma}_{21}(\boldsymbol{\Sigma}_{11} - \boldsymbol{\Sigma}_{12}\boldsymbol{\Sigma}_{22}^{-1}\boldsymbol{\Sigma}_{21})^{-1}\boldsymbol{\Sigma}_{12}\boldsymbol{\Sigma}_{22}^{-1})\boldsymbol{\Sigma}_{21}\boldsymbol{\Sigma}_{11}^{-1}) \\
= \; & \text{tr}((\boldsymbol{\Sigma}_{11} - \boldsymbol{\Sigma}_{12}\boldsymbol{\Sigma}_{22}^{-1}\boldsymbol{\Sigma}_{21})\boldsymbol{\Sigma}_{11}^{-1}\boldsymbol{\Sigma}_{12}\boldsymbol{\Sigma}_{22}^{-1}\boldsymbol{\Sigma}_{21}\boldsymbol{\Sigma}_{11}^{-1}) \\
+ \; & \text{tr}((\boldsymbol{\Sigma}_{11} - \boldsymbol{\Sigma}_{12}\boldsymbol{\Sigma}_{22}^{-1}\boldsymbol{\Sigma}_{21})\boldsymbol{\Sigma}_{11}^{-1}\boldsymbol{\Sigma}_{12}\boldsymbol{\Sigma}_{22}^{-1}\boldsymbol{\Sigma}_{21} \\
\times \; & (\boldsymbol{\Sigma}_{11} - \boldsymbol{\Sigma}_{12}\boldsymbol{\Sigma}_{22}^{-1}\boldsymbol{\Sigma}_{21})^{-1}\boldsymbol{\Sigma}_{12}\boldsymbol{\Sigma}_{22}^{-1}\boldsymbol{\Sigma}_{21}\boldsymbol{\Sigma}_{11}^{-1}) \\
= \; & \text{tr}(\boldsymbol{\Sigma}_{12}\boldsymbol{\Sigma}_{22}^{-1}\boldsymbol{\Sigma}_{21}\boldsymbol{\Sigma}_{11}^{-1})
\end{aligned}
$$

Finally the equation (3.12) becomes

$$
\begin{aligned}
& \text{ADR}(\widehat{\boldsymbol{\beta}}_1^{\text{SM}}; \boldsymbol{W}) - \text{ADR}(\widehat{\boldsymbol{\beta}}_1^{\text{FM}}; \boldsymbol{W}) \\
= \; & \boldsymbol{\delta}^\top\boldsymbol{\Sigma}_{21}\boldsymbol{\Sigma}_{11}^{-1}\boldsymbol{\Sigma}_{11.2}\boldsymbol{\Sigma}_{11}^{-1}\boldsymbol{\Sigma}_{12}\boldsymbol{\delta} - \text{tr}(\boldsymbol{\Sigma}_{12}\boldsymbol{\Sigma}_{22}^{-1}\boldsymbol{\Sigma}_{21}\boldsymbol{\Sigma}_{11}^{-1}) = \boldsymbol{M}_1 - \boldsymbol{M}_2, \qquad (3.13)
\end{aligned}
$$

where $\boldsymbol{M}_1 = \boldsymbol{\delta}^\top\boldsymbol{M}^*\boldsymbol{\delta}$ and $\boldsymbol{M}_2 = \text{tr}(\boldsymbol{\Sigma}_{12}\boldsymbol{\Sigma}_{22}^{-1}\boldsymbol{\Sigma}_{21}\boldsymbol{\Sigma}_{11}^{-1})$ with $\boldsymbol{M}^* = \boldsymbol{\Sigma}_{21}\boldsymbol{\Sigma}_{11}^{-1}\boldsymbol{\Sigma}_{11.2}\boldsymbol{\Sigma}_{11}^{-1}\boldsymbol{\Sigma}_{12}$.

Clearly, if the sparsity assumption is nearly true, $\widehat{\boldsymbol{\beta}}_1^{\text{SM}}$ is more efficient than $\widehat{\boldsymbol{\beta}}_1^{\text{FM}}$. As mentioned earlier, the risk of $\widehat{\boldsymbol{\beta}}_1^{\text{SM}}$ is unbounded function of $||\boldsymbol{\delta}||$, and for large values of $||\boldsymbol{\delta}||$ the full model estimator $\widehat{\boldsymbol{\beta}}_1^{\text{FM}}$ performs better than the submodel estimator.

3.4.2 Risk Comparison of $\hat{\beta}_1^{\text{FM}}$ and $\hat{\beta}_1^{\text{S}}$

The risk difference of $\widehat{\boldsymbol{\beta}}_1^{S}$ and $\widehat{\boldsymbol{\beta}}_1^{\text{FM}}$ is

$$
\begin{aligned}
& \text{ADR}\left(\widehat{\boldsymbol{\beta}}_1^{S}; \boldsymbol{W}\right) - \text{ADR}\left(\widehat{\boldsymbol{\beta}}_1^{\text{FM}}; \boldsymbol{W}\right) \\
= \; & [\nu^2\text{E}(\Xi_1^{-2}) - 2\nu\text{E}(\Xi_1^{-1})]\text{tr}(\boldsymbol{M}_2), \quad \text{where } \boldsymbol{W} = \boldsymbol{\Sigma}_{11.2} \\
+ \; & [\nu^2\text{E}(\Xi_2^{-2}) + 2\nu\text{E}(\Xi_1^{-1}) - 2\nu\text{E}(\Xi_2^{-1})]\text{tr}(\boldsymbol{M}_1) \qquad (3.14)
\end{aligned}
$$

We know that

$$
\begin{aligned}
\text{E}(\Xi_1^{-1}) - \text{E}(\Xi_2^{-1}) &= 2\text{E}(\Xi_2^{-2}), \qquad (3.15) \\
\text{E}(\Xi_1^{-1}) - \nu\text{E}(\Xi_1^{-2}) &= 2\Delta\text{E}(\Xi_2^{-2}). \qquad (3.16)
\end{aligned}
$$

Using (3.14) and (3.15), (3.16) can be written as

$$
\begin{aligned}
& \text{ADR}\left(\widehat{\boldsymbol{\beta}}_1^{S}; \boldsymbol{W}\right) - \text{ADR}\left(\widehat{\boldsymbol{\beta}}_1^{\text{FM}}; \boldsymbol{W}\right) \\
= \; & \left[\nu^2\text{E}(\Xi_1^{-2}) - 2\nu\left(\nu\text{E}(\Xi_1^{-2}) + \Delta\text{E}(\Xi_2^{-2})\right)\right]\text{tr}(\boldsymbol{M}_2)
\end{aligned}
$$

$$
\begin{aligned}
& + \quad \nu(\nu+4)\mathrm{E}(\Xi_2^{-2})\mathrm{tr}(\boldsymbol{M}_1), \quad \text{where } \boldsymbol{W} = \boldsymbol{\Sigma}_{11.2} \\
&= \quad -\nu^2\mathrm{E}(\Xi_1^{-2})\mathrm{tr}(\boldsymbol{M}_2) - 2\Delta\nu\mathrm{E}(\Xi_2^{-2})\mathrm{tr}(\boldsymbol{M}_2) + \nu(\nu+4)\mathrm{E}(\Xi_2^{-2})\mathrm{tr}(\boldsymbol{M}_1) \\
&= \quad -\nu\mathrm{tr}(\boldsymbol{M}_1)\left[\nu\mathrm{E}(\Xi_2^{-2}) + \left(1 - \frac{(\nu+4)\mathrm{tr}(\boldsymbol{M}_1)}{2\Delta\mathrm{tr}(\boldsymbol{M}_2)}\right)2\Delta\mathrm{E}(\Xi_1^{-2})\right]
\end{aligned}
$$

The above risk difference is non-negative, that is, $\mathrm{ADR}\left(\widehat{\boldsymbol{\beta}}_1^S; \boldsymbol{W}\right) - \mathrm{ADR}\left(\widehat{\boldsymbol{\beta}}_1^{\mathrm{FM}}; \boldsymbol{W}\right) \leq 0$ for $\nu > 1$, or $p_2 > 3$, and $\Delta > 0$ when

$$
\begin{aligned}
& 1 - \frac{(\nu+4)\mathrm{tr}(\boldsymbol{M}_1)}{2\Delta\mathrm{tr}(\boldsymbol{M}_2)} \geq 0 \\
& \frac{(\nu+4)\mathrm{tr}(\boldsymbol{M}_1)}{2\Delta\mathrm{tr}(\boldsymbol{M}_2)} \leq 1 \\
& \frac{(p_2+2)\mathrm{Ch}_{\max}(\boldsymbol{M}_2)}{2\mathrm{tr}(\boldsymbol{M}_2)} \leq 1, \qquad \text{by Courant Theorem} \\
& \frac{\mathrm{Ch}_{\max}(\boldsymbol{M}_2)}{\mathrm{tr}(\boldsymbol{M}_2)} \leq \frac{(p_2+2)}{2}, \qquad p_2 > 3
\end{aligned}
$$

Under the above conditions, the risk of $\widehat{\boldsymbol{\beta}}_1^S$ is smaller than the risk of $\widehat{\boldsymbol{\beta}}_1^{\mathrm{FM}}$ in the entire parameter space and the upper limit is attained when $\Delta \to \infty$. It clearly indicates the asymptotic inferiority of $\widehat{\boldsymbol{\beta}}_1^{\mathrm{FM}}$ and the largest gain in risk is achieved when the sparsity assumption is true.

3.4.3 Risk Comparison of $\hat{\boldsymbol{\beta}}_1^{\mathrm{S}}$ and $\hat{\boldsymbol{\beta}}_1^{\mathrm{SM}}$

The risk difference between $\widehat{\boldsymbol{\beta}}_1^{\mathrm{SM}}$ and $\widehat{\boldsymbol{\beta}}_1^S$ is given by

$$
\begin{aligned}
& \mathrm{ADR}\left(\widehat{\boldsymbol{\beta}}_1^{\mathrm{SM}}; \boldsymbol{W}\right) - \mathrm{ADR}\left(\widehat{\boldsymbol{\beta}}_1^S; \boldsymbol{W}\right) \\
= \; & \mathrm{tr}(\boldsymbol{W}\boldsymbol{\Gamma}_{12}) + \boldsymbol{\gamma}^\top\boldsymbol{W}\boldsymbol{\gamma} \\
& -\mathrm{ADR}\left(\widehat{\boldsymbol{\beta}}_1^{\mathrm{FM}}; \boldsymbol{W}\right) - \left(\nu^2\mathrm{E}\left(\Xi_1^{-2}\right) - 2\nu\mathrm{E}\left(\Xi_1^{-1}\right)\right)\mathrm{tr}\left(\boldsymbol{W}\boldsymbol{\Gamma}_{13}\right) \\
& - \left(\nu^2\mathrm{E}\left(\Xi_2^{-2}\right) - 2\nu\mathrm{E}\left(\Xi_1^{-1}\right) + 2\nu\mathrm{E}\left(\Xi_2^{-1}\right)\right)\mathrm{tr}\left(\boldsymbol{\gamma}^\top\boldsymbol{W}\boldsymbol{\gamma}\right) \\
= \; & \mathrm{ADR}\left(\widehat{\boldsymbol{\beta}}_1^{\mathrm{FM}}; \boldsymbol{W}\right) - \mathrm{tr}(\boldsymbol{W}\boldsymbol{\Sigma}_{11}^{-1}\boldsymbol{\Sigma}_{12}\boldsymbol{\Sigma}_{22.1}^{-1}\boldsymbol{\Sigma}_{21}\boldsymbol{\Sigma}_{11}^{-1}) \\
& +\boldsymbol{\delta}^\top\boldsymbol{\Sigma}_{21}\boldsymbol{\Sigma}_{11}^{-1}\boldsymbol{W}\boldsymbol{\Sigma}_{11}^{-1}\boldsymbol{\Sigma}_{12}\boldsymbol{\delta} \\
& -\mathrm{ADR}\left(\widehat{\boldsymbol{\beta}}_1^{\mathrm{FM}}; \boldsymbol{W}\right) - \left(\nu^2\mathrm{E}\left(\Xi_1^{-2}\right) - 2\nu\mathrm{E}\left(\Xi_1^{-1}\right)\right)\mathrm{tr}\left(\boldsymbol{W}\boldsymbol{\Gamma}_{13}\right) \\
& - \left(\nu^2\mathrm{E}\left(\Xi_2^{-2}\right) - 2\nu\mathrm{E}\left(\Xi_1^{-1}\right) + 2\nu\mathrm{E}\left(\Xi_2^{-1}\right)\right)\mathrm{tr}\left(\boldsymbol{\gamma}^\top\boldsymbol{W}\boldsymbol{\gamma}\right) \\
= \; & \mathrm{tr}(\boldsymbol{M}_1) - \mathrm{tr}(\boldsymbol{M}_2) - \left[\nu^2\mathrm{E}(\Xi_1^{-2}) - 2\nu\left(\nu\mathrm{E}(\Xi_1^{-2}) + \Delta\mathrm{E}(\Xi_2^{-2})\right)\right]\mathrm{tr}(\boldsymbol{M}_2) \\
& -\nu(\nu+4)\mathrm{E}(\Xi_2^{-2})\mathrm{tr}(\boldsymbol{M}_1), \quad \text{where } \boldsymbol{W} = \boldsymbol{\Sigma}_{11.2} \\
= \; & \mathrm{tr}(\boldsymbol{M}_1) - \mathrm{tr}(\boldsymbol{M}_2) + \nu^2\mathrm{E}(\Xi_1^{-2})\mathrm{tr}(\boldsymbol{M}_2) \\
& -2\Delta\nu\mathrm{E}(\Xi_2^{-2})\mathrm{tr}(\boldsymbol{M}_2) - \nu(\nu+4)\mathrm{E}(\Xi_2^{-2})\mathrm{tr}(\boldsymbol{M}_1) \\
= \; & (1 - (p^2 - 4)\mathrm{E}(\Xi_2^{-2}))\mathrm{tr}(\boldsymbol{M}_1) \\
& -(1 - (p_2-2)^2\mathrm{E}(\Xi_1^{-2}) + 2(p_2-2)\Delta\mathrm{E}(\Xi_2^{-2}))\mathrm{tr}(\boldsymbol{M}_2)
\end{aligned}
$$

Noting,

$$
\frac{\boldsymbol{\delta}^\top\boldsymbol{M}^*\boldsymbol{\delta}}{\boldsymbol{\Delta}} = \frac{\boldsymbol{\delta}^\top\boldsymbol{M}^*\boldsymbol{\delta}}{\boldsymbol{\delta}^\top\boldsymbol{\Sigma}_{22.1}\boldsymbol{\delta}} \leq \mathrm{Ch}_{\max}(\boldsymbol{M}^*\boldsymbol{\Sigma}_{22.1}) = \mathrm{Ch}_{\max}(\boldsymbol{M}_2) = g\mathrm{tr}(\boldsymbol{M}_2),
$$

where $g = \text{Ch}_{\max}(\boldsymbol{M}_2)/\text{tr}(\boldsymbol{M}_2)$ and $\boldsymbol{M}_1 = \boldsymbol{\delta}^\top \boldsymbol{M}^* \boldsymbol{\delta}$. Thus we have

$$\text{ADR}\left(\widehat{\boldsymbol{\beta}}_1^{\text{SM}}; \boldsymbol{W}\right) - \text{ADR}\left(\widehat{\boldsymbol{\beta}}_1^{S}; \boldsymbol{W}\right) \leq (1 - (p^2 - 4)\text{E}(\Xi_2^{-2}))g\Delta \text{tr}(\boldsymbol{M}_2)$$
$$-(1 - (p_2 - 2)^2 \text{E}(\Xi_1^{-2}) + 2(p_2 - 2)\Delta \text{E}(\Xi_2^{-2}))\text{tr}(\boldsymbol{M}_2) \tag{3.17}$$

The right side of (3.17) is negative if Δ is near zero and $p_2 \geq 3$. Thus $\widehat{\boldsymbol{\beta}}_1^{\text{SM}}$ perform better than $\widehat{\boldsymbol{\beta}}_1^{S}$ when the sparsity assumption is nearly true. When Δ increases the risk difference is positive indicating poor performance of the submodel estimator. Again, the validity of the sparsity assumption is fatal to a submodel estimator.

3.4.4 Risk Comparison of $\hat{\beta}_1^{\text{PS}}$ and $\hat{\beta}_1^{\text{FM}}$

The risk difference between $\widehat{\boldsymbol{\beta}}_1^{PS}$ and $\widehat{\boldsymbol{\beta}}_1^{\text{FM}}$ is given by

$$\begin{aligned}
& \text{ADR}\left(\widehat{\boldsymbol{\beta}}_1^{PS}; \boldsymbol{W}\right) - \text{ADR}\left(\widehat{\boldsymbol{\beta}}_1^{\text{FM}}; \boldsymbol{W}\right) \\
= \; & \text{ADR}\left(\widehat{\boldsymbol{\beta}}_1^{S}; \boldsymbol{W}\right) - \text{ADR}\left(\widehat{\boldsymbol{\beta}}_1^{\text{FM}}; \boldsymbol{W}\right) \\
& -\text{E}\left((1 - \nu\Xi_1^{-1})^2 I\left(\Xi_1 < \nu\right)\right) \text{tr}\left(\boldsymbol{W}\boldsymbol{\Gamma}_{13}\right) \\
& + \left(2\Psi_{\nu+4}(\nu, \Delta) - 2\nu \text{E}\left(\Xi_1^{-1} I\left(\Xi_1 < \nu\right)\right)\right. \\
& \left. - \text{E}\left(\left(1 - \nu\Xi_2^{-1}\right)^2 I\left(\Xi_2 < \nu\right)\right)\right) \text{tr}\left(\boldsymbol{\gamma}^\top \boldsymbol{W} \boldsymbol{\gamma}\right).
\end{aligned}$$

We know that from the risk comparison between $\widehat{\boldsymbol{\beta}}_1^{S}$ and $\widehat{\boldsymbol{\beta}}_1^{\text{FM}}$ that $\text{ADR}\left(\widehat{\boldsymbol{\beta}}_1^{S}; \boldsymbol{W}\right) - \text{ADR}\left(\widehat{\boldsymbol{\beta}}_1^{\text{FM}}; \boldsymbol{W}\right) \leq 0$. Also from Risk comparison of $\widehat{\boldsymbol{\beta}}_1^{PS}$ and $\widehat{\boldsymbol{\beta}}_1^{S}$ shows that the third term in the above expression is negative. That is, we can say for all Δ and $p_2 \geq 3$ that

$$\text{ADR}\left(\widehat{\boldsymbol{\beta}}_1^{PS}; \boldsymbol{W}\right) \leq \text{ADR}\left(\widehat{\boldsymbol{\beta}}_1^{\text{FM}}; \boldsymbol{W}\right)$$

3.4.5 Risk Comparison of $\hat{\beta}_1^{\text{PS}}$ and $\hat{\beta}_1^{\text{S}}$

The risk difference of $\widehat{\boldsymbol{\beta}}_1^{S}$ and $\widehat{\boldsymbol{\beta}}_1^{PS}$ is

$$\begin{aligned}
& \text{ADR}\left(\widehat{\boldsymbol{\beta}}_1^{PS}; \boldsymbol{W}\right) - \text{ADR}\left(\widehat{\boldsymbol{\beta}}_1^{S}; \boldsymbol{W}\right) \\
= \; & \text{E}\left((1 - \nu\Xi_1^{-1})^2 I\left(\Xi_1 < \nu\right)\right) \text{tr}\left(\boldsymbol{W}\boldsymbol{\Gamma}_{13}\right) \\
& - \left(2\Psi_{\nu+4}(\nu, \Delta) - 2\nu \text{E}\left(\Xi_1^{-1} I\left(\Xi_1 < \nu\right)\right)\right. \\
& \left. + \text{E}\left(\left(1 - \nu\Xi_2^{-1}\right)^2 I\left(\Xi_2 < \nu\right)\right)\right) \text{tr}\left(\boldsymbol{\gamma}^\top \boldsymbol{W} \boldsymbol{\gamma}\right). \\
= \; & \text{E}\left((1 - (p_2 - 2)\Xi_1^{-1})^2 I\left(\Xi_1 < (p_2 - 2)\right)\right) \text{tr}\left(\boldsymbol{W}\boldsymbol{\Gamma}_{13}\right) \\
& - \left(2\Psi_{p_2+2}((p_2 - 2), \Delta) - 2(p_2 - 2)\text{E}\left(\Xi_1^{-1} I\left(\Xi_1 < (p_2 - 2)\right)\right)\right. \\
& \left. + \text{E}\left(\left(1 - (p_2 - 2)\Xi_2^{-1}\right)^2 I\left(\Xi_2 < (p_2 - 2)\right)\right)\right) \text{tr}\left(\boldsymbol{\gamma}^\top \boldsymbol{W} \boldsymbol{\gamma}\right), \quad \nu = p_2 - 2.
\end{aligned}$$

The right-hand side of the above expression is positive, since the expectation of a positive random variable is positive. By thedefinition of an indicator function since

$$\begin{aligned}
& (1 - \nu\Xi_1^{-1})^2 I\left(\Xi_1 < \nu\right) \geq 0, \quad \left(1 - (p_2 - 2)\Xi_2^{-1}\right)^2 I\left(\Xi_2 < (p_2 - 2)\right) \geq 0, \\
& \text{and } 2\Psi_{p_2+2}((p_2 - 2), \Delta) - 2(p_2 - 2)\text{E}\left(\Xi_1^{-1} I\left(\Xi_1 < (p_2 - 2)\right)\right) \leq 0,
\end{aligned}$$

where $\Psi_{p_2+2}((p_2 - 2), \Delta)$ lies between 0 and 1. Thus, for all Δ and $p_2 \geq 3$

$$\text{ADR}\left(\widehat{\boldsymbol{\beta}}_1^{PS}; \boldsymbol{W}\right) \leq \text{ADR}\left(\widehat{\boldsymbol{\beta}}_1^{S}; \boldsymbol{W}\right) \leq \text{ADR}\left(\widehat{\boldsymbol{\beta}}_1^{\text{FM}}; \boldsymbol{W}\right),$$

with strict inequality for some Δ. Hence we can conclude that the proposed estimator $\widehat{\boldsymbol{\beta}}_1^{PS}$ is asymptotically superior to $\widehat{\boldsymbol{\beta}}_1^{S}$ and hence to $\widehat{\boldsymbol{\beta}}_1^{\text{FM}}$, as well.

Based on the above findings, we can safely conclude that the submodel estimator dominates the full model and a shrinkage estimator if the sparsity assumption is nearly correct. Thus, the performance of the submodel estimator heavily depends on the sparsity assumption, that is, $\boldsymbol{\beta} = \mathbf{0}$. The risk of this estimator may become unbounded when the sparsity assumption does not hold. The shrinkage estimators outperform the full model estimator of the regression parameters vector in the entire parameter space induced by the sparsity assumption. We suggest to use $\widehat{\boldsymbol{\beta}}_1^{PS}$ over all other estimators in the class when sparsity assumption may judiciously satisfied.

3.4.6 Mean Squared Prediction Error

Let $(y_i, \boldsymbol{x}_i)$ be the training data that will be used to fit the multiple linear regression model (MLR) and $(y_i^*, \boldsymbol{x}_i^*)$ be the testing data on which predictions will be made. Mean squared prediction error (MSPE)focuses on the prediction errors of a model. It can be derive as using the testing data. Based on this data the MLR becomes $\boldsymbol{y}^* = \boldsymbol{X}_1^*\boldsymbol{\beta}_1 + \boldsymbol{X}_2^*\boldsymbol{\beta}_2 + \boldsymbol{\varepsilon}^*$. When we suspect that $\boldsymbol{\beta}_2 = \mathbf{0}$, then the MLR model becomes $\boldsymbol{y}^* = \boldsymbol{X}_1^*\boldsymbol{\beta}_1 + \boldsymbol{\varepsilon}^*$. Let $\widehat{\boldsymbol{y}}^* = \boldsymbol{X}_1^*\widehat{\boldsymbol{\beta}}_1^{\text{FM}}$ denote the predicted values based on the test data set. Then

$$\begin{aligned}
\text{E}||\boldsymbol{y}^* - \widehat{\boldsymbol{y}}^*||^2 &= \text{E}||\boldsymbol{X}_1^*\boldsymbol{\beta}_1 + \boldsymbol{\varepsilon}^* - \boldsymbol{X}_1^*\widehat{\boldsymbol{\beta}}_1^{\text{FM}}||^2 \\
&= \text{E}||\boldsymbol{X}_1^*(\boldsymbol{\beta}_1 - \widehat{\boldsymbol{\beta}}_1^{\text{FM}}) + \boldsymbol{\varepsilon}^*||^2 \\
&= \text{E}||\boldsymbol{X}_1^*(\boldsymbol{\beta}_1 - \widehat{\boldsymbol{\beta}}_1^{\text{FM}})||^2 + \text{E}||\boldsymbol{\varepsilon}^*||^2 \\
&= \text{E}||(\boldsymbol{\beta}_1 - \widehat{\boldsymbol{\beta}}_1^{\text{FM}})^\top \boldsymbol{X}_1^{*\top}\boldsymbol{X}_1(\boldsymbol{\beta}_1 - \widehat{\boldsymbol{\beta}}_1^{\text{FM}})|| + n^*\sigma *^2 \\
&= \text{tr}\left(\boldsymbol{X}_1^{*\top}\boldsymbol{X}_1^* \cdot \text{E}(\boldsymbol{\beta}_1 - \widehat{\boldsymbol{\beta}}_1^{\text{FM}})^\top(\boldsymbol{\beta}_1 - \widehat{\boldsymbol{\beta}}_1^{\text{FM}})\right) + n^*\sigma *^2 \\
&= \text{tr}\left(\boldsymbol{X}_1^{*\top}\boldsymbol{X}_1^* \cdot \boldsymbol{\Sigma}_{11}\right) + n^*\sigma*^2,
\end{aligned}$$

where $\boldsymbol{\Sigma}_{11}$ is the variance-covariance matrix of $\widehat{\boldsymbol{\beta}}_1^{\text{FM}}$ from the training process, n^* is the size of the testing set, and $\sigma*^2$ is the error variance from testing set. For practical reasons and to illustrate the properties of the theoretical results, we conducted a simulation study, reported in the next section, to compare the performance of the proposed estimators for moderate and large sample sizes.

3.5 Simulation Experiments

In this section, we conduct Monte Carlo simulation experiments to examine the quadratic risk (namely MSE) performance of the proposed estimators. Our simulation is based on a multiple linear regression model with different numbers of predictors, sample sizes, and degrees of sparsity. The response variable is centered and the predictors are standardized so that the intercept term can be left out.

The performance of an estimator is evaluated by using the relative mean squared error (RMSE) criterion. The RMSE of an estimator $\widehat{\boldsymbol{\beta}}_1^*$ with respect to $\widehat{\boldsymbol{\beta}}_1^{\mathrm{FM}}$ is defined as follows

$$\mathrm{RMSE}\left(\widehat{\boldsymbol{\beta}}_1^*\right) = \frac{\mathrm{MSE}\left(\widehat{\boldsymbol{\beta}}_1^{\mathrm{FM}}\right)}{\mathrm{MSE}\left(\widehat{\boldsymbol{\beta}}_1^*\right)}, \tag{3.18}$$

where $\widehat{\boldsymbol{\beta}}_1^*$ is one of the listed estimators. Keeping in mind that the amount by which a RMSE is larger than one indicates the degree of superiority of the estimator $\widehat{\boldsymbol{\beta}}_1^{\mathrm{FM}}$. In our simulation experiment, each realization was repeated 1000 times, as a further increase in the number of realizations did not significantlychange the result, and we report the average RMSE based on 1000 replications.

We divide our simulation into two subsections: the first one deals with situations when there are no weak signals in the model, and the second one includes weak signals as well.

3.5.1 Strong Signals and Noises

In this section, we consider the case when a model is sparse, that is, it has a few strong signals and the rest are zero signals. The following are some details of the simulation study:

- The regression coefficients are set $\boldsymbol{\beta} = \left(\boldsymbol{\beta}_1^\top, \boldsymbol{\beta}_2^\top\right)^\top$, where $\boldsymbol{\beta}_1$ is defined as the vector of 1 with dimension p_1 and $\boldsymbol{\beta}_2$ is defined as the vector of zeros with dimension p_2.

- In order to investigate the behavior of the estimators when $\boldsymbol{\beta}_2 = \mathbf{0}$ is violated, we define $\Delta = \|\boldsymbol{\beta} - \boldsymbol{\beta}_0\|$, where $\boldsymbol{\beta}_0 = \left(\boldsymbol{\beta}_1^\top, \mathbf{0}_{p_2}^\top\right)^\top$ and $\|\cdot\|$ is the Euclidean norm. We considered Δ values between 0 and 4. If $\Delta = 0$, then it means that we will have $\boldsymbol{\beta} = (\underbrace{1,\ldots,1}_{p_1},\underbrace{0,\ldots,0}_{p_2})^\top$. If $\Delta > 0$, then it indicates that the selected submodel may not be the right one. We are interested in quantifying the performance of the suggested shrinkage estimators in the real setting, that is, when the selected submodel may not be a correct one.

Based on Figures 3.1–3.3, we summarize the findings as follows:

- The submodel estimator outshines all the other estimators when the restriction is at or near $\Delta = 0$. By contrast, when Δ is larger than zero, the estimated RMSE of $\widehat{\boldsymbol{\beta}}_1^{\mathrm{SM}}$ increases and becomes unbounded, whereas the estimated RMSEs of all other estimators remain bounded and approach one. It can be safely concluded that the departure from the restriction is fatal to $\widehat{\boldsymbol{\beta}}_1^{\mathrm{SM}}$. This is consistent with our asymptotic theory.

- With increasing Δ, the RMSE of the shrinkage estimators with respect to the MLE decreases and converges to one, regardless of p_1, p_2, or n. In other words, shrinkage estimators outperform the full model estimator, regardless of the correctness of the selected submodel at hand.

- Further, the shrinkage estimators work better in cases with large p_2. Thus, the shrinkage estimators are preferable in high-dimensional cases.

- The $\widehat{\boldsymbol{\beta}}_1^{\mathrm{PS}}$ performs better than $\widehat{\boldsymbol{\beta}}_1^{\mathrm{S}}$ in the entire parameter space induced by Δ.

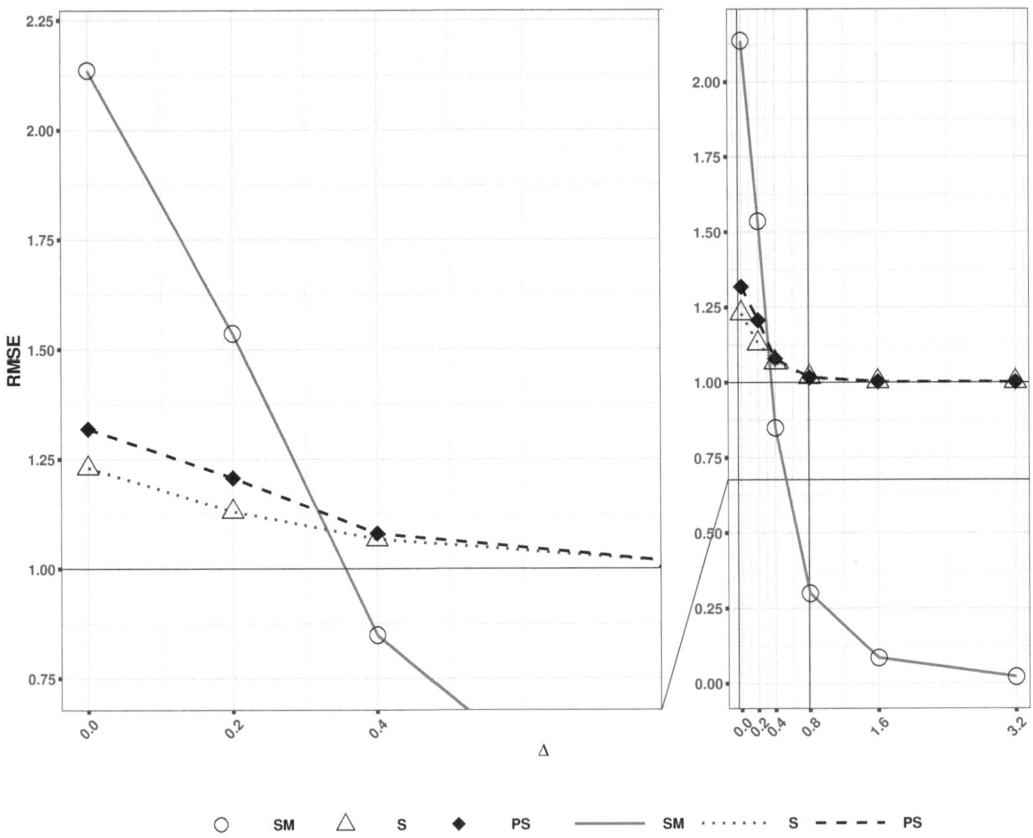

FIGURE 3.1: RMSE of the Estimators for $n = 30$, $p_1 = 3$, and $p_2 = 3$.

3.5.2 Signals and Noises

Now, we consider a more realistic situation when models contain all three signals, that is, strong, weak, and zero signals. Such predictors with a small amount of influence on the response variable are often ignored incorrectly in variable selection methods. If we borrow information from those predictors using the proposed shrinkage methods, the prediction performance based on the selected submodel can be improved substantially. However, weak signalsmay be embedded either in strong signals or zero signals. Thus, both zero and weak signals coexist in our simulation settings to provide a fair comparison.

We simulate the response from the following model:

$$y_i = x_{1i}\beta_1 + x_{2i}\beta_2 + ... + x_{pi}\beta_p + \varepsilon_i \ , \ i = 1, 2, ..., n, \tag{3.19}$$

where x_i and ε_i are i.i.d. $N(0, 1)$. We consider the regression coefficients are set$\boldsymbol{\beta} = \left(\boldsymbol{\beta}_1^\top, \boldsymbol{\beta}_2^\top, \boldsymbol{\beta}_3^\top\right)^\top$,with dimensions p_1, p_2 and p_3, respectively.

Further, $\boldsymbol{\beta}_1$ represent strong signals, that is, $\boldsymbol{\beta}_1$ is a vector of 1 values, $\boldsymbol{\beta}_2$ is a vector of 0.1 values, and $\boldsymbol{\beta}_3$ means no signals, that is, $\boldsymbol{\beta}_3 = \mathbf{0}$. In this simulation setting, we simulated 1000 data sets consisting of $n = 30, 50, 100$, with $p_1 = 3, 5$, $p_2 = 0, 3, 6, 9$ and $p_3 = 3, 6, 9, 12$. We also consider two cases: in case 1, the weak signals are not combined with strong signals in the calculation of the MSE of the submodel estimator, and in case 2, the weak signals are combined with strong signals.

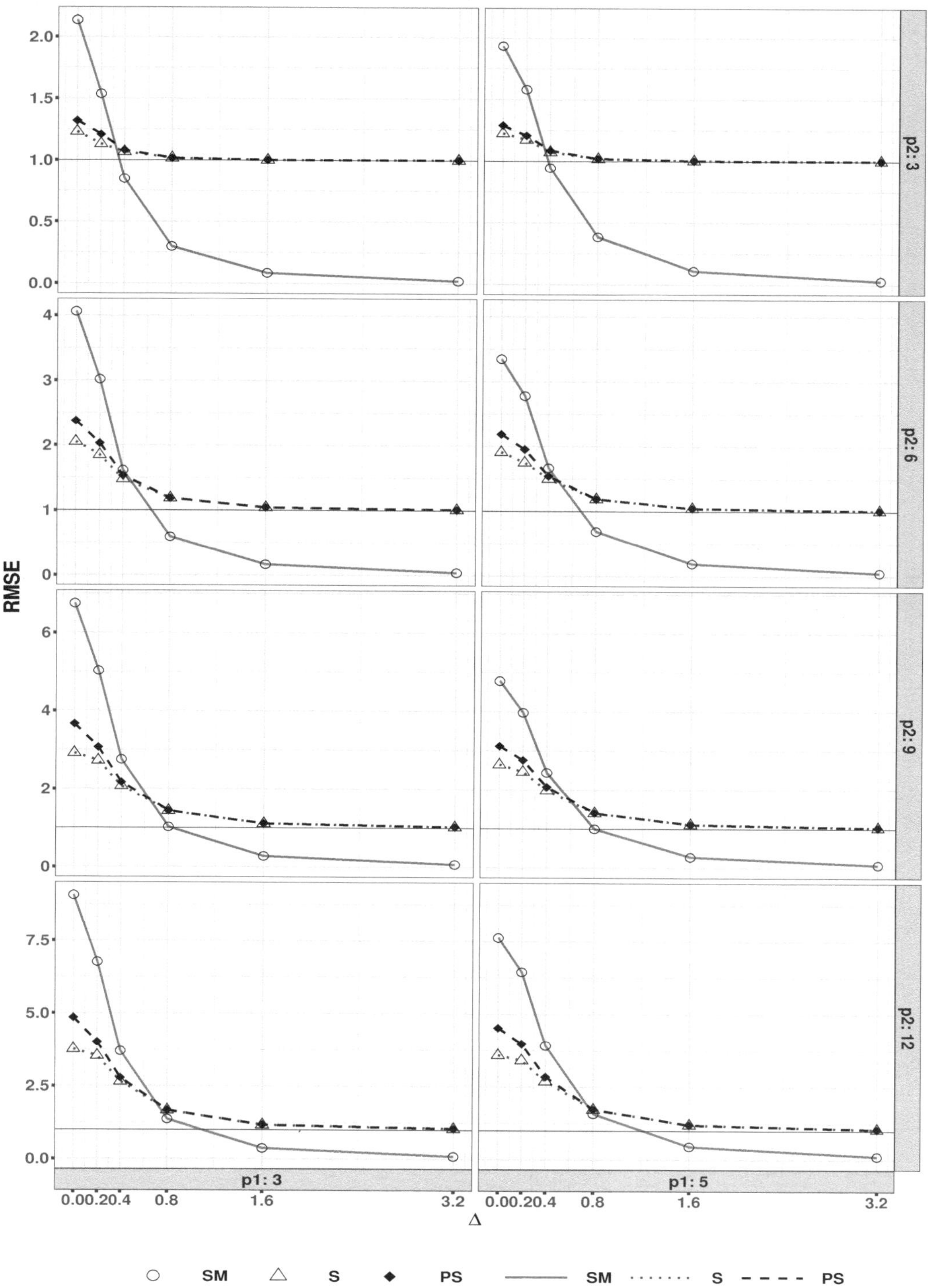

FIGURE 3.2: RMSE of the Estimators for $n = 30$ and Different Combinations of p_1 and p_2.

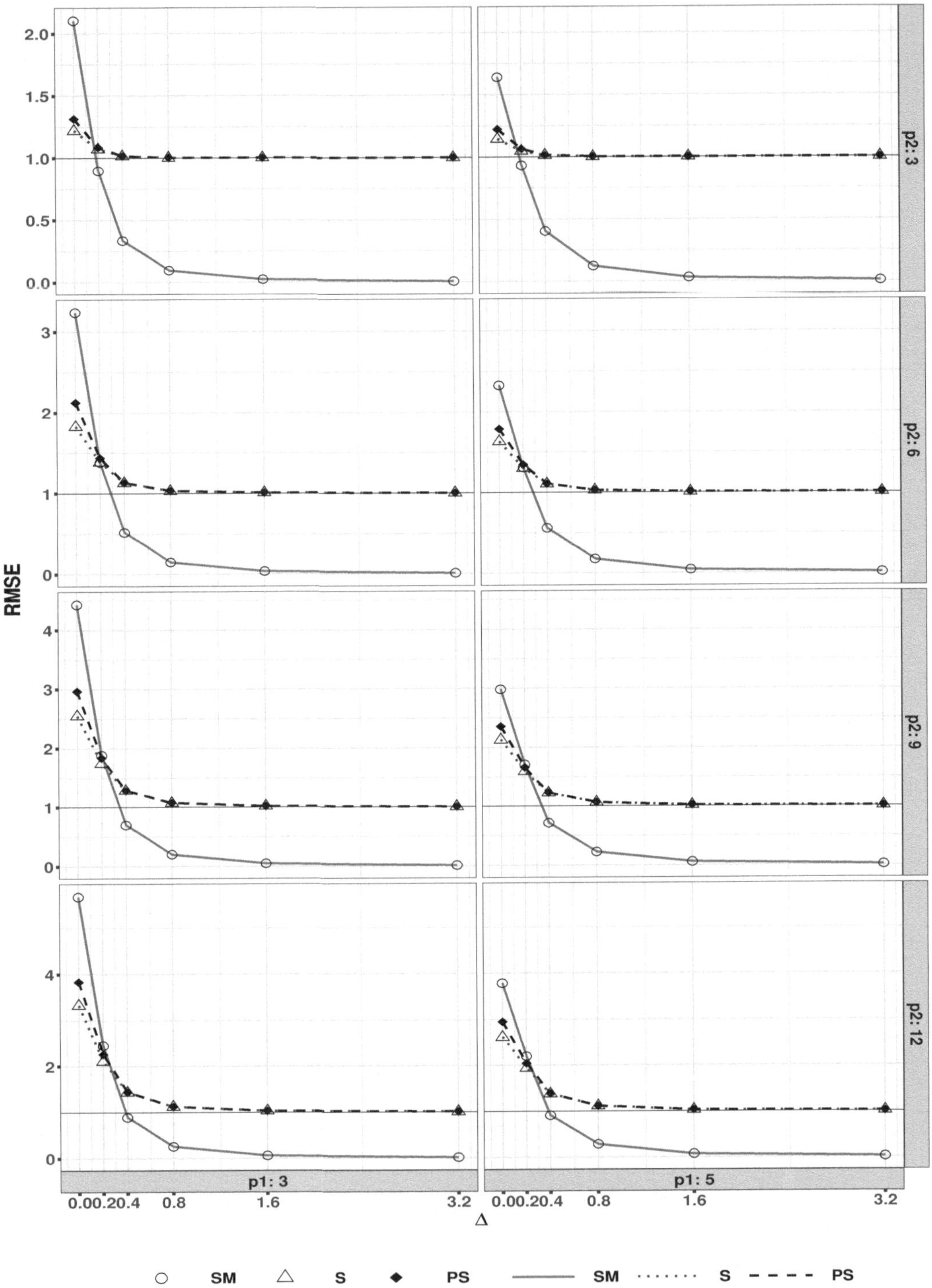

FIGURE 3.3: RMSE of the Estimators for $n = 100$ and Different Combinations of p_1 and p_2.

First, we present the result when weak signals coexist with the strong signals in Figures 3.8–3.11. As evident from these figures, the performance of the estimators is the same as in the case when weak signals were not present. This makes sense since including weak signals with strong signals is the same since weak signals are part of the true submodel. Thus, the submodel estimators continue to perform better than shrinkage a estimators for a range of Δ values. More importantly, shrinkage estimators dominate the full model estimators for all the values of Δ. When weak signals are combined with zero signals, the picture becomes completely different.In this case, as expected, the performance of the submodel becomes worse, and shrinkage estimators perform better than the submodel estimators. Figures 3.4–3.7 clearly display such characteristics.

A simple explanation is that shrinkage estimators are shrinking toward full model estimators, which seem to have a lower MSE than that of submodel estimators in this scenario. This shows the beauty and power of shrinkage estimators. In a sense, the shrinkage estimators are robust in the presence of weak coefficients.

In an effort to easily quantify the amount of improvement of SM and PS estimators over the full model estimator, we provide the following tables: We discarded shrinkage estimators in this study since they suffer the over-shrinking problem and are dominated by the PS. The Table 3.1 showcases the RMSE when there are no weak signals in the model and reveals that at $\Delta = 0$ with $n = 30, p_1 = 3, p_3 = 12$ the RMSE of the SM and PS are 9.068 and 4.849, respectively. However, as Δ increases, the RMSE of SM decreases and converges to zero. On the other hand, the RMSE of the PS is always more or equal to 1. This cleanly demonstrates the superiority of the PS over the full model estimator.

Tables 3.2–3.4 include various values of Δ in the simulation study and are combined with the strong signals in computing the MSE of the submodels.

Simile to graphical analysis the RMSE of the SM increases as the number of weak signals increases.Consequently, the RMSE of the PS also increases.

Tables 3.5–3.7 report the RMSE of the estimators when the weak signals are combined with the zero signals for some configurations of (n, p_1, p_2, p_3). In this scenario the submodel estimators perform badly relative to PS estimators even at $\Delta = 0$ in most cases.

3.5.3 Comparing Shrinkage Estimators with Penalty Estimators

In this section, we compare shrinkage estimations with the three penalized likelihood methods, namely, ENET, LASSO,and ALASSO using RMSE.

Further, for the data, we calculate the prediction errors and relative prediction errors of the respective estimator as follows:

The PE of an estimator $\widehat{\boldsymbol{\beta}}_1^*$ is defined as

$$\text{PE}\left(\widehat{\boldsymbol{\beta}}_1^*\right) = \left\|\boldsymbol{y}_{\text{test}} - (\boldsymbol{X}_1)_{\text{test}}\widehat{\boldsymbol{\beta}}_1^*\right\|^2 \tag{3.20}$$

where $(\boldsymbol{X}_1)_{\text{test}}$ is the design matrix of main effects. Finally, relative prediction error is defined as

$$\text{RPE}\left(\widehat{\boldsymbol{\beta}}_1^*\right) = \frac{\text{PE}\left(\widehat{\boldsymbol{\beta}}_1^{\text{FM}}\right)}{\text{PE}\left(\widehat{\boldsymbol{\beta}}_1^*\right)}. \tag{3.21}$$

If RPE is greater than 1, it means that the suggested estimator is better than $\widehat{\boldsymbol{\beta}}_1^{\text{FM}}$.

We randomly split the data into two equal groups of observations. The first part is the training set, and the other part is for the test set. The listed estimators are obtained from the training set only. In variable selection strategies, the Bayesian information criterion (BIC) is used to choose all tuning parameters.

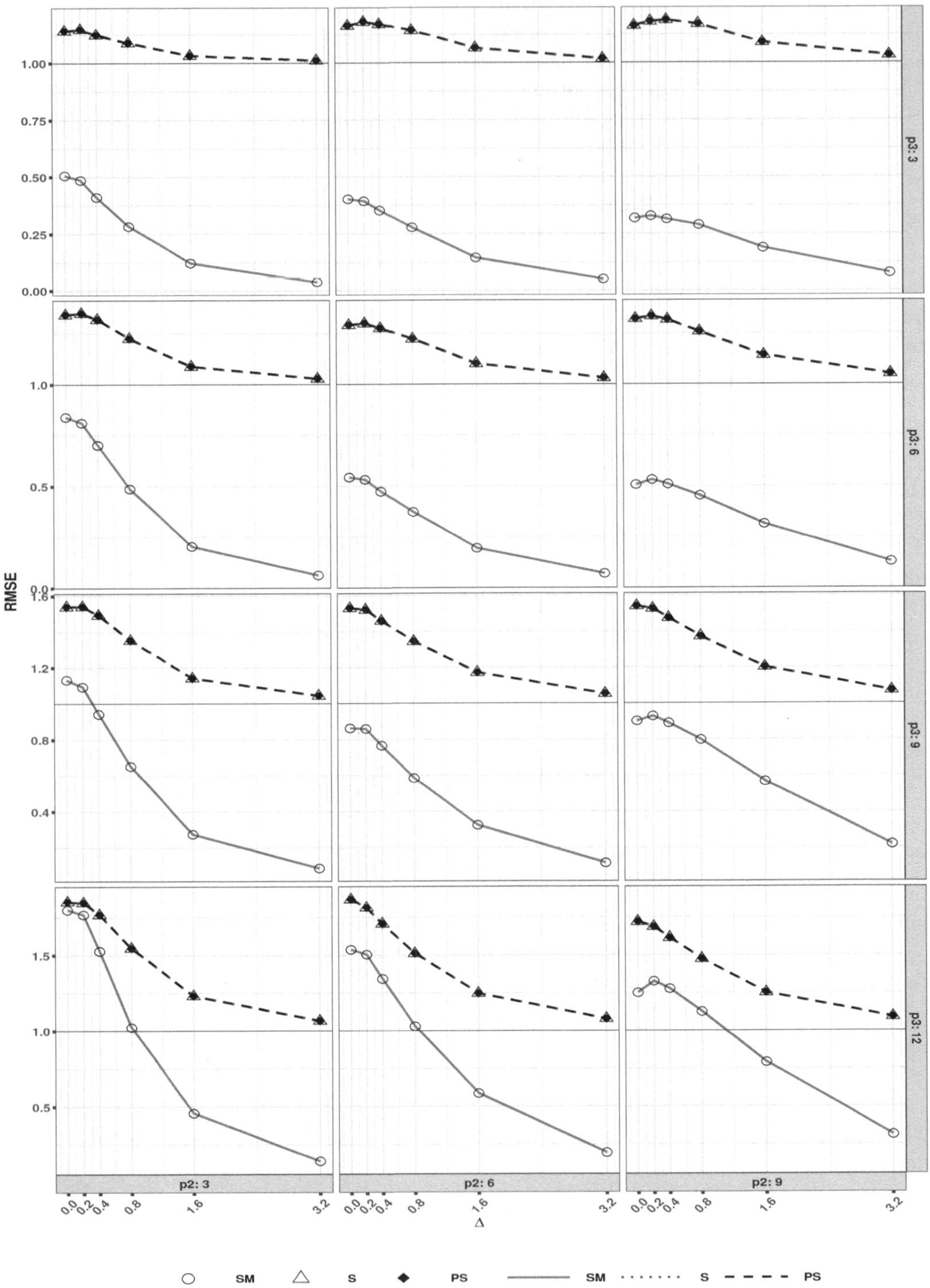

FIGURE 3.4: RMSE of the Estimators for Case 1, and $n = 30$ and $p_1 = 3$.

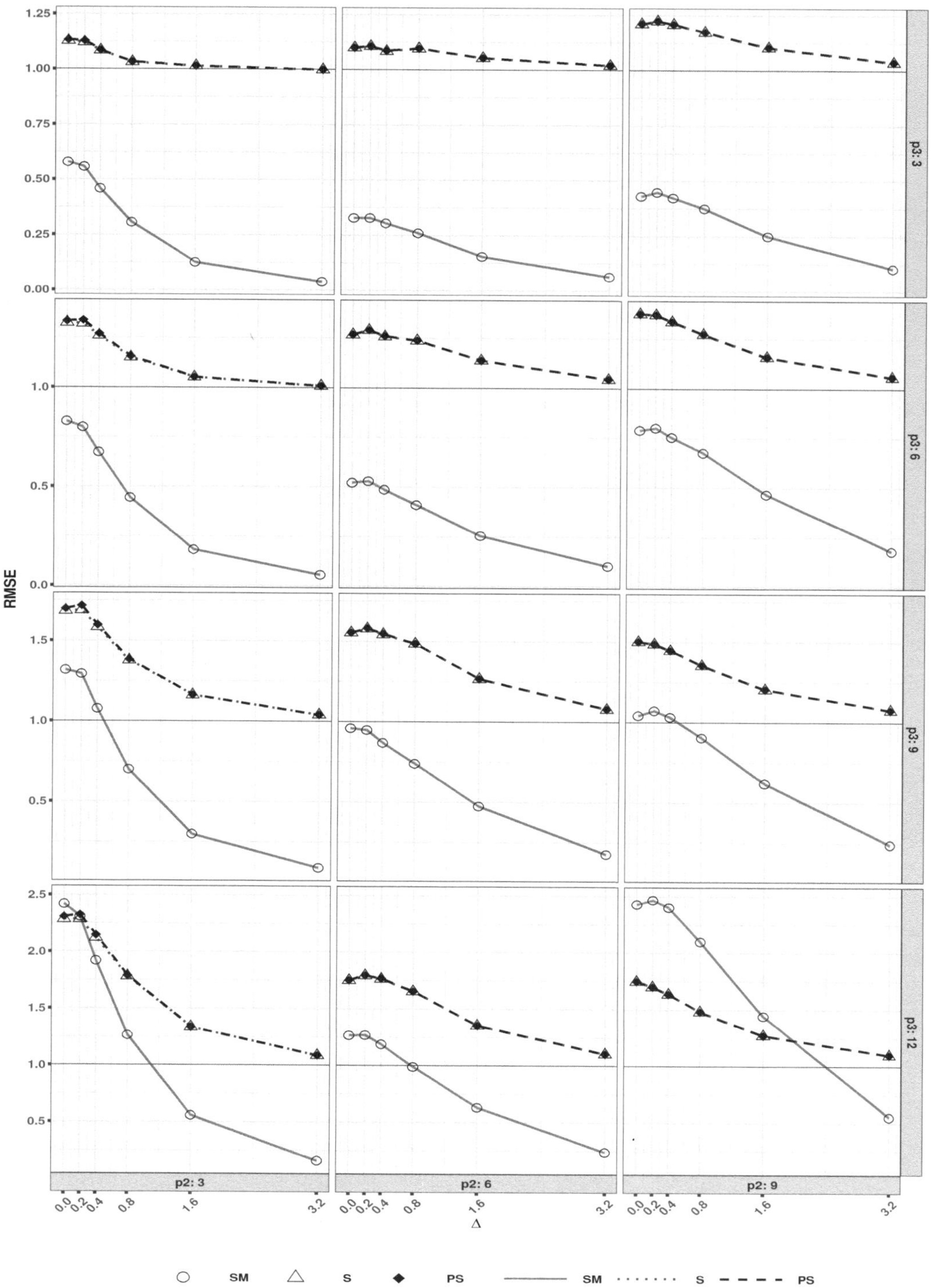

FIGURE 3.5: RMSE of the Estimators for Case 1, and $n = 30$ and $p_1 = 5$.

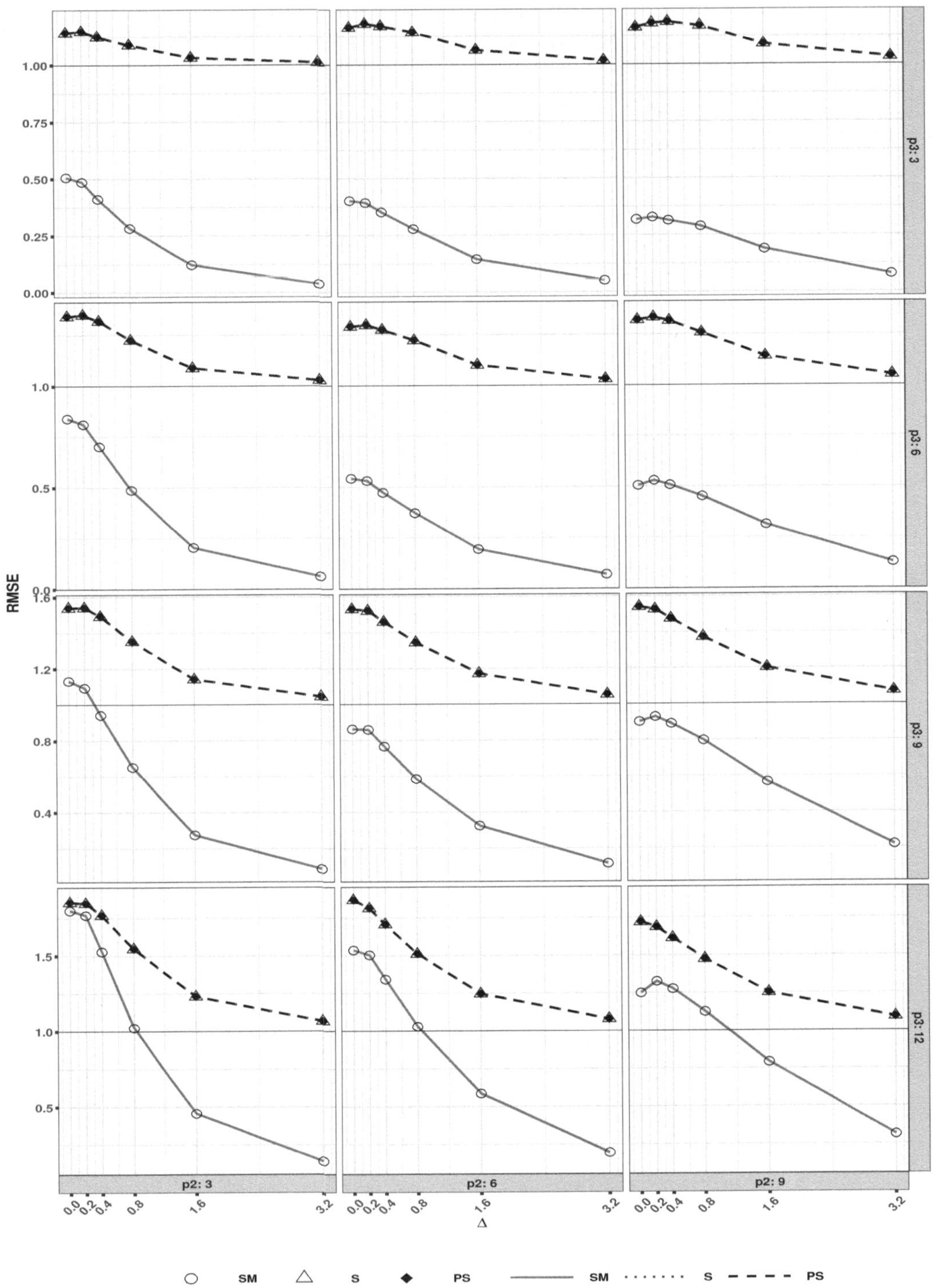

FIGURE 3.6: RMSE of the Estimators for Case 1, and $n = 100$ and $p_1 = 3$.

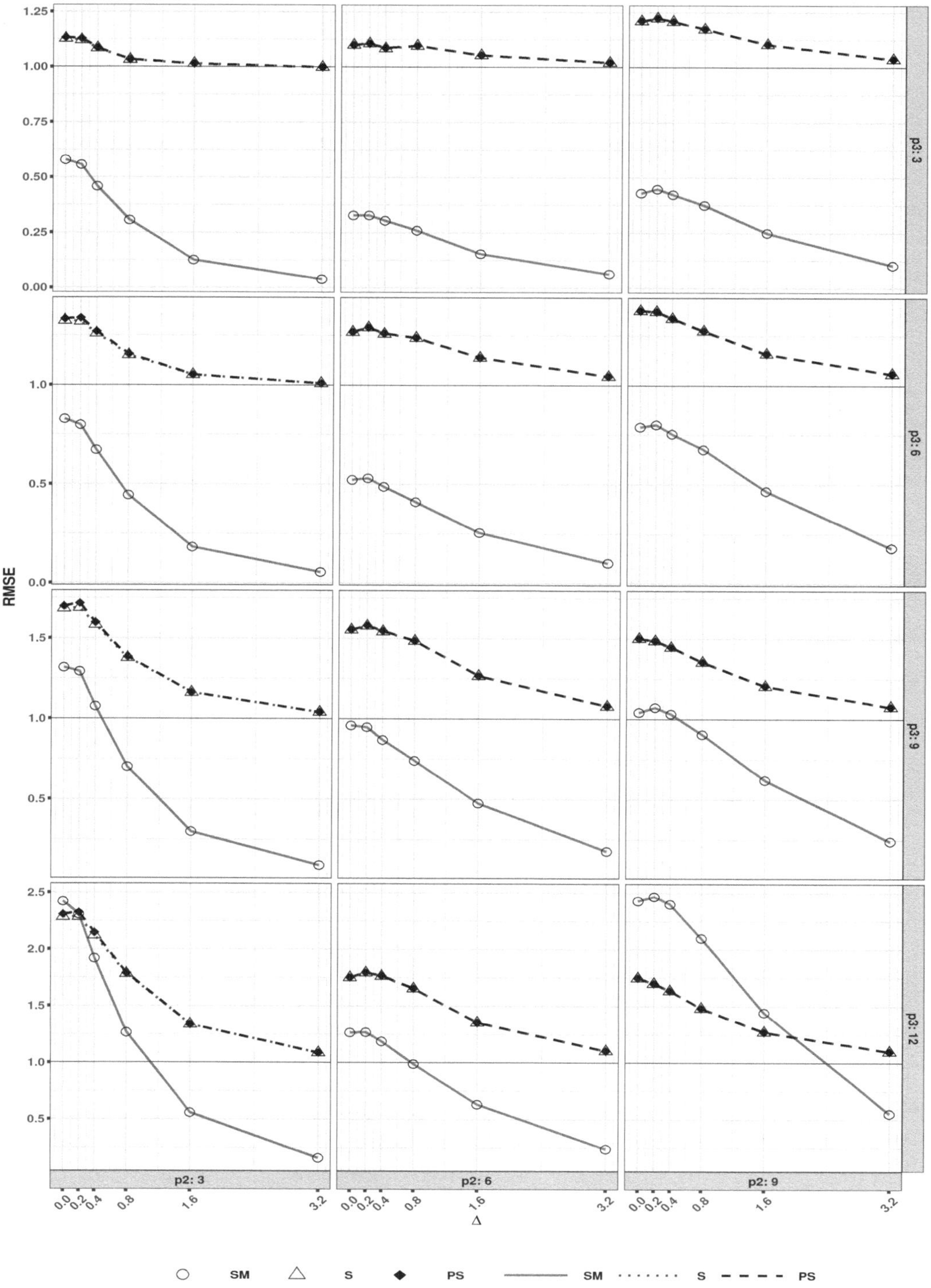

FIGURE 3.7: RMSE of the Estimators for Case 1, and $n = 100$ and $p_1 = 5$.

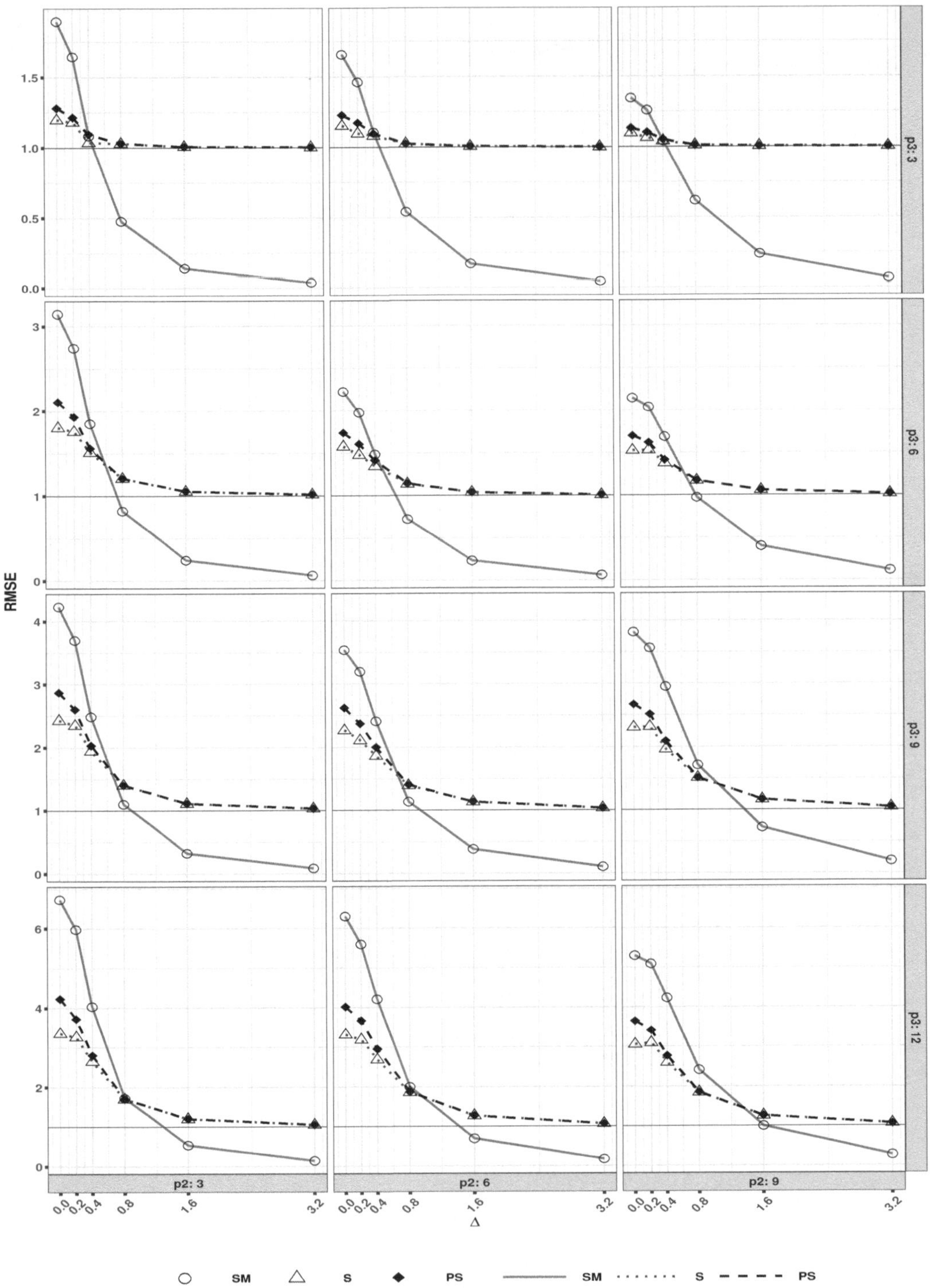

FIGURE 3.8: RMSE of the Estimators for Case 2, and $n = 30$ and $p_1 = 3$.

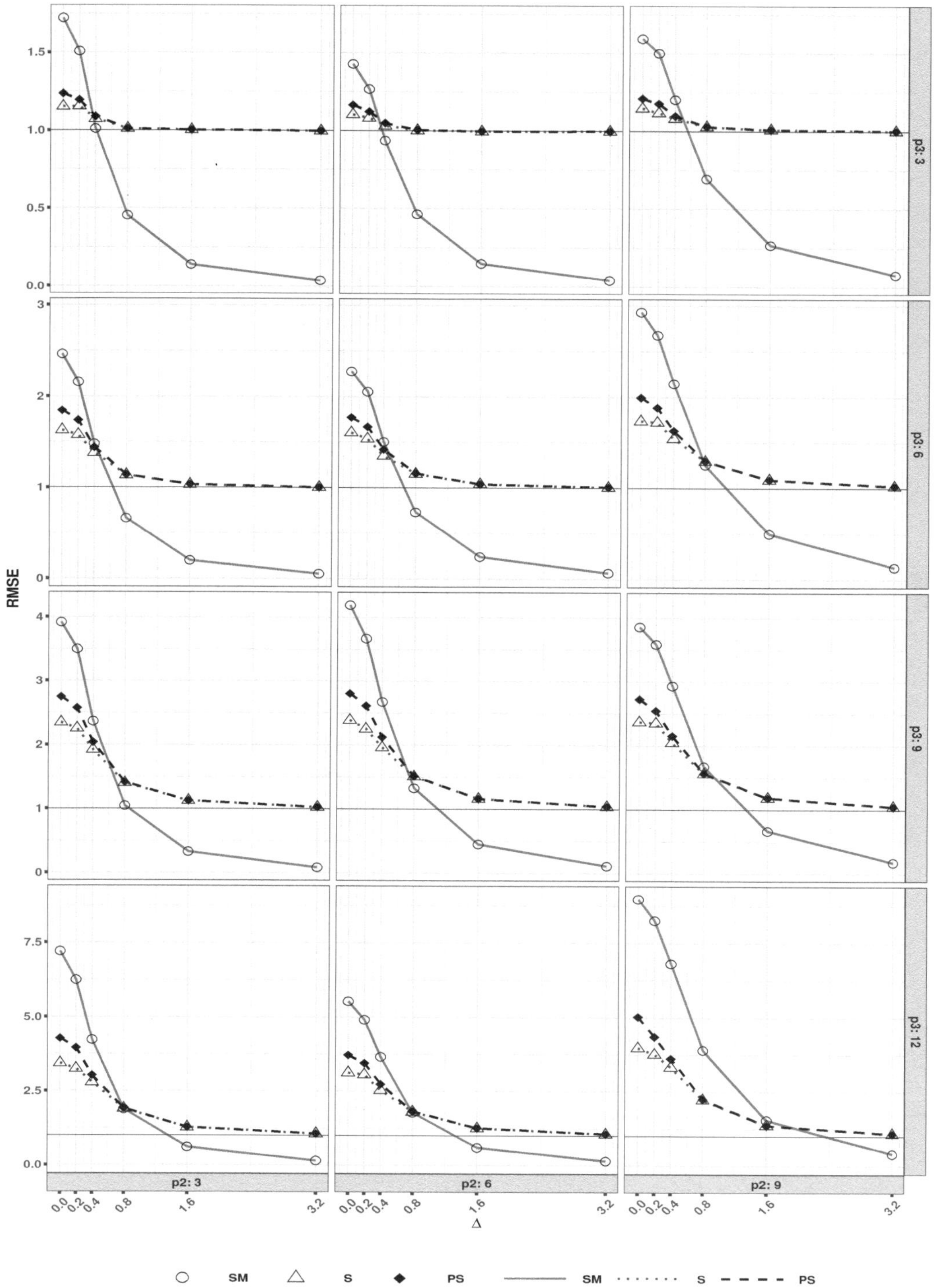

FIGURE 3.9: RMSE of the Estimators for Case 2, and $n = 30$ and $p_1 = 5$.

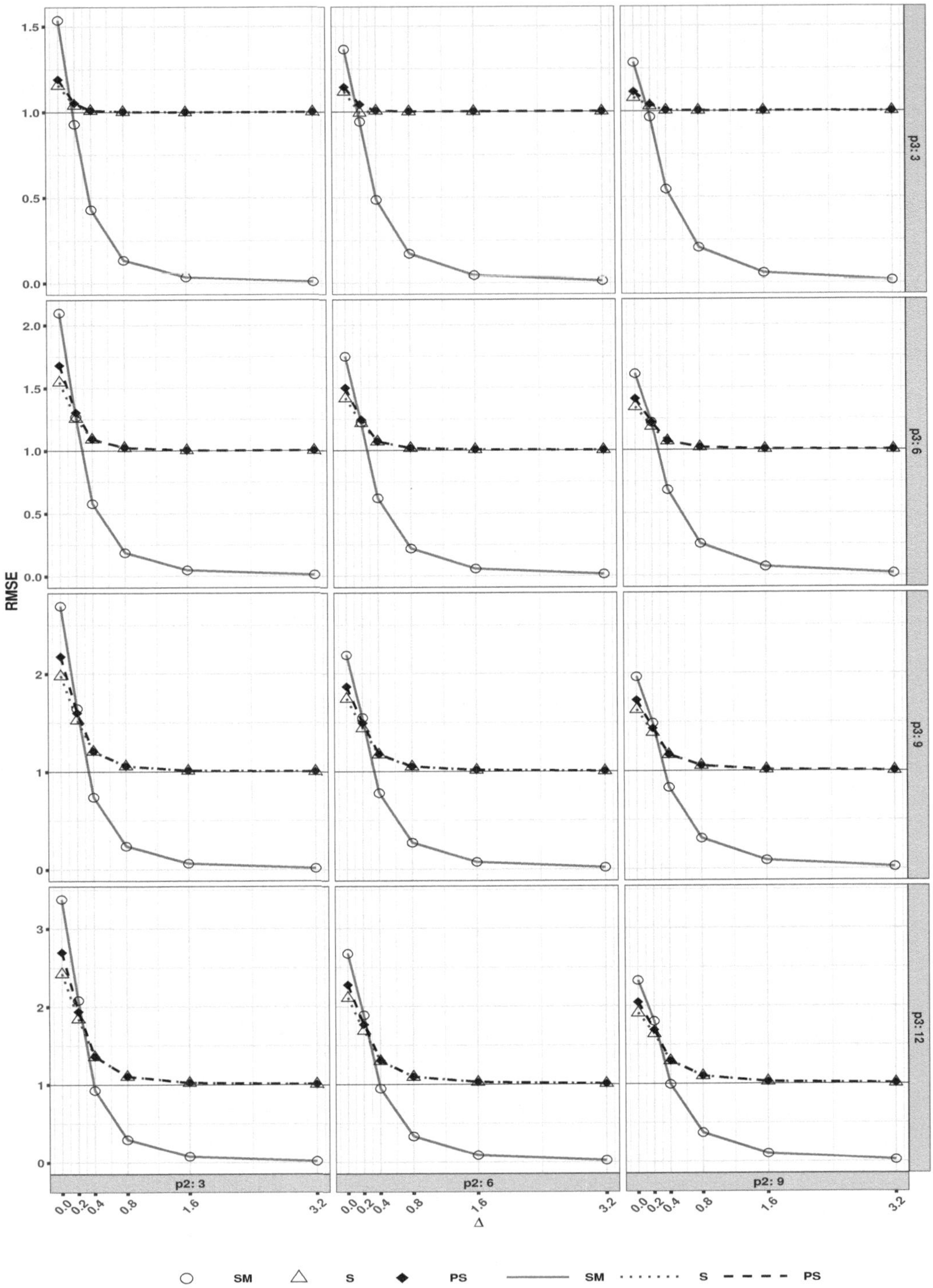

FIGURE 3.10: RMSE of the Estimators for Case 2, and $n = 100$ and $p_1 = 3$.

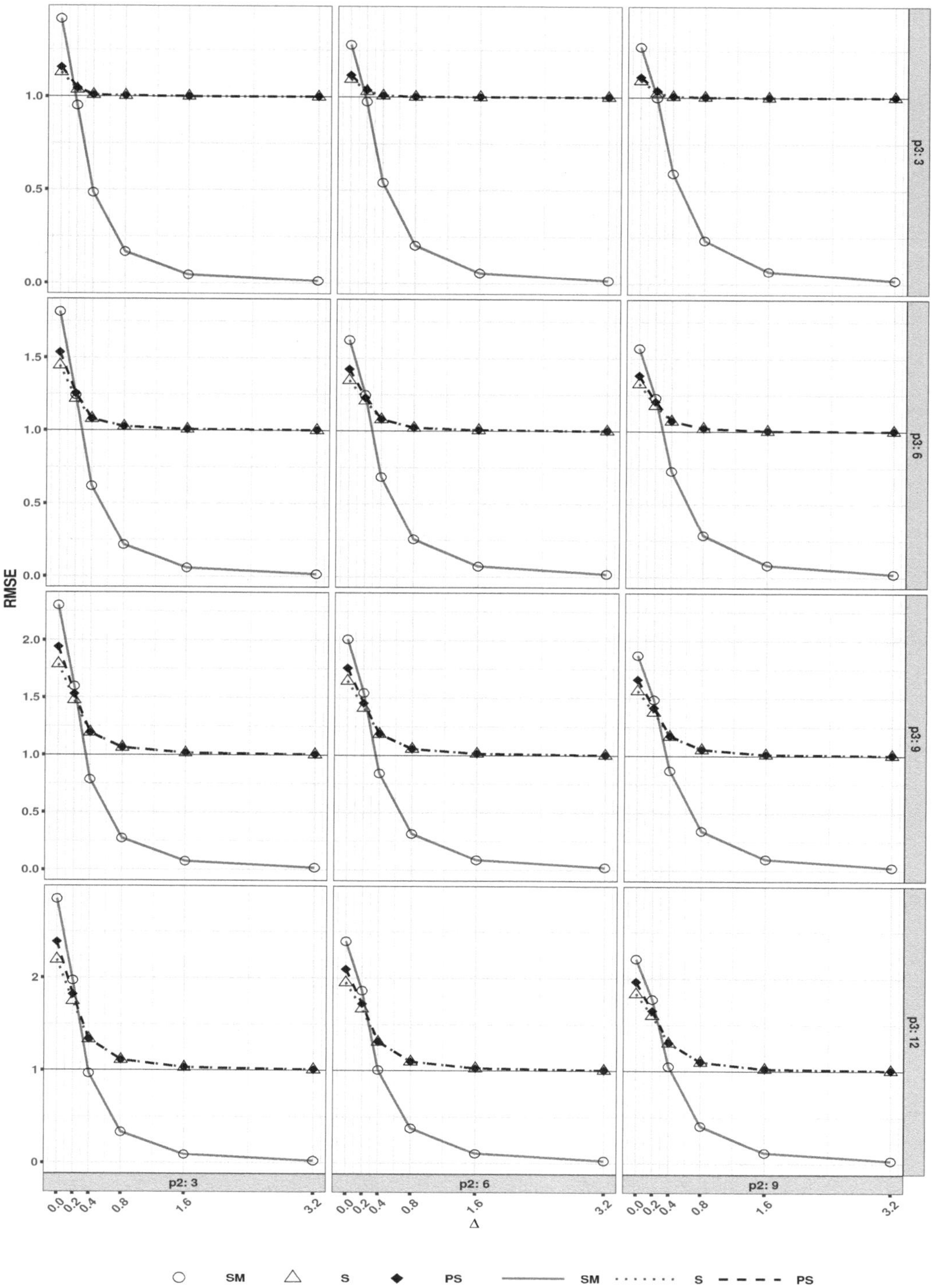

FIGURE 3.11: RMSE of the Estimators for Case 2, $n = 100$ and $p_1 = 5$.

TABLE 3.1: The RMSE of the Estimators for $p_2 = 0$.

		$p_1 = 3$				$p_1 = 5$			
		$n = 30$		$n = 100$		$n = 30$		$n = 100$	
p_3	Δ	SM	PS	SM	PS	SM	PS	SM	PS
3	0.0	2.136	1.319	2.104	1.315	1.934	1.291	1.640	1.222
	0.2	1.536	1.207	0.895	1.085	1.581	1.208	0.932	1.064
	0.4	0.849	1.080	0.334	1.016	0.946	1.088	0.403	1.014
	0.8	0.300	1.017	0.095	1.003	0.386	1.021	0.124	1.002
	1.6	0.085	1.002	0.024	1.001	0.111	1.004	0.032	1.001
	3.2	0.021	1.001	0.006	1.000	0.029	1.001	0.008	1.000
6	0.0	4.062	2.377	3.238	2.125	3.347	2.184	2.327	1.784
	0.2	3.019	2.029	1.382	1.438	2.780	1.952	1.307	1.342
	0.4	1.616	1.533	0.518	1.132	1.667	1.544	0.562	1.108
	0.8	0.593	1.186	0.148	1.032	0.687	1.189	0.176	1.028
	1.6	0.165	1.043	0.037	1.009	0.194	1.048	0.045	1.008
	3.2	0.042	1.012	0.010	1.001	0.051	1.012	0.012	1.002
9	0.0	6.751	3.665	4.430	2.962	4.788	3.123	2.980	2.350
	0.2	5.028	3.065	1.884	1.837	3.982	2.763	1.709	1.658
	0.4	2.755	2.170	0.698	1.284	2.445	2.077	0.717	1.229
	0.8	1.023	1.432	0.204	1.075	0.999	1.404	0.226	1.065
	1.6	0.276	1.107	0.051	1.020	0.279	1.107	0.058	1.017
	3.2	0.071	1.028	0.013	1.003	0.074	1.028	0.015	1.005
12	0.0	9.068	4.849	5.683	3.832	7.622	4.521	3.782	2.945
	0.2	6.766	4.006	2.447	2.262	6.449	3.974	2.193	2.038
	0.4	3.700	2.777	0.891	1.439	3.920	2.848	0.909	1.385
	0.8	1.367	1.668	0.261	1.124	1.577	1.729	0.287	1.114
	1.6	0.370	1.168	0.066	1.032	0.458	1.199	0.074	1.030
	3.2	0.096	1.044	0.017	1.007	0.120	1.051	0.019	1.008

In Tables 3.8–3.11, we give RMSE and RPE of shrinkage and three penalty-type estimators with respect to the FM for selected configurations of n, p_1, p_2, p_3. We only do the comparison when $\Delta = 0$ because the penalty estimator we consider here does not take advantage of the fact that regression parameter is partitioned into important parameters and nuisance parameters, and thus are at a disadvantage when $\Delta > 0$. We see that, when p_2, is relatively small, the penalty estimators perform better than our shrinkage methods. On the other hand, the shrinkage strategies perform better when p_2 is large, which is consistent with the asymptotic theory of the shrinkage estimators. Thus, we recommend using the positive-part shrinkage estimator when p_2 is large, which is the case in practice.

TABLE 3.2: The RMSE of the Estimators for Case 2 and $p_2 = 3$.

		$p_1 = 3$				$p_1 = 5$			
		$n = 30$		$n = 100$		$n = 30$		$n = 100$	
p_3	Δ	SM	PS	SM	PS	SM	PS	SM	PS
3	0.0	1.898	1.281	1.537	1.191	1.722	1.236	1.418	1.156
	0.2	1.645	1.215	0.931	1.053	1.507	1.195	0.952	1.045
	0.4	1.084	1.096	0.430	1.009	1.011	1.088	0.485	1.008
	0.8	0.478	1.030	0.135	1.002	0.454	1.011	0.167	1.004
	1.6	0.144	1.007	0.036	1.000	0.139	1.003	0.044	1.001
	3.2	0.038	1.002	0.009	1.000	0.036	0.999	0.012	1.000
6	0.0	3.142	2.107	2.097	1.682	2.462	1.842	1.814	1.536
	0.2	2.743	1.931	1.269	1.304	2.157	1.733	1.244	1.254
	0.4	1.852	1.561	0.579	1.092	1.477	1.437	0.619	1.078
	0.8	0.823	1.206	0.187	1.024	0.661	1.138	0.216	1.026
	1.6	0.241	1.053	0.048	1.004	0.200	1.032	0.057	1.007
	3.2	0.064	1.014	0.013	1.002	0.053	1.002	0.015	1.002
9	0.0	4.229	2.870	2.694	2.177	3.917	2.747	2.302	1.941
	0.2	3.698	2.601	1.647	1.602	3.500	2.574	1.596	1.529
	0.4	2.488	2.031	0.739	1.210	2.365	2.037	0.785	1.193
	0.8	1.100	1.399	0.239	1.059	1.044	1.410	0.273	1.063
	1.6	0.323	1.108	0.062	1.013	0.328	1.123	0.073	1.017
	3.2	0.087	1.029	0.016	1.005	0.086	1.024	0.019	1.005
12	0.0	6.733	4.232	3.378	2.697	7.207	4.263	2.854	2.387
	0.2	5.975	3.725	2.084	1.939	6.248	3.947	1.971	1.821
	0.4	4.036	2.806	0.927	1.359	4.218	3.021	0.967	1.335
	0.8	1.725	1.703	0.297	1.106	1.884	1.911	0.334	1.110
	1.6	0.539	1.201	0.078	1.025	0.614	1.278	0.090	1.030
	3.2	0.141	1.051	0.020	1.008	0.150	1.064	0.023	1.008

3.6 Prostrate Cancer Data Example

Prostrate cancer accounts for 1 in 5 new diagnoses of cancer in men, and with an aging population, this number is also expected to rise. Studies suggest rapidly increasing prevalence rates after the age of 66. The American Cancer Society expects that almost 250,000 new cases of prostrate cancer are expected to be diagnosed in the US during 2021, and approximately 34,000 men are expected to die. As cancer researchers conduct large studies to advance our understanding of why cancer occurs, the long-term goal remains to identify men

TABLE 3.3: The RMSE of the Estimators for Case 2 and $p_2 = 6$.

		$p_1 = 3$				$p_1 = 5$			
		$n = 30$		$n = 100$		$n = 30$		$n = 100$	
p_3	Δ	SM	PS	SM	PS	SM	PS	SM	PS
3	0.0	1.657	1.228	1.363	1.144	1.428	1.166	1.279	1.117
	0.2	1.462	1.173	0.943	1.042	1.267	1.122	0.975	1.038
	0.4	1.105	1.097	0.486	1.006	0.938	1.049	0.540	1.008
	0.8	0.540	1.025	0.172	1.001	0.464	1.005	0.203	1.002
	1.6	0.172	1.005	0.046	1.001	0.147	0.997	0.056	1.001
	3.2	0.046	1.000	0.012	1.000	0.040	0.998	0.015	1.000
6	0.0	2.225	1.738	1.750	1.500	2.274	1.768	1.624	1.425
	0.2	1.976	1.606	1.224	1.245	2.054	1.665	1.249	1.229
	0.4	1.485	1.413	0.621	1.071	1.501	1.418	0.685	1.079
	0.8	0.722	1.140	0.220	1.018	0.732	1.157	0.257	1.022
	1.6	0.231	1.036	0.059	1.006	0.242	1.040	0.071	1.006
	3.2	0.063	1.006	0.015	1.002	0.065	1.007	0.019	1.001
9	0.0	3.540	2.622	2.190	1.867	4.187	2.804	2.012	1.758
	0.2	3.195	2.370	1.549	1.496	3.667	2.613	1.543	1.453
	0.4	2.408	1.999	0.778	1.178	2.676	2.124	0.843	1.187
	0.8	1.133	1.401	0.273	1.052	1.322	1.515	0.315	1.056
	1.6	0.384	1.131	0.074	1.016	0.451	1.159	0.088	1.016
	3.2	0.102	1.030	0.019	1.004	0.113	1.037	0.023	1.003
12	0.0	6.298	4.021	2.677	2.278	5.529	3.713	2.395	2.096
	0.2	5.593	3.670	1.890	1.769	4.912	3.441	1.865	1.725
	0.4	4.213	2.953	0.947	1.306	3.657	2.736	1.009	1.312
	0.8	1.998	1.871	0.333	1.097	1.772	1.817	0.379	1.098
	1.6	0.695	1.274	0.091	1.029	0.600	1.258	0.106	1.027
	3.2	0.174	1.066	0.023	1.006	0.151	1.064	0.028	1.006

who are at risk of cancer and the preventative measures that can be taken to reduce this risk. Epidemiological research continues to investigate long-term survivorship, treatment and prevention, and policies and guidelines.

Epidemiological research is based on statistical and machine learning methods, which help us make predictions that are both accurate and easy to understand. To gain a better understanding of how the methods can be applied to a context such as prostrate cancer research, we will analyze a simple prostate cancer data set. Stamey et al. (1989) conducted a study that showed that a prostrate specific antigen is useful as a preoperative marker as it strongly correlated with the volume of prostrate cancer. The data set is publicly available

TABLE 3.4: The RMSE of the Estimators for Case 2 and $p_2 = 9$.

		$p_1 = 3$				$p_1 = 5$			
		$n = 30$		$n = 100$		$n = 30$		$n = 100$	
p_3	Δ	SM	PS	SM	PS	SM	PS	SM	PS
3	0.0	1.346	1.133	1.284	1.114	1.592	1.210	1.270	1.109
	0.2	1.260	1.103	0.966	1.040	1.500	1.177	0.998	1.036
	0.4	1.040	1.051	0.544	1.005	1.203	1.098	0.591	1.006
	0.8	0.619	1.010	0.204	1.002	0.694	1.030	0.233	1.002
	1.6	0.239	1.004	0.057	1.000	0.268	1.011	0.066	1.000
	3.2	0.069	1.001	0.015	1.000	0.076	1.003	0.017	1.000
6	0.0	2.144	1.697	1.608	1.410	2.929	1.994	1.571	1.386
	0.2	2.037	1.618	1.221	1.220	2.677	1.881	1.232	1.206
	0.4	1.686	1.413	0.683	1.070	2.148	1.628	0.728	1.072
	0.8	0.971	1.169	0.254	1.021	1.254	1.297	0.286	1.022
	1.6	0.397	1.055	0.071	1.005	0.501	1.093	0.081	1.005
	3.2	0.113	1.014	0.018	1.002	0.132	1.023	0.021	1.002
9	0.0	3.819	2.670	1.964	1.724	3.859	2.726	1.872	1.663
	0.2	3.569	2.514	1.490	1.435	3.588	2.547	1.488	1.418
	0.4	2.952	2.090	0.830	1.168	2.935	2.156	0.871	1.172
	0.8	1.712	1.509	0.310	1.055	1.678	1.570	0.344	1.056
	1.6	0.718	1.157	0.087	1.014	0.666	1.183	0.098	1.013
	3.2	0.192	1.039	0.022	1.005	0.177	1.045	0.026	1.004
12	0.0	5.298	3.647	2.327	2.047	8.990	5.029	2.211	1.966
	0.2	5.088	3.414	1.804	1.691	8.265	4.351	1.777	1.654
	0.4	4.232	2.771	0.993	1.292	6.819	3.616	1.052	1.307
	0.8	2.415	1.851	0.372	1.099	3.901	2.251	0.409	1.099
	1.6	1.003	1.259	0.104	1.026	1.553	1.369	0.117	1.025
	3.2	0.276	1.065	0.027	1.008	0.405	1.090	0.031	1.008

as a built-in data frame in R. The data frame consists of 97 men who were due to receive a radical prostatectomy and 8 columns of different biomarkers such as age and gleason score to predict the prostrate specific antigen levels (log(psa)).

We will analyze the data using both proposed shrinkage strategies and penalized methods.We use AIC, BIC, and BSS techniques to obtain respective submodels to construct shrinkage estimators. Further, we will apply three machine learning methods, namely, neural network, random forest, and K-nearest neighbours. We will compare the models and their prediction errors.

TABLE 3.5: The RMSE of the Estimators for Case 1 and $p_2 = 3$.

		$p_1 = 3$				$p_1 = 5$			
		$n = 30$		$n = 100$		$n = 30$		$n = 100$	
p_3	Δ	SM	PS	SM	PS	SM	PS	SM	PS
3	0.0	0.507	1.142	0.126	1.027	0.578	1.132	0.149	1.023
	0.2	0.486	1.148	0.117	1.027	0.556	1.127	0.138	1.025
	0.4	0.411	1.124	0.104	1.027	0.459	1.085	0.123	1.018
	0.8	0.283	1.088	0.069	1.013	0.305	1.032	0.085	1.015
	1.6	0.122	1.032	0.029	1.002	0.126	1.014	0.035	1.006
	3.2	0.038	1.012	0.009	1.002	0.038	0.998	0.011	1.003
6	0.0	0.839	1.341	0.172	1.062	0.828	1.334	0.190	1.053
	0.2	0.811	1.348	0.160	1.060	0.798	1.337	0.181	1.056
	0.4	0.702	1.318	0.140	1.057	0.672	1.269	0.157	1.043
	0.8	0.488	1.225	0.095	1.032	0.444	1.156	0.109	1.033
	1.6	0.205	1.088	0.039	1.010	0.181	1.052	0.046	1.014
	3.2	0.063	1.028	0.012	1.005	0.055	1.009	0.014	1.006
9	0.0	1.132	1.539	0.220	1.101	1.317	1.698	0.242	1.094
	0.2	1.094	1.540	0.207	1.098	1.291	1.716	0.232	1.096
	0.4	0.942	1.493	0.179	1.089	1.077	1.597	0.199	1.075
	0.8	0.652	1.353	0.122	1.054	0.701	1.383	0.138	1.056
	1.6	0.275	1.141	0.050	1.019	0.298	1.161	0.059	1.024
	3.2	0.086	1.044	0.015	1.008	0.090	1.040	0.018	1.009
12	0.0	1.800	1.854	0.276	1.145	2.421	2.306	0.299	1.139
	0.2	1.768	1.848	0.261	1.141	2.307	2.322	0.286	1.140
	0.4	1.528	1.769	0.225	1.126	1.920	2.145	0.245	1.114
	0.8	1.023	1.547	0.151	1.081	1.264	1.789	0.169	1.082
	1.6	0.458	1.233	0.063	1.030	0.557	1.336	0.072	1.036
	3.2	0.139	1.068	0.019	1.011	0.157	1.087	0.022	1.012

3.6.1 Classical Strategy

Whenever one has a dataset with multiple numeric variables, it is a good idea to look at the correlations among these variables. One reason is that if you have a dependent variable, you can easily see which independent variables correlate with that dependent variable. A second reason is that if you will be constructing a multiple regression model, adding an independent variable that is strongly correlated with an independent variable already in the model is unlikely to improve the model much, and you may have good reason to choose one variable over another. Figure 3.12 demonstrates that although we have some correlation, the values do not exceed a correlation value of 0.6, a reasonable cut-off value for empirical data.

TABLE 3.6: The RMSE of the Estimators for Case 1 and $p_2 = 6$.

		$p_1 = 3$				$p_1 = 5$			
		$n = 30$		$n = 100$		$n = 30$		$n = 100$	
p_3	Δ	SM	PS	SM	PS	SM	PS	SM	PS
3	0.0	0.402	1.164	0.089	1.030	0.328	1.099	0.097	1.022
	0.2	0.394	1.181	0.084	1.031	0.328	1.106	0.094	1.026
	0.4	0.352	1.170	0.078	1.030	0.304	1.086	0.086	1.024
	0.8	0.279	1.143	0.062	1.022	0.259	1.096	0.070	1.019
	1.6	0.145	1.064	0.031	1.011	0.155	1.054	0.037	1.012
	3.2	0.052	1.018	0.011	1.005	0.063	1.020	0.013	1.004
6	0.0	0.542	1.291	0.114	1.050	0.522	1.268	0.123	1.044
	0.2	0.531	1.299	0.109	1.050	0.530	1.291	0.120	1.047
	0.4	0.473	1.274	0.100	1.047	0.487	1.260	0.109	1.042
	0.8	0.372	1.223	0.079	1.036	0.410	1.239	0.089	1.034
	1.6	0.195	1.101	0.040	1.020	0.254	1.139	0.047	1.020
	3.2	0.070	1.031	0.014	1.007	0.102	1.045	0.017	1.006
9	0.0	0.863	1.531	0.143	1.073	0.959	1.553	0.152	1.067
	0.2	0.859	1.523	0.138	1.072	0.947	1.580	0.148	1.071
	0.4	0.766	1.460	0.125	1.067	0.868	1.546	0.135	1.064
	0.8	0.585	1.346	0.099	1.052	0.739	1.484	0.109	1.051
	1.6	0.324	1.171	0.051	1.029	0.475	1.265	0.058	1.029
	3.2	0.114	1.054	0.018	1.010	0.178	1.078	0.020	1.009
12	0.0	1.536	1.869	0.174	1.096	1.265	1.752	0.181	1.092
	0.2	1.504	1.815	0.168	1.095	1.267	1.796	0.179	1.098
	0.4	1.343	1.706	0.152	1.088	1.188	1.768	0.161	1.086
	0.8	1.030	1.511	0.120	1.070	0.989	1.655	0.131	1.066
	1.6	0.586	1.247	0.062	1.039	0.632	1.353	0.070	1.039
	3.2	0.195	1.080	0.021	1.013	0.238	1.104	0.025	1.013

It is also worthwhile to look at the distribution of the numeric variables. If the distributions differ greatly, using Kendall or Spearman correlations may be more appropriate. Also, if independent variables differ in distribution from the dependent variable, the independent variables may need to be transformed. In this case, our dependent variable is normally distributed. Next, it is important to evaluate the regression diagnostics and see that the assumptions of multiple regression are held true. Figure 3.13 demonstrates that the assumptions are upheld.

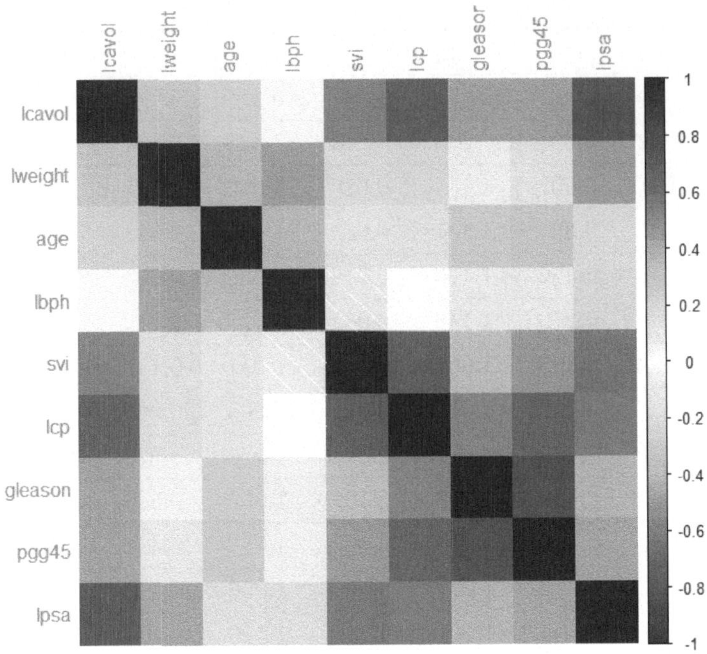

FIGURE 3.12: Correlation Plot.

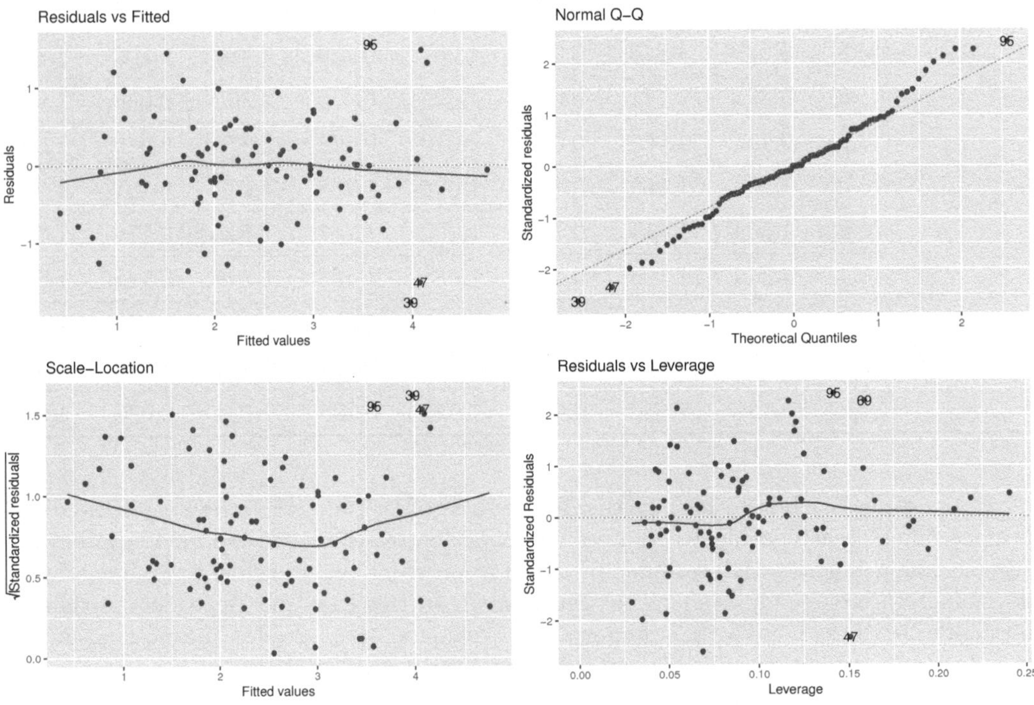

FIGURE 3.13: Regression Diagnostics.

TABLE 3.7: The RMSE of the Estimators for Case 1 and $p_2 = 9$.

		$p_1 = 3$				$p_1 = 5$			
		$n = 30$		$n = 100$		$n = 30$		$n = 100$	
p_3	Δ	SM	PS	SM	PS	SM	PS	SM	PS
3	0.0	0.318	1.163	0.075	1.031	0.429	1.210	0.087	1.030
	0.2	0.328	1.181	0.073	1.033	0.448	1.223	0.086	1.032
	0.4	0.314	1.187	0.069	1.034	0.423	1.208	0.080	1.029
	0.8	0.289	1.170	0.060	1.029	0.375	1.173	0.069	1.028
	1.6	0.188	1.088	0.035	1.015	0.250	1.103	0.040	1.014
	3.2	0.079	1.032	0.013	1.009	0.103	1.037	0.015	1.007
6	0.0	0.506	1.320	0.094	1.048	0.789	1.376	0.108	1.046
	0.2	0.531	1.333	0.093	1.049	0.801	1.371	0.106	1.050
	0.4	0.508	1.316	0.087	1.049	0.754	1.337	0.098	1.045
	0.8	0.453	1.257	0.074	1.042	0.676	1.274	0.084	1.041
	1.6	0.313	1.142	0.044	1.023	0.466	1.159	0.049	1.023
	3.2	0.128	1.050	0.017	1.012	0.180	1.058	0.019	1.011
9	0.0	0.902	1.540	0.115	1.064	1.041	1.501	0.129	1.065
	0.2	0.929	1.527	0.113	1.067	1.072	1.487	0.128	1.069
	0.4	0.891	1.475	0.106	1.065	1.032	1.446	0.118	1.062
	0.8	0.798	1.373	0.091	1.056	0.904	1.353	0.101	1.054
	1.6	0.566	1.200	0.054	1.033	0.621	1.204	0.060	1.031
	3.2	0.218	1.071	0.020	1.015	0.241	1.075	0.023	1.014
12	0.0	1.250	1.720	0.136	1.081	2.428	1.749	0.152	1.084
	0.2	1.327	1.686	0.137	1.085	2.467	1.701	0.152	1.089
	0.4	1.277	1.613	0.126	1.082	2.400	1.637	0.142	1.083
	0.8	1.125	1.474	0.109	1.071	2.101	1.483	0.120	1.071
	1.6	0.790	1.252	0.064	1.043	1.444	1.272	0.071	1.043
	3.2	0.313	1.091	0.025	1.019	0.552	1.100	0.028	1.018

3.6.2 Shrinkage and Penalty Strategies

In order to make shrinkage estimators, we first choose submodels using the AIC, BIC, and BSS methods for choosing variables, and then we combine those submodels with the full model estimator. We also apply four penalized likelihood methods, LASSO, ALASSO, SCAD, and ENET, to the data set.

For $i = 1, \ldots, 97$, the full and three submodels based on variable selection methods, and four models based on penalized likelihood methods are given as follows:

TABLE 3.8: The RMSE of the Estimators for $p_1 = 3$.

			Case 1		Case 2				
n	p_2	p_3	SM	PS	SM	PS	ENET	LASSO	ALASSO
30	0	4	4.066	1.867	4.066	1.867	1.611	1.673	3.273
		8	7.707	3.596	7.707	3.596	2.158	2.351	6.049
		16	27.533	8.356	27.533	8.356	7.317	7.840	21.717
	5	4	0.609	1.300	1.594	1.309	1.224	1.052	0.767
		8	1.012	1.619	2.646	2.069	1.734	1.562	1.189
		16	3.878	2.854	10.183	5.784	5.461	5.202	4.131
50	0	4	2.821	1.687	2.821	1.687	1.289	1.282	2.587
		8	4.817	2.984	4.817	2.984	1.736	1.767	4.410
		16	11.659	6.431	11.659	6.431	3.145	3.392	10.700
	15	4	0.191	1.121	1.409	1.217	0.973	0.833	0.690
		8	0.260	1.170	1.916	1.653	1.009	0.814	0.725
		16	0.548	1.305	4.044	3.225	1.392	1.203	1.179
100	0	4	2.612	1.632	2.612	1.632	1.187	1.100	2.292
		8	4.150	2.804	4.150	2.804	1.412	1.372	3.640
		12	6.027	4.024	6.027	4.024	1.692	1.724	5.288
		16	8.094	5.320	8.094	5.320	2.021	2.116	7.101
	25	10	0.091	1.068	1.532	1.444	1.032	0.917	0.915
		20	0.135	1.099	2.289	2.126	1.171	1.020	1.124
		30	0.210	1.134	3.535	3.216	1.358	1.166	1.310
		40	0.333	1.174	5.626	4.945	1.700	1.481	1.619
	50	10	0.131	1.079	1.518	1.423	1.084	0.984	0.720
		20	0.213	1.100	2.461	2.240	1.304	1.146	0.866
		30	0.369	1.123	4.278	3.743	1.766	1.586	1.251
		40	1.161	1.151	13.481	9.235	4.876	4.447	3.497

$$
\begin{aligned}
\text{Full Model: } \text{lpsa}_i &= \beta_0 + \beta_1 \text{lcavol}_i + \beta_2 \text{lweight}_i + \beta_3 \text{svi}_i \\
&+ \beta_4 \text{lbph}_i + \beta_5 \text{age}_i + \beta_6 \text{lcp}_i \\
&+ \beta_7 \text{gleason}_i + \beta_8 \text{pgg45}_i + \epsilon \\
\text{Sub Model (AIC): } \text{lpsa}_i &= \beta_0 + \beta_1 \text{lcavol}_i + \beta_2 \text{lweight}_i + \beta_3 \text{svi}_i \\
&+ \beta_4 \text{lbph}_i + \beta_5 \text{age}_i + \epsilon \\
\text{Sub Model (BIC): } \text{lpsa}_i &= \beta_0 + \beta_1 \text{lcavol}_i + \beta_2 \text{lweight}_i + \beta_3 \text{svi}_i + \epsilon \\
\text{Sub Model (BSS): } \text{lpsa}_i &= \beta_0 + \beta_1 \text{lcavol}_i + \beta_2 \text{lweight}_i + \epsilon \\
\text{(LASSO): } \text{lpsa}_i &= \beta_0 + \beta_1 \text{lcavol}_i + \beta_2 \text{lweight}_i + \beta_3 \text{svi}_i + \epsilon
\end{aligned}
$$

TABLE 3.9: The RPE of the Estimators for $p_1 = 3$.

n	p_2	p_3	Case 1 SM	Case 1 PS	Case 2 SM	Case 2 PS	ENET	LASSO	ALASSO
30	0	4	1.122	1.072	1.122	1.072	0.962	0.950	1.072
		8	1.307	1.240	1.307	1.240	1.059	1.045	1.248
		16	2.254	2.032	2.254	2.032	1.715	1.715	2.152
	5	4	0.816	1.056	1.126	1.075	1.030	0.965	0.904
		8	0.989	1.173	1.366	1.282	1.171	1.110	1.069
		16	2.097	1.852	2.902	2.492	2.233	2.185	2.166
50	0	4	1.087	1.053	1.087	1.053	1.083	1.070	1.098
		8	1.190	1.156	1.190	1.156	1.175	1.163	1.202
		16	1.492	1.440	1.492	1.440	1.446	1.433	1.508
	15	4	0.352	1.047	1.130	1.074	0.997	0.910	0.816
		8	0.405	1.076	1.301	1.234	1.012	0.884	0.807
		16	0.632	1.185	2.031	1.864	1.250	1.100	1.052
100	0	4	1.029	1.018	1.029	1.018	0.958	0.933	0.994
		8	1.056	1.047	1.056	1.047	0.947	0.921	1.020
		12	1.090	1.080	1.090	1.080	0.949	0.930	1.053
		16	1.125	1.114	1.125	1.114	0.961	0.947	1.087
	25	10	0.301	1.036	1.107	1.094	0.982	0.960	0.984
		20	0.342	1.055	1.260	1.240	1.020	0.986	1.053
		30	0.405	1.080	1.490	1.459	1.077	1.028	1.116
		40	0.520	1.116	1.915	1.860	1.228	1.170	1.270
	50	10	0.248	1.047	1.193	1.163	1.000	0.960	0.823
		20	0.331	1.069	1.591	1.527	1.118	1.022	0.886
		30	0.476	1.096	2.292	2.168	1.376	1.270	1.119
		40	1.197	1.137	5.772	4.929	3.094	2.880	2.539

$$
\begin{aligned}
\text{(ENET)}: \text{lpsa}_i &= \beta_0 + \beta_1 \text{lcavol}_i + \beta_2 \text{lweight}_i + \beta_3 \text{svi}_i \\
&+ \beta_4 \text{lbph}_i + \beta_5 \text{age}_i \\
&+ \beta_7 \text{gleason}_i + \beta_8 \text{pgg45}_i + \epsilon \\
\text{(ALASSO)}: \text{lpsa}_i &= \beta_0 + \beta_1 \text{lcavol}_i + \beta_2 \text{lweight}_i + \beta_3 \text{svi}_i \\
+\epsilon & \\
\text{(SCAD)}: \text{lpsa}_i &= \beta_0 + \beta_1 \text{lcavol}_i + \beta_2 \text{lweight}_i + \epsilon
\end{aligned}
$$

TABLE 3.10: The RMSE of the Estimators for $p_1 = 3$.

			Case 1		Case 2				
n	p_3	p_2	SM	PS	SM	PS	ENET	LASSO	ALASSO
100	4	0	2.447	1.590	2.447	1.590	1.189	1.063	2.184
		4	0.121	1.036	1.698	1.344	1.065	0.941	1.018
		8	0.090	1.044	1.440	1.237	1.039	0.933	0.994
		12	0.076	1.043	1.313	1.179	1.008	0.910	0.940
	8	0	4.024	2.666	4.024	2.666	1.457	1.411	3.587
		4	0.179	1.090	2.419	1.942	1.202	1.094	1.311
		8	0.115	1.069	1.899	1.666	1.084	0.956	1.212
		12	0.095	1.057	1.639	1.494	1.042	0.907	1.048
200	4	0	2.388	1.595	2.388	1.595	1.164	1.030	1.924
		4	0.055	1.020	1.586	1.317	1.034	0.904	1.206
		8	0.040	1.020	1.392	1.218	1.001	0.870	1.148
		12	0.034	1.016	1.279	1.166	0.997	0.893	1.066
	8	0	3.903	2.709	3.903	2.709	1.335	1.301	3.116
		4	0.078	1.040	2.226	1.864	1.129	0.987	1.475
		8	0.051	1.025	1.796	1.603	1.041	0.860	1.369
		12	0.044	1.023	1.573	1.446	1.015	0.871	1.240

3.6.3 Prediction Error via Bootstrapping

In real data examples, the prediction error (PE) is used to evaluate the performance of an estimator. In this step, we split data into train and test sets while resampling the input B bootstraps times. Each time, a new random split of the data is performed and samples are drawn (with replacement) on each side of the split to construct the training and test sets. Before beginning analysis, we center all variables based on the training data set. A constant term is therefore not counted as a parameter.

The general idea is as follows for each iteration:

1- Pick a number of samples, say $\text{Sample}^1, \ldots, \text{Sample}^{\text{B}}$.

2- Randomly and correspondingly divide the samples into train and test sets. For instance, if 20% of the dataset is designated as the test set, 20% of samples will be selected at random and the remaining 80% will become the training set. This step is to acquire Train. In this step, obtain $\text{Train}^1, \ldots, \text{Train}^{\text{B}}$ and $\text{Test}^1, \ldots, \text{Test}^{\text{B}}$.

3- Calculate $\bar{\boldsymbol{X}}_{\text{Train}^{\text{b}}} = (\bar{X}_{1,\text{Train}^{\text{b}}}, \ldots, \bar{X}_{\text{p},\text{Train}^{\text{b}}})$ and $\bar{Y}_{\text{Train}^{\text{b}}}$ for $b = 1, \ldots, B$.

4- Fit model on the training set. Obtain $\widehat{\boldsymbol{\beta}}^1, \ldots, \widehat{\boldsymbol{\beta}}^B$.

TABLE 3.11: The RMSE of the Estimators for $p_1 = 3$.

n	p_3	p_2	Case 1 SM	Case 1 PS	Case 2 SM	Case 2 PS	ENET	LASSO	ALASSO
30	4	0	4.066	1.867	4.066	1.867	1.611	1.673	3.273
		4	0.667	1.296	1.896	1.415	1.214	1.002	0.822
		8	0.550	1.273	1.577	1.297	1.250	1.123	0.885
		12	0.799	1.319	2.267	1.493	1.982	1.763	1.305
	8	0	7.707	3.596	7.707	3.596	2.158	2.351	6.049
		4	1.053	1.605	2.991	2.249	1.471	1.339	1.179
		8	1.244	1.495	3.576	2.397	2.342	2.156	1.751
		12	1.454	1.459	4.124	2.660	2.735	2.418	1.963
100	4	0	2.612	1.632	2.612	1.632	1.187	1.100	2.292
		4	0.111	1.021	1.586	1.320	1.016	0.819	0.696
		8	0.083	1.040	1.453	1.249	1.034	0.921	1.088
		12	0.073	1.042	1.340	1.192	1.026	0.928	1.005
	8	0	4.150	2.804	4.150	2.804	1.412	1.372	3.640
		4	0.162	1.066	2.305	1.927	1.084	0.852	0.803
		8	0.111	1.065	1.950	1.689	1.095	0.947	1.309
		12	0.094	1.061	1.718	1.538	1.082	0.961	1.163
500	4	0	2.390	1.573	2.390	1.573	1.073	0.975	1.492
		4	0.022	1.006	1.603	1.313	0.998	0.860	1.216
		8	0.015	1.006	1.379	1.212	0.985	0.866	1.115
		12	0.013	1.006	1.282	1.160	0.983	0.880	1.070
	8	0	3.825	2.614	3.825	2.614	1.267	1.213	2.391
		4	0.030	1.013	2.212	1.861	1.083	0.942	1.672
		8	0.019	1.009	1.768	1.574	1.039	0.896	1.417
		12	0.016	1.009	1.563	1.435	1.019	0.888	1.300

5- Calculate each testing set's response vectors,

$$\widehat{\boldsymbol{Y}}_{\text{Test}^1} = \boldsymbol{X}_{\text{Test}^1}\widehat{\boldsymbol{\beta}}^1$$
$$\vdots$$
$$\widehat{\boldsymbol{Y}}_{\text{Test}^{\text{B}}} = \boldsymbol{X}_{\text{Test}^{\text{B}}}\widehat{\boldsymbol{\beta}}^B.$$

For machine learning strategies in this book, this step is obtained directly, skipping step 4.

TABLE 3.12: PE of estimators for Prostrate Data.

Shrinkage Estimation				
	AIC	BIC	BSS	LASSO
FM	0.511(0.005)	0.511(0.005)	0.511(0.005)	0.511(0.005)
SM	0.5(0.004)	0.511(0.004)	0.559(0.004)	0.511(0.004)
S	0.506(0.005)	0.499(0.005)	0.501(0.005)	0.499(0.005)
PS	0.504(0.005)	0.498(0.005)	0.501(0.005)	0.498(0.005)
Penalized Methods				
ENET	LASSO	RIDGE	ALASSO	SCAD
0.493(0.004)	0.563(0.005)	0.500(0.005)	0.57(0.005)	0.538(0.005)
Machine Learnings				
NN	RF	KNN		
0.577(0.006)	0.398(0.005)	0.646(0.007)		

The numbers in parenthesis are the corresponding standard errors of the prediction errors.

6- Calculate PE for each sample,

$$\mathrm{PE}^b = \frac{1}{n_{\mathrm{Test^b}}}(\boldsymbol{r}^b)^\top \boldsymbol{r}^b, b = 1, \ldots, B,$$

where $\boldsymbol{r}^b = \widehat{\boldsymbol{Y}}_{\mathrm{Test^b}} - \bar{Y}_{\mathrm{Train^b}} - (\boldsymbol{X}_{\mathrm{Test^b}} - \bar{\boldsymbol{X}}_{\mathrm{Train^b}})\widehat{\boldsymbol{\beta}}^b$.

7- Calculate average of $\mathrm{PE}^1, \ldots, \mathrm{PE}^\mathrm{B}$.

To calculate the prediction error of estimators based on the above strategies, we randomly split the data into a training set that has 70% of the observations and a testing set that has the remaining 30% of the observations. Since the splitting of the data is a random procedure, to account for the random variation, we repeat the process $B = 250$ times and estimate the average prediction errors along with their standard deviations. The number of repetitions was initially varied, and we settled on this number as no noticeable variations in the standard deviations were observed for higher numbers of repetitions. The results are given in Table 3.12. Further, the respective prediction errors based on the machine learning strategies are reported as well.

Table 3.12 reveals that the submodel estimator based on the AIC criteria outperforms the penalty methods in terms of prediction error. This indicates the inactive predictors that are deleted from the submodel are indeed irrelevant, or nearly so, for the response. Thus, the shrinkage estimators based on AIC produce the smallest prediction error as compared to all the other listed estimators. The penalized methods are comparable, except SCAD is showing a little bit higher prediction error. On the other hand, the machine learning method of random forest gives the smallest prediction error but comes at a cost in interpretability and computational power. We suggest the use of a shrinkage estimator using an AIC selection method for this data set. Comparing shrinkage strategies with penalized and machine learning methods, shrinkage estimators are in closed form, computationally attractive, and free from tuning parameters. However, there may be situations when other estimators may perform better than shrinkage estimators. These results are consistent with our analytical and simulated findings.

3.6.4 Machine Learning Strategies

We use selected machine learning strategies to analyze this data. In order to implement the neural network method in R using the neuralnet package, the data must be normalized between 0 and 1. We scale the data using the minimum and maximum scaling methods. We use one hidden layer for this analysis since our data is low-dimensional. Using many hidden layers in a neural network is also known as "deep learning" and is used when the data is high-dimensional and complex. But for the purpose of this example, we train the data using one hidden layer with 8 inputs, which are our explanatory variables, and produce a value for our target variable. A graphical representation of the neural network is depicted next in Figure 3.14. Visualizing what the neural network's architecture looks like helps us see how the 8 inputs align with the hidden layer and the nonlinear bias B1 to adjust the weights before another non-linear bias B2 is imposed when producing the prediction for the output O1, lpsa.

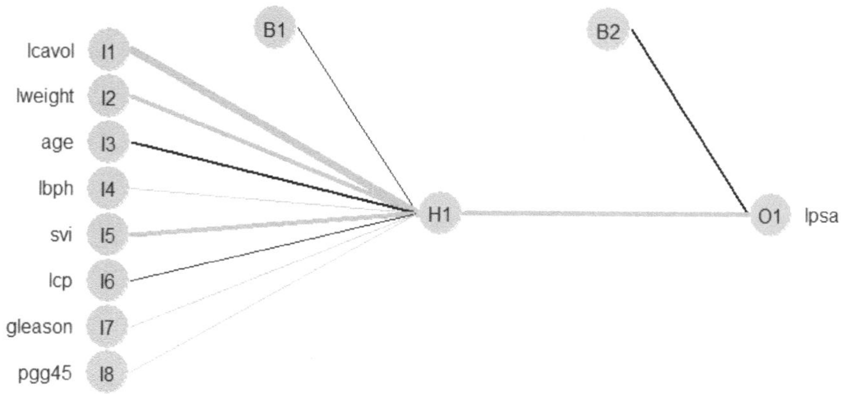

FIGURE 3.14: Neural Network Architecture.

Figure 3.15 allows us to visualize which variables are most important in our neural network. Here we see prostate weight, seminal vesicle invasion, and cancer volume are the most important. Although this information is helpful, we still do not know the sign of the relationship, only the magnitude. We then employ the Olden method to calculate the data's variable importance to see the magnitude and sign of variable importance, which the Garson method does not. Figure 3.16 allows us to see that age and capsular penetration not only have lesser importance in the neural network but in fact have a negative relationship with our dependent variable.

Random forest involves the process of creating multiple decision trees and the combining of their results. For the prostrate data, the only parameter required for this analysis is to find the optimal number of trees, which is found to be 15 using the which.min function from the randomForest package. In order to visualize the variable importance of this data, we use minimal depth, and it is presented in the 3.17 because our data is low-dimensional and

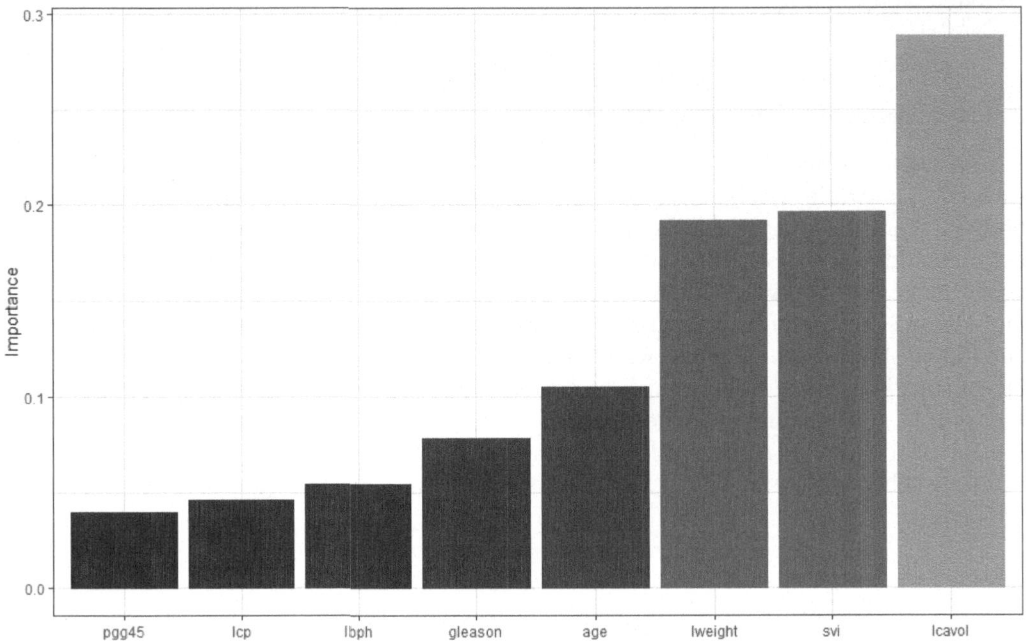

FIGURE 3.15: Variable Importance Chart via Garson's Algorithm.

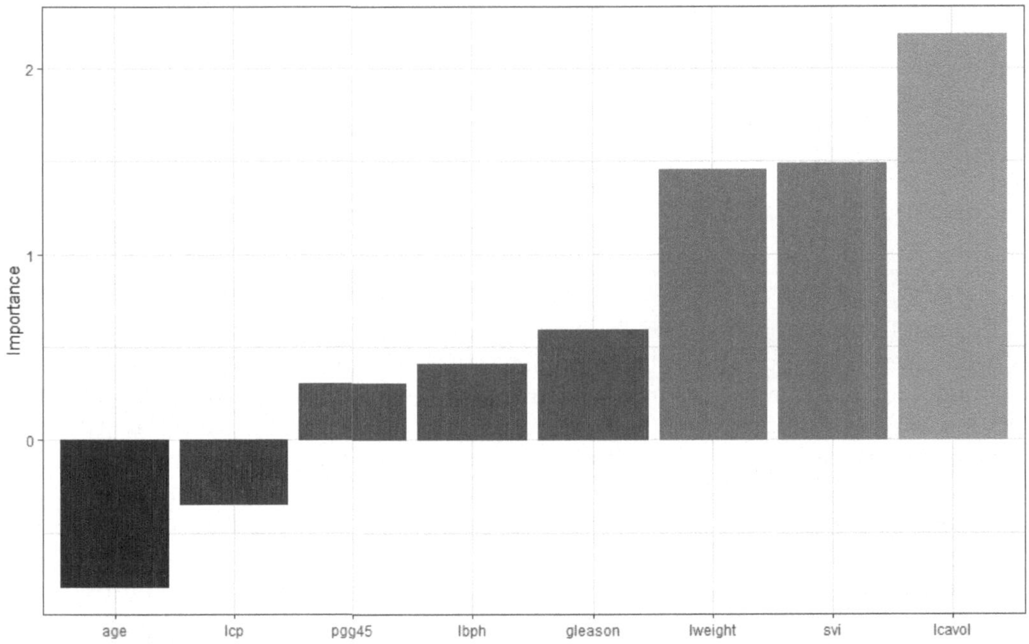

FIGURE 3.16: Variable Importance Chart via Olden's Algorithm.

comparison can be easily ascertained. This figure shows our 8 inputs on the y-axis, and their respective rainbow gradient reveals the distribution of minimal depth as the number of trees increases. The lower the mean of the minimal depth, as indicated by the black vertical bar of each variable in the data, the greater the importance. For our data, we see that cancer volume and prostate weight hold higher importance again, with mean minimal depths of 1.51 and 1.86, respectively. We also see that capsular penetration is of greater importance than we saw in our neural network analysis.

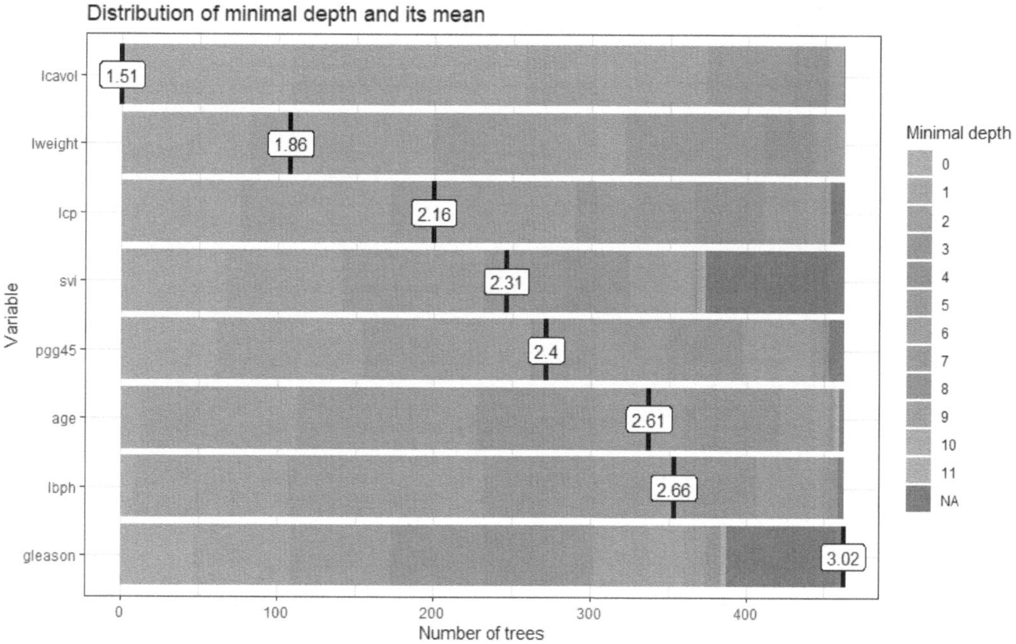

FIGURE 3.17: Random Forest Distribution of Minimal Depth and Mean.

Another visualization we have employed to evaluate variable importance is through a multi-way importance plot seen in Figure 3.18. This figure allows us to plot the number of trees in which the root is split on each of our variables against the mean minimal depth. We see that the variable cancer volume's mean minimal tree depth of 1.51 is split 94 times. This confirms that cancer volume has the highest importance, followed by prostrate weight and capsular penetration.

Next, we employ the K-nearest neighbours method and see how it performs for the prostrate data. We use the squareroot of the number of observations in the training set to determine the number of neighbours in our training set, which is 15. To confirm if this is the optimal number, we calculate the RMSE over different values of k.

Figure 3.19 demonstrates the RMSE values over different numbers of k and shows that the lowest RMSE of 0.678 is when k is set to 13. Using 13 as k, we run KNN to find our prediction values for lpsa.

Below in Figure 3.20 we plot the actual versus predicted values of the prostate specific antigen and their respective prediction errors. We can see that multiple regression outperforms machine learning techniques for this low-dimensional, prostrate dataset!

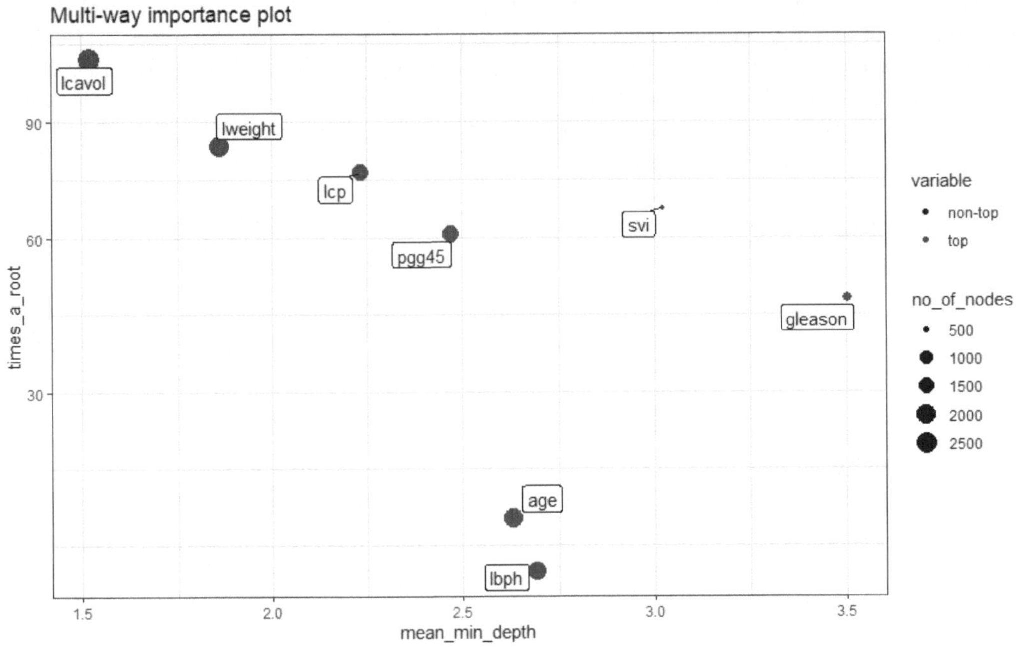

FIGURE 3.18: Random Forest Multiway Importance Plot.

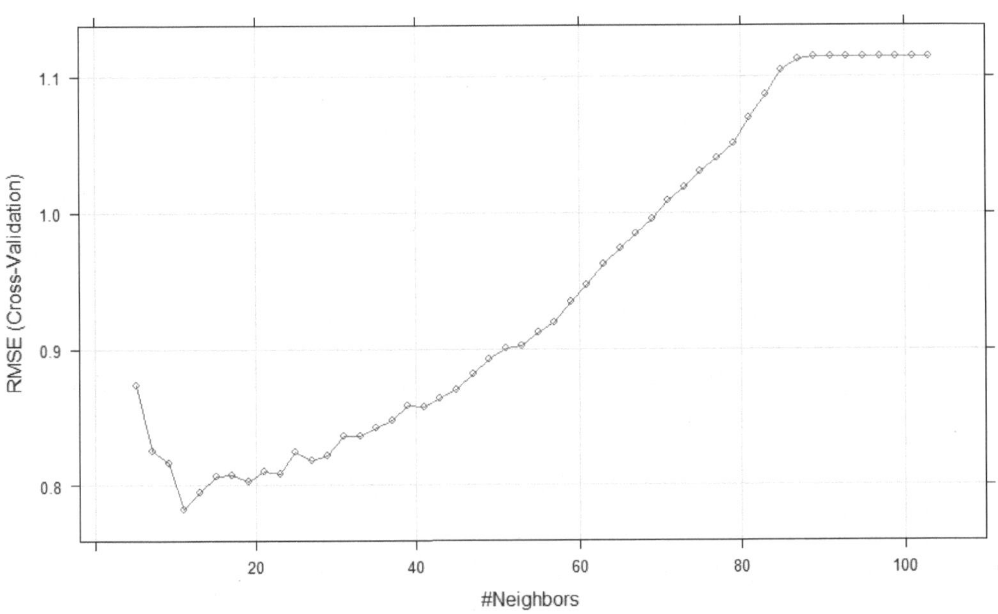

FIGURE 3.19: RMSE versus Number of Nearest Neighbours from KNN

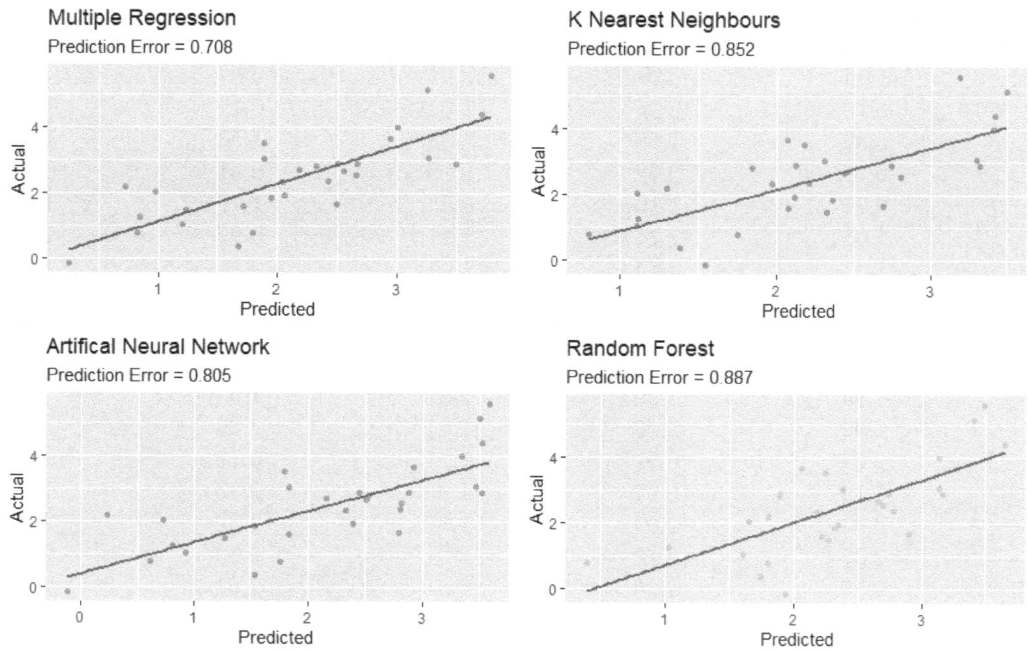

FIGURE 3.20: Prediction Results

3.7 R-Codes

```
> library('MASS')    # It is for 'mvrnorm' function
> library('lmtest') # It is for 'lrtest' function
> library('caret')   # It is for 'split' function
> set.seed(2500)
> # Defining Shrinkage and Positive Shrinkage estimation functions
> Shrinkage_Est <- function(beta_FM,beta_SM,test_stat,p2){
+   return(beta_FM - ((beta_FM - beta_SM)*(p2-2)/test_stat))
+ }
> PShrinkage_Est <- function(beta_FM,beta_SM,test_stat,p2){
+   return(beta_SM + max(0,(1-(p2-2)/test_stat))*(beta_FM-beta_SM))
+ }
> # The function of prediction error
> Prediction_Error <- function(y,yhat){mean((y-yhat)^2)}
> # The function of MSE
> MSE <- function(beta,beta_hat){mean((beta-beta_hat)^2)}
> n  <- 100 # The number of sample
> p1 <- 3   # The number of significant covariates
> p2 <- 5   # The number of insignificant covariates
> p <- p1+p2
> beta_true <- rep(1,p1)
> beta2_true <- rep(0,p2)
> beta_true <- c(beta_true,beta2_true) # The ture value of covariates
> # The design matrix from multivariate normal distribution with zero
> # mean and the pairwise correlation between Xi and Xj was set to
```

TABLE 3.13: Prediction Values.

Actual	MR	KNN	NN	RF
-0.16	0.24	1.54	-0.12	1.94
0.37	1.68	1.38	1.53	1.73
0.77	0.82	0.80	0.61	0.40
0.77	1.80	1.75	1.76	1.80
1.05	1.21	1.11	0.92	1.60
1.27	0.84	1.11	0.81	1.03
1.47	1.23	2.33	1.27	2.26
1.60	1.72	2.07	1.84	2.21
1.66	2.50	2.70	2.80	2.90
1.85	1.95	2.37	1.53	2.31
1.92	2.06	2.11	2.39	2.34
2.05	0.97	1.11	0.72	1.64
2.19	0.72	1.30	0.23	1.79
2.33	2.04	1.97	2.32	2.17
2.37	2.42	2.21	2.80	2.78
2.57	2.66	2.81	2.81	2.70
2.66	2.56	2.44	2.51	2.66
2.68	2.18	2.45	2.16	2.51
2.79	2.33	1.84	2.53	1.89
2.88	3.49	3.32	3.52	3.15
2.88	2.51	2.13	2.44	1.88
2.89	2.66	2.75	2.88	2.74
3.04	1.88	2.31	1.83	2.38
3.08	3.26	3.30	3.47	3.13
3.52	1.89	2.18	1.79	2.24
3.68	2.94	2.07	2.92	2.05
3.99	3.00	3.42	3.34	3.13
4.39	3.71	3.43	3.51	3.66
5.14	3.25	3.50	3.50	3.41
5.58	3.78	3.19	3.58	3.49

```
> # be corr(Xi,Xj)=0.5^|i-j|.
> MU <- rep(0,p)
> SIGMA <- matrix(0,p,p)
> for (i in 1:p){
+   for (j in 1:p){
+     SIGMA[i,j] = 0.5^{abs(i-j)}
+   }
+ }
> X = mvrnorm(n, mu=MU, Sigma=SIGMA)
> ## assigning colnames of X to "X1","X2",...,"X10".
> v<-NULL
> for(i in 1:p){
+   v[i] <- paste("X",i,sep="")
+   assign(v[i],X[,i])
+ }
> #X_scale <-  scale(X)      # Centering and Scaling the X
> epsilon <- rnorm(n) # The errors
```

```
> sigma <- 1
> y<-X%*%beta_true+sigma*epsilon # The response
> # Split data into train and test set
> all.folds <- split (sample (1 : n), rep (1 : 2, length = n))
> train_ind <- all.folds$'1'
> test_ind  <- all.folds$'2'
> y_train <- y[train_ind]
> X_train <- X[train_ind,]
> ### Centering train data of y and X
> X_train_mean  <- apply(X_train,2,mean)
> X_train_scale <- scale(X_train,X_train_mean,F)
> y_train_mean  <- mean(y_train)
> y_train_scale <- scale(y_train,y_train_mean,F)
> train_scale_df <- data.frame(y_train_scale,X_train_scale)
> #test data
> y_test <- y[test_ind]
> X_test <- X[test_ind,]
> ### Centering test data of y and X based on train means
> X_test_scale <- scale(X_test,X_train_mean,F)
> y_test_scale <- scale(y_test,y_train_mean,F)
> # Formula of the Full model
> xcount.FM <- c(0,paste("X", 1:p, sep=""))
> Formula_FM<- as.formula(paste("y_train_scale~",
+                               paste(xcount.FM, collapse= "+")))
> # Formula of the Sub model
> xcount.SM <- c(0,paste("X", 1:p1, sep=""))
> Formula_SM<- as.formula(paste("y_train_scale~",
+                               paste(xcount.SM, collapse= "+")))
> # The full model fit
> fit_FM   <-lm(Formula_FM,data = train_scale_df)
> beta.FM  <-fit_FM$coef
> # The sub model fit
> fit_SM   <-lm(Formula_SM,data = train_scale_df)
> beta.SM  <-c(fit_SM$coef,rep(0,p2))
> # Likelihood ratio test
> test_LR <- lrtest(fit_SM, fit_FM)
> test_stat <- test_LR$Chisq[2]
> #
> beta.S  <- Shrinkage_Est(beta.FM,beta.SM,test_stat,p2)
> beta.PS <- PShrinkage_Est(beta.FM,beta.SM,test_stat,p2)
> # Estimate Prediction Errors based on test data
> yhat_beta.FM <- X_test_scale%*%beta.FM
> yhat_beta.SM <- X_test_scale%*%beta.SM
> yhat_beta.S  <- X_test_scale%*%beta.S
> yhat_beta.PS <- X_test_scale%*%beta.PS
> # Calculte prediction errors of estimators based on test data
> PE_values <- c(FM=Prediction_Error(y_test_scale,yhat_beta.FM),
+                SM=Prediction_Error(y_test_scale,yhat_beta.SM),
+                S =Prediction_Error(y_test_scale,yhat_beta.S),
+                PS=Prediction_Error(y_test_scale,yhat_beta.PS))
> # print and sort the results
> sort(PE_values)
       SM         S        PS        FM
0.6930014 0.7619520 0.7619520 0.8311779
> # Calculte MSEs of estimators
```

```
> MSE_values <- c(FM=MSE(beta_true,beta.FM ),
+                 SM=MSE(beta_true,beta.SM ),
+                 S = MSE(beta_true,beta.S ),
+                 PS=MSE(beta_true,beta.PS ))
> # print and sort the results
> sort(MSE_values)
         SM           S          PS          FM
0.004994425 0.032718440 0.032718440 0.053856227
> ###
> # An Example of the Real data
> library('ElemStatLearn')  # It is for 'prostate' data
> library('dplyr')   # for data cleaning
> library('leaps')      # model selection function of 'regsubsets'
> data("prostate")
> # Center y andX will be standardized
> y <- prostate %>% dplyr::select(lpsa) %>%
+   scale(center = TRUE, scale = FALSE) %>% as.matrix()
> X <- prostate %>% dplyr::select(-c(lpsa,train)) %>%
+   scale(center = TRUE, scale = TRUE) %>% as.matrix()
> raw_data <-  data.frame(y,X)
> p <- ncol(X)
> # perform best subset selection
> best_subset <- regsubsets(lpsa~., raw_data, nvmax = p)
> results <- summary(best_subset)
> # significant variable by BIC
> sub_names <- names(coef(best_subset,which.min(results$bic))[-1])
> full_names <- colnames(X)
> # indexes of significant variables
> p1_indx <- which(full_names %in% sub_names)
> p1 <- length(p1_indx) # the value of p1
> p2 <- p-p1 # the value of p2
> # Split by train and test set by using train column
> train_set <- prostate %>% subset(train==TRUE)  %>%
   dplyr::select(-train)
> test_set  <- prostate %>% subset(train==FALSE) %>%
   dplyr::select(-train)
> # Center y and X for the train data
> # For y
> y_train      <- train_set %>% dplyr::select(lpsa)  %>% as.matrix()
> y_train.mean <- mean(y_train)
> y_train_scale <- y_train- y_train.mean
> # For X
> X_train        <- train_set %>% dplyr::select(-lpsa) %>% as.matrix()
> X_train.mean  <- apply (X_train, 2, mean)
> X_train_scale <- scale(X_train, X_train.mean, F)
> # Center y on the y_train.mean in the test set
> y_test_scale <- test_set %>% dplyr::select(lpsa) %>%
+   scale(y_train.mean, scale = FALSE) %>% as.matrix()
> # Center y on the y_train.mean in the test set
> X_test_scale <- test_set %>% dplyr::select(-lpsa) %>%
+   scale(center=X_train.mean,scale = F ) %>% as.matrix()
> # data frame based on train data
> df_train <- data.frame(y_train_scale,X_train_scale)
> # Formula of the Full model
> xcount.FM <- c(0,full_names)
```

```
> Formula_FM<- as.formula(paste("lpsa~",
+                               paste(xcount.FM, collapse= "+")))
> # Formula of the Sub model
> xcount.SM <- c(0,sub_names)
> Formula_SM<- as.formula(paste("lpsa~",
+                               paste(xcount.SM, collapse= "+")))
> # The full model fit based on train data
> fit_FM <-lm(Formula_FM,df_train)
> beta.FM  <-fit_FM$coef
> # The sub model fit based on train data
> fit_SM <-lm(Formula_SM,df_train)
> beta.SM <- rep(0,p)
> beta.SM[p1_indx]  <-fit_SM$coef
> # Likelihood ratio test
> test_LR   <- lrtest(fit_SM, fit_FM)
> test_stat <- test_LR$Chisq[2]
> #
> beta.S  <- Shrinkage_Est(beta.FM,beta.SM,test_stat,p2)
> beta.PS <- PShrinkage_Est(beta.FM,beta.SM,test_stat,p2)
> # Estimate Prediction Errors based on test data
> yhat_beta.FM <- X_test_scale%*%beta.FM
> yhat_beta.SM <- X_test_scale%*%beta.SM
> yhat_beta.S  <- X_test_scale%*%beta.S
> yhat_beta.PS <- X_test_scale%*%beta.PS
> # Calculte prediction errors of estimators
> PE_values <- c(FM=Prediction_Error(y_test_scale,yhat_beta.FM),
+                SM=Prediction_Error(y_test_scale,yhat_beta.SM),
+                S =Prediction_Error(y_test_scale,yhat_beta.S),
+                PS=Prediction_Error(y_test_scale,yhat_beta.PS))
> # print and sort the results
> sort(PE_values)
       SM        PS         S        FM
0.4005308 0.4759919 0.4759919 0.5212740
> library('MASS')   # It is for 'mvrnorm' function
> library('lmtest') # It is for 'lrtest' function
> library('caret')  # It is for 'split' function
> set.seed(2500)
> # Defining Shrinkage and Positive Shrinkage estimation functions
> Shrinkage_Est<-function(beta_FM,beta_SM,test_stat,p2){
+   return(beta_FM - ((beta_FM - beta_SM)*(p2-2)/test_stat))
+ }
> PShrinkage_Est<-function(beta_FM,beta_SM,test_stat,p2){
+   return(beta_SM + max(0,(1-(p2-2)/test_stat))*(beta_FM-beta_SM))
+ }
> # The function of prediction error
> Prediction_Error<-function(y,yhat){mean((y-yhat)^2)}
> # The function of MSE
> MSE<-function(beta,beta_hat){mean((beta-beta_hat)^2)}
> #
> n <-100 # The number of sample
> p1<-3  # The number of significant covariates
> p2<-5  # The number of insignificant covariates
> p<-p1+p2
> beta_true<-rep(1,p1)
> beta2_true<-rep(0,p2)
```

```
> beta_true<-c(beta_true,beta2_true) # The ture value of covariates
> # The design matrix from multivariate normal distribution with zero
> # mean and the pairwise correlation between Xi and Xj was set to
> # be corr(Xi,Xj)=0.5^|i-j|.
> MU<-rep(0,p)
> SIGMA<-matrix(0,p,p)
> for (i in 1:p){
+   for (j in 1:p){
+     SIGMA[i,j] = 0.5^{abs(i-j)}
+   }
+ }
> X = mvrnorm(n, mu=MU, Sigma=SIGMA)
> ## assigning colnames of X to "X1","X2",...,"X10".
> v<-NULL
> for(i in 1:p){
+   v[i]<-paste("X",i,sep="")
+   assign(v[i],X[,i])
+ }
> epsilon<-rnorm(n) # The errors
> sigma<-1
> y<-X%*%beta_true+sigma*epsilon # The response
> # Split data into train and test set
> all.folds<-split (sample (1 : n), rep (1 : 2, length = n))
> train_ind<-all.folds$'1'
> test_ind <-all.folds$'2'
> y_train<-y[train_ind]
> X_train<-X[train_ind,]
> ### Centering train data of y and X
> X_train_mean <-apply(X_train,2,mean)
> X_train_scale<-scale(X_train,X_train_mean,F)
> y_train_mean <-mean(y_train)
> y_train_scale<-scale(y_train,y_train_mean,F)
> train_scale_df<-data.frame(y_train_scale,X_train_scale)
> #test data
> y_test<-y[test_ind]
> X_test<-X[test_ind,]
> ### Centering test data of y and X based on train means
> X_test_scale<-scale(X_test,X_train_mean,F)
> y_test_scale<-scale(y_test,y_train_mean,F)
> # Formula of the Full model
> xcount.FM<-c(0,paste("X", 1:p, sep=""))
> Formula_FM<- as.formula(paste("y_train_scale~",
+                               paste(xcount.FM, collapse= "+")))
> # Formula of the Sub model
> xcount.SM<-c(0,paste("X", 1:p1, sep=""))
> Formula_SM<- as.formula(paste("y_train_scale~",
+                               paste(xcount.SM, collapse= "+")))
> # The full model fit
> fit_FM   <-lm(Formula_FM,data = train_scale_df)
> beta.FM  <-fit_FM$coef
> # The sub model fit
> fit_SM   <-lm(Formula_SM,data = train_scale_df)
> beta.SM  <-c(fit_SM$coef,rep(0,p2))
> # Likelihood ratio test
> test_LR<-lrtest(fit_SM, fit_FM)
```

```
> test_stat<-test_LR$Chisq[2]
> # Shrinkage Estimation
> beta.S <-Shrinkage_Est(beta.FM,beta.SM,test_stat,p2)
> beta.PS<-PShrinkage_Est(beta.FM,beta.SM,test_stat,p2)
> # Estimate Prediction Errors based on test data
> yhat_beta.FM<-X_test_scale%*%beta.FM
> yhat_beta.SM<-X_test_scale%*%beta.SM
> yhat_beta.S <-X_test_scale%*%beta.S
> yhat_beta.PS<-X_test_scale%*%beta.PS
> # Calculte prediction errors of estimators based on test data
> PE_values<-
+c(FM=Prediction_Error(y_test_scale,yhat_beta.FM),
+SM=Prediction_Error(y_test_scale,yhat_beta.SM),
+S =Prediction_Error(y_test_scale,yhat_beta.S),
+PS=Prediction_Error(y_test_scale,yhat_beta.PS))
> # print and sort the results
> sort(PE_values)
       SM         S        PS        FM
0.6930014 0.7619520 0.7619520 0.8311779
> # Calculte MSEs of estimators
> MSE_values<-c(FM=MSE(beta_true,beta.FM ),
+                 SM=MSE(beta_true,beta.SM ),
+                 S = MSE(beta_true,beta.S ),
+                 PS=MSE(beta_true,beta.PS ))
> # print and sort the results
> sort(MSE_values)
         SM           S          PS          FM
0.004994425 0.032718440 0.032718440 0.053856227
> # An Example of the Real data
> library('ElemStatLearn') # It is for 'prostate' data
> library('dplyr')   # for data cleaning
> library('leaps')      # model selection function of 'regsubsets'
> data("prostate")
> # Center y andX will be standardized
> y<-prostate %>% dplyr::select(lpsa) %>%
+   scale(center = TRUE, scale = FALSE) %>% as.matrix()
> X<-prostate %>% dplyr::select(-c(lpsa,train)) %>%
+   scale(center = TRUE, scale = TRUE) %>% as.matrix()
> raw_data<- data.frame(y,X)
> p<-ncol(X)
> # perform best subset selection
> best_subset<-regsubsets(lpsa~., raw_data, nvmax = p)
> results<-summary(best_subset)
> # significant variable by BIC
> sub_names<-
+names(coef(best_subset,which.min(results$bic))[-1])
> full_names<-colnames(X)
> # indexes of significant variables
> p1_indx<-which(full_names %in% sub_names)
> p1<-length(p1_indx) # the value of p1
> p2<-p-p1 # the value of p2
> # Split by train and test set by using train column
> train_set<-prostate %>% subset(train==TRUE)  %>%
   dplyr::select(-train)
```

```
> test_set <-prostate %>% subset(train==FALSE) %>%
   dplyr::select(-train)
> # Center y and X for the train data
> # For y
> y_train     <-train_set %>% dplyr::select(lpsa)  %>% as.matrix()
> y_train.mean<-mean(y_train)
> y_train_scale<-y_train- y_train.mean
> # For X
> X_train      <-train_set %>% dplyr::select(-lpsa) %>% as.matrix()
> X_train.mean <-apply (X_train, 2, mean)
> X_train_scale<-scale(X_train, X_train.mean, F)
> # Center y on the y_train.mean in the test set
> y_test_scale<-test_set %>% dplyr::select(lpsa) %>%
+   scale(y_train.mean, scale = FALSE) %>% as.matrix()
> # Center y on the y_train.mean in the test set
> X_test_scale<-test_set %>% dplyr::select(-lpsa) %>%
+   scale(center=X_train.mean,scale = F ) %>% as.matrix()
> # data frame based on train data
> df_train<-data.frame(y_train_scale,X_train_scale)
> # Formula of the Full model
> xcount.FM<-c(0,full_names)
> Formula_FM<- as.formula(paste("lpsa~",
+                                 paste(xcount.FM, collapse= "+")))
> # Formula of the Sub model
> xcount.SM<-c(0,sub_names)
> Formula_SM<- as.formula(paste("lpsa~",
+                                 paste(xcount.SM, collapse= "+")))
> # The full model fit based on train data
> fit_FM <-lm(Formula_FM,df_train)
> beta.FM  <-fit_FM$coef
> # The sub model fit based on train data
> fit_SM <-lm(Formula_SM,df_train)
> beta.SM<-rep(0,p)
> beta.SM[p1_indx]  <-fit_SM$coef
> # Likelihood ratio test
> test_LR  <-lrtest(fit_SM, fit_FM)
> test_stat<-test_LR$Chisq[2]
> #Shrinkage Estimation
> beta.S <-Shrinkage_Est(beta.FM,beta.SM,test_stat,p2)
> beta.PS<-PShrinkage_Est(beta.FM,beta.SM,test_stat,p2)
> # Estimate Prediction Errors based on test data
> yhat_beta.FM<-X_test_scale%*%beta.FM
> yhat_beta.SM<-X_test_scale%*%beta.SM
> yhat_beta.S <-X_test_scale%*%beta.S
> yhat_beta.PS<-X_test_scale%*%beta.PS
> # Calculte prediction errors of estimators
> PE_values<-
> c(FM=Prediction_Error(y_test_scale,yhat_beta.FM),
+SM=Prediction_Error(y_test_scale,yhat_beta.SM),
+S =Prediction_Error(y_test_scale,yhat_beta.S),
+PS=Prediction_Error(y_test_scale,yhat_beta.PS))
> # print and sort the results
> sort(PE_values)
       SM        PS         S        FM
0.4005308 0.4759919 0.4759919 0.5212740
```

3.8 Concluding Remarks

In this chapter, we consider some high-dimensional post-selection shrinkage estimators for the commonly used multiple regression models in low and high-dimensional settings.We develop the asymptotic risk functions of the suggested estimators in a low-dimensional case and provide apairwise comparison of the listed estimators. Finally, we also give a global dominance picture under certain conditions. The continuing use of least squares and/or likelihood strategies seems inexplicable, and the suggested shrinkage strategy demonstrates this convincingly, especially when many regression parameters are in the model. Importantly, the shrinkage estimator is computationally elementary, under-demanding, and can be easily implemented in a host of statistical models. Interestingly, the shrinkage approach is free from any tuning parameters, and numerical work is not iterative. Finally, the simulation results and the real data example strongly corroborate the contention that the shrinkage strategy is superior to classical estimation.We suggest using positive-part shrinkage estimators because they outperform the listed estimators in the entire parameter space in low-dimensional cases.

We also included some penalized strategies and compared them with the post-shrinkage estimation method through extensive numerical studies. We consider both sparse and weak sparse models in our simulation study. The simulation results demonstrate that the post-selection shrinkage estimators have favorable performance and are a good and safe alternative to penalized estimators, especially in the presence of weak signals. They continue to perform better than full model, submodel, and penalized estimators when the validity of the sparsity may not hold. They save the loss of efficiency of the penalized estimators due to the effect of imprecise variable selection at the first stage.The post-selection shrinkage strategy has a superior estimation and prediction performance over other penalized regression estimators in numerous scenarios.

The same result holds for the high-dimensional cases. However, in such cases, we use two penalized strategies: one less aggressive to select an overfittedmodel (full model) and one aggressive one to select an underfitted model (submodel), and then we combine both models in the usual way to construct a shrinkage strategy.

The real-world data example illustrated the benefit of the shrinkage strategy. After using some machine learning techniques to analyze real data, we recommend using a shrinkage strategy to minimize bias, especially when the assumption of complete sparsity on the model cannot be judiciously satisfied.

The most important message in this chapter is that when there are a large number of inactive predictors included in the model at the initial stage of model building, a substantial gain in precision may be achieved by judiciously exploiting information that suggests restrictions on the parameter space. Our numerical results indicate that using the shrinkage strategy, a significant reduction of the MSE seems quite possible in many situations. Thus, it seems desirable to pay attention to these considerations in the development of statistical models and inference theory. It may be worth mentioning that this is one of the two areas Professor Efron predicted for the early 21st century (RSS News, January 1995). When it comes to combining estimation problems, shrinkage and likelihood-based methods are still very useful.

4

Shrinkage Strategies in High-Dimensional Regression Models

4.1 Introduction

Model selection, post-estimation, and prediction is imperative for anyone conducting an analysis. As we try to advance business practices, being able to predict financial, operational, transactional, etc. information is a lucrative skill. For example, many retailers want to provide their customers with the appropriate advertisements in their emails. In order to target these customers, the analytics team must collect and analyze the data to monitor consumer behavior. Based on their customers purchasing history, the retailer can predict their next purchases and provide a personalized flyer/coupon. However, prediction is only as good as its model. Bias in the model can cost a business to make ill-informed decisions. Statistical bias is a systematic partiality that is present in the data collection process, resulting in misleading results. Just how social biases can affect our personal decisions, statistical bias affects analytical modeling. For example, gender-biased employers often overlook women compared to their male counterparts, resulting in women being passed over for positions and promotions. Such biases can translate into the statistical world. For instance, if the sample collected has more men than women it results an unbalanced model for the success rate of a CEO .

In this chapter, we consider model selection, parameter estimation, and prediction problems in high-dimensional regression model, that is, when the sample size is smaller than the size of data elements associated with each observation. One of the important objectives of regression analysis is to identify important predictors that are associated with the response variable and for estimation and prediction purposes. These tasks are more challenging when the number of predictors is relatively large compared to the sample size. A vast literature is available focusing on the development of methodological and numerical methods for regression analysis on high-dimensional data (Tibshirani (1996); Fan and Li (2001); Zhang (2010), among others).

Representation and modeling of high-dimensional data is an important feature in a host of social sciences, medical, environmental, engineering, financial studies, social network modeling, clinical, genetic, and phenotypic data among others. For example, genomics data is large and vast accounting for every gene in the body and every genes phenotypic expressions. As another example, The Cancer Genome Atlas (TCGA) (http://cancergenome.nih.gov/) has generated more than 2.5 petabytes of clinical and genomic data, including DNA alterations, methylation profiles, and the expression levels of RNA and protein, for over 11,000 cancer samples with 33 types of cancer. The rapid growth in the size and scope of data sets in several unrelated disciplines has created a need for innovative statistical strategies to provide insights on such data. Among many problems arisen from big data in the realm of statistics, many researchers are interested in data sets containing larger number of predictors (regression parameters) than the number of observations (sample size). There is an increasing demand for variable selection procedures and then prediction strategies for

DOI: 10.1201/9781003170259-4

analyzing HDD. Developing innovative statistical learning algorithms and data analytic techniques play a fundamental role for the future of research in these fields. More public and private sectors are now acknowledging the importance of statistical tools and its critical role in analyzing HDD.

The challenges are to find novel statistical methods to extract meaningful insights and interpretable results from HDD. The classical statistical strategies do not provide solutions to such problems. Traditionally, for low-dimensional cases, practitioners and professionals used best subset selection or other variable selection methods to select predictors that are highly correlated with the response variable. Based on the selected predictors, statisticians employed classical statistical methods to analyze HDD. However, with a huge number of predictors, implementing a best subset selection is already computationally burdensome. On top of that, these variable selection techniques suffer from high variability due to their nature. To resolve such issues, a class of penalized estimation methods have been proposed in the literature. They are referred to as penalized estimation methods since they share the idea to estimate parameters in a model using classical maximum likelihood or least squares approach with an additional penalty term. Some of these methods perform variables selection and parameter estimation simultaneously. Existing penalized techniques that deal with high-dimensional data mostly rely on various L_1 penalty regularizes. Due to the trade-off between model complexity and model prediction, the statistical inference of model selection becomes an extremely important and challenging problem in high-dimensional data analysis. However, penalized methods assume that a model is sparse, where it contains a few strong predictors and the rest have no influence on prediction. Clearly, this assumption may not be true in many applications where weak signals may have a joint impact on prediction. In most published studies on HDD, it has been assumed that the strong signals and noises are well separated. Penalization methods, including LASSO, group LASSO, and SCAD, typically focus on selecting variables with strong effects while ignoring weak signals. This may result in biased prediction, especially when weak signals outnumber strong signals. Again, the conventional penalized theory relies on a strong assumption on the minimum signal strength, which aims to estimate the coefficients for strong signals without considering weak signals as weak signals tend to be shrunk toward zero. For example, weak signals exist in the HIV-1 drug resistance study Qu and Shi (2016) and present quite ubiquitously in general. Donoho and Jin (2008) illustrated that a single weak signal might not contribute significantly to the response, but all of them combined together could have significant influence toward scientific discovery. In general, identification of weak signals could facilitate us to demystify the entire scope of studies.

Shrinkage analysis has been one of first author main research fields for many years. Previously, the focus was to shrink a full estimator in the direction of an estimator under a submodel. However, in a high-dimensional setting, there is no unique solution for a full estimator. Thus, it becomes an interesting, but very challenging problem to study shrinkage analysis in a high-dimensional setting. Gao et al. (2017a) and Gao et al. (2017b) suggested an idea of using ridge regression to produce a useful full estimator, and using any existing penalized method such as LASSO or MCP to select a good submodel. Thus, a shrinkage strategy could be adopted to improve the prediction efficiency of Lasso-type submodels. Eventually, they successfully provided a framework for high-dimensional shrinkage analysis when both strong signals (with only a small number of candidates) and weak signals (with a very large amount of members) co-exist. The idea of borrowing the joint strength of a large number of weak signals to improve the prediction efficiency of strong signals adopting a ridge estimator is new in shrinkage analysis. The theoretical investigation of the research is sophisticated, the work is original, and can be extended in many model settings. In this chapter we integrate two submodels based on penalized methods to construct the post-selection shrinkage estimator.

We consider the estimation problem of regression parameters when there are many potential predictors in the initial/working model and:

1. most of them may not have any influence (sparse signals) on the response of interest
2. some of the predictors may have strong influence (strong signals) on the response of interest
3. some of them may have weak-moderate influence (weak-moderate signals) on the response of interest

The model and some estimators are introduced in Section 4.2. In Section 4.3 we showcase our suggested estimation strategy. The results of a simulation study includes comparison of suggested estimators with the penalty estimators are reported in Section 4.4. Application to real data sets is given in Section 4.5. The R codes are available in Section 4.6. Finally, we offer concluding remarks in Section 4.7

4.2 Estimation Strategies

We consider a high-dimensional linear regression sparse model:

$$y_i = \sum_{j=1}^{d} x_{ij}\beta_j + \varepsilon_i, \quad 1 \le i \le n << p \tag{4.1}$$

where y_i observed response variable with predictors x_is, and β_j are the regression parameters. Further, ε_is are independent and identically distributed (iid) random errors with center 0 and variance σ^2.

Regularization techniques are helpful in dealing with the problem when the number of features, p are much larger than the data points, n. However, the solution is based upon the stringent assumption that the true underlying model is sparse. Most LASSO-type penalties make assumptions on both the true model and the designed covariates. For example, the true model is often assumed to be sparse such that

- Most of the regression coefficients are zeros except for a few ones
- All non-zero β_j's are larger than noise level, $c\sigma\sqrt{(2/n)\log(p)}$ with $c \ge 1/2$

The minimum signal strength condition is critical for theoretical derivation. However, it is rather difficult to verify this condition in a finite sample size. Further, there is no guarantee that it will be satisfied in analysing real datasets. On the contrary, if this condition is not satisfied, or in other words, if weak signals exist, then the method suggested by Zhao et al. (2018) would fail in theory and the weak coefficients would shrink toward zero. Recently, Gao et al. (2017a), and Qu and Shi (2016) and Li et al. (2019) provide strategies for weak signal detection and estimation.

Additional assumptions made regarding the designed covariates include the adaptive irrepresentable condition and the restricted eigenvalue conditions. We refer to Zhao and Yu (2006), Huang et al. (2008), and Bickel et al. (2009) for some insights. The LASSO and ENET penalized methods tend to select an overfitted model since it penalizes all the regression coefficients equally and subject to inherited model selection bias and lacks an oracle property, see Leng et al. (2006). In reviewed literature, several modifications and methodologies have been suggested to improve model selection and prediction accuracy for the LASSO strategy. For that purpose, we consider SCAD, ALASSO, and the minimax concave penalty (MCP). The MCP strategy was introduced by Zhang (2010). These methods select a submodel by shrinking some regression coefficients to zero and provide shrinkage estimators of the remaining coefficients. However, these methods may force the relatively more weak coefficients toward zeros as compared to LASSO, resulting in underfitted models subject to a much larger selection bias, particularly in the presence of significant number of weak signals.

In this chapter, we consider the model selection, estimation and prediction problems for the sparse regression models when there are many potential predictors that may have no and/or weak influence on the response of interest. This leads to the consideration of two models: the overfitted model that includes predictors with strong signals and possibly some predictors with weak signals selected by LASSO or ENET. We select an underfitted model that possibly includes the predictors with strong signals only by using SCAD or ALASSO. One way to deal with this framework is to use Stein-type shrinkage estimators, where the estimated regression coefficient vector is shrunk in the direction of the underfitted model. This "soft threshold" modification of the shrinkage method has been shown to be efficient in a host frameworks. Ahmed et al. (2012), among others have examined the properties of Stein-type shrinkage estimation strategies for a number of interesting statistical models.

Let us rewrite our model in the following matrix model

$$\boldsymbol{Y} = \boldsymbol{X}\boldsymbol{\beta} + \boldsymbol{\varepsilon}, \tag{4.2}$$

where $\boldsymbol{Y} = (y_1, y_2, \ldots, y_n)^\top$ is a vector of responses, $\boldsymbol{X}$ is an $n \times p$ fixed design matrix, $\boldsymbol{\beta} = (\beta_1, \ldots, \beta_p)^\top$ is an unknown vector of parameters, $\boldsymbol{\varepsilon} = (\varepsilon_1, \varepsilon_2, \ldots, \varepsilon_n)^\top$ is the vector of unobservable random errors. We do not make any distributional assumptions about the errors except that $\boldsymbol{\varepsilon}$ has a cumulative distribution function $F(\boldsymbol{\varepsilon})$ with $E(\boldsymbol{\varepsilon}) = \mathbf{0}$, and $E(\boldsymbol{\varepsilon}\boldsymbol{\varepsilon}^\top) = \sigma^2\boldsymbol{I}$, where σ^2 is finite.

For $n > p$ the FM of $\boldsymbol{\beta}$ is given by

$$\widehat{\boldsymbol{\beta}}^{\mathrm{FM}} = (\mathbf{X}^\top\mathbf{X})^{-1}\mathbf{X}^\top\mathbf{Y}.$$

Since we are dealing with a high-dimensional situation, i.e. $n < p$ the inverse of the Gram matrix, $(\mathbf{X}^\top\mathbf{X})^{-1}$ does not exist and there will be infinitely many solutions for the least squares minimization, hence there is no well-defined solution. In fact, even in the case $p \leq n$ and p close to n, the LSE is generally not considered very useful because standard deviations of estimators are usually very high. In other words, LSE may not be stable.

The main goal of this chapter to improve the estimation and prediction accuracy of the important set of the regression parameters by combining an overfitted model estimators with an underfitted one. As stated earlier, the LASSO and ENET produce an overfitted model, respectively as compared with SCAD, ALASSO and other penalized methods. The LASSO strategy retains some regression coefficients with strong signals and as well as some with weak signals in the resulted model. On the other hand, aggressive penalized strategies may force some, if not all, moderate and weak signals toward zero, resulting in underfitted models with a fewer predictors of strong signals. Thus, we combine estimators from an underfitted model with an overfitted model leading to a non-linear shrinkage strategy or integrated estimation strategy for the regression parameters estimation.

4.3 Integrating Submodels

In this section we show how to combine two models produced by two distinct variable selection methods. The idea is to start with an intimal model that may have all the possible predictors, and then apply two different variable selection procedures to obtain two submodels, respectively. Ahmed and Yüzbaşı (2016) and Ahmed and Yüzbaşı (2017) suggested to combine the estimates from two submodels to improve post-estimation and prediction performances of the estimators, respectively.

4.3.1 Sparse Regression Model

Consider the high-dimensional sparse regression model as mentioned earlier

$$\boldsymbol{Y} = \boldsymbol{X}\boldsymbol{\beta} + \boldsymbol{\varepsilon}, \quad p > n. \tag{4.3}$$

Suppose we can divide the index set $\{1, \cdots, p\}$ into three disjoint subsets: S_1, S_2 and S_3. In particular, S_1 includes indexes of non-zero β_i's which are large and easily detectable. The set S_2, being the intermediate, includes indexes of those non-zero β_j with weak-to-moderate signals but strictly non-zero signals. By the assumption of sparsity, S_3 includes indexes with only zero coefficients and can be comfortably discarded by penalized methods. Thus, S_1 and S_3 are able to be retained and discarded by using penalized techniques, respectively. However, it is possible that the S_2 may be covertly hiding either in S_2 or S_3 depending on penalized methods being used. For the case when S_2 may not be separated from S_3; we refer to Zhang and Zhang (2014) and others. Hansen (2016) has showed using simulation studies that such a LASSO estimate often performs worse than the post-selection least square estimate. To improve the prediction error of a LASSO-type variable selection approach, some (modified) post least squares estimators are studied in Belloni and Chernozhukov (2013) and Liu and Yu (2013).

However, we are interested in cases when predictors in S_1 are kept in the model, and some or all predictors in S_2 are also included in S_1, which may or may not be useful for prediction purposes. It is possible that one variable selection strategies may produce an overfitted model, that is retaining predictors from S_1 and S_2. On the other hand, other methods may produce an underfitted model keeping only predictors from S_1. Thus, the predictors in S_2 should be subject to further scrutiny to improve the prediction error.

We partition the design matrix such that $\mathbf{X} = (\mathbf{X}_{S_1}|\mathbf{X}_{S_2}|\mathbf{X}_{S_3})$, Further, $\mathbf{X}_1$ is $n \times p_1$, $\mathbf{X}_2$ is $n \times p_2$, and $\mathbf{X}_3$ is $n \times p_3$ submatrix of predictors, respectively; and $p = p_1 + p_2 + p_3$. Here we make the usual assumption that $p_1 \leq p_2 < n$ and $p_3 > n$.

Thus, sparse regression model is:

$$\boldsymbol{Y} = \boldsymbol{X}_1\boldsymbol{\beta}_1 + \boldsymbol{X}_2\boldsymbol{\beta}_2 + \boldsymbol{X}_3\boldsymbol{\beta}_3 + \boldsymbol{\varepsilon}, \quad p > n, \quad p_1 + p_2 < n. \tag{4.4}$$

4.3.2 Overfitted Regression Model

Suppose a penalized method which keeps both strong and weak-moderate signals as follows:

$$\boldsymbol{Y} = \boldsymbol{X}_1\boldsymbol{\beta}_1 + \boldsymbol{X}_2\boldsymbol{\beta}_2 + \varepsilon, \quad p_1 \leq p_2 < n, p_1 + p_2 < n. \tag{4.5}$$

The LASSO and ENET strategies which usually eliminate the sparse signals and retains weak-moderate and strong signals in the resulting model, thus are useful to obtain an overfitted model. Keeping in mind that there is no guarantee this outcome will always be achieved in real situations.

4.3.3 Underfitted Regression Model

Using more aggressive penalized method which keeps only strong signals and eliminates all other signals in the resulting model:

$$\boldsymbol{Y} = \boldsymbol{X}_1\boldsymbol{\beta}_1 + \varepsilon, \quad p_1 < n. \tag{4.6}$$

One can use the SCAD or ALASSO strategy which usually retains the strong signals that may produce a lower dimensional model as compared with LASSO. This model may be deemed as an underfitted model.

We are interested in estimating $\boldsymbol{\beta}_1$, when $\boldsymbol{\beta}_2$ may or may not be a set of nuisance parameters. We suggest a non-linear shrinkage strategy based on the Stein-rule for estimating $\boldsymbol{\beta}_1$ suspecting $\boldsymbol{\beta}_2 = \mathbf{0}$. In sum, we are combining estimates of the overfitted with the estimates of underfitted models to improve the performance of an underfitted model.

4.3.4 Non-Linear Shrinkage Estimation Strategies

In the spirit of Ahmed (2014), the shrinkage estimator of $\boldsymbol{\beta}_1$ defined by combining overfitted model estimate $\widehat{\boldsymbol{\beta}}_1^{\mathrm{OF}}$ with the underfitted $\widehat{\boldsymbol{\beta}}_1^{\mathrm{UF}}$ as follows

$$\widehat{\boldsymbol{\beta}}_1^{\mathrm{S}} = \widehat{\boldsymbol{\beta}}_1^{\mathrm{UF}} + \left(\widehat{\boldsymbol{\beta}}_1^{\mathrm{OF}} - \widehat{\boldsymbol{\beta}}_1^{\mathrm{UF}}\right)\left(1 - (p_2 - 2)W_n^{-1}\right), \, p_2 \geq 3,$$

where the weight function W_n is defined by

$$W_n = \frac{1}{\widehat{\sigma}^2}(\widehat{\boldsymbol{\beta}}_2^{\mathrm{LSE}})^\top(\boldsymbol{X}_{S_2}^\top \boldsymbol{M}_1 \boldsymbol{X}_{S_2})\widehat{\boldsymbol{\beta}}_2^{\mathrm{LSE}},$$

and $\boldsymbol{M}_1 = \boldsymbol{I}_n - \boldsymbol{X}_{S_1}\left(\boldsymbol{X}_{S_1}^\top \boldsymbol{X}_{S_1}\right)^{-1}\boldsymbol{X}_{S_1}^\top$, $\widehat{\boldsymbol{\beta}}_2^{\mathrm{LSE}} = \left(\boldsymbol{X}_{S_2}^\top \boldsymbol{M}_1 \boldsymbol{X}_{S_2}\right)^{-1}\boldsymbol{X}_{S_2}^\top \boldsymbol{M}_1 \boldsymbol{y}$ and

$$\widehat{\sigma}^2 = \frac{1}{n - p_1}(\boldsymbol{y} - \boldsymbol{X}_{S_1}\widehat{\boldsymbol{\beta}}_1^{\mathrm{LSE}})^\top(\boldsymbol{y} - \boldsymbol{X}_{S_1}\widehat{\boldsymbol{\beta}}_1^{\mathrm{LSE}}).$$

The $\widehat{\boldsymbol{\beta}}_1^{\mathrm{UF}}$ maybe based on SCAD or ALASSO and $\widehat{\boldsymbol{\beta}}_1^{\mathrm{OF}}$ is obtained by LASSO or ENET methods, respectively.

In an effort to avoid the over-shrinking problem inherited by $\widehat{\boldsymbol{\beta}}_1^{\mathrm{S}}$, we suggest to use the positive part of the shrinkage estimator of $\boldsymbol{\beta}_1$ defined by

$$\widehat{\boldsymbol{\beta}}_1^{\mathrm{PS}} = \widehat{\boldsymbol{\beta}}_1^{\mathrm{UF}} + \left(\widehat{\boldsymbol{\beta}}_1^{\mathrm{OF}} - \widehat{\boldsymbol{\beta}}_1^{\mathrm{UF}}\right)\left(1 - (p_2 - 2)W_n^{-1}\right)^+.$$

In the following section, we conduct a Monte Carlo simulation study to appraise the performance of the listed estimators.

4.4 Simulation Experiments

Here we present the details of the Monte Carlo simulation study. We simulate the response from the following model:

$$y_i = x_{1i}\beta_1 + x_{2i}\beta_2 + ... + x_{pi}\beta_p + \varepsilon_i \, , \; i = 1, 2, ..., n, \tag{4.7}$$

where x_i and ε_i are i.i.d. $N(0, 1)$. We consider the regression coefficients are set $\boldsymbol{\beta} = \left(\boldsymbol{\beta}_1^\top, \boldsymbol{\beta}_2^\top, \boldsymbol{\beta}_3^\top\right)^\top$, with dimensions p_1, p_2 and p_3, respectively. $\boldsymbol{\beta}_1$ represent the strong signals, that is, $\boldsymbol{\beta}_1$ is a vector of 1 values, $\boldsymbol{\beta}_2$ stands for the weak signals, with the signal strength $\kappa = 0, 0.237, 0.475, 0.712, 0.950$, and $\boldsymbol{\beta}_3$ means no signals, namely $\boldsymbol{\beta}_3 = \mathbf{0}$. If $\kappa = 0$, then it indicates that the selected submodel is the right one. In this simulation setting, we simulated 100 data sets consisting of $n = 50, 100$, with $p_1 = 3, 9$, $p_2 = 10, 30$ and $p_3 = 100, 1000, 10000$.

We use ENET, an overfitted (loosely speaking full) model and subsequentially use ALASSO to generate a underfitted (loosely speaking submodel) model to construct shrinkage estimators. We calculate the RMSE of the listed estimators with respect to full model and the results are presented in Tables 4.1–4.4 for the selected values simulation parameters. As expected, the submodel estimator (in this case ALASSO) outperforms its competitors in many cases, since its MSE was calculated under the assumption of model accuracy. However, for large values of κ its performance is not satisfactory with RMSE less than 1. We also observe that SCAD yields larger RMSE than both ridge and LASSO at $\kappa = 0$. However, as expected, RMSE of all penalty estimators converge to zero for larger values of κ. The performance of the ridge estimator is rather poor and we suggest not to use it for high-dimensional cases. Generally speaking, ridge estimators do well in the presence of multicollinearity mostly in fixed low-dimensional cases.

The performance of both shrinkage estimators are impressive. More importantly, the RMSEs of the estimators based on shrinkage principle are bounded in κ. The suggested shrinkage estimators outperform the penalty estimators for almost all values κ. Thus, the performance of the shrinkage estimators remain similar as in fixed low-dimensional cases as reported in Chapter 3.

4.5 Real Data Examples

In this section, we apply the proposed post-selection shrinkage strategies to three real data sets. In an effort to construct post-estimation shrinkage strategies, we consider models based on ENET and LASSO procedures as a full model (large number of predictors). Conversely, we implement ALASSO, SCAD, and MCP penalized methods to produce respective submodels (fewer predictors than ENET). We then combine full model estimators with their respective submodel estimators to obtain shrinkage estimators. In other words, we first obtain submodels from three variable selection techniques: ALASSO, SCAD and MCP. Then, the full models are selected based on ENET and LASSO. Finally, we combine selected full and submodel one at a time to construct the suggested shrinkage post-selection estimators. We also include ridge regression and three machine learning strategies in our data analysis. The average number of selected predictors and non zeros of penalized methods for the data sets are reported in Tables 4.5 and 4.6.

As described in Section 3.6.3, we evaluate the performance of estimators based on the PE of the estimators with B=100 bootstrap replications. In order to facilitate comparisons, we also compute $\mathrm{RPE}(\widehat{\boldsymbol{\beta}}^*) = \frac{\mathrm{PE}(\widehat{\boldsymbol{\beta}}^{\mathrm{FM}})}{\mathrm{PE}(\widehat{\boldsymbol{\beta}}^*)}$. If the RPE is greater than one, this indicates that the method is superior to the full model estimator.

4.5.1 Eye Data

The eye data data set of Scheetz et al. (2006), which contains gene expression of mammalian eye tissue samples. The format consists of a list containing the design matrix which represents the data of 120 rats with 200 gene probes, and the response vector with 120 dimensions which represents the expression level of TRIM32 gene. Thus, for this data set we

TABLE 4.1: The RMSE of the Estimators for $n = 50$ and $p_1 = 3$.

p_2	p_3	κ	SM	S	PS	LASSO	SCAD	Ridge
10	100	0.000	2.570	1.855	2.349	1.068	0.326	0.256
		0.238	1.336	1.150	1.150	1.105	0.407	0.429
		0.475	0.985	1.010	1.010	1.000	0.469	0.572
		0.712	0.974	1.003	1.003	1.000	0.477	0.582
		0.950	1.153	1.005	1.005	1.000	0.316	0.471
	1000	0.000	4.173	1.702	3.341	2.216	1.879	1.000
		0.238	1.572	1.254	1.254	1.414	0.934	1.000
		0.475	0.879	1.015	1.015	1.002	0.517	1.009
		0.712	0.891	1.005	1.005	1.005	0.482	1.005
		0.950	0.902	1.001	1.001	1.008	0.521	0.996
	10000	0.000	1.176	1.100	1.214	1.221	0.900	1.000
		0.238	0.840	1.025	1.025	1.001	0.521	1.000
		0.475	0.756	1.000	1.000	0.893	0.471	1.000
		0.712	0.763	1.000	1.000	0.878	0.465	1.000
		0.950	0.782	1.001	1.001	0.892	0.476	1.000
30	100	0.000	3.073	2.835	2.929	1.238	0.413	0.314
		0.238	0.967	1.030	1.030	1.035	0.558	0.795
		0.475	0.869	0.987	0.987	1.000	0.574	1.077
		0.712	0.882	0.994	0.994	1.000	0.556	1.142
		0.950	0.898	0.997	0.997	1.000	0.599	1.112
	1000	0.000	3.873	3.385	3.657	2.167	1.928	1.000
		0.238	1.024	1.102	1.102	1.098	0.680	1.000
		0.475	0.821	0.999	0.999	0.920	0.536	1.033
		0.712	0.856	0.997	0.997	0.937	0.600	1.103
		0.950	0.919	0.998	0.998	0.994	0.663	1.195
	10000	0.000	1.146	1.156	1.188	1.198	0.813	1.000
		0.238	0.810	0.977	0.977	0.944	0.549	1.000
		0.475	0.823	0.996	0.996	0.919	0.577	1.000
		0.712	0.831	0.999	0.999	0.905	0.583	1.001
		0.950	0.832	1.000	1.000	0.894	0.577	1.000

TABLE 4.2: The RMSE of the Estimators for $n = 50$ and $p_1 = 9$.

p_2	p_3	κ	SM	S	PS	LASSO	SCAD	Ridge
10	100	0.000	1.537	1.439	1.454	1.000	0.848	0.436
		0.238	1.143	1.088	1.088	1.000	0.713	0.582
		0.475	0.947	1.003	1.003	1.000	0.578	0.822
		0.712	0.893	0.997	0.997	1.000	0.488	0.901
		0.950	0.884	0.998	0.998	1.000	0.460	0.937
	1000	0.000	0.876	0.857	0.931	0.992	0.511	0.989
		0.238	0.894	1.015	1.015	1.002	0.527	1.011
		0.475	0.879	1.007	1.007	0.979	0.522	1.035
		0.712	0.873	0.998	0.998	0.973	0.546	1.101
		0.950	0.861	0.998	0.998	0.962	0.572	1.136
	10000	0.000	0.786	0.774	0.844	0.896	0.478	1.000
		0.238	0.801	0.970	0.970	0.901	0.524	1.000
		0.475	0.802	0.997	0.997	0.896	0.524	1.000
		0.712	0.810	0.999	0.999	0.895	0.548	1.000
		0.950	0.810	1.000	1.000	0.887	0.547	1.000
30	100	0.000	1.296	1.306	1.294	1.000	0.893	0.520
		0.238	0.905	0.979	0.979	1.000	0.563	0.969
		0.475	0.910	0.993	0.993	1.000	0.612	1.088
		0.712	0.912	0.997	0.997	1.000	0.593	1.126
		0.950	0.909	0.998	0.998	1.000	0.583	1.144
	1000	0.000	0.873	0.877	0.914	0.993	0.511	0.996
		0.238	0.848	0.994	0.994	0.955	0.563	1.025
		0.475	0.886	1.001	1.001	0.963	0.628	1.048
		0.712	0.919	0.998	0.998	0.995	0.609	1.120
		0.950	0.927	0.999	0.999	0.999	0.601	1.159
	10000	0.000	0.785	0.781	0.826	0.895	0.482	1.000
		0.238	0.828	0.981	0.981	0.917	0.575	1.000
		0.475	0.834	0.997	0.997	0.906	0.569	1.000
		0.712	0.825	0.999	0.999	0.889	0.565	1.000
		0.950	0.828	1.000	1.000	0.887	0.564	1.001

TABLE 4.3: The RMSE of the Estimators for $n = 100$ and $p_1 = 3$.

p_2	p_3	κ	SM	S	PS	LASSO	SCAD	Ridge
10	100	0.000	3.762	2.444	2.886	1.000	1.512	0.087
		0.238	1.375	1.075	1.075	1.000	0.609	0.294
		0.475	1.616	1.028	1.028	1.000	0.626	0.247
		0.712	2.885	1.020	1.020	1.000	0.743	0.164
		0.950	3.743	1.013	1.013	1.000	1.527	0.112
	1000	0.000	31.192	2.163	9.515	7.855	2.469	1.000
		0.238	4.132	1.235	1.235	2.991	1.431	1.000
		0.475	2.020	1.041	1.041	1.803	1.426	1.000
		0.712	2.095	1.019	1.019	1.671	1.659	0.958
		0.950	2.378	1.011	1.011	1.605	2.636	0.776
	10000	0.000	27.900	3.662	8.512	5.947	3.001	1.000
		0.238	3.553	1.220	1.220	2.424	1.588	1.000
		0.475	1.296	1.024	1.024	1.332	0.794	1.000
		0.712	0.919	1.004	1.004	1.023	0.539	1.000
		0.950	0.872	1.001	1.001	0.968	0.526	1.000
30	100	0.000	3.758	3.156	3.461	1.000	0.856	0.092
		0.238	1.038	1.024	1.024	1.000	0.519	0.520
		0.475	0.958	1.002	1.002	1.000	0.535	0.586
		0.712	1.025	1.002	1.002	1.000	0.437	0.503
		0.950	1.642	1.007	1.007	1.000	0.312	0.314
	1000	0.000	32.122	9.380	17.070	7.813	2.382	1.000
		0.238	1.673	1.148	1.148	1.653	0.899	1.000
		0.475	0.931	1.007	1.007	1.043	0.533	0.979
		0.712	0.900	0.999	0.999	1.000	0.501	0.999
		0.950	0.912	0.999	0.999	1.000	0.518	0.990
	10000	0.000	25.790	9.078	14.350	5.810	3.027	1.000
		0.238	1.459	1.124	1.124	1.462	0.878	1.000
		0.475	0.905	1.008	1.008	0.999	0.534	1.000
		0.712	0.840	1.002	1.002	0.925	0.509	1.000
		0.950	0.843	1.001	1.001	0.913	0.511	1.000

TABLE 4.4: The RMSE of the Estimators for $n = 100$ and $p_1 = 9$.

p_2	p_3	κ	SM	S	PS	LASSO	SCAD	Ridge
10	100	0.000	3.773	2.488	2.960	1.000	1.597	0.092
		0.238	1.492	1.088	1.088	1.000	0.660	0.168
		0.475	1.572	1.025	1.025	1.000	0.852	0.195
		0.712	2.731	1.018	1.018	1.000	1.173	0.157
		0.950	3.444	1.012	1.012	1.000	2.620	0.116
	1000	0.000	32.728	2.604	8.788	5.443	5.742	0.915
		0.238	6.633	1.289	1.289	3.375	3.258	0.941
		0.475	1.883	1.037	1.037	1.643	1.762	0.844
		0.712	1.075	1.005	1.005	1.082	0.653	0.736
		0.950	1.008	1.001	1.001	1.006	0.618	0.685
	10000	0.000	2.653	1.824	2.266	1.675	3.584	1.000
		0.238	1.248	1.083	1.083	1.230	0.802	1.000
		0.475	0.857	1.005	1.005	0.963	0.502	1.000
		0.712	0.783	1.000	1.000	0.887	0.494	1.000
		0.950	0.816	1.001	1.001	0.916	0.487	1.000
30	100	0.000	3.730	3.460	3.403	1.000	1.151	0.096
		0.238	1.115	1.043	1.043	1.000	0.563	0.277
		0.475	0.936	1.001	1.001	1.000	0.529	0.471
		0.712	0.912	0.999	0.999	1.000	0.350	0.529
		0.950	0.923	1.000	1.000	1.000	0.314	0.484
	1000	0.000	31.734	10.027	16.718	5.504	5.673	0.907
		0.238	2.274	1.198	1.198	1.753	1.503	0.760
		0.475	0.956	1.002	1.002	1.012	0.470	0.794
		0.712	0.898	0.999	0.999	1.000	0.491	0.980
		0.950	0.902	0.999	0.999	1.000	0.497	1.048
	10000	0.000	2.615	2.464	2.456	1.675	3.740	1.000
		0.238	1.052	1.061	1.061	1.133	0.602	1.000
		0.475	0.840	1.006	1.006	0.948	0.461	1.000
		0.712	0.816	1.002	1.002	0.910	0.470	1.000
		0.950	0.826	1.002	1.002	0.912	0.474	1.002

TABLE 4.5: The Average Number of Selected Predictors.

		Eye Data		Lu2004		Riboflavin	
FM	SM	S	PS	S	PS	S	PS
ENET	ALASSO	21	12	22	13	35	33
	SCAD	21	13	22	15	37	34
	MCP	20	11	22	13	35	32
LASSO	ALASSO	13	11	13	10	17	16
	SCAD	15	12	14	12	18	16
	MCP	13	9	12	9	15	12

TABLE 4.6: The Number of the Predicting Variables of Penalized Methods.

	Eye Data $(n, p) = (120, 200)$	Lu 2004 $(n, p) = (30, 403)$	Riboflavin $(n, p) = (71, 4088)$
ENET	22	22	32
LASSO	19	20	28
ALASSO	6	9	9
SCAD	9	9	16
MCP	9	9	16
Ridge	200	403	4088

have $(n, p) = (120, 200)$. Aldahmani and Dai (2015) analyzed this date set and applied some penalty estimation procedures, and found that the LASSO method identifies 24 influential predictors. However, we use both LASSO and ENET penalty methods to form so-called full models, and three other penalty methods to obtain the respective submodels. To evaluate the prediction accuracy of the listed estimators, we randomly divided the data into two parts: 70% of the dataset was designated as the training set, and 30% was designated as the test set. The results are given in Table 4.7.

TABLE 4.7: RPE of the Estimators for Eye Data.

FM	SM	SM	S	PS	Ridge	RF	NN	kNN
ENET	ALASSO	1.061	1.034	1.075	0.933	0.914	0.616	0.756
	SCAD	0.998	0.963	1.015				
	MCP	0.970	0.880	0.996				
LASSO	ALASSO	1.210	1.166	1.218	1.013	1.000	0.673	0.827
	SCAD	1.130	1.062	1.131				
	MCP	1.082	0.975	1.098				

The results are consistent with the findings of our simulation study. We observe that the suggested positive-part shrinkage estimator outperforms both submodels and full models estimators in all cases. For this data, ENET performs relatively better than three selected penalized methods used to construct the shrinkage estimators, perhaps due to an inherited large amount of bias being more aggressive in variable selection. Table 4.6 shows that ENET is selecting 22 predictors, whereas ALASSO, SCAD, and MCP select 6, 9, and 9 predictors respectively. Interestingly, all these three penalized methods are superior to LASSO. Nevertheless, the positive-part shrinkage estimator is outperforming the listed penalty estimators either we use ENET or LASSO to construct it.

More importantly, the suggested shrinkage estimator is superior to all three listed machine learning estimators. As a matter of fact, all statistical learning estimators perform much better than selected machine learning estimators for this data set.

We suggest to combine two penalized based on shrinkage principle to improve the prediction error, a clear winner!.

4.5.2 Expression Data

The expression data is obtained from the microarray study of Lu et al. (2004). This data set contains measurements of the gene expression of 403 genes from 30 human brain samples. The response variable is the age of each patient is provided, thus $(n, p) = (30, 403)$. Zuber and Strimmer (2011) reported that the LASSO method selects 36 predictors. Again, we construct shrinkage estimators by combining two penalty estimators at a time. To evaluate the prediction accuracy of the listed estimators, we randomly divided the data into two parts: 90% of the dataset was designated as the training set, and 10% was designated as the test set.

TABLE 4.8: RPE of the Estimators for Expression Data.

FM	SM	SM	S	PS	Ridge	RF	NN	kNN
ENET	ALASSO	1.104	1.072	1.112	0.963	0.937	0.632	0.776
	SCAD	1.004	0.983	1.018				
	MCP	1.048	0.949	1.055				
LASSO	ALASSO	1.215	1.185	1.221	1.033	1.004	0.678	0.831
	SCAD	1.093	1.105	1.108				
	MCP	1.143	1.044	1.152				

Table 4.8 clearly reveals that the positive part estimator dominates the all the estimators in the class, that is, both estimators based on penalized and machine learning strategies.

4.5.3 Riboflavin Data

In this example, we use a data set of riboavin production by bacillus subtilis containing 71 observations of 4088 predictors (gene expressions) and a one-dimensional response. In this data set, the response variable is the Log-transformed riboavin production rate and the predictor variables measure the logarithm of the expression level of 4088 genes Bühlmann et al. (2014).

Similarly, we first obtain models from the three variable selection techniques: ALASSO, SCAD and MCP. Then, we select two models based on ENET and LASSO. Finally, we combine two selected penalized models at a time to construct the suggested shrinkage estimators. To evaluate the prediction accuracy of the listed estimators, we randomly divided the data into two parts: 70% of the dataset was designated as the training set, and 30% was designated as the test set. We report the RPE in the Table 4.9.

TABLE 4.9: RPE of the Estimators for Riboflavin Data.

FM	SM	SM	S	PS	Ridge	RF	NN	kNN
ENET	ALASSO	1.066	1.132	1.133	0.894	0.796	0.485	0.553
	SCAD	0.910	0.971	0.986				
	MCP	0.826	0.759	0.930				
LASSO	ALASSO	1.214	1.286	1.286	1.015	0.901	0.546	0.628
	SCAD	1.036	1.084	1.102				
	MCP	0.940	0.802	1.047				

Once again, our suggested positive-part estimator is outperforms other estimators in the class. However, when we combine we use ENET as a full model estimator, the post-selection under-performs (RPE is little less than one) when combined with SCAD and MCP respectively, this is perhaps due to a sampling error caused by outlying observations and needs to be further investigated. The RPE of the post-selection should be at least one theoretically speaking. There maybe issues with variable selection by PS (34, 32) as it selects relatively more of the predictors as compared with ENET (32).

4.6 R-Codes

```
> library('MASS')    # It is for 'mvrnorm' function
> library('lmtest') # It is for 'lrtest' function
> library('caret')  # It is for 'split' function
> library('ncvreg') # It is for 'cv.ncvreg' function
> library('glmnet') # It is for 'cv.glmnet' function
> set.seed(2500)
> # Defining Shrinkage and Positive Shrinkage estimation functions
> Shrinkage_Est<-function(beta_FM,beta_SM,test_stat,p2){
+   return(beta_FM - ((beta_FM - beta_SM)*(p2-2)/test_stat))
+ }
> PShrinkage_Est<-function(beta_FM,beta_SM,test_stat,p2){
+   return(beta_SM + max(0,(1-(p2-2)/test_stat))*(beta_FM-beta_SM))
+ }
> # The function of prediction error
> Prediction_Error<-function(y,yhat){mean((y-yhat)^2)}
> # The function of MSE
> MSE<-function(beta,beta_hat){mean((beta-beta_hat)^2)}
> n <-100   # The number of sample
> p1<-5    # The number of strong signals
> p2<-10    # The number of weak signals
> p3<-300   # The number of zeros, no signal
> p<-p1+p2+p3
> beta1_true<-rep(5   ,p1)
> beta2_true<-rep(0.5,p2)
> beta3_true<-rep(0   ,p3)
> # The ture value of covariates
```

```
> beta_true<-c(beta1_true,beta2_true,beta3_true)
> # The design matrix
> X<-matrix(0, n, p)
> for (i in 1:p){
+   X[,i]= rnorm(n)
+ }
> ## assigning colnames of X to "X1","X2",...,"X10".
> v<-NULL
> for(i in 1:p){
+   v[i]<-paste("X",i,sep="")
+   assign(v[i],X[,i])
+ }
> epsilon<-rnorm(n) # The errors
> sigma<-1
> y<-X%*%beta_true+sigma*epsilon # The response
> # Split data into train and test set
> all.folds<-split (sample (1 : n), rep (1 : 2, length = n))
> train_ind<-all.folds$‘1‘
> test_ind <-all.folds$‘2‘
> y_train<-y[train_ind]
> X_train<-X[train_ind,]
> ### Centering train data of y and X
> X_train_mean <-apply(X_train,2,mean)
> X_train_scale<-scale(X_train,X_train_mean,F)
> y_train_mean <-mean(y_train)
> y_train_scale<-scale(y_train,y_train_mean,F)
> train_scale_df<-data.frame(y_train_scale,X_train_scale)
> #test data
> y_test<-y[test_ind]
> X_test<-X[test_ind,]
> ### Centering test data of y and X based on train means
> X_test_scale<-scale(X_test,X_train_mean,F)
> y_test_scale<-scale(y_test,y_train_mean,F)
> # Formula of the Full model
> xcount.FM<-c(0,paste("X", 1:(p1+p2), sep=""))
> Formula_FM<- as.formula(paste("y_train_scale~",
+                             paste(xcount.FM, collapse= "+")))
> # Formula of the Sub model
> xcount.SM<-c(0,paste("X", 1:p1, sep=""))
> Formula_SM<- as.formula(paste("y_train_scale~",
+                             paste(xcount.SM, collapse= "+")))
> # The full model fit
> alphas<-seq(0.1,0.9,length.out = 9)
> for (ind in 1:length(alphas)){
+   fits.enet[[ind]]<-glmnet(X_train_scale, y_train_scale,
+                            alpha = alphas[ind],intercept = F,
+                            standardize = F)}
> cvs.enet<-sapply(fits.enet, function(x) {
+   BIC = deviance(x) + log(NROW(X_train_scale))*x$df
+   beta.enet = coef(x)[-1,which.min(BIC)]
+   return(list(PE=mean((y_test_scale - X_test_scale%*%beta.enet)^2),
+               beta.enet = beta.enet))})
> ind_opt <-which.min(do.call(rbind,cvs.enet[1,]))
> b.enet<-do.call(rbind,cvs.enet[2,])[ind_opt,]
> #LASSO based on BIC
```

```
> lasso.fit<-glmnet(X_train_scale, y_train_scale, alpha = 1,
+                   intercept = F, standardize = F)
> lasso.bic = deviance(lasso.fit) +
+   log(NROW(X_train_scale))*lasso.fit$df
> b.lasso = coef(lasso.fit)[-1,which.min(lasso.bic)]
> #Ridge based on BIC
> ridge.fit<-glmnet(X_train_scale, y_train_scale, alpha = 0,
+                   intercept = F, standardize = F)
> ridge.bic = deviance(ridge.fit) +
+   log(NROW(X_train_scale))*ridge.fit$df
> b.ridge = coef(ridge.fit)[-1,which.min(ridge.bic)]
> # Estimation of SCAD
> scad.fit<-ncvreg(X_train_scale, y_train_scale, penalty=c("SCAD"))
> lam<-scad.fit$lambda[which.min(BIC(scad.fit))]
> b.scad<-coef(scad.fit, lambda=lam)[-1]
> # The Sub model fit
> #ALASSO based on BIC
> weight<-b.lasso
> weight<-ifelse(weight == 0, 0.00001, weight)
> alasso.fit<-glmnet(X_train_scale, y_train_scale,
+                    alpha = 1, penalty.factor=1/abs(weight),
+                    intercept = F, standardize = F)
> alasso.bic = deviance(alasso.fit) +
+   log(NROW(X_train_scale))*alasso.fit$df
> b.alasso = coef(alasso.fit)[-1,which.min(alasso.bic)]
> # Likelihood ratio test
> fit_FM   <-lm(Formula_FM,data = train_scale_df)
> fit_SM   <-lm(Formula_SM,data = train_scale_df)
> test_LR<-lrtest(fit_SM, fit_FM)
> test_stat<-test_LR$Chisq[2]
> #Shrinkage Estimation
> beta.S <-Shrinkage_Est(b.enet,b.alasso,test_stat,p2)
> beta.PS<-PShrinkage_Est(b.enet,b.alasso,test_stat,p2)
> # Estimate Prediction Errors based on test data
> yhat_beta.FM  <-X_test_scale%*%b.enet
> yhat_beta.SM  <-X_test_scale%*%b.alasso
> yhat_beta.S   <-X_test_scale%*%beta.S
> yhat_beta.PS  <-X_test_scale%*%beta.PS
> yhat_beta.lasso<- X_test_scale%*%b.lasso
> yhat_beta.scad<-X_test_scale%*%b.scad
> yhat_beta.ridge<- X_test_scale%*%b.ridge
> # Calculte prediction errors of estimators based on test data
> PE_values<-c(FM   =Prediction_Error(y_test_scale,yhat_beta.FM),
+              SM   =Prediction_Error(y_test_scale,yhat_beta.SM),
+              S    =Prediction_Error(y_test_scale,yhat_beta.S),
+              PS   =Prediction_Error(y_test_scale,yhat_beta.PS),
+
   LASSO=Prediction_Error(y_test_scale,yhat_beta.lasso),
+              SCAD =Prediction_Error(y_test_scale,yhat_beta.scad),
+
   RIDGE=Prediction_Error(y_test_scale,yhat_beta.ridge))
> # print and sort the results
> round(sort(PE_values),3)
   SM     PS      S  LASSO   SCAD     FM  RIDGE
 4.107  4.234  4.234  4.483  4.742  4.803 99.588
```

```
> # Calculte MSEs of estimators
> MSE_values<-c(FM= MSE(beta_true,b.enet ),
+                 SM= MSE(beta_true,b.alasso ),
+                 S = MSE(beta_true,beta.S ),
+                 PS= MSE(beta_true,beta.PS ),
+                 LASSO= MSE(beta_true,b.lasso ),
+                 SCAD = MSE(beta_true,b.scad ),
+                 RIDGE= MSE(beta_true,b.ridge ))
> # print and sort the results
> round(sort(MSE_values),4)
    SM   SCAD     PS      S  LASSO     FM  RIDGE
0.0106 0.0119 0.0152 0.0152 0.0162 0.0175 0.3615
```

4.7 Concluding Remarks

In this chapter, we suggested integrating two submodels by incorporating shrinkage estimation strategies in high-dimensional sparse models. Our suggested post-selection procedure are constructed by integration with one submodel, which may be either obtained by LASSO or ENET variable selection methods with a model based on more aggressive variable selection methods such as ALASSO, SCAD or MCP. Our Monte Carlo simulation studies suggest that post-selection shrinkage strategies perform better than penalty estimators for both variable selection and prediction purposes in many instances. The reported results of the three data sets are consistent with the our simulation study, demonstrating the superior performance of suggested post-selection shrinkage strategies. However, we fall short in providing the theoretical justification now, which is an ongoing work and will be investigated in further studies. More importantly, the numerical work corroborates with the theoretical results presented in Chapter 3 for low-dimensional cases.

For future work, one may consider combining all the estimators produced by by multiple (more than two) variable selection techniques into a single estimator to improve the overall prediction error. In another study, it would be interesting to include penalty estimators that correspond to Bayes procedures based on priors with polynomial tails.

5

Shrinkage Estimation Strategies in Partially Linear Models

5.1 Introduction

The partially linear regression model (PLM) is defined as follows.

$$y_i = \boldsymbol{x}_i^\top \boldsymbol{\beta} + f(t_i) + \varepsilon_i, \quad i = 1, \ldots, n, \tag{5.1}$$

where y_is are observed values of the variable of interest, $\boldsymbol{x}_i^\top = (x_{i1}, \ldots, x_{ip})$ is the ith observed vector of predicting variables, $\boldsymbol{\beta} = (\beta_1, \ldots, \beta_p)^\top$ is an unknown $p-$dimensional vector of regression coefficients with $p \leq n$, t_is are values of an extra univariate variable satisfying $t_1 \leq \ldots \leq t_n$, $f(\cdot)$ is an unknown smooth function, and ε_is are random noises assumed to be as $\mathcal{N}(0, \sigma^2)$. We treat the vector $\boldsymbol{\beta}$ as the parametric part and $f(\cdot)$ is the non-parametric part of the model. The model (5.1) can be referred to as a semi-parametric model that includes both parametric and nonparametric parts. We can rewrite the PLM in vector-matrix form in the following way,

$$\boldsymbol{y} = \boldsymbol{X}\boldsymbol{\beta} + \boldsymbol{f} + \boldsymbol{\varepsilon}, \tag{5.2}$$

where $\boldsymbol{y} = (y_1, \ldots, y_n)^\top$, $\boldsymbol{X} = (\boldsymbol{x}_1, \ldots, \boldsymbol{x}_n)^\top$, $\boldsymbol{f} = (f(t_1), \ldots, f(t_n))^\top$, and $\boldsymbol{\varepsilon} = (\varepsilon_1, \ldots, \varepsilon_n)^\top$ is random vector with $\mathrm{E}(\boldsymbol{\varepsilon}) = \boldsymbol{0}$ and $\mathrm{Var}(\boldsymbol{\varepsilon}) = \sigma^2 \boldsymbol{I}_n$.

The model (5.1) was first applied by Engle et al. (1986) in analyzing the relationship between the weather and electricity sales. PLMs since then have had a plethora of applications. Speckman (1988), Eubank (1986), Schimek (2000), Liang (2006) and Ahmed (2014), amongst others, have investigated PLMs. Hossain et al. (2016) developed marginal analysis methods for longitudinal data under partially linear models. Recently, Ahmed et al. (2021) introduced a modified kernel-type ridge estimator for partially linear models under randomly-right censored data.

For PLMs, Ahmed et al. (2007) considered shrinkage estimation methods based on least squares estimation, and the non-parametric component is estimated by using a kernel smoothing function. Raheem et al. (2012) extended this study via B-spline-based estimates of the non-parametric component. Further, Phukongtong et al. (2020) estimated the regression parameters using the profile likelihood, where the non-parametric component is approximated by using smoothing splines. In the above work, the shrinkage estimations are compared with some penalized likelihood estimators.

In real-life applications, the variance of the least squares estimator may be very large due to multicollinearity in the model; we refer to (5.1). There are many methods available in the reviewed literature to deal with multicollinearity. One of the most popular strategies is ridge regression, suggested by Hoerl and Kennard (1970). Yüzbaşı and Ahmed (2016) suggested the shrinkage ridge estimation in the presence of multicollinearity and the model may be sparse or restricted to a subspace. In their study, the non-parametric component is estimated using kernel smoothing. Yüzbaşı et al. (2020) extended this work by using

DOI: 10.1201/9781003170259-5

smoothing splines. Gao and Ahmed (2014) developed shrinkage estimation strategies in high-dimensional partially linear regression models. The paper establishes the consistency and asymptotic normality of the estimator. Additionally, the author derived the asymptotic distribution of the risk via a quadratic loss function, with simulations and data analysis to complement the theoretical work. This paper provides an interesting alternative to classical penalization methods through the use of shrinkage. The numerical results support the theoretical results established, improve the performance of the proposed methods, and contain high quality statistical methodology.

Based on Yüzbaşı and Ahmed (2016) work, this chapter presents a shrinkage ridge regression strategy using a kernel smoothing method for estimation of the nonparametric component in a PLM and is compared with penalized strategies.

The chapter is structured as follows. In Section 5.1.1, we provide a synopsis of the situation. In Section 5.2, the full, submodel, and shrinkage estimators are presented. Section 5.3 provides the estimators' asymptotic bias and risk. A Monte Carlo simulation is used to evaluate the performance of the estimators, including a comparison with the penalized estimators described in Section 5.4. The examples of real data are provided in Section 5.5. Section 5.6 demonstrates uses for a high-dimensional model. The R codes can be found in Section 5.7. The chapter concludes with Section 5.8.

5.1.1 Statement of the Problem

We are mainly interested in the estimation of the regression parameters vector $\boldsymbol{\beta}$ when restricted to a subspace. Let us consider the model (5.1)

$$y_i = \boldsymbol{x}_i^\top \boldsymbol{\beta} + f(t_i) + \varepsilon_i \quad \text{subject to } \boldsymbol{\beta}^\top \boldsymbol{\beta} \leq \phi \text{ and } \boldsymbol{R}\boldsymbol{\beta} = \boldsymbol{r}, \tag{5.3}$$

where ϕ is a tuning parameter, $\boldsymbol{R}$ is an $m \times p$ restriction matrix, and $\boldsymbol{r}$ is an $m \times 1$ vector of constants. Suppose the model is sparse and we can partition the data matrix $\boldsymbol{X} = (\boldsymbol{X_1}, \boldsymbol{X_2})$, where $\boldsymbol{X_1}$ is an $n \times p_1$ sub-matrix containing the regressors of interest and $\boldsymbol{X_2}$ is an $n \times p_2$ sub-matrix that may or may not be relevant in the analysis of the main regressors. Similarly, let $\boldsymbol{\beta} = \left(\boldsymbol{\beta}_1^\top, \boldsymbol{\beta}_2^\top\right)^\top$ be the vector of parameters, where $\boldsymbol{\beta}_1$ and $\boldsymbol{\beta}_2$ have dimensions p_1 and p_2, respectively, with $p_1 + p_2 = p$. Hence model (5.2) can be written as:

$$\boldsymbol{y} = \boldsymbol{X}_1\boldsymbol{\beta}_1 + \boldsymbol{X}_2\boldsymbol{\beta}_2 + \boldsymbol{f} + \boldsymbol{\varepsilon}. \tag{5.4}$$

We are interested in the estimation of $\boldsymbol{\beta}_1$ when it is suspected that $\boldsymbol{\beta}_2$ is close to $\boldsymbol{0}$. In other words, $\boldsymbol{R} = [\boldsymbol{0}, \boldsymbol{I}]$, and $\boldsymbol{r} = \boldsymbol{0}$, where $\boldsymbol{0}$ is a $p_2 \times p_1$ matrix of zeroes and $\boldsymbol{I}$ is the identity matrix of order $p_2 \times p_2$, which means $\boldsymbol{\beta}_2 = \boldsymbol{0}$. Essentially, we are considering the conditional estimation of $\boldsymbol{\beta}_1$ in a submodel:

$$\boldsymbol{y} = \boldsymbol{X}_1\boldsymbol{\beta}_1 + \boldsymbol{f} + \boldsymbol{\varepsilon} \tag{5.5}$$

5.2 Estimation Strategies

Let us briefly describe the estimation of the non-parametric component of the model. Assume that $\left(\boldsymbol{x}_i^\top, t_i, y_i; i = 1, 2, ..., n\right)$ satisfies model (5.1). Since $E(\varepsilon_i) = 0$, we have $f(t_i) = E\left(y_i - \boldsymbol{x}_i^\top \boldsymbol{\beta}\right)$ for $i = 1, 2, ..., n$. Hence, if we know $\boldsymbol{\beta}$, a natural non-parametric estimator of $f(\cdot)$, is

$$\widehat{f}(t, \boldsymbol{\beta}) = \sum_{i=1}^{n} W_{ni}(t)\left(y_i - \boldsymbol{x}_i^\top \boldsymbol{\beta}\right),$$

where the positive weight functions $W_{ni}(\cdot)$ satisfies the following conditions:

(i) $\max_{1\leq i\leq n}\sum_{j=1}^{n} W_{ni}(t_j) = O(1)$,

(ii) $\max_{1\leq i,j\leq n} W_{ni}(t_j) = O\left(n^{-2/3}\right)$,

(iii) $\max_{1\leq i\leq n}\sum_{j=1}^{n} W_{ni}(t_j) I(|t_i - t_j| > c_n) = O(d_n)$, where I is an indicator function, c_n satisfies $\limsup_{n\to\infty} nc_n^3 < \infty$, and d_n satisfies $\limsup_{n\to\infty} nd_n^3 < \infty$.

Hence, a full model ridge estimator is readily obtained by minimizing:

$$\arg\min_{\boldsymbol{\beta}} \left(\tilde{\boldsymbol{y}} - \tilde{\boldsymbol{X}}\boldsymbol{\beta}\right)^\top \left(\tilde{\boldsymbol{y}} - \tilde{\boldsymbol{X}}\boldsymbol{\beta}\right) \text{ subject to } \boldsymbol{\beta}^\top\boldsymbol{\beta} \leq \phi,$$

where $\tilde{\boldsymbol{y}} = (\tilde{y}_1, \tilde{y}_2, ..., \tilde{y}_n)^\top$, $\tilde{\boldsymbol{X}} = (\tilde{\boldsymbol{x}}_1\tilde{\boldsymbol{x}}_2 \ldots \tilde{\boldsymbol{x}}_n)^\top$, $\tilde{y}_i = y_i - \sum_{j=1}^{n} W_{nj}(t_i) y_j$ and $\tilde{\boldsymbol{x}}_i = \boldsymbol{x}_i - \sum_{j=1}^{n} W_{nj}(t_i)\boldsymbol{x}_j$ for $i = 1, 2, ..., n$.

The ridge estimator of $\boldsymbol{\beta}$ based on full model is given as follows:

$$\widehat{\boldsymbol{\beta}}^{\text{FM}} = \left(\tilde{\boldsymbol{X}}^\top\tilde{\boldsymbol{X}} + k\boldsymbol{I}_p\right)^{-1} \tilde{\boldsymbol{X}}^\top\tilde{\mathbf{y}},$$

where $k \geq 0$ is the tuning parameter. As a special case for $k = 1$, we get the LSE of the $\boldsymbol{\beta}$.

However, we are interested in the estimation of $\boldsymbol{\beta}_1$, the estimation of strong signals in the model. Let $\widehat{\boldsymbol{\beta}}_1^{\text{FM}}$ be the unrestricted or full model ridge estimator of $\boldsymbol{\beta}_1$, which is given by

$$\widehat{\boldsymbol{\beta}}_1^{\text{FM}} = \left(\tilde{\boldsymbol{X}}_1^\top\tilde{\boldsymbol{M}}_2(k)\tilde{\boldsymbol{X}}_1 + k\boldsymbol{I}_{p_1}\right)^{-1} \tilde{\boldsymbol{X}}_1^\top\tilde{\boldsymbol{M}}_2(k)\tilde{\boldsymbol{y}},$$

where $\tilde{\boldsymbol{M}}_2(k) = \boldsymbol{I}_n - \tilde{\boldsymbol{X}}_2\left(\tilde{\boldsymbol{X}}_2^\top\tilde{\boldsymbol{X}}_2 + k\boldsymbol{I}_{p_2}\right)^{-1}\tilde{\boldsymbol{X}}_2^\top$. For $k = 1$, we can obtain the LSE of the $\boldsymbol{\beta}$ in the full model.

Now, assuming the model is sparse, such that $\boldsymbol{\beta}_2 = \mathbf{0}$, then we have the following partial linear submodel:

$$\boldsymbol{y} = \boldsymbol{X}_1\boldsymbol{\beta}_1 + \boldsymbol{f} + \boldsymbol{\varepsilon} \quad \text{subject to } \boldsymbol{\beta}_1^\top\boldsymbol{\beta}_1 \leq \phi^2. \tag{5.6}$$

Let us denote $\widehat{\boldsymbol{\beta}}_1^{\text{SM}}$ as the submodel or the restricted ridge estimator of $\boldsymbol{\beta}_1$, then

$$\widehat{\boldsymbol{\beta}}_1^{\text{SM}} = \left(\tilde{\boldsymbol{X}}_1^\top\tilde{\boldsymbol{X}}_1 + k\boldsymbol{I}_{p_1}\right)^{-1} \tilde{\boldsymbol{X}}_1^\top\tilde{\boldsymbol{y}}.$$

Intuitively, $\widehat{\boldsymbol{\beta}}_1^{\text{SM}}$ will perform better than $\widehat{\boldsymbol{\beta}}_1^{\text{FM}}$ in terms of MSE when $\boldsymbol{\beta}_2$ is close to $\mathbf{0}$. However, when $\boldsymbol{\beta}_2$ is far away from $\mathbf{0}$, $\widehat{\boldsymbol{\beta}}_1^{\text{SM}}$ can be inefficient. We suggest shrinkage estimation in order to improve the performance of the submodel estimator.

The shrinkage ridge estimator $\widehat{\boldsymbol{\beta}}_1^{\text{S}}$ of $\boldsymbol{\beta}_1$ is defined by

$$\widehat{\boldsymbol{\beta}}_1^{\text{S}} = \widehat{\boldsymbol{\beta}}_1^{\text{SM}} + \left(\widehat{\boldsymbol{\beta}}_1^{\text{FM}} - \widehat{\boldsymbol{\beta}}_1^{\text{SM}}\right)\left(1 - (p_2 - 2)\mathcal{T}_n^{-1}\right), p_2 \geq 3,$$

where $\mathcal{T}_n$ is a standardized distance measure defined as:

$$\mathcal{T}_n = \frac{1}{\widehat{\sigma}^2}(\widehat{\boldsymbol{\beta}}_2^{\text{LSE}})^\top\left(\tilde{\boldsymbol{X}}_2^\top\mathbf{M}_1\tilde{\boldsymbol{X}}_2\right)\widehat{\boldsymbol{\beta}}_2^{\text{LSE}},$$

where

$$\widehat{\sigma}^2 = \frac{1}{n-p}\left(\tilde{\boldsymbol{y}} - \tilde{\boldsymbol{X}}_1\widehat{\boldsymbol{\beta}}_1^{\text{SM}}\right)^\top\left(\tilde{\boldsymbol{y}} - \tilde{\boldsymbol{X}}_1\widehat{\boldsymbol{\beta}}_1^{\text{SM}}\right)$$

and

$$\widehat{\boldsymbol{\beta}}_2^{\mathrm{LSE}} = \left(\tilde{\boldsymbol{X}}_2^\top \tilde{\mathbf{M}}_1 \tilde{\boldsymbol{X}}_2\right)^{-1} \tilde{\boldsymbol{X}}_2^\top \tilde{\mathbf{M}}_1 \tilde{\boldsymbol{y}},$$

with $\tilde{\mathbf{M}}_1 = \boldsymbol{I}_n - \tilde{\boldsymbol{X}}_1 \left(\tilde{\boldsymbol{X}}_1^\top \tilde{\boldsymbol{X}}_1\right)^{-1} \tilde{\boldsymbol{X}}_1^\top$.

By design of the shrinkage estimator, it is possible that it may have a different sign from the full model estimator due to an over-shrinking problem. As a remedy, we consider the positive part of the shrinkage ridge estimator $\widehat{\boldsymbol{\beta}}_1^{\mathrm{PS}}$ of $\boldsymbol{\beta}_1$ defined by

$$\widehat{\boldsymbol{\beta}}_1^{\mathrm{PS}} = \widehat{\boldsymbol{\beta}}_1^{\mathrm{SM}} + \left(\widehat{\boldsymbol{\beta}}_1^{\mathrm{FM}} - \widehat{\boldsymbol{\beta}}_1^{\mathrm{SM}}\right)\left(1 - (p_2 - 2)\mathcal{T}_n^{-1}\right)^+,$$

where $z^+ = max\,(0, z)$.

In the following section we provide some important large-sample properties of the estimators.

5.3 Asymptotic Properties

In this section, we define expressions for asymptotic distributional bias (ADB), asymptotic covariance matrices, and asymptotic distributional risk (ADR) of the listed estimators. For this purpose, we consider a sequence $\{K_n\}$ is given by

$$K_n : \boldsymbol{\beta}_2 = \boldsymbol{\beta}_{2(n)} = \frac{\boldsymbol{w}}{\sqrt{n}}, \quad \boldsymbol{w} = (w_1, \ldots, w_{p_2})^\top \in \mathbb{R}^{p_2}.$$

Now, we define a quadratic loss function using a positive definite matrix (p.d.m) $\mathbf{W}$, by

$$L\left(\boldsymbol{\beta}_1^*\right) = n\left(\boldsymbol{\beta}_1^* - \boldsymbol{\beta}_1\right)^\top \mathbf{W}\left(\boldsymbol{\beta}_1^* - \boldsymbol{\beta}_1\right),$$

where $\boldsymbol{\beta}_1^*$ is anyone of suggested estimators. Under $\{K_n\}$, we can write the asymptotic distribution function of $\boldsymbol{\beta}_1^*$ as

$$F(\boldsymbol{x}) = \lim_{n\to\infty} P\left(\sqrt{n}\left(\boldsymbol{\beta}_1^* - \boldsymbol{\beta}_1\right) \le \boldsymbol{x} | K_n\right),$$

where $F(\boldsymbol{x})$ is non-degenerate. Then the ADR of $\boldsymbol{\beta}_1^*$ is defined as:

$$\mathrm{ADR}\left(\boldsymbol{\beta}_1^*\right) = \mathrm{tr}\left(\mathbf{W}\int_{\mathbb{R}^{p_1}} \int \boldsymbol{x}\boldsymbol{x}^\top dF(\boldsymbol{x})\right) = \mathrm{tr}\left(\mathbf{W}\mathbf{V}\right),$$

where $\mathbf{V}$ is the dispersion matrix for the distribution $F(\boldsymbol{x})$.

Asymptotic distributional bias of an estimator $\boldsymbol{\beta}_1^*$ is defined as

$$\mathrm{ADB}\left(\boldsymbol{\beta}_1^*\right) = \mathrm{E}\left\{\lim_{n\to\infty} \sqrt{n}\left(\boldsymbol{\beta}_1^* - \boldsymbol{\beta}_1\right)\right\}.$$

We consider the following two regularity conditions:

(i) $\frac{1}{n} \max_{1\le i\le n} \tilde{\boldsymbol{x}}_i^\top (\tilde{\boldsymbol{X}}^\top \tilde{\boldsymbol{X}})^{-1} \tilde{\boldsymbol{x}}_i \to 0$ as $n \to \infty$, where $\tilde{\boldsymbol{x}}_i^\top$ is the ith row of $\tilde{\boldsymbol{X}}$,

(ii) $\frac{1}{n}\sum_{i=1}^n \tilde{\boldsymbol{X}}^\top \tilde{\boldsymbol{X}} \to \tilde{\boldsymbol{Q}}$, where $\tilde{\boldsymbol{Q}}$ is a finite positive-definite matrix.

Theorem 5.1 *Under assumed regularity conditions and $\{K_n\}$, the ADBs of the estimators are:*

$$\begin{aligned}
\mathrm{ADB}\left(\widehat{\boldsymbol{\beta}}_1^{\mathrm{FM}}\right) &= -\boldsymbol{\eta}_{11.2},\\
\mathrm{ADB}\left(\widehat{\boldsymbol{\beta}}_1^{\mathrm{SM}}\right) &= -\boldsymbol{\xi},\\
\mathrm{ADB}\left(\widehat{\boldsymbol{\beta}}_1^{\mathrm{S}}\right) &= -\boldsymbol{\eta}_{11.2} - (p_2-2)\boldsymbol{\delta}\mathrm{E}\left(\chi_{p_2+2}^{-2}(\Delta)\right),\\
\mathrm{ADB}\left(\widehat{\boldsymbol{\beta}}_1^{\mathrm{PS}}\right) &= -\boldsymbol{\eta}_{11.2} - \boldsymbol{\delta}\mathrm{H}_{p_2+2}\left(\chi_{p_2,\alpha}^{2};\Delta\right),\\
&\quad -(p_2-2)\boldsymbol{\delta}\mathrm{E}\left\{\chi_{p_2+2}^{-2}(\Delta)\mathrm{I}\left(\chi_{p_2+2}^{2}(\Delta) > p_2-2\right)\right\},
\end{aligned}$$

where $\tilde{\boldsymbol{Q}} = \begin{pmatrix} \tilde{\boldsymbol{Q}}_{11} & \tilde{\boldsymbol{Q}}_{12} \\ \tilde{\boldsymbol{Q}}_{21} & \tilde{\boldsymbol{Q}}_{22} \end{pmatrix}$, $\Delta = \left(\boldsymbol{w}^{\intercal}\tilde{\boldsymbol{Q}}_{22.1}^{-1}\boldsymbol{w}\right)\sigma^{-2}$, $\tilde{\boldsymbol{Q}}_{22.1} = \tilde{\boldsymbol{Q}}_{22} - \tilde{\boldsymbol{Q}}_{21}\tilde{\boldsymbol{Q}}_{11}^{-1}\tilde{\boldsymbol{Q}}_{12}$, $\boldsymbol{\eta} = \begin{pmatrix} \boldsymbol{\eta}_{11.2} \\ \boldsymbol{\eta}_{22.1} \end{pmatrix} = -\lambda_0\tilde{\boldsymbol{Q}}^{-1}\boldsymbol{\beta} = \begin{pmatrix} -\lambda_0\tilde{\boldsymbol{Q}}_{11.2}^{-1}\boldsymbol{\beta}_1 + \frac{\lambda_0\boldsymbol{\omega}}{\sqrt{n}}\tilde{\boldsymbol{Q}}_{11}^{-1}\tilde{\boldsymbol{Q}}_{12}\tilde{\boldsymbol{Q}}_{22}^{-1} \\ \lambda_0\tilde{\boldsymbol{Q}}_{22}^{-1}\tilde{\boldsymbol{Q}}_{21}\tilde{\boldsymbol{Q}}_{11.2}^{-1}\boldsymbol{\beta}_1 - \frac{\lambda_0\boldsymbol{\omega}}{\sqrt{n}}\tilde{\boldsymbol{Q}}_{22}^{-1} \end{pmatrix}$, $\boldsymbol{\xi} = \boldsymbol{\eta}_{11.2} - \boldsymbol{\delta}$, $\boldsymbol{\delta} = \tilde{\boldsymbol{Q}}_{11}^{-1}\tilde{\boldsymbol{Q}}_{12}\boldsymbol{\omega}$ *and* $\mathrm{H}_v(x,\Delta)$ *be the cumulative distribution function of the non-central chi-squared distribution with non-centrality parameter* Δ, v *degrees of freedom, and*

$$\mathrm{E}\left(\chi_v^{-2j}(\Delta)\right) = \int_0^\infty x^{-2j}d\mathrm{H}_v(x,\Delta).$$

Proof *See Appendix.*

Since the bias expressions for all estimators are not in scalar form, we convert them to quadratic forms. We define the quadratic asymptotic distributional bias (QADB) of an estimator $\boldsymbol{\beta}_1^*$ as:

$$\mathrm{QADB}(\boldsymbol{\beta}_1^*) = \left(\mathrm{ADB}(\boldsymbol{\beta}_1^*)\right)^\top \tilde{\boldsymbol{Q}}_{11.2}\left(\mathrm{ADB}(\boldsymbol{\beta}_1^*)\right), \tag{5.7}$$

where $\tilde{\boldsymbol{Q}}_{11.2} = \tilde{\boldsymbol{Q}}_{11} - \tilde{\boldsymbol{Q}}_{12}\tilde{\boldsymbol{Q}}_{22}^{-1}\tilde{\boldsymbol{Q}}_{21}$. Thus,

$$\begin{aligned}
\mathrm{QADB}\left(\widehat{\boldsymbol{\beta}}_1^{\mathrm{FM}}\right) &= \boldsymbol{\eta}_{11.2}^\top\tilde{\boldsymbol{Q}}_{11.2}\boldsymbol{\eta}_{11.2},\\
\mathrm{QADB}\left(\widehat{\boldsymbol{\beta}}_1^{\mathrm{SM}}\right) &= \boldsymbol{\xi}^\top\tilde{\boldsymbol{Q}}_{11.2}\boldsymbol{\xi},\\
\mathrm{QADB}\left(\widehat{\boldsymbol{\beta}}_1^{\mathrm{S}}\right) &= \boldsymbol{\eta}_{11.2}^\top\tilde{\boldsymbol{Q}}_{11.2}\boldsymbol{\eta}_{11.2} + (p_2-2)\boldsymbol{\eta}_{11.2}^\top\tilde{\boldsymbol{Q}}_{11.2}\boldsymbol{\delta}\mathrm{E}\left(\chi_{p_2+2}^{-2}(\Delta)\right)\\
&\quad +(p_2-2)\boldsymbol{\delta}^\top\tilde{\boldsymbol{Q}}_{11.2}\boldsymbol{\eta}_{11.2}\mathrm{E}\left(\chi_{p_2+2}^{-2}(\Delta)\right)\\
&\quad +(p_2-2)^2\boldsymbol{\delta}^\top\tilde{\boldsymbol{Q}}_{11.2}\boldsymbol{\delta}\left(\mathrm{E}\left(\chi_{p_2+2}^{-2}(\Delta)\right)\right)^2,\\
\mathrm{QADB}\left(\widehat{\boldsymbol{\beta}}_1^{\mathrm{PS}}\right) &= \boldsymbol{\eta}_{11.2}^\top\tilde{\boldsymbol{Q}}_{11.2}\boldsymbol{\eta}_{11.2} + \left(\boldsymbol{\delta}^\top\tilde{\boldsymbol{Q}}_{11.2}\boldsymbol{\eta}_{11.2} + \boldsymbol{\eta}_{11.2}^\top\tilde{\boldsymbol{Q}}_{11.2}\boldsymbol{\delta}\right)\\
&\quad \cdot\left[\mathrm{H}_{p_2+2}\left((p_2-2);\Delta\right)\right.\\
&\quad \left.+(p_2-2)\mathrm{E}\left\{\chi_{p_2+2}^{-2}(\Delta)\mathrm{I}\left(\chi_{p_2+2}^{-2}(\Delta) > p_2-2\right)\right\}\right]\\
&\quad +\boldsymbol{\delta}^\top\tilde{\boldsymbol{Q}}_{11.2}\boldsymbol{\delta}\left[\mathrm{H}_{p_2+2}\left((p_2-2);\Delta\right)\right.\\
&\quad \left.+(p_2-2)\mathrm{E}\left\{\chi_{p_2+2}^{-2}(\Delta)\mathrm{I}\left(\chi_{p_2+2}^{-2}(\Delta) > p_2-2\right)\right\}\right]^2.
\end{aligned}$$

Quadratic Bias and Analysis:

Assuming that $\tilde{\boldsymbol{Q}}_{12} \neq \mathbf{0}$, in the following discussion.

(i) The QADB of $\widehat{\boldsymbol{\beta}}_1^{\mathrm{FM}}$ is an constant with $\boldsymbol{\eta}_{11.2}^\top\tilde{\boldsymbol{Q}}_{11.2}\boldsymbol{\eta}_{11.2}$ and independent of $\boldsymbol{\xi}^\top\tilde{\boldsymbol{Q}}_{11.2}\boldsymbol{\xi}$.

(ii) The QADB of $\widehat{\boldsymbol{\beta}}_1^{\mathrm{SM}}$ is an unbounded function of $\boldsymbol{\xi}^\top\tilde{\boldsymbol{Q}}_{11.2}\boldsymbol{\xi}$.

(iii) The QADB of $\widehat{\boldsymbol{\beta}}_1^{\mathrm{S}}$ starts from $\boldsymbol{\eta}_{11.2}^\top\tilde{\boldsymbol{Q}}_{11.2}\boldsymbol{\eta}_{11.2}$ at $\Delta = 0$, and it increases to a point then decreases toward zero for non-zero Δ values. This is due to the impact of $\mathrm{E}\left(\chi_{p_2+2}^{-2}(\Delta)\right)$ being a non-increasing log convex function of Δ. Lastly, for all Δ values, the shrinkage strategy can be viewed as a bias reducing and controlling technique.

(iv) The performance of the QADB of $\widehat{\boldsymbol{\beta}}_1^{\mathrm{PS}}$ is almost the same as that of $\widehat{\boldsymbol{\beta}}_1^{\mathrm{S}}$. However, the quadratic bias curve of $\widehat{\boldsymbol{\beta}}_1^{\mathrm{PS}}$ remains below the curve of $\widehat{\boldsymbol{\beta}}_1^{\mathrm{S}}$ in the entire parameter space induced by Δ.

We now present the asymptotic covariance matrices of the proposed estimators which are given in the following theorem.

Theorem 5.2 *Under assumed regularity conditions and $\{K_n\}$, asymptotic covariance matrices of the estimators are:*

$$
\begin{aligned}
\mathrm{Cov}\left(\widehat{\boldsymbol{\beta}}_1^{\mathrm{FM}}\right) &= \sigma^2\tilde{\boldsymbol{Q}}_{11.2}^{-1} + \boldsymbol{\eta}_{11.2}\boldsymbol{\eta}_{11.2}^{\top},\\
\mathrm{Cov}\left(\widehat{\boldsymbol{\beta}}_1^{\mathrm{SM}}\right) &= \sigma^2\tilde{\boldsymbol{Q}}_{11}^{-1} + \boldsymbol{\xi}\boldsymbol{\xi}^{\top},\\
\mathrm{Cov}\left(\widehat{\boldsymbol{\beta}}_1^{\mathrm{S}}\right) &= \sigma^2\tilde{\boldsymbol{Q}}_{11.2}^{-1} + \boldsymbol{\eta}_{11.2}\boldsymbol{\eta}_{11.2}^{\top} + 2(p_2-2)\boldsymbol{\delta}\boldsymbol{\eta}_{11.2}^{\top}\mathrm{E}\left(\chi_{p_2+2}^{-2}(\Delta)\right)\\
&\quad -(p_2-2)\sigma^2\tilde{\boldsymbol{Q}}_{11}^{-1}\tilde{\boldsymbol{Q}}_{12}\tilde{\boldsymbol{Q}}_{22.1}^{-1}\tilde{\boldsymbol{Q}}_{21}\tilde{\boldsymbol{Q}}_{11}^{-1}\{2\mathrm{E}\left(\chi_{p_2+2}^{-2}(\Delta)\right)\\
&\quad -(p_2-2)\mathrm{E}\left(\chi_{p_2+2}^{-4}(\Delta)\right)\}\\
&\quad +(p_2-2)\boldsymbol{\delta}\boldsymbol{\delta}^{\top}\{2\mathrm{E}\left(\chi_{p_2+2}^{-2}(\Delta)\right)\\
&\quad -2\mathrm{E}\left(\chi_{p_2+4}^{-2}(\Delta)\right) - (p_2-2)\mathrm{E}\left(\chi_{p_2+4}^{-4}(\Delta)\right)\},\\
\mathrm{Cov}\left(\widehat{\boldsymbol{\beta}}_1^{\mathrm{PS}}\right) &= \mathrm{Cov}\left(\widehat{\boldsymbol{\beta}}_1^{\mathrm{S}}\right)\\
&\quad -2\boldsymbol{\delta}\boldsymbol{\eta}_{11.2}^{\top}\mathrm{E}\left(\{1-(p_2-2)\chi_{p_2+2}^{-2}(\Delta)\}\mathrm{I}\left(\chi_{p_2+2}^{2}(\Delta)\le p_2-2\right)\right)\\
&\quad +(p_2-2)\sigma^2\tilde{\boldsymbol{Q}}_{11}^{-1}\tilde{\boldsymbol{Q}}_{12}\tilde{\boldsymbol{Q}}_{22.1}^{-1}\tilde{\boldsymbol{Q}}_{21}\tilde{\boldsymbol{Q}}_{11}^{-1}\\
&\quad \cdot\left[2\mathrm{E}\left(\chi_{p_2+2}^{-2}(\Delta)\mathrm{I}\left(\chi_{p_2+2}^{2}(\Delta)\le p_2-2\right)\right)\right.\\
&\quad \left. -(p_2-2)\mathrm{E}\left(\chi_{p_2+2}^{-4}(\Delta)\mathrm{I}\left(\chi_{p_2+2}^{2}(\Delta)\le p_2-2\right)\right)\right]\\
&\quad -\sigma^2\tilde{\boldsymbol{Q}}_{11}^{-1}\tilde{\boldsymbol{Q}}_{12}\tilde{\boldsymbol{Q}}_{22.1}^{-1}\tilde{\boldsymbol{Q}}_{21}\tilde{\boldsymbol{Q}}_{11}^{-1}\mathrm{H}_{p_2+2}\left((p_2-2);\Delta\right)\\
&\quad +\boldsymbol{\delta}\boldsymbol{\delta}^{\top}\left[2\mathrm{H}_{p_2+2}\left((p_2-2);\Delta\right) - \mathrm{H}_{p_2+4}\left((p_2-2);\Delta\right)\right]\\
&\quad -(p_2-2)\boldsymbol{\delta}\boldsymbol{\delta}^{\top}\left[2\mathrm{E}\left(\chi_{p_2+2}^{-2}(\Delta)\mathrm{I}\left(\chi_{p_2+2}^{2}(\Delta)\le p_2-2\right)\right)\right.\\
&\quad -2\mathrm{E}\left(\chi_{p_2+4}^{-2}(\Delta)\mathrm{I}\left(\chi_{p_2+4}^{2}(\Delta)\le p_2-2\right)\right)\\
&\quad \left. +(p_2-2)\mathrm{E}\left(\chi_{p_2+2}^{-4}(\Delta)\mathrm{I}\left(\chi_{p_2+2}^{2}(\Delta)\le p_2-2\right)\right)\right].
\end{aligned}
$$

Proof *See Appendix.*

Finally, we obtain the ADRs of the estimators under $\{K_n\}$ given as:

Theorem 5.3

$$
\begin{aligned}
\mathrm{ADR}\left(\widehat{\boldsymbol{\beta}}_1^{\mathrm{FM}}\right) &= \sigma^2\mathrm{tr}\left(\boldsymbol{W}\tilde{\boldsymbol{Q}}_{11.2}^{-1}\right) + \boldsymbol{\eta}_{11.2}^{\top}\boldsymbol{W}\boldsymbol{\eta}_{11.2},\\
\mathrm{ADR}\left(\widehat{\boldsymbol{\beta}}_1^{\mathrm{SM}}\right) &= \sigma^2\mathrm{tr}\left(\boldsymbol{W}\tilde{\boldsymbol{Q}}_{11}^{-1}\right) + \boldsymbol{\xi}^{\top}\boldsymbol{W}\boldsymbol{\xi},\\
\mathrm{ADR}\left(\widehat{\boldsymbol{\beta}}_1^{\mathrm{S}}\right) &= \mathrm{ADR}\left(\widehat{\boldsymbol{\beta}}_1^{\mathrm{FM}}\right) + 2(p_2-2)\boldsymbol{\eta}_{11.2}^{\top}\boldsymbol{W}\boldsymbol{\delta}\mathrm{E}\left(\chi_{p_2+2}^{-2}(\Delta)\right)\\
&\quad -(p_2-2)\sigma^2\mathrm{tr}\left(\tilde{\boldsymbol{Q}}_{21}\tilde{\boldsymbol{Q}}_{11}^{-1}\boldsymbol{W}\tilde{\boldsymbol{Q}}_{11}^{-1}\tilde{\boldsymbol{Q}}_{12}\tilde{\boldsymbol{Q}}_{22.1}^{-1}\right)\{2\mathrm{E}\left(\chi_{p_2+2}^{-2}(\Delta)\right)\\
&\quad -(p_2-2)\mathrm{E}\left(\chi_{p_2+2}^{-4}(\Delta)\right)\}\\
&\quad +(p_2-2)\boldsymbol{\delta}^{\top}\boldsymbol{W}\boldsymbol{\delta}\{2\mathrm{E}\left(\chi_{p_2+2}^{-2}(\Delta)\right)\\
&\quad -2\mathrm{E}\left(\chi_{p_2+4}^{-2}(\Delta)\right) - (p_2-2)\mathrm{E}\left(\chi_{p_2+4}^{-4}(\Delta)\right)\},\\
\mathrm{ADR}\left(\widehat{\boldsymbol{\beta}}_1^{\mathrm{PS}}\right) &= \mathrm{ADR}\left(\widehat{\boldsymbol{\beta}}_1^{\mathrm{S}}\right)\\
&\quad -2\boldsymbol{\eta}_{11.2}^{\top}\boldsymbol{W}\boldsymbol{\delta}\mathrm{E}\\
&\quad \times\left(\{1-(p_2-2)\chi_{p_2+2}^{-2}(\Delta)\}\mathrm{I}\left(\chi_{p_2+2}^{2}(\Delta)\le p_2-2\right)\right)
\end{aligned}
$$

$$
\begin{aligned}
&+(p_2-2)\sigma^2 \mathrm{tr}\left(\tilde{\boldsymbol{Q}}_{21}\tilde{\boldsymbol{Q}}_{11}^{-1}\boldsymbol{W}\tilde{\boldsymbol{Q}}_{11}^{-1}\tilde{\boldsymbol{Q}}_{12}\tilde{\boldsymbol{Q}}_{22.1}^{-1}\right) \\
&\cdot\left[2\mathrm{E}\left(\chi_{p_2+2}^{-2}(\Delta)\mathrm{I}\left(\chi_{p_2+2}^{2}(\Delta)\le p_2-2\right)\right)\right. \\
&\left.-(p_2-2)\mathrm{E}\left(\chi_{p_2+2}^{-4}(\Delta)\mathrm{I}\left(\chi_{p_2+2}^{2}(\Delta)\le p_2-2\right)\right)\right] \\
&-\sigma^2 \mathrm{tr}\left(\tilde{\boldsymbol{Q}}_{21}\tilde{\boldsymbol{Q}}_{11}^{-1}\boldsymbol{W}\tilde{\boldsymbol{Q}}_{11}^{-1}\tilde{\boldsymbol{Q}}_{12}\tilde{\boldsymbol{Q}}_{22.1}^{-1}\right)\mathrm{H}_{p_2+2}\left((p_2-2);\Delta\right) \\
&+\boldsymbol{\delta}^\top\boldsymbol{W}\boldsymbol{\delta}\left[2\mathrm{H}_{p_2+2}\left((p_2-2);\Delta\right)-\mathrm{H}_{p_2+4}\left((p_2-2);\Delta\right)\right] \\
&-(p_2-2)\boldsymbol{\delta}^\top\boldsymbol{W}\boldsymbol{\delta}\left[2\mathrm{E}\left(\chi_{p_2+2}^{-2}(\Delta)\mathrm{I}\left(\chi_{p_2+2}^{2}(\Delta)\le p_2-2\right)\right)\right. \\
&-2\mathrm{E}\left(\chi_{p_2+4}^{-2}(\Delta)\mathrm{I}\left(\chi_{p_2+4}^{2}(\Delta)\le p_2-2\right)\right) \\
&\left.+(p_2-2)\mathrm{E}\left(\chi_{p_2+2}^{-4}(\Delta)\mathrm{I}\left(\chi_{p_2+2}^{2}(\Delta)\le p_2-2\right)\right)\right].
\end{aligned}
$$

Proof *See Appendix.*

If $\tilde{\boldsymbol{Q}}_{12}=\mathbf{0}$, then $\boldsymbol{\delta}=\mathbf{0}$, $\boldsymbol{\xi}=\boldsymbol{\eta}_{11.2}$ and $\tilde{\boldsymbol{Q}}_{11.2}=\tilde{\boldsymbol{Q}}_{11}$, all the ADRs reduce to a common value $\sigma^2\mathrm{tr}\left(\mathbf{W}\tilde{\boldsymbol{Q}}_{11}^{-1}\right)+\boldsymbol{\eta}_{11.2}^\top\mathbf{W}\boldsymbol{\eta}_{11.2}$ for all $\boldsymbol{\omega}$ and nothing to compare. Next we consider $\tilde{\boldsymbol{Q}}_{12}\neq\mathbf{0}$ and summarize our findings as follows.

(i) The ADR of the full model estimator is constant since it does not depend on the sparsity condition, $\Delta=0$.

(ii) The ADR of the submodel estimator is the smallest compared to the listed estimators for smaller values of $\Delta=0$. Conversely, as Δ moves away from 0 the ADR $\left(\widehat{\boldsymbol{\beta}}_1^{\mathrm{SM}}\right)$ increases without bounds.

(iii) When $\Delta=0$, the ADR of the both shrinkage estimators are smaller than the ADR of the full model estimator. Further, it can be seen that ADR $\left(\widehat{\boldsymbol{\beta}}_1^{\mathrm{PS}}\right)\le$ ADR $\left(\widehat{\boldsymbol{\beta}}_1^{\mathrm{S}}\right)$.

(iv) Consider the situation when $\Delta\neq 0$, it can be shown that for all $\mathbf{W}$ and $\boldsymbol{\delta}$ ADR $\left(\widehat{\boldsymbol{\beta}}_1^{\mathrm{S}}\right)\le$ ADR $\left(\widehat{\boldsymbol{\beta}}_1^{\mathrm{FM}}\right)$, if

$$
\frac{\mathrm{tr}\left(\tilde{\boldsymbol{Q}}_{21}\tilde{\boldsymbol{Q}}_{11}^{-1}\mathbf{W}\tilde{\boldsymbol{Q}}_{11}^{-1}\tilde{\boldsymbol{Q}}_{12}\tilde{\boldsymbol{Q}}_{22.1}^{-1}\right)}{ch_{max}\left(\tilde{\boldsymbol{Q}}_{21}\tilde{\boldsymbol{Q}}_{11}^{-1}\mathbf{W}\tilde{\boldsymbol{Q}}_{11}^{-1}\tilde{\boldsymbol{Q}}_{12}\tilde{\boldsymbol{Q}}_{22.1}^{-1}\right)}\ge\frac{p_2+2}{2},
$$

where $ch_{max}(\cdot)$ is the maximum characteristic root.

(v) Comparing $\widehat{\boldsymbol{\beta}}_1^{\mathrm{PS}}$ and $\widehat{\boldsymbol{\beta}}_1^{\mathrm{S}}$, it is observed that the ADR $\left(\widehat{\boldsymbol{\beta}}_1^{\mathrm{PS}}\right)$ is equal to or less than the ADR $\left(\widehat{\boldsymbol{\beta}}_1^{\mathrm{S}}\right)$ for all the values of Δ, and a strict inequality holds at $\Delta=0$. Thus, positive part shrinkage estimator dominates the usual shrinkage estimator in the entire parameter space generated by Δ.

(vi) Finally, we establish that ADR $\left(\widehat{\boldsymbol{\beta}}_1^{\mathrm{PS}}\right)\le$ ADR $\left(\widehat{\boldsymbol{\beta}}_1^{\mathrm{S}}\right)\le$ ADR $\left(\widehat{\boldsymbol{\beta}}_1^{\mathrm{FM}}\right)$ for all $\mathbf{W}$ and $||\boldsymbol{\delta}||$, with a strict inequality for some $||\boldsymbol{\delta}||$ close to zero. Thus, both shrinkage estimators outshine the full model estimator in estimating the parameter vector $\boldsymbol{\beta}_1$.

It is important to note that the performance of a submodel estimator heavily depends on the correctness of the sparsity condition. However, one is seldom sure of the reliability of this information. In an effort to resolve this issue, shrinkage estimators are used to increase the precision of the submodel estimators.

5.4 Simulation Experiments

Consider a Monte Carlo simulation to evaluate the relative quadratic risk performance of the listed estimators. The main purpose of this simulation is to examine the quality of statistical estimation based on a large-sample methodology in moderate sample situations, with varying levels of sparsity and degrees of collinearity in the model.

We simulate the response from the following model:

$$y_i = x_{1i}\beta_1 + x_{2i}\beta_2 + ... + x_{pi}\beta_p + f(t_i) + \varepsilon_i \ , \ i = 1, \ldots, n. \tag{5.8}$$

In an effort to inject different degrees of collinearity among the predictors, we employ the following equation

$$x_{ij} = (1-\gamma^2)^{1/2} z_{ij} + \gamma z_{ip}$$

where z_{ij} are random numbers following a standard normal distribution such that $i = 1, 2, \ldots n$, $j = 1, 2, \ldots p$, where n is the sample size and p is the number of regressors Gibbons (1981). Three degrees of correlation γ are compared: $0.3, 0.6$ and 0.9.

Furthermore, we consider $\beta_j = 0$ for $j = p_1 + 1, p_1 + 2, ..., p$, with $p = p_1 + p_2$. Hence, we can partition the regression coefficients as $\boldsymbol{\beta} = (\boldsymbol{\beta}_1, \boldsymbol{\beta}_2) = (\boldsymbol{\beta}_1, \mathbf{0})$ with $\boldsymbol{\beta}_1 = (2, 0.5, -1, 3)$.

In (5.8), the non-parametric function $f(t_i) = \sqrt{t_i(1-t_i)} \sin\left(\frac{2.1\pi}{t_i + 0.05}\right)$, called the Doppler function for $t_i = (i - 0.5)/n$, is used to generate the variable of interest y_i.

Regarding the non-parametric component, there are a number of methods available for bandwidth selection for a PLM in reviewed literature, we refer to Aydın et al. (2016); Yilmaz et al. (2021). Here we use generalized cross-validation (GCV) to select the optimal value of k and h_n values. The GCV score function of kernel smoothing can be procured by

$$\text{GCV}(k, h_n) = \frac{n\left\|\boldsymbol{y} - \widehat{\boldsymbol{y}}\right\|^2}{\{\text{tr}(\boldsymbol{I}_n - \mathbf{H}_{h_n})\}^2},$$

where $\boldsymbol{H} = \tilde{\boldsymbol{X}}\left(\tilde{\boldsymbol{X}}^\top\tilde{\boldsymbol{X}} + k\boldsymbol{I}_p\right)^{-1}\tilde{\boldsymbol{X}}^\top$ and $\boldsymbol{H}_{h_n} = \boldsymbol{W} + (\boldsymbol{I}_n - \boldsymbol{W})\boldsymbol{H}$. We use the following weight function.

$$W_{ni}(t) = \frac{K\left(\frac{t-t_i}{h_n}\right)}{\sum_i K\left(\frac{t-t_i}{h_n}\right)},$$

where $K(u) = \frac{1}{\sqrt{2\pi}}\exp\left\{-\frac{1}{2}u^2\right\}$.

One thousand simulations for each set of observations were determined to be adequate. Initially the number of simulations were varied, but it was observed that a further increase in the number did not significantly change the result.

We define $\Delta = \|\boldsymbol{\beta} - \boldsymbol{\beta}_0\|$, where $\boldsymbol{\beta}_0 = (\boldsymbol{\beta}_1, \mathbf{0})$, and $\|\cdot\|$ is the Euclidean norm. In an effort to investigate the behavior of the estimators for $\Delta > 0$, more data sets were generated from distributions based on the selected values Δ.

The performance of an estimator was evaluated by using MSE. For ease in comparison, we also calculate the relative mean squared efficiency (RMSE) of the $\widehat{\boldsymbol{\beta}}_1^*$ to the $\widehat{\boldsymbol{\beta}}_1^{\text{FM}}$ given by

$$\text{RMSE}\left(\widehat{\boldsymbol{\beta}}_1^*\right) = \frac{\text{MSE}\left(\widehat{\boldsymbol{\beta}}_1^{\text{FM}}\right)}{\text{MSE}\left(\widehat{\boldsymbol{\beta}}_1^*\right)},$$

where $\widehat{\boldsymbol{\beta}}_1^*$ is one of the listed estimators. It is important to note that a value of RMSE greater than 1 indicates the degree of superiority of the selected estimator to the full model estimator.

Tables 5.1 and 5.2 provide the RMSE for the listed estimates over the full model estimator for $n = 60, 120$ and $p_2 = 4, 8, 12$. It is apparent from these tables that the submodel estimator dominates the shrinkage estimators when the assumption of sparsity is nearly correct, i.e. values of the sparsity parameter Δ are small. Alternatively, when the value of Δ increases, the performance of the submodel estimator becomes worse making it a desirable strategy. On the other hand, the performance of shrinkage estimators are stable in such cases, i.e. it achieves a maximum RMSE at $\Delta = 0$ which monotonically decreases then tends to the MSE from the above to the full model estimator. More importantly, the shrinkage estimators are superior to the full model estimator for all values of Δ, and the strict inequality holds for some values of Δ near zero. Further, the positive part shrinkage estimator dominates the usual shrinkage estimator. In short, Tables 5.1 and 5.2 reveal that for Δ close to 0, all the proposed estimators are highly efficient relative to the full model estimator.

Finally, for all values of Δ the numerical-based performance of the estimators is similar to the asymptotic analysis provided in Section 5.3. It is also observed that the shrinkage estimators are relatively more efficient than the full model estimator as n and p_2 increase, as is consistent with the theory presented.

5.4.1 Comparing with Penalty Estimators

In this section, we compare the full model, submodel, and shrinkage estimators with four penalized estimators: ENET, LASSO, ALASSO and SCAD. In this simulation study, we consider the true values of the regression coefficients as $\boldsymbol{\beta} = (\boldsymbol{\beta}_1, \boldsymbol{\beta}_2) = (\boldsymbol{\beta}_1, \mathbf{0})$ with $\boldsymbol{\beta} = (\underbrace{1, \ldots, 1}_{p_1})^\top$. We calculate the RMSE and RPE to provide a demonstration of dominance of the listed estimators. The data is randomly split into two equal parts of observations to calculate these quantities. The first part is the training set and the other is the test data. The suggested estimators are obtained from the training set only.

In this simulation study where $\boldsymbol{\beta}_1 \neq 0$ and $\boldsymbol{\beta}_2 = 0$, the penalty estimation methods are anticipated to estimate $\boldsymbol{\beta}_1$ efficiently by adaptively choosing the tuning parameter λ and produce sparse solutions, where the components of $\boldsymbol{\beta}_2$ are set to zero as compared to full model ridge regression estimator. The shrinkage estimators are expected to perform well by adaptively placing more weights on the submodel ridge estimator. Here we compare these two types of procedures with respect to the full model ridge estimator.

In Tables 5.3–5.6, we give the simulated RMSE and RPE of the submodel, positive shrinkage, and penalty estimators with respect to the full model ridge estimator for $n = 60, 100$ when $p_1 = 3, 6$ – non-zero coefficients, and $p_2 = 3, 6, 9, 12$ – zero coefficients.

As expected, Tables 5.3–5.6 show that the submodel estimator performs best since it is assumed to be correct. The positive shrinkage estimator performs better than penalty estimators in almost all cases.

In Table 5.7, we provide RMSE and RPE for larger values of p_2. We found that as p_2 increased, the performance of the positive shrinkage was relatively better. Finally, in Table 5.8 we use the full model estimator based on LSE. However, it is suggested not to use LSE when a high level of multicollinearity is presented in the data.

TABLE 5.1: RMSE of the Estimator for $n = 60$ and $p_1 = 4$.

		$\gamma = 0.3$			$\gamma = 0.6$			$\gamma = 0.9$		
p_2	Δ	SM	S	PS	SM	S	PS	SM	S	PS
4	0.0	1.925	1.309	1.401	1.883	1.227	1.410	1.481	1.062	1.263
	0.5	0.853	1.042	1.042	1.026	1.078	1.078	1.220	0.874	1.121
	1.0	0.325	0.998	0.998	0.449	1.008	1.008	0.839	1.028	1.030
	1.5	0.164	0.988	0.988	0.244	0.995	0.995	0.586	1.011	1.011
	2.0	0.098	0.986	0.986	0.150	0.989	0.989	0.426	0.996	0.996
	2.5	0.069	0.987	0.987	0.109	0.989	0.989	0.338	0.993	0.993
	3.0	0.052	0.987	0.987	0.085	0.989	0.989	0.279	0.990	0.990
	3.5	0.042	0.988	0.988	0.070	0.989	0.989	0.245	0.990	0.990
	4.0	0.036	0.989	0.989	0.060	0.991	0.991	0.214	0.990	0.990
8	0.0	2.674	1.746	2.009	2.512	1.607	1.937	1.667	1.248	1.480
	0.5	1.169	1.247	1.249	1.362	1.305	1.313	1.423	1.188	1.327
	1.0	0.452	1.043	1.043	0.597	1.072	1.072	0.984	1.127	1.131
	1.5	0.227	0.995	0.995	0.320	1.011	1.011	0.679	1.048	1.048
	2.0	0.131	0.984	0.984	0.191	0.991	0.991	0.483	1.008	1.008
	2.5	0.096	0.983	0.983	0.143	0.988	0.988	0.380	0.993	0.993
	3.0	0.072	0.978	0.978	0.108	0.982	0.982	0.315	0.982	0.982
	3.5	0.053	0.977	0.977	0.083	0.980	0.980	0.262	0.978	0.978
	4.0	0.044	0.976	0.976	0.070	0.980	0.980	0.232	0.977	0.977
12	0.0	3.507	2.424	2.747	3.513	2.288	2.737	2.080	1.573	1.843
	0.5	1.622	1.539	1.540	1.897	1.664	1.676	1.761	1.518	1.608
	1.0	0.612	1.119	1.119	0.780	1.175	1.175	1.181	1.259	1.270
	1.5	0.299	1.031	1.031	0.399	1.055	1.055	0.777	1.121	1.121
	2.0	0.184	0.996	0.996	0.253	1.006	1.006	0.574	1.042	1.042
	2.5	0.126	0.984	0.984	0.179	0.988	0.988	0.451	1.005	1.005
	3.0	0.093	0.977	0.977	0.136	0.979	0.979	0.372	0.985	0.985
	3.5	0.075	0.976	0.976	0.113	0.977	0.977	0.321	0.977	0.977
	4.0	0.062	0.972	0.972	0.096	0.974	0.974	0.285	0.972	0.972

TABLE 5.2: RMSE of the Estimator for $n = 120$ and $p_1 = 4$.

		$\gamma = 0.3$			$\gamma = 0.6$			$\gamma = 0.9$		
p_2	Δ	SM	S	PS	SM	S	PS	SM	S	PS
4	0.0	1.809	1.158	1.374	2.131	1.049	1.489	1.532	0.864	1.292
	0.5	0.726	1.039	1.039	1.108	1.081	1.081	1.311	1.104	1.122
	1.0	0.248	1.007	1.007	0.434	1.022	1.022	0.893	1.048	1.048
	1.5	0.117	1.001	1.001	0.215	1.009	1.009	0.583	1.020	1.020
	2.0	0.066	0.998	0.998	0.127	1.003	1.003	0.401	1.007	1.007
	2.5	0.044	0.997	0.997	0.087	1.000	1.000	0.303	1.002	1.002
	3.0	0.034	0.996	0.996	0.067	0.999	0.999	0.242	0.999	0.999
	3.5	0.027	0.995	0.995	0.053	0.998	0.998	0.200	0.997	0.997
	4.0	0.022	0.995	0.995	0.044	0.998	0.998	0.170	0.996	0.996
8	0.0	2.262	1.722	1.857	2.278	1.656	1.876	1.606	1.276	1.466
	0.5	0.952	1.165	1.165	1.190	1.234	1.234	1.345	1.248	1.261
	1.0	0.322	1.039	1.039	0.462	1.064	1.064	0.879	1.112	1.112
	1.5	0.156	1.012	1.012	0.234	1.024	1.024	0.578	1.049	1.049
	2.0	0.090	1.000	1.000	0.141	1.007	1.007	0.403	1.018	1.018
	2.5	0.056	0.994	0.994	0.092	0.999	0.999	0.295	1.002	1.002
	3.0	0.043	0.991	0.991	0.071	0.995	0.995	0.236	0.994	0.994
	3.5	0.036	0.990	0.990	0.058	0.993	0.993	0.198	0.990	0.990
	4.0	0.030	0.989	0.989	0.049	0.991	0.991	0.171	0.988	0.988
12	0.0	2.873	2.189	2.348	3.198	2.250	2.517	2.074	1.631	1.835
	0.5	1.137	1.279	1.280	1.461	1.395	1.401	1.628	1.403	1.479
	1.0	0.362	1.072	1.072	0.497	1.120	1.120	0.981	1.220	1.220
	1.5	0.177	1.022	1.022	0.249	1.047	1.047	0.608	1.101	1.101
	2.0	0.104	1.003	1.003	0.149	1.018	1.018	0.410	1.044	1.044
	2.5	0.071	0.995	0.995	0.103	1.005	1.005	0.297	1.015	1.015
	3.0	0.050	0.989	0.989	0.075	0.996	0.996	0.228	0.997	0.997
	3.5	0.040	0.986	0.986	0.059	0.992	0.992	0.185	0.989	0.989
	4.0	0.034	0.985	0.985	0.051	0.989	0.989	0.162	0.985	0.985

TABLE 5.3: RMSE and RPE of the Estimators for $n = 60$ and $p_1 = 3$.

		$\gamma = 0.3$		$\gamma = 0.6$		$\gamma = 0.9$	
p_2	Method	RMSE	RPE	RMSE	RPE	RMSE	RPE
3	SM	2.272	1.111	2.812	1.111	3.856	1.088
	S	1.302	1.042	1.307	1.035	0.106	0.544
	PS	1.344	1.046	1.459	1.050	1.715	1.048
	ENET	1.049	1.002	1.003	0.993	0.746	0.955
	LASSO	1.074	1.009	1.033	0.997	0.743	0.958
	ALASSO	1.151	1.033	1.106	1.014	0.583	0.909
	SCAD	1.120	1.026	1.080	1.009	0.501	0.908
6	SM	2.896	1.242	2.983	1.242	3.927	1.181
	S	1.849	1.159	1.741	1.136	1.730	1.089
	PS	2.087	1.177	2.245	1.180	3.086	1.145
	ENET	1.161	1.032	0.956	1.024	0.548	0.974
	LASSO	1.281	1.056	1.125	1.054	0.590	0.990
	ALASSO	1.312	1.112	1.079	1.083	0.475	0.957
	SCAD	1.084	1.079	0.710	0.996	0.284	0.878
9	SM	4.311	1.394	4.363	1.357	6.124	1.237
	S	2.743	1.311	2.828	1.283	3.376	1.197
	PS	3.166	1.328	3.352	1.303	4.280	1.207
	ENET	1.469	1.119	1.287	1.084	0.863	1.011
	LASSO	1.761	1.159	1.494	1.124	0.824	1.008
	ALASSO	1.490	1.187	1.354	1.164	0.633	0.971
	SCAD	1.391	1.187	0.948	1.083	0.279	0.823
12	SM	5.442	1.560	5.683	1.482	7.140	1.288
	S	3.561	1.472	3.540	1.407	4.198	1.250
	PS	3.776	1.475	4.036	1.418	5.010	1.254
	ENET	1.395	1.148	1.067	1.090	0.502	0.965
	LASSO	1.746	1.231	1.470	1.171	0.711	1.017
	ALASSO	1.529	1.231	1.345	1.169	0.495	0.952
	SCAD	1.682	1.264	1.180	1.155	0.303	0.852

TABLE 5.4: RMSE and RPE of the Estimators for $n = 60$ and $p_1 = 6$.

		$\gamma = 0.3$		$\gamma = 0.6$		$\gamma = 0.9$	
p_2	Method	RMSE	RPE	RMSE	RPE	RMSE	RPE
3	SM	1.696	1.070	2.812	1.111	1.867	1.127
	S	1.233	1.022	1.307	1.035	0.523	0.864
	PS	1.241	1.028	1.459	1.050	1.402	1.070
	ENET	1.064	1.002	1.003	0.993	0.791	0.972
	LASSO	1.084	0.997	1.033	0.997	0.793	0.977
	ALASSO	1.242	1.021	1.106	1.014	0.718	0.958
	SCAD	1.370	1.040	1.080	1.009	0.569	0.948
6	SM	2.067	1.143	2.983	1.242	2.516	1.231
	S	1.503	1.071	1.741	1.136	1.803	1.180
	PS	1.667	1.101	2.245	1.180	2.018	1.188
	ENET	1.093	1.015	0.956	1.024	0.875	0.987
	LASSO	1.102	1.015	1.125	1.054	0.838	0.979
	ALASSO	1.345	1.054	1.079	1.083	0.829	1.002
	SCAD	1.461	1.068	0.710	0.996	0.564	0.879
9	SM	3.089	1.226	4.363	1.357	3.034	1.382
	S	2.128	1.153	2.828	1.283	2.258	1.307
	PS	2.447	1.184	3.352	1.303	2.432	1.320
	ENET	1.291	1.052	1.287	1.084	0.904	1.019
	LASSO	1.370	1.067	1.494	1.124	1.018	1.048
	ALASSO	1.576	1.087	1.354	1.164	0.896	1.037
	SCAD	2.008	1.149	0.948	1.083	0.448	0.842
12	SM	3.018	1.284	5.683	1.482	3.686	1.544
	S	2.333	1.221	3.540	1.407	2.879	1.458
	PS	2.489	1.248	4.036	1.418	3.064	1.472
	ENET	1.327	1.082	1.067	1.090	0.929	1.053
	LASSO	1.383	1.087	1.470	1.171	1.044	1.115
	ALASSO	1.692	1.124	1.345	1.169	0.958	1.128
	SCAD	2.308	1.211	1.180	1.155	0.554	0.881

TABLE 5.5: RMSE and RPE of the Estimators for $n = 100$ and $p_1 = 3$.

		$\gamma = 0.3$		$\gamma = 0.6$		$\gamma = 0.9$	
p_2	Method	RMSE	RPE	RMSE	RPE	RMSE	RPE
3	SM	2.318	1.059	3.574	1.061	5.670	1.055
	S	1.307	1.025	1.278	1.018	0.789	0.965
	PS	1.394	1.030	1.591	1.032	1.948	1.032
	ENET	1.110	1.008	1.156	1.008	0.900	0.995
	LASSO	1.200	1.012	1.346	1.015	0.928	0.998
	ALASSO	1.479	1.034	1.582	1.027	0.782	0.978
	SCAD	1.614	1.045	1.752	1.038	0.562	0.954
6	SM	2.981	1.138	4.101	1.126	7.364	1.098
	S	1.929	1.095	2.195	1.092	2.178	1.067
	PS	2.094	1.101	2.523	1.096	3.569	1.082
	ENET	1.236	1.023	1.267	1.027	1.092	1.012
	LASSO	1.398	1.040	1.475	1.035	1.092	1.014
	ALASSO	1.689	1.073	1.832	1.066	1.022	1.001
	SCAD	2.269	1.116	2.467	1.098	0.589	0.943
9	SM	3.728	1.198	5.048	1.181	8.718	1.131
	S	2.504	1.147	2.675	1.132	2.804	1.090
	PS	2.750	1.166	3.287	1.153	4.696	1.116
	ENET	1.397	1.058	1.428	1.053	1.108	1.017
	LASSO	1.518	1.074	1.597	1.068	1.150	1.027
	ALASSO	1.850	1.105	2.010	1.098	1.097	1.012
	SCAD	2.761	1.164	2.555	1.126	0.605	0.949
12	SM	5.474	1.274	6.507	1.245	9.983	1.179
	S	3.006	1.203	3.129	1.183	3.296	1.135
	PS	3.785	1.235	4.299	1.213	6.011	1.162
	ENET	1.825	1.117	1.703	1.101	1.079	1.036
	LASSO	1.954	1.126	1.857	1.111	1.094	1.037
	ALASSO	1.992	1.136	1.882	1.122	0.935	1.017
	SCAD	3.119	1.223	2.208	1.153	0.481	0.935

TABLE 5.6: RMSE and RPE of the Estimators for $n = 100$ and $p_1 = 6$.

		$\gamma = 0.3$		$\gamma = 0.6$		$\gamma = 0.9$	
p_2	Method	RMSE	RPE	RMSE	RPE	RMSE	RPE
3	SM	1.696	1.070	2.611	1.069	4.525	1.064
	S	1.233	1.022	0.714	0.938	1.145	1.001
	PS	1.241	1.028	1.555	1.035	2.051	1.040
	ENET	1.064	1.002	1.178	1.002	0.968	0.984
	LASSO	1.084	0.997	1.258	0.997	0.852	0.964
	ALASSO	1.242	1.021	1.417	1.006	0.548	0.898
	SCAD	1.370	1.040	1.513	1.018	0.554	0.920
6	SM	2.067	1.143	3.007	1.136	4.769	1.116
	S	1.503	1.071	1.514	1.043	1.452	1.019
	PS	1.667	1.101	2.178	1.100	3.059	1.092
	ENET	1.093	1.015	1.141	1.012	0.879	0.983
	LASSO	1.102	1.015	1.174	1.005	0.838	0.977
	ALASSO	1.345	1.054	1.302	1.023	0.526	0.905
	SCAD	1.461	1.068	1.316	1.028	0.361	0.870
9	SM	3.089	1.226	3.732	1.198	5.157	1.169
	S	2.128	1.153	2.312	1.130	2.277	1.101
	PS	2.447	1.184	2.890	1.164	3.820	1.146
	ENET	1.291	1.052	1.231	1.044	0.853	0.995
	LASSO	1.370	1.067	1.325	1.052	0.820	0.999
	ALASSO	1.576	1.087	1.389	1.059	0.567	0.936
	SCAD	2.008	1.149	1.465	1.074	0.227	0.849
12	SM	3.018	1.284	3.815	1.259	6.449	1.204
	S	2.333	1.221	2.616	1.192	3.351	1.147
	PS	2.489	1.248	2.966	1.227	4.291	1.182
	ENET	1.327	1.082	1.365	1.069	1.003	1.006
	LASSO	1.383	1.087	1.489	1.083	1.069	1.009
	ALASSO	1.692	1.124	1.699	1.090	0.814	0.937
	SCAD	2.308	1.211	1.641	1.094	0.225	0.811

TABLE 5.7: RMSE and RPE of the Estimators for $n = 100$ and $p_1 = 3$.

		$p_2 = 20$		$p_2 = 40$		$p_2 = 60$		$p_2 = 80$	
γ	Method	RMSE	RPE	RMSE	RPE	RMSE	RPE	RMSE	RPE
0.3	SM	6.643	1.221	13.474	1.488	21.115	1.767	63.387	3.161
	S	3.977	1.196	9.520	1.465	15.650	1.747	45.868	3.101
	PS	4.950	1.205	10.299	1.470	16.312	1.748	46.458	3.099
	LSE	0.831	0.980	0.649	0.872	0.381	0.648	0.348	0.522
	ENET	2.211	1.127	3.457	1.338	4.419	1.548	11.917	2.733
	LASSO	2.570	1.143	4.045	1.365	5.160	1.586	14.107	2.797
	ALASSO	2.528	1.151	3.128	1.319	3.440	1.492	22.687	3.008
	SCAD	4.898	1.215	11.879	1.485	16.260	1.751	40.849	3.137
0.6	SM	7.830	1.206	15.193	1.430	23.328	1.642	66.441	2.657
	S	4.290	1.182	10.349	1.410	16.882	1.627	49.736	2.621
	PS	5.609	1.192	11.292	1.417	17.827	1.629	51.371	2.620
	LSE	0.641	0.967	0.460	0.835	0.251	0.602	0.217	0.440
	ENET	2.140	1.118	3.270	1.290	4.031	1.450	10.213	2.324
	LASSO	2.681	1.137	3.895	1.320	4.788	1.482	11.789	2.364
	ALASSO	2.757	1.142	3.493	1.300	3.748	1.432	23.434	2.564
	SCAD	4.403	1.195	12.348	1.427	16.512	1.627	42.132	2.652
0.9	SM	14.040	1.155	22.134	1.251	26.877	1.324	46.127	1.583
	S	5.663	1.139	13.106	1.240	18.864	1.318	37.035	1.575
	PS	8.592	1.147	15.359	1.245	20.647	1.319	38.360	1.575
	LSE	0.349	0.920	0.194	0.725	0.094	0.483	0.057	0.261
	ENET	1.516	1.069	2.220	1.136	2.229	1.172	3.297	1.388
	LASSO	2.085	1.088	2.584	1.154	2.615	1.194	3.918	1.418
	ALASSO	2.269	1.095	2.477	1.149	2.065	1.158	6.069	1.486
	SCAD	1.562	1.088	2.709	1.179	2.636	1.216	3.990	1.436

TABLE 5.8: RMSE and RPE of the Estimators for $n = 100$ and $p_1 = 3$ – FM is based on LSE.

		$p_2 = 20$		$p_2 = 40$		$p_2 = 60$		$p_2 = 80$	
γ	Method	RMSE	RPE	RMSE	RPE	RMSE	RPE	RMSE	RPE
0.3	SM	7.667	1.255	22.757	1.717	57.254	2.742	158.864	6.084
	S	4.607	1.222	13.507	1.676	31.746	2.685	90.068	5.960
	PS	5.560	1.236	16.188	1.692	35.416	2.695	95.692	5.974
	Ridge	1.207	1.021	1.535	1.145	2.568	1.527	2.896	1.917
	ENET	2.627	1.147	5.394	1.535	11.750	2.396	36.813	5.287
	LASSO	3.221	1.174	6.189	1.566	14.365	2.455	43.818	5.389
	ALASSO	2.867	1.169	4.679	1.496	9.238	2.296	71.765	5.801
	SCAD	5.954	1.240	18.045	1.700	45.116	2.714	111.789	5.987
0.6	SM	10.332	1.254	33.790	1.720	86.420	2.733	238.259	6.091
	S	5.458	1.220	17.089	1.678	40.630	2.678	112.239	5.972
	PS	6.780	1.235	21.343	1.696	45.962	2.688	120.886	5.987
	Ridge	1.557	1.034	2.174	1.194	3.927	1.650	4.648	2.286
	ENET	3.348	1.155	7.235	1.550	16.108	2.410	46.331	5.300
	LASSO	4.306	1.180	8.499	1.580	21.013	2.480	59.196	5.433
	ALASSO	4.192	1.189	7.634	1.551	16.402	2.404	118.649	5.887
	SCAD	8.308	1.244	27.331	1.713	68.007	2.711	182.635	6.036
0.9	SM	19.163	1.254	63.084	1.725	145.180	2.727	392.049	6.045
	S	7.058	1.221	22.884	1.683	51.937	2.676	140.592	5.929
	PS	9.335	1.236	29.991	1.700	59.778	2.685	153.346	5.943
	Ridge	2.865	1.085	5.205	1.377	10.727	2.073	17.771	3.832
	ENET	4.440	1.165	11.449	1.569	23.208	2.412	59.654	5.311
	LASSO	6.032	1.185	13.416	1.594	30.567	2.487	73.914	5.461
	ALASSO	5.921	1.188	11.593	1.570	24.148	2.427	114.187	5.724
	SCAD	6.235	1.193	14.418	1.618	31.987	2.539	64.449	5.443

5.5 Real Data Examples

In this section, we present two real data examples.

5.5.1 Housing Prices (HP) Data

We implement the proposed strategies on a dataset comprised of housing attributes and the associated hedonic prices, as used by Ho (1997) by the method of partial least squares. The data contains 92 detached homes in the region of Ottawa, Ontario. Following Roozbeh (2016), the response variable y is sale price; the predicting variables include lot size (LA), square footage of housing (US), average neighborhood income (AI), distance to highway (DH), presence of garage (GR), fireplace (FP). We first consider the parametric model:

$$y_i = \beta_0 + \beta_1\text{LA}_i + \beta_2\text{US}_i + \beta_3\text{AI}_i + \beta_4\text{DH}_i + \beta_5\text{GR}_i + \beta_6\text{FP}_i + \epsilon_i.$$

In Table 5.9, the correlation among the predictors are presented. As shown, multicollinearity potentially exists between DH & AI and FP & US, among others. The condition number (CN) is used to test the multicollinearity, which is defined as the ratio of the largest eigenvalue to the smallest eigenvalue of the matrix $\boldsymbol{X}^\top\boldsymbol{X}$. If the CN is larger than 30, it implies the existence of multicollinearity in the data set, referring to Belsley (2014). Here the eigenvalues of $\boldsymbol{X}^\top\boldsymbol{X}$ are $\lambda_1 = 238468.999$, $\lambda_2 = 228.036$, $\lambda_3 = 23.826$, $\lambda_4 = 18.739$, $\lambda_5 = 15.501$ and $\lambda_6 = 6.163$. Hence, the CN is approximately equal to 38693.56 implying the design matrix $\boldsymbol{X}$ has a multicollinearity problem. Thus, the ridge regression method will be a good option for modeling the data.

TABLE 5.9: Correlation Matrix for HP Data.

	y	LA	US	DH	GR	FP
LA	-0.229					
US	0.550	-0.257				
DH	-0.694	-0.154	-0.480			
GR	0.142	-0.088	-0.044	-0.227		
FP	0.246	-0.265	0.477	-0.310	-0.282	
AI	0.344	-0.178	0.221	-0.612	-0.382	0.392

We consider AI as a nonparametric part of the model since there exists a non-linear relationship between itself and the response. The full model is thus given by

$$y_i = \beta_0 + \beta_1\text{LA}_i + \beta_2\text{US}_i + \beta_3\text{FP}_i + \beta_4\text{DH}_i + \beta_5\text{GR}_i + f(\text{AI}_i) + \epsilon_i.$$

In order to apply the proposed methods, we implement a two-step approach since prior information is not available here. In the first step, we apply the usual variable selection methods to select the best possible submodel. We use the forward AIC method via the *olsrr* package in R. It is observed that LA, FP and DH may be ignored since they do not seem to be significantly important. The submodel is then given by

$$y_i = \beta_0 + \beta_2\text{US}_i + \beta_5\text{GR}_i + f(\text{AI}_i) + \epsilon_i.$$

The response variable is centered and the predictors are standardized for analysis purposes; therefore, a constant term is not counted as a parameter. To assess the prediction accuracy of the listed estimators, we randomly split the data into two equal parts of observations; the first part is the training set, and the other is the test set. We fitted the model on the training set only. We consider the following measure to assess the performance of the estimators.

$$\text{PE}(\widehat{\boldsymbol{\beta}}^*) = \left\| \boldsymbol{y}_{\text{test}} - \boldsymbol{X}_{\text{test}}\widehat{\boldsymbol{\beta}}^* \right\|^2, \tag{5.9}$$

where $\widehat{\boldsymbol{\beta}}^*$ is the one of the listed estimators. This process is repeated 1000 times, and the mean values are reported. For the ease of comparison, we calculate the $\text{RPE}(\widehat{\boldsymbol{\beta}}^*) = \frac{\text{PE}(\widehat{\boldsymbol{\beta}}^{\text{FM}})}{\text{PE}(\widehat{\boldsymbol{\beta}}^*)}$. If the RPE is larger than one, then this indicates the superiority of that method over the full model estimator. The results are given in Table 5.10. We also consider three machine learning techniques, and the RPE of the machine learning estimators with the full model estimator is reported in Table 5.10. Looking at the RPE values in Table 5.10, it is clear that PS has an edge over on all other estimators. Although the RPE of SM is the highest, it is only efficient when the selected submodel is the correct one, otherwise its RPE may converge to zero. On the other hand, the RPE of the shrinkage will never go below one! Interestingly, machine learning methods are not doing well either for this data set. We suggest trying a host of statistical and machine leaning strategies for the data at hand, then to selecting accordingly.

TABLE 5.10: The RPE of Estimators for HD Data.

Estimation Strategy	RPE
SM	1.047
Shrinkage Methods	
S	0.970
PS	**1.023**
Penalized Methods	
ENET	0.996
LASSO	0.985
ALASSO	0.982
SCAD	0.947
Machine Learning Methods	
NN	0.862
RF	0.967
KNN	0.856

5.5.2 Investment Data of Turkey

We apply the listed estimation strategies to an economic dataset regarding Turkey's attraction for foreign direct investment (FDI). The data was collected over the period between 1983 and 2018, and the response variable FDI, the net inflows (% of GDP), is given by y. The predictor variables include GDP per capita growth (annual %) as GROWTH, inflation, GDP deflator (annual %) as DEFLATOR, exports of goods and services (% of GDP) as EXPORTS, imports of goods and services (% of GDP) as IMPORTS, general government final consumption expenditure (% of GDP) as GGFCE, total reserves (includes gold, current US$) / GDP (current US$) as RESERVES, personal remittances, received (% of GDP) as PREM, current account balance (% of GDP) as BALANCA. Here $(n, p) = (36, 9)$. This data is part of the study of Yüzbaşı et al. (2020), and the raw data is available from the World Bank [1].

We first consider the parametric model:

$$\begin{aligned} y_i &= \beta_0 + \beta_1 \text{GROWTH}_i + \beta_2 \text{DEFLATOR}_i + \beta_3 \text{EXPORTS}_i \\ &+ \beta_4 \text{IMPORTS}_i + \beta_5 \text{SAVINGS}_i + \beta_6 \text{GGFCE}_i \\ &+ \beta_7 \text{RESERVES}_i + \beta_8 \text{BALANCE}_i + \beta_9 \text{PREM}_i + \epsilon_i. \end{aligned}$$

In Table 5.11, we present the variance inflation factor (VIF) values and eigenvalues of the predicting variables. Since EXPORTS, IMPORTS, GGFCE, PREM, and BALANCE have a VIF above 10, it indicates high correlation and should be cause for concern. The CN is approximately equal to 29336658 that implies the data has the serious problem of multicollinearity. Thus, the ridge regression methods is a good option for modeling this data.

TABLE 5.11: Diagnostics for Multicollinearity in Investment Data.

	VIFs	Eigenvalues
GROWTH	3.23	137579.46
DEFLATOR	4.10	21074.61
EXPORTS	22.72	1075.70
IMPORTS	23.72	661.18
SAVINGS	2.82	120.07
GGFCE	10.25	59.83
RESERVES	9.05	19.79
PREM	12.88	3.46
BALANCE	12.95	0.01

In order to identify the parametric and non-parametric components of the model, we investigate plots of each predictor versus the response variable. This suggests that PREM may be considered as a non-parametric part of the model. Hence, the semi-parametric full model is given by:

$$\begin{aligned} y_i &= \beta_0 + \beta_1 \text{GROWTH}_i + \beta_2 \text{DEFLATOR}_i + \beta_3 \text{EXPORTS}_i \\ &+ \beta_4 \text{IMPORTS}_i + \beta_5 \text{SAVINGS}_i + \beta_6 \text{GGFCE}_i \\ &+ \beta_7 \text{RESERVES}_i + \beta_8 \text{BALANCE}_i + f(\text{PREM}_i) + \epsilon_i. \end{aligned}$$

Using the forward AIC method, we find that DEFLATOR, IMPORTS, SAVING, GGFCES, and RESERVES may be deleted from the model. The new submodel is given by

$$\begin{aligned} y_i &= \beta_0 + \beta_1 \text{GROWTH}_i + \beta_3 \text{EXPORTS}_i + \beta_8 \text{BALANCE}_i \\ &+ f(\text{PREM}_i) + \epsilon_i. \end{aligned}$$

As usual, the response variable is centered and the predictors are standardized. To assess the prediction accuracy of the listed estimators, we randomly split the data into two equal groups of observations. The first part is the training set where the models are fitted, and the other is the testing set.

[1] https://data.worldbank.org

We calculate the prediction error together with their respective standard error (SE) based on 1000 repetitions, and the mean values are reported. The results are given in Table 5.12. We consider ridge regression and LSE as the full model estimator. We also report the RPE of each of the listed estimators relative to the full model estimator. If the RPE is larger than one, this is indicative of the superiority of the selected estimator over the full model estimator.

TABLE 5.12: PE and RPE of the Investment Data.

The FM is Ridge			The FM is LSE		
Method	PE(SE)	RPE	Method	PE(SE)	RPE
FM	0.127(0.003)	1	FM	0.367(0.036)	1
SM	0.108(0.002)	1.173	SM	0.108(0.003)	3.391
S	0.120(0.004)	1.054	S	0.241(0.042)	1.525
PS	0.113(0.002)	**1.122**	PS	0.190(0.014)	**1.930**
LSE	0.376(0.031)	0.337	Ridge	0.143(0.007)	2.566
ENET	0.143(0.006)	0.888	ENET	0.134(0.006)	2.742
LASSO	0.135(0.007)	0.938	LASSO	0.124(0.004)	2.963
ALASSO	0.138(0.006)	0.919	ALASSO	0.140(0.008)	2.632
SCAD	0.132(0.006)	0.963	SCAD	0.134(0.008)	2.750
NN	0.159(0.003)	0.797	NN	0.151(0.002)	2.433
RF	0.100(0.002)	**1.264**	RF	0.096(0.002)	3.810
KNN	0.118(0.002)	1.073	KNN	0.112(0.002)	3.284

The positive shrinkage strategy is the clear winner in the class when the ridge estimator is being used as a full model estimator. However, RF shows the highest efficiency among all the estimators. We suggest using shrinkage strategy for meaningful interpretation and for further statistical analysis. The results are different when we use LSE as a full model estimator, but such findings can be misleading when ignoring the multicollinearity in data. This is a classical example of not using the right initial model. Before choosing an initial model, it's important to do all the diagnostics and safety checks that are needed.

5.6 High-Dimensional Model

In this section, we will perform a numerical study to investigate the performance of the shrinkage estimators in high-dimensional cases. For brevity's sake, we also consider the case when both n and p are large enough. Our aim is to examine the relative performance of the positive shrinkage estimator with four penalized methods: LASSO, ALASSO, SCAD, and ENET. The ridge estimator is used as a benchmark (full model) estimator, and submodel estimators are obtained via the above penalized methods. They are then combined to build shrinkage estimators. Monte Carlo simulation experiments are conducted to evaluate the relative performance of the listed estimators with respect to the ridge estimator. We conveniently partition $\boldsymbol{\beta} = \boldsymbol{\beta_1} + \boldsymbol{\beta_2} + \boldsymbol{\beta_3}$, where $\boldsymbol{\beta_1}$ is p_1 vector for the strong signals in the model, $\boldsymbol{\beta_2}$ is p_2 vector for the weak signals, and $\boldsymbol{\beta_3}$ is p_3 vector for the sparse signals. The strength of the weak signals and level of multicollinearity are denoted by Greek letters κ and γ, respectively. We select values $\gamma = 0.3, 0.6, 0.9$ and $\kappa = 0, 0.05, 0.1$ for illustrative purposes. When $\kappa = 0$, there are no weak signals in the simulated model, and the model consists of strong and sparse signals.

We simulate the response from the following model:

$$y_i = x_{1i}\beta_1 + x_{2i}\beta_2 + ... + x_{pi}\beta_p + f(t_i) + \varepsilon_i \ , \ i = 1, \ldots, n. \tag{5.10}$$

where $x_{ij} = (1 - \gamma^2)^{1/2} z_{ij} + \gamma z_{ip}$ and z_{ij} are random numbers following a standard normal distribution such that $i = 1, 2, \ldots n$, $j = 1, 2, \ldots p$. The degree of correlation is given by $\gamma = 0.3, 0.6, 0.9$. In (5.10), we considered the non-parametric function $f(t_i) = \sqrt{t_i(1 - t_i)} \sin\left(\frac{2.1\pi}{t_i + 0.05}\right)$, called the doppler function for $t_i = (i - 0.5)/n$ to generate the variable on interest y_i.

The regression coefficients are set $\boldsymbol{\beta} = \left(\boldsymbol{\beta}_1^\top, \boldsymbol{\beta}_2^\top, \boldsymbol{\beta}_3^\top\right)^\top$ with dimensions p_1, p_2 and p_3, respectively. $\boldsymbol{\beta}_1$ represent strong signals, i.e. $\boldsymbol{\beta}_1$ is a vector of 1 values, $\boldsymbol{\beta}_2$ stands for the weak signals of $\boldsymbol{\kappa} = 0, 0.05, 0.10$, and $\boldsymbol{\beta}_3$ means no signals, $\boldsymbol{\beta}_3 = \mathbf{0}$. In this simulation setting, 50 datasets are simulated consisting of $n = 75, 150, 750$, with $p_1 = 4$, $p_2 = 50, 100, 500$ and $p_3 = 1000$.

The performance of an estimator was evaluated by using mean squared error (MSE). The relative mean squared error (RMSE) of the listed estimators and the ridge estimators is also calculated. Keeping in mind that a value of RMSE greater than 1 indicates the degree of superiority of the selected estimator to the ridge estimator. Simulation results are reported in Tables 5.13 and 5.14, using $p_1 = 3$, $p_3 = 1000$, and varying values of n, γ, κ and p_2. We observe that the performance of the positive shrinkage estimator is superior to all the other estimators in almost all simulation configurations. For example, the performance of the positive shrinkage estimator is relatively steady as the values of p_2 and κ increase individually and simultaneously. The penalized methods perform poorly in such cases. As expected, when the level of multicolinearity increases the RMSE of the penalized methods approaches to zero. Some instances reported in Table 5.13 show very large RMSE values for all the estimators, possibly due to the poor performance of the ridge estimator. The simulation results re-establish the fact that penalized methods are not suitable to handle a large number of weak signals, and the impact on RMSE is rather dramatic. In this sense, the shrinkage strategy performs like a robust one and successfully reduces the impact of weak signals.

5.6.1 Real Data Example

We utilize the Berndt (1991) data set to demonstrate the applicability and performance of the high-dimensional shrinkage and penalty estimators. The data presents the wage information of 534 workers, as well as their education, living region, gender, race, occupation, marital status, and years of experience. Xie and Huang (2009); Gao et al. (2017a) assume a nonlinear link between years of experience and wage level and propose a partial linear regression model. However, the primary concern is the significance of other variables for wages. Specifically, we evaluate:

$$y_i = \beta_0 + \sum_{j=1}^{14} x_{ij}\beta_j + f(t_i) + \epsilon_i, i = 1, \ldots, 534,$$

where y_i is the ith worker's wage, t_i is their years of experience, x_{ij} is the jth variable and ϵ_i's are i.i.d variables with mean 0 and finite variance. 14 factors are considered in Table 5.15.

Since there is limited information, we employ a two-step procedure to implement the proposed methods. In the first stage, we select an appropriate submodel using standard

TABLE 5.13: The RMSE of the Estimators for $p_1 = 4$ and $p_3 = 1000$.

γ	n	p_2	κ	SM	PS	ENET	LASSO	ALASSO	SCAD
0.3	75	50	0.00	61.15	19.14	4.11	5.17	9.70	20.39
			0.05	13.44	4.89	3.21	3.48	4.63	5.63
			0.10	3.75	1.84	2.04	1.93	1.92	1.45
	150	100	0.00	78.31	29.47	10.06	11.97	43.30	66.79
			0.05	5.72	2.11	2.68	2.41	2.15	2.14
			0.10	1.55	1.26	0.96	0.82	0.57	0.66
	750	500	0.00	283.56	107.19	61.79	74.12	490.41	439.15
			0.05	0.31	1.08	0.15	0.14	0.09	0.10
			0.10	0.11	1.02	0.06	0.05	0.04	0.04
0.6	75	50	0.00	56.84	19.85	3.15	3.90	7.29	9.48
			0.05	4.60	5.42	2.46	2.22	2.17	1.52
			0.10	1.22	1.93	1.35	0.95	0.50	0.44
	150	100	0.00	84.02	32.55	7.55	8.98	41.17	66.58
			0.05	1.35	2.08	1.31	0.93	0.49	0.59
			0.10	0.38	1.27	0.38	0.29	0.13	0.13
	750	500	0.00	361.18	126.90	49.68	58.50	426.24	477.25
			0.05	0.06	1.09	0.04	0.04	0.03	0.03
			0.10	0.02	1.03	0.02	0.01	0.01	0.01
0.9	75	50	0.00	37.46	18.28	1.42	1.21	1.10	0.74
			0.05	2.73	11.05	1.32	0.61	0.38	0.17
			0.10	0.75	5.36	0.87	0.25	0.12	0.10
	150	100	0.00	57.62	29.82	2.55	2.85	4.28	0.90
			0.05	0.76	7.63	0.86	0.30	0.11	0.10
			0.10	0.22	2.08	0.19	0.10	0.05	0.05
	750	500	0.00	301.58	127.48	16.77	19.02	91.67	1.03
			0.05	0.03	1.32	0.01	0.01	0.01	0.01
			0.10	0.01	1.09	0.01	0.01	0.01	0.01

TABLE 5.14: The PE of the Estimators for $p_1 = 4$ and $p_3 = 1000$.

γ	n	p_2	κ	SM	PS	ENET	LASSO	ALASSO	SCAD
0.3	75	50	0.00	3.98	3.59	2.38	2.60	3.20	3.63
			0.05	2.72	2.28	2.11	2.25	2.59	2.90
			0.10	1.50	1.42	1.79	1.90	2.27	2.39
	150	100	0.00	3.71	3.52	3.03	3.12	3.60	3.66
			0.05	1.41	1.51	2.57	2.66	3.04	3.14
			0.10	0.57	1.12	1.99	2.07	2.24	2.37
	750	500	0.00	3.85	3.79	3.72	3.75	3.87	3.87
			0.05	0.12	1.03	2.47	2.46	1.29	1.84
			0.10	0.05	1.00	1.63	1.58	0.94	1.02
0.6	75	50	0.00	3.11	2.91	1.93	2.07	2.51	2.64
			0.05	1.86	2.13	1.78	1.83	2.00	2.26
			0.10	0.90	1.44	1.54	1.62	1.68	1.64
	150	100	0.00	2.96	2.86	2.43	2.50	2.90	2.94
			0.05	0.82	1.47	2.09	2.19	2.14	2.30
			0.10	0.29	1.14	1.74	1.80	1.13	1.30
	750	500	0.00	3.19	3.16	3.10	3.12	3.22	3.22
			0.05	0.06	1.05	1.98	1.96	0.90	0.90
			0.10	0.02	1.01	1.24	1.20	0.84	0.85
0.9	75	50	0.00	1.64	1.61	1.14	1.15	1.15	0.87
			0.05	1.30	1.55	1.13	1.09	1.00	0.99
			0.10	0.83	1.48	1.04	1.05	0.95	0.99
	150	100	0.00	1.59	1.58	1.32	1.34	1.41	1.15
			0.05	0.77	1.44	1.21	1.25	1.01	0.99
			0.10	0.33	1.24	1.14	1.21	0.98	0.99
	750	500	0.00	1.68	1.68	1.64	1.65	1.69	1.30
			0.05	0.07	1.10	1.31	1.17	0.93	0.97
			0.10	0.02	1.03	0.97	0.88	0.88	0.96

TABLE 5.15: The Description of Wage Data.

Variable	Description
south	1 = southern region, 0 = other
fe	1 = Female, 0 = Male
union	1 = union member, 0 = nonmember
nonwh	1 = black, 0 = other
hisp	1 = Hispanic, 0 = other
manag	1 = management, 0 = other
sales	1 = sales, 0 = other
cler	1 = clerical, 0 = other
serv	1 = service, 0 = other
prof	1 = professional, 0 = other
manuf	1 = manufacturing, 0 = other
constr	1 = construction, 0 = other
marr	1 = married, 0 = other

variable selection methods. We employ the forward AIC approach of the R package *olsrr*. It is observed that nonwh, sales, and constr may be disregarded as they do not appear to be of significant importance. The submodel is then given by

$$\begin{aligned} y_i &= \beta_0 + \beta_1 \text{south}_i + \beta_2 \text{fe}_i + \beta_3 \text{union}_i + \beta_5 \text{hisp}_i + \beta_6 \text{manag}_i + \beta_8 \text{cler}_i \\ &+ \beta_9 \text{serv}_i + \beta_{10} \text{prof}_i + \beta_{11} \text{manuf}_i + \beta_{13} \text{marr}_i + f(t_i) + \epsilon_i \end{aligned}$$

For analysis purposes, the response variable is centered and the predictors are standardized. Consequentially, a constant term is not considered a parameter. To evaluate the prediction accuracy of the given estimators, we randomly divide the data into two equal parts, one for the training set where the model is fitted, and the other for the test set. We consider the measure of PE in (5.9) to evaluate the estimators' performance. The average values are reported after 500 repetitions of this process. We also compute the RPE values for comparison purposes. If the RPE is greater than one, this indicates that the method is superior to the full model estimator. The results are given in Table 5.16. Three machine learning methods are also implemented for comparative purposes with this data set. Looking at RPE values in Table 5.16, PS has an advantage over all other estimators, despite SM having the highest RPE. However, SM has the highest RPE only when the selected submodel is the correct one; otherwise, its RPE may converge to zero. In contrast, the RPE of the shrinkage will never be less than one! The machine learning methods perform poorly with this data set as well. It is again recommended to implement a variety of statistical and machine learning strategies on a set of data to make an informed decision.

5.7 R-Codes

```
> library('MASS')  # It is for 'mvrnorm' and 'lm.ridge' function
> library('dplyr') # for data cleaning
> library('lmtest')# It is for 'lrtest' function
```

TABLE 5.16: Prediction Performance of Methods.

	PE(SE)	RPE
FM	0.1978(0.00063)	1
SM	0.19688(0.00064)	1.00470
PS	0.19725(0.00063)	1.00278
ENET	0.19854(0.00069)	0.99627
LASSO	0.1993(0.00069)	0.99247
ALASSO	0.20108(0.00068)	0.98371
SCAD	0.19978(0.00072)	0.99008
NN	0.24736(0.00102)	0.79965
RF	0.21412(0.00066)	0.92377
KNN	0.25577(0.00087)	0.77337

```
> library('caret') # It is for 'split' function
> library('psych') # It is for 'tr' function
> library('glmnet')# It is for 'glmnet' function
> library('gdata')# It is for Read Excel files
> set.seed(2500)
> # Defining Shrinkage and Positive Shrinkage estimation functions
> Shrinkage_Est<-function(beta_FM,beta_SM,test_stat,p2){
+   return(beta_FM - ((beta_FM - beta_SM)*(p2-2)/test_stat))
+ }
> PShrinkage_Est<-function(beta_FM,beta_SM,test_stat,p2){
+   return(beta_SM + max(0,(1-(p2-2)/test_stat))*
+            (beta_FM-beta_SM))
+ }
> # The function of prediction error
> Prediction_Error<-function(y,yhat){mean((y-yhat)^2)}
> # The function of MSE
> MSE<-function(beta,beta_hat){mean((beta-beta_hat)^2)}
> n <-100 # The number of sample
> p1<-4  # The number of significant covariates
> p2<-4  # The number of insignificant covariates
> p<-p1+p2
> beta_true<-rep(1,p1)
> beta2_true<-rep(0,p2)
> # The ture value of covariates
> beta_true<-c(beta_true,beta2_true)
> #Generate the design matrix
> Phi = 0.5 # the magnitude of the multicollinearity
> X = matrix(0,n,p)
> w=matrix(rnorm(n*p,mean=0,sd=1), n, p)
> for (i in 1:n){
+   for (j in 1:p){
+     X[i,j]=sqrt(1-Phi^2)*w[i,j]+(Phi)*w[i,p];
+   }
+ }
> ## assigning colnames of X to "X1","X2",....
> v<-NULL
> for(i in 1:p){
+   v[i]<-paste("X",i,sep="")
+   assign(v[i],X[,i])
```

```
+ }
> ############ Nonparametric Part ############
> t<-c()
> for(i in 1:n){
+   z = (i-0.5)/n
+   t<-c(t,z)}
> f<-sqrt(t*(1-t))*sin(2.1*pi/t+.05)
> ############ ############ ############
> epsilon<-rnorm(n) # The errors
> sigma<-1
> y<-X%*%beta_true+f+sigma*epsilon # The response
> # kernel function
> kernel <-function(u) {(1/sqrt(2*pi))*exp(-.5*u^2)}
> # Grid of kernel tuning parameter
> a<-as.matrix(abs(outer(t, t, "-")))
> for (i in 1:n) {
+   a[i, i]<--1000
+ }
> a<-as.vector(a[a != -1000])
> b.min<-quantile(a, 0.05)
> b.max<-(max(t) - min(t))
> Lp <-seq(b.min, b.max, length.out = 100) * 0.25
> m<-length(Lp)
> ### Estimation of nonparametric part
> GCV<- rep(0,m)
> W<-array(0,dim = c(n,n,m))
> for(z in 1:m){
+   Kx<-rbind(sapply(t, function(Xi) kernel((t - Xi) / Lp[z]) /
+                    n*Lp[z]))
+   # Weights
+   W[,,z]<-Kx / rowSums(Kx) # Column recycling!
+   Xt=(diag(n)-W[,,z])%*%X; yt=(diag(n)-W[,,z])%*%y
+   kl<- lm.ridge(yt~Xt)$kHKB  ## Another option could be '$kLW'
+   H=Xt%*%solve(t(Xt)%*%Xt+kl*diag(p))%*%t(Xt)
+   H1 = W[,,z]+(diag(n)-W[,,z])%*%H; N1=y-H1%*%y;
+   GCV[z]=(n*(norm(N1,"F")^2))/((tr(diag(n)-H1))^2)
+ }
> #plot(GCV)
> W.opt<-W[,,which.min(GCV)];
> # Transform semiprametric model to parametric model
> XT<-(diag(n)-W.opt)%*%X;
> yT<-(diag(n)-W.opt)%*%y;
> # Split data into train and test set
> all.folds<-split (sample (1 : n), rep (1 : 2, length = n))
> train_ind<-all.folds$`1`
> test_ind <-all.folds$`2`
> y_train<-yT[train_ind]
> X_train<-XT[train_ind,]
> ### Centering train data of y and X
> X_train_mean <-apply(X_train,2,mean)
> X_train_scale<-scale(X_train,X_train_mean,F)
> y_train_mean <-mean(y_train)
> y_train_scale<-scale(y_train,y_train_mean,F)
> train_scale_df<-data.frame(y_train_scale,X_train_scale)
> #test data
```

```
> y_test<-yT[test_ind]
> X_test<-XT[test_ind,]
> ### Centering test data of y and X based on train means
> X_test_scale<-scale(X_test,X_train_mean,F)
> y_test_scale<-scale(y_test,y_train_mean,F)
> # Formula of the Full model
> xcount.FM<-c(0,paste("X", 1:p, sep=""))
> Formula_FM<- as.formula(paste("y_train_scale~",
+                              paste(xcount.FM, collapse= "+")))
> # Formula of the Sub model
> xcount.SM<-c(0,paste("X", 1:p1, sep=""))
> Formula_SM<- as.formula(paste("y_train_scale~",
+                              paste(xcount.SM, collapse= "+")))
> # Likelihood ratio test
> fit_FM    <-lm(Formula_FM,data = train_scale_df)
> fit_SM    <-lm(Formula_SM,data = train_scale_df)
> test_LR<-lrtest(fit_SM, fit_FM)
> test_stat<-test_LR$Chisq[2]
> cv.ridge.full <-cv.glmnet(X_train_scale, y_test_scale,
+                          alpha = 0, nfold = 5,
+                          intercept = F, standardize = F)
> beta.FM <-coef(cv.ridge.full, s = "lambda.min")[-1]
> X1_train_scale<-X_train_scale[,1:p1]
> cv.ridge.sub <-cv.glmnet(X1_train_scale, y_test_scale,
+                         alpha = 0, nfold = 5,
+                         intercept = F, standardize = F)
> beta.SM<-rep(0,p)
> beta.SM[1:p1] <-coef(cv.ridge.sub, s = "lambda.min")[-1]
> beta.S <-Shrinkage_Est(beta.FM,beta.SM,test_stat,p2)
> beta.PS<-PShrinkage_Est(beta.FM,beta.SM,test_stat,p2)
> # Estimate Prediction Errors based on test data
> yhat_beta.FM<-X_test_scale%*%beta.FM
> yhat_beta.SM<-X_test_scale%*%beta.SM
> yhat_beta.S <-X_test_scale%*%beta.S
> yhat_beta.PS<-X_test_scale%*%beta.PS
> # Calculte prediction errors of estimators based on test data
> PE_values<-
+c(FM=Prediction_Error(y_test_scale,yhat_beta.FM),
+SM=Prediction_Error(y_test_scale,yhat_beta.SM),
+S =Prediction_Error(y_test_scale,yhat_beta.S),
+PS=Prediction_Error(y_test_scale,yhat_beta.PS))
> # print and sort the results
> sort(PE_values)
      SM        S       PS       FM
4.724918 4.872494 4.872494 4.983623
> # Calculte MSEs of estimators
> MSE_values<-c(FM=MSE(beta_true,beta.FM ),
+               SM=MSE(beta_true,beta.SM ),
+               S = MSE(beta_true,beta.S ),
+               PS=MSE(beta_true,beta.PS ))
> # print and sort the results
> sort(MSE_values)
        S        PS        FM        SM
0.4950663 0.4950663 0.4963238 0.5062714
> # An Example of the Real data
```

```
> set.seed(2500)
> # read data for xls
> HP<-read.xls("~/Downloads/HousingPrices.xlsx",sheet=1,
> +header= T) # This data can be requested from authors
> y<-HP %>% dplyr::select(sellprix)%>% as.matrix()
> X<-HP %>% dplyr::select(c(lotarea,usespace,disthwy,garage,
+                           fireplac,avginc)) %>% as.matrix()
> n<-dim(X)[1]
> p<-dim(X)[2]
> # conditon test
> ev<-eigen(t(X)%*%X)
> round(ev$values,5)
[1] 238468.99865    228.03567     23.82619     18.73857     15.50144
        6.16302
> sqrt(max(ev$values)/min(ev$values))
[1] 196.7068
> #df_scale<-data.frame(y,X)
> # Model Selection
> # scale X, center y
> Xs<-scale(X)
> ys<-scale(y,T,F)
> df_scale<-data.frame(ys,Xs)
> # perform bacward AIC
> require(olsrr)
> model_parametric<-lm(sellprix~., data = df_scale)
> aic<-ols_step_backward_aic(model_parametric)
> sub_ind  <-names(aic$ model$ coefficients)[-1]
> sub_ind
[1] "usespace" "garage"   "avginc"
> # select the nonparameteric predictor
> t <-HP$avginc
> # Sort t for plotting
> s<-sort(unique(t))
> # update X and Xs, p
> X <-X %>% as.data.frame() %>%
+   dplyr::select(-avginc) %>% as.matrix()
> Xs<-scale(X)
> p<-dim(X)[2]
> #####
> FM<-colnames(X)
> SM<-sub_ind[-3] # 'avginc' is non parametric part
> sub_indx<-which(FM %in% SM)
> p2<-p-length(sub_indx)
> # Grid of kernel tuning parameter
> a<-as.matrix(abs(outer(t, t, "-")))
> for (i in 1:n) {
+   a[i, i]<--1000
+ }
> a<-as.vector(a[a != -1000])
> b.min<-quantile(a, 0.05)
> b.max<-(max(t) - min(t)) * 0.25
> Lp <-seq(b.min, b.max, length.out = 100)
> m<-length(Lp)
> #Nadaraya-Watson
> kernel<-function(u) {(15/16)*(1-u^2)^2*I(abs(u)<=1)}
```

```
> # Gaussian is one of another option
> #kernel <-function(u) {(1/sqrt(2*pi))*exp(-.5*u^2)}
> ### Estimation of nonparametric part
> GCV<- rep(0,m)
> W<-array(0,dim = c(n,n,m))
> for(z in 1:m){
+   Kx<-rbind(sapply(t, function(Xi) kernel((t - Xi) / Lp[z]) /
+                         n*Lp[z]))
+   # Weights
+   W[,,z]<-Kx / rowSums(Kx) # Column recycling!
+   Xt=(diag(n)-W[,,z])%*%X; yt=(diag(n)-W[,,z])%*%y
+   kl<- lm.ridge(yt~Xt)$kHKB  ## Another option could be '$kLW'
+   H=Xt%*%solve(t(Xt)%*%Xt+kl*diag(p))%*%t(Xt)
+   H1 = W[,,z]+(diag(n)-W[,,z])%*%H; N1=y-H1%*%y;
+   GCV[z]=(n*(norm(N1,"F")^2))/((tr(diag(n)-H1))^2)
+ }
> #plot(GCV)
> W.opt<-W[,,which.min(GCV)];
> # Transform semiprametric model to parametric model
> XT<-(diag(n)-W.opt)%*%X;
> yT<-(diag(n)-W.opt)%*%y;
> # Split data into train and test set
> set.seed(1)
> all.folds<-split (sample (1 : n), rep (1 : 2, length = n))
> train_ind<-all.folds$'1'
> test_ind <-all.folds$'2'
> y_train<-yT[train_ind]
> X_train<-XT[train_ind,]
> ### Centering train data of y and X
> X_train_mean <-apply(X_train,2,mean)
> X_train_scale<-scale(X_train,X_train_mean,F)
> y_train_mean <-mean(y_train)
> y_train_scale<-scale(y_train,y_train_mean,F)
> train_scale_df<-data.frame(y_train_scale,X_train_scale)
> #test data
> y_test<-yT[test_ind]
> X_test<-XT[test_ind,]
> ### Centering test data of y and X based on train means
> X_test_scale<-scale(X_test,X_train_mean,F)
> y_test_scale<-scale(y_test,y_train_mean,F)
> # Formula of the Full model
> Formula_FM<-as.formula(paste("y_train_scale ~ ",
+                             paste(FM, collapse= "+")))
> # Formula of the Sub model
> Formula_SM<-as.formula(paste("y_train_scale ~ ",
+                             paste(SM, collapse= "+")))
> # Likelihood ratio test
> fit_FM   <-lm(Formula_FM,data = train_scale_df)
> fit_SM   <-lm(Formula_SM,data = train_scale_df)
> test_LR<-lrtest(fit_SM, fit_FM)
> test_stat<-test_LR$Chisq[2]
> #FM Ridge based on BIC
> ridge.fit_FM<-glmnet(X_train_scale, y_train_scale,
+                       alpha = 0,intercept = F,
+                       standardize = F)
```

```
> ridge.bic_FM = deviance(ridge.fit_FM) +
+   log(NROW(X_train_scale))*ridge.fit_FM$df
> beta.FM = coef(ridge.fit_FM)[-1,which.min(ridge.bic_FM)]
> #SM Ridge based on BIC
> X1_train_scale<-X_train_scale[,sub_indx]
> ridge.fit_SM<-glmnet(X1_train_scale, y_train_scale,
+                       alpha = 0,intercept = F,
+                       standardize = F)
> ridge.bic_SM = deviance(ridge.fit_SM) +
+   log(NROW(X1_train_scale))*ridge.fit_SM$df
> beta.SM<-rep(0,p)
> beta.SM[sub_indx]<-
+coef(ridge.fit_SM)[-1,which.min(ridge.bic_SM)]
> beta.S <-Shrinkage_Est(beta.FM,beta.SM,test_stat,p2)
> beta.PS<-PShrinkage_Est(beta.FM,beta.SM,test_stat,p2)
> # Estimate Prediction Errors based on test data
> yhat_beta.FM<-X_test_scale%*%beta.FM
> yhat_beta.SM<-X_test_scale%*%beta.SM
> yhat_beta.S <-X_test_scale%*%beta.S
> yhat_beta.PS<-X_test_scale%*%beta.PS
> # Calculte prediction errors of estimators based on test data
> PE_values<-
+c(FM=Prediction_Error(y_test_scale,yhat_beta.FM),
+SM=Prediction_Error(y_test_scale,yhat_beta.SM),
+S =Prediction_Error(y_test_scale,yhat_beta.S),
+PS=Prediction_Error(y_test_scale,yhat_beta.PS))
> # print and sort the results
> sort(PE_values)
      SM        S       PS       FM
554.3848 609.0107 609.0107 636.1945
> ### Plot data for whole data set
> XT_scale<-scale(XT)
> yT_scale<-scale(yT,T,F)
> scale_df<-data.frame(yT_scale,XT_scale)
> # Formula of the Full model
> Formula_FM<-as.formula(paste("sellprix ~ ",
+                              paste(FM, collapse= "+")))
> # Formula of the Sub model
> Formula_SM<-as.formula(paste("sellprix ~ ",
+                              paste(SM, collapse= "+")))
> # Likelihood ratio test
> fit_FM   <-lm(Formula_FM,data = scale_df)
> fit_SM   <-lm(Formula_SM,data = scale_df)
> test_LR<-lrtest(fit_SM, fit_FM)
> test_stat<-test_LR$Chisq[2]
> #FM Ridge based on BIC
> ridge.fit_FM<-glmnet(XT_scale, yT_scale, alpha = 0,
+                      intercept = F, standardize = F)
> ridge.bic_FM = deviance(ridge.fit_FM) +
+   log(NROW(XT_scale))*ridge.fit_FM$df
> beta.FM = coef(ridge.fit_FM)[-1,which.min(ridge.bic_FM)]
> #SM Ridge based on BIC
> XT1_scale<-XT_scale[,sub_indx]
> ridge.fit_SM<-glmnet(XT1_scale, yT_scale, alpha = 0,
+                      intercept = F, standardize = F)
```

```
> ridge.bic_SM = deviance(ridge.fit_SM) +
+   log(NROW(XT1_scale))*ridge.fit_SM$df
> beta.SM<-rep(0,p)
> min_index_SM<-which.min(ridge.bic_SM)
> beta.SM[sub_indx]<-coef(ridge.fit_SM)[-1,min_index_SM]
> beta.S <-Shrinkage_Est(beta.FM,beta.SM,test_stat,p2)
> beta.PS<-PShrinkage_Est(beta.FM,beta.SM,test_stat,p2)
> FM_fhat <-W.opt%*%(ys-Xs%*%beta.FM)
> SM_fhat <-W.opt%*%(ys-Xs%*%beta.SM)
> PS_fhat <-W.opt%*%(ys-Xs%*%beta.PS)
> ox     <-match(s, t)
> # Plot for estimated functions
> plot(t,ys,pch=".", cex=3, ylab="f(AI)",xlab="AI",cex.lab = 1.25)
> lines(s,FM_fhat[ox],type='l', lty = 2, col = "2",lwd=3)
> lines(s,SM_fhat[ox],type='l', lty = 4, col = "3",lwd=3)
> lines(s,PS_fhat[ox],type='l', lty = 4, col = "4",lwd=3)
> legend("topleft", cex=1.2 ,c("Data", "FM", "SM","PS"), lwd=3,
+        col = c(1,2:4),text.col = "black" , lty= c(0,2:4),
+        pch = c(19,NA,NA,NA))
```

5.8 Concluding Remarks

In this chapter, we consider shrinkage, penalty estimation, and machine learning methods for partially linear models. The parametric component is estimated by using the ridge regression approach, and the non-parametric component is estimated by using a kernel smoothing approach. The properties of some estimators are studied both theoretically and numerically. As expected, the submodel estimator outperforms shrinkage, penalty, and other estimators when the selected model is assumed to be true. Conversely, as the sparsity assumption is violated, the efficiency of the submodel estimator gradually decreases, even more so than the full model estimator, and may also converge to zero.

However, the shrinkage estimators perform better than the full model estimator in the entire parameter space induced by the sparsity assumption. Our asymptotic theory is well supported by numerical analyses.

Regularization methods, including LASSO, ALASSO, SCAD, and others typically focus on selecting predictors with strong signals while ignoring weak signals. This may result in a biased prediction, especially when weak signals outnumber strong signals. In high-dimensional cases, when p is much larger than n, it is reasonable to assume that model sparsity exists when most predictors do not contribute. However, some predictors may still make a small but non-trivial contribution to the response. Existing penalized strategies usually lead to a sparse model but tend to miss the possible small contributions from some predictors, resulting in excessive prediction errors or inefficient estimation. In other words, most regularization methods assume the restrictive condition, which requires that the strength of a non-zero regression coefficient be larger than a certain noise level. Hence, regularization methods may fail to detect weak signals with non-zero but small regression coefficients, resulting in biased estimates and inaccurate predictions, especially when weak signals outnumber strong signals. The post-shrinkage strategy, which accounts for possible contributions of predictors with weak and/or moderate signals, has dominant estimation and prediction performances over submodel estimates generated from penalized methods.

Our simulation results demonstrate that a shrinkage strategy produces much lower MSE and prediction errors than penalized methods. The shrinkage estimation provides a better trade-off between bias and variance when selected submodels underfit the true model. Additionally, we estimate strong signals with a larger magnitude, indicating that the estimation of the strong signal could be attenuated if the weak signals are ignored. The sparsity assumption is not realistic in real life situations as it may not hold frequently, one needs to be careful when applying penalized methods and should check for weak sparsity. The machine learning methods are subject to heavy bias, lack interpretation, and should be used cautiously.

Appendix

We present the following lemma below, which will enable us to derive the results of Theorems 5.1 and 5.3 in this chapter.

Lemma 5.4 *If $k/\sqrt{n} \to \lambda_0 \geq 0$ and $\tilde{\boldsymbol{Q}}$ is non-singular, then*

$$\sqrt{n}\left(\widehat{\boldsymbol{\beta}}^{\mathrm{FM}} - \boldsymbol{\beta}\right) \xrightarrow{d} \mathcal{N}\left(-\lambda_0 \tilde{\boldsymbol{Q}}^{-1}\boldsymbol{\beta}, \, \sigma^2 \tilde{\boldsymbol{Q}}^{-1}\right),$$

where "$\xrightarrow{d}$" denotes convergence in distribution.

Proof *Let define $V_n(\boldsymbol{u})$ as follows:*

$$\sum_{i=1}^{n}\left[\left(\tilde{\varepsilon}_i - \boldsymbol{u}^\top \tilde{\boldsymbol{x}}_i/\sqrt{n}\right)^2 - \tilde{\varepsilon}_i^2\right] + k\sum_{j=1}^{p}\left[\left|\beta_j + u_j/\sqrt{n}\right|^2 - |\beta_j|^2\right]$$

where $\boldsymbol{u} = (u_1, \ldots, u_p)^\top$. Following Knight and Fu (2000), it can be shown that

$$\sum_{i=1}^{n}\left[\left(\tilde{\varepsilon}_i - \boldsymbol{u}^\top \tilde{\boldsymbol{x}}_i/\sqrt{n}\right)^2 - \tilde{\varepsilon}_i^2\right] \xrightarrow{d} -2\boldsymbol{u}^\top \boldsymbol{D} + \boldsymbol{u}^\top \tilde{\boldsymbol{Q}}\boldsymbol{u}$$

where $\mathbf{D} \sim \mathcal{N}\left(\mathbf{0}, \sigma^2 \mathbf{I}_p\right)$ with finite-dimensional convergence holding trivially. Hence,

$$k\sum_{j=1}^{p}\left[\left|\beta_j + u_j/\sqrt{n}\right|^2 - |\beta_j|^2\right] \xrightarrow{d} \lambda_0 \sum_{j=1}^{p} u_j sgn(\beta_j)|\beta_j|.$$

Hence, $V_n(\boldsymbol{u}) \xrightarrow{d} V(\boldsymbol{u})$. Because V_n is convex and V has a unique minimum, by following Geyer (1996), it yields

$$arg\min(V_n) = \sqrt{n}\left(\widehat{\boldsymbol{\beta}}^{\mathrm{FM}} - \boldsymbol{\beta}\right) \xrightarrow{d} arg\min(V).$$

Hence,

$$\sqrt{n}\left(\widehat{\boldsymbol{\beta}}^{\mathrm{FM}} - \boldsymbol{\beta}\right) \xrightarrow{d} \tilde{\boldsymbol{Q}}^{-1}\left(\boldsymbol{D} - \lambda_0\boldsymbol{\beta}\right) \sim \mathcal{N}\left(-\lambda_0 \tilde{\boldsymbol{Q}}^{-1}\boldsymbol{\beta}, \, \sigma^2 \tilde{\boldsymbol{Q}}^{-1}\right).$$

We further consider the following proposition for proving theorems.

Proposition 5.5 *Under local alternative $\{K_n\}$ as $n \to \infty$, we have*

$$\begin{pmatrix} \vartheta_1 \\ \vartheta_3 \end{pmatrix} \sim \mathcal{N}\left[\begin{pmatrix} -\boldsymbol{\eta}_{11.2} \\ \boldsymbol{\delta} \end{pmatrix}, \begin{pmatrix} \sigma^2\tilde{\boldsymbol{Q}}_{11.2}^{-1} & \boldsymbol{\Phi}_* \\ \boldsymbol{\Phi}_* & \boldsymbol{\Phi}_* \end{pmatrix}\right],$$
$$\begin{pmatrix} \vartheta_3 \\ \vartheta_2 \end{pmatrix} \sim \mathcal{N}\left[\begin{pmatrix} \boldsymbol{\delta} \\ -\boldsymbol{\xi} \end{pmatrix}, \begin{pmatrix} \boldsymbol{\Phi}_* & \boldsymbol{0} \\ \boldsymbol{0} & \sigma^2\tilde{\boldsymbol{Q}}_{11}^{-1} \end{pmatrix}\right],$$

where $\vartheta_1 = \sqrt{n}\left(\widehat{\boldsymbol{\beta}}_1^{\mathrm{FM}} - \boldsymbol{\beta}_1\right)$, $\vartheta_2 = \sqrt{n}\left(\widehat{\boldsymbol{\beta}}_1^{\mathrm{SM}} - \boldsymbol{\beta}_1\right)$ and $\vartheta_3 = \vartheta_1 - \vartheta_2$.

Proof *Under the light of Lemma 5.4 and Lemma 3.2, it can easily be obtained*

$$\vartheta_1 \xrightarrow{d} \mathcal{N}\left(-\boldsymbol{\eta}_{11.2}, \sigma^2\tilde{\boldsymbol{Q}}_{11.2}^{-1}\right).$$

Define $\boldsymbol{y}^* = \boldsymbol{y} - \tilde{\boldsymbol{X}}_2\widehat{\boldsymbol{\beta}}_2$, *and*

$$\begin{aligned}
\widehat{\boldsymbol{\beta}}_1^{\mathrm{FM}} &= \underset{\boldsymbol{\beta}_1}{\arg\min}\left\{\left\|\boldsymbol{y}^* - \tilde{\boldsymbol{X}}_1\boldsymbol{\beta}_1\right\| + k\left\|\boldsymbol{\beta}_1\right\|^2\right\} \\
&= \left(\tilde{\boldsymbol{X}}_1^\top\tilde{\boldsymbol{X}}_1 + k\boldsymbol{I}_{p_1}\right)^{-1}\tilde{\boldsymbol{X}}_1^\top\boldsymbol{y}^* \\
&= \left(\tilde{\boldsymbol{X}}_1^\top\tilde{\boldsymbol{X}}_1 + k\boldsymbol{I}_{p_1}\right)^{-1}\tilde{\boldsymbol{X}}_1^\top\boldsymbol{y} - \left(\tilde{\boldsymbol{X}}_1^\top\tilde{\boldsymbol{X}}_1 + k\boldsymbol{I}_{p_1}\right)^{-1}\tilde{\boldsymbol{X}}_1^\top\tilde{\boldsymbol{X}}_2\widehat{\boldsymbol{\beta}}_2^{\mathrm{LSE}} \\
&= \widehat{\boldsymbol{\beta}}_1^{\mathrm{SM}} - \left(\tilde{\boldsymbol{X}}_1^\top\tilde{\boldsymbol{X}}_1 + k\boldsymbol{I}_{p_1}\right)^{-1}\tilde{\boldsymbol{X}}_1^\top\tilde{\boldsymbol{X}}_2\widehat{\boldsymbol{\beta}}_2^{\mathrm{LSE}}. \qquad (5.11)
\end{aligned}$$

By using equation (5.11),

$$\begin{aligned}
\mathrm{E}\left\{\lim_{n\to\infty}\sqrt{n}\left(\widehat{\boldsymbol{\beta}}_1^{\mathrm{SM}} - \boldsymbol{\beta}_1\right)\right\} &= \mathrm{E}\left\{\lim_{n\to\infty}\sqrt{n}\left(\widehat{\boldsymbol{\beta}}_1^{\mathrm{FM}} + \tilde{\boldsymbol{Q}}_{11}^{-1}\tilde{\boldsymbol{Q}}_{12}\widehat{\boldsymbol{\beta}}_2^{\mathrm{LSE}} - \boldsymbol{\beta}_1\right)\right\} \\
&= \mathrm{E}\left\{\lim_{n\to\infty}\sqrt{n}\left(\widehat{\boldsymbol{\beta}}_1^{\mathrm{FM}} - \boldsymbol{\beta}_1\right)\right\} \\
&\quad + \mathrm{E}\left\{\lim_{n\to\infty}\sqrt{n}\left(\tilde{\boldsymbol{Q}}_{11}^{-1}\tilde{\boldsymbol{Q}}_{12}\widehat{\boldsymbol{\beta}}_2^{\mathrm{LSE}}\right)\right\} \\
\text{by Lemma 3.2,} \quad &= -\boldsymbol{\eta}_{11.2} + \tilde{\boldsymbol{Q}}_{11}^{-1}\tilde{\boldsymbol{Q}}_{12}\boldsymbol{\omega} \\
&= -\left(\boldsymbol{\eta}_{11.2} - \boldsymbol{\delta}\right) \\
&= -\boldsymbol{\xi}.
\end{aligned}$$

Hence, $\vartheta_2 \xrightarrow{d} \mathcal{N}\left(-\boldsymbol{\xi}, \sigma^2\tilde{\boldsymbol{Q}}_{11}^{-1}\right)$.
Using the equation (5.11), *we can obtain* $\boldsymbol{\Phi}_*$ *as follows:*

$$\begin{aligned}
\boldsymbol{\Phi}_* &= \mathrm{Cov}\left(\widehat{\boldsymbol{\beta}}_1^{\mathrm{FM}} - \widehat{\boldsymbol{\beta}}_1^{\mathrm{SM}}\right) \\
&= \mathrm{E}\left[\left(\widehat{\boldsymbol{\beta}}_1^{\mathrm{FM}} - \widehat{\boldsymbol{\beta}}_1^{\mathrm{SM}}\right)\left(\widehat{\boldsymbol{\beta}}_1^{\mathrm{FM}} - \widehat{\boldsymbol{\beta}}_1^{\mathrm{SM}}\right)^\top\right] \\
&= \mathrm{E}\left[\left(\tilde{\boldsymbol{Q}}_{11}^{-1}\tilde{\boldsymbol{Q}}_{12}\widehat{\boldsymbol{\beta}}_2\right)\left(\tilde{\boldsymbol{Q}}_{11}^{-1}\tilde{\boldsymbol{Q}}_{12}\widehat{\boldsymbol{\beta}}_2^{\mathrm{LSE}}\right)^\top\right] \\
&= \tilde{\boldsymbol{Q}}_{11}^{-1}\tilde{\boldsymbol{Q}}_{12}\mathrm{E}\left[\widehat{\boldsymbol{\beta}}_2^{\mathrm{LSE}}\left(\widehat{\boldsymbol{\beta}}_2^{\mathrm{LSE}}\right)^\top\right]\tilde{\boldsymbol{Q}}_{21}\tilde{\boldsymbol{Q}}_{11}^{-1} \\
&= \sigma^2\tilde{\boldsymbol{Q}}_{11}^{-1}\tilde{\boldsymbol{Q}}_{12}\tilde{\boldsymbol{Q}}_{22.1}^{-1}\tilde{\boldsymbol{Q}}_{21}\tilde{\boldsymbol{Q}}_{11}^{-1}.
\end{aligned}$$

We also know that

$$\boldsymbol{\Phi}_* = \sigma^2\tilde{\boldsymbol{Q}}_{11}^{-1}\tilde{\boldsymbol{Q}}_{12}\tilde{\boldsymbol{Q}}_{22.1}^{-1}\tilde{\boldsymbol{Q}}_{21}\tilde{\boldsymbol{Q}}_{11}^{-1} = \sigma^2\left(\tilde{\boldsymbol{Q}}_{11.2}^{-1} - \tilde{\boldsymbol{Q}}_{11}^{-1}\right).$$

Hence, it is obtained $\vartheta_3 \xrightarrow{d} \mathcal{N}\left(\boldsymbol{\delta}, \boldsymbol{\Phi}_*\right)$.

Theorem 5.1 ADB$\left(\widehat{\boldsymbol{\beta}}_1^{\mathrm{FM}}\right)$ *and* ADB$\left(\widehat{\boldsymbol{\beta}}_1^{\mathrm{SM}}\right)$ *are directly obtained from Proposition 5.5. Also, the ADBs of S and PS are obtained as follows:*

$$\begin{aligned}
\mathrm{ADB}\left(\widehat{\boldsymbol{\beta}}_1^{\mathrm{S}}\right) &= \mathrm{E}\left\{\lim_{n\to\infty}\sqrt{n}\left(\widehat{\boldsymbol{\beta}}_1^{\mathrm{S}}-\boldsymbol{\beta}_1\right)\right\} \\
&= \mathrm{E}\left\{\lim_{n\to\infty}\sqrt{n}\left(\widehat{\boldsymbol{\beta}}_1^{\mathrm{FM}}-\left(\widehat{\boldsymbol{\beta}}_1^{\mathrm{FM}}-\widehat{\boldsymbol{\beta}}_1^{\mathrm{SM}}\right)(p_2-2)\,\mathcal{T}_n^{-1}-\boldsymbol{\beta}_1\right)\right\} \\
&= \mathrm{E}\left\{\lim_{n\to\infty}\sqrt{n}\left(\widehat{\boldsymbol{\beta}}_1^{\mathrm{FM}}-\boldsymbol{\beta}_1\right)\right\} \\
&\quad -\mathrm{E}\left\{\lim_{n\to\infty}\sqrt{n}\left(\left(\widehat{\boldsymbol{\beta}}_1^{\mathrm{FM}}-\widehat{\boldsymbol{\beta}}_1^{\mathrm{SM}}\right)(p_2-2)\,\mathcal{T}_n^{-1}\right)\right\} \\
&= -\boldsymbol{\eta}_{11.2}-(p_2-2)\,\boldsymbol{\delta}\mathrm{E}\left(\chi_{p_2+2}^{-2}(\Delta)\right).
\end{aligned}$$

$$\begin{aligned}
\mathrm{ADB}\left(\widehat{\boldsymbol{\beta}}_1^{\mathrm{PS}}\right) &= \mathrm{E}\left\{\lim_{n\to\infty}\sqrt{n}\left(\widehat{\boldsymbol{\beta}}_1^{\mathrm{PS}}-\boldsymbol{\beta}_1\right)\right\} \\
&= \mathrm{E}\left\{\lim_{n\to\infty}\sqrt{n}\left(\widehat{\boldsymbol{\beta}}_1^{\mathrm{SM}}+\left(\widehat{\boldsymbol{\beta}}_1^{\mathrm{FM}}-\widehat{\boldsymbol{\beta}}_1^{\mathrm{SM}}\right)\right.\right. \\
&\quad \times\left.\left.\left(1-(p_2-2)\,\mathcal{T}_n^{-1}\right)\mathrm{I}\left(\mathcal{T}_n>p_2-2\right)-\boldsymbol{\beta}_1\right)\right\} \\
&= \mathrm{E}\left\{\lim_{n\to\infty}\sqrt{n}\left[\widehat{\boldsymbol{\beta}}_1^{\mathrm{SM}}+\left(\widehat{\boldsymbol{\beta}}_1^{\mathrm{FM}}-\widehat{\boldsymbol{\beta}}_1^{\mathrm{SM}}\right)\left(1-\mathrm{I}\left(\mathcal{T}_n\leq p_2-2\right)\right)\right.\right. \\
&\quad \left.\left.-\left(\widehat{\boldsymbol{\beta}}_1^{\mathrm{FM}}-\widehat{\boldsymbol{\beta}}_1^{\mathrm{SM}}\right)(p_2-2)\,\mathcal{T}_n^{-1}\mathrm{I}\left(\mathcal{T}_n>p_2-2\right)-\boldsymbol{\beta}_1\right]\right\} \\
&= \mathrm{E}\left\{\lim_{n\to\infty}\sqrt{n}\left(\widehat{\boldsymbol{\beta}}_1^{\mathrm{FM}}-\boldsymbol{\beta}_1\right)\right\} \\
&\quad -\mathrm{E}\left\{\lim_{n\to\infty}\sqrt{n}\left(\widehat{\boldsymbol{\beta}}_1^{\mathrm{FM}}-\widehat{\boldsymbol{\beta}}_1^{\mathrm{SM}}\right)\mathrm{I}\left(\mathcal{T}_n\leq p_2-2\right)\right\} \\
&\quad -\mathrm{E}\left\{\lim_{n\to\infty}\sqrt{n}\left(\widehat{\boldsymbol{\beta}}_1^{\mathrm{FM}}-\widehat{\boldsymbol{\beta}}_1^{\mathrm{SM}}\right)(p_2-2)\,\mathcal{T}_n^{-1}\mathrm{I}\left(\mathcal{T}_n>p_2-2\right)\right\} \\
&= -\boldsymbol{\eta}_{11.2}-\boldsymbol{\delta}\mathrm{H}_{p_2+2}\left(p_2-2;(\Delta)\right) \\
&\quad -\boldsymbol{\delta}(p_2-2)\,\mathrm{E}\left\{\chi_{p_2+2}^{-2}(\Delta)\,\mathrm{I}\left(\chi_{p_2+2}^{2}(\Delta)>p_2-2\right)\right\}.
\end{aligned}$$

The asymptotic covariance of an estimator $\boldsymbol{\beta}_1^*$ is defined as follows:

$$\mathrm{Cov}\left(\boldsymbol{\beta}_1^*\right) = \mathrm{E}\left\{\lim_{n\to\infty}n\left(\boldsymbol{\beta}_1^*-\boldsymbol{\beta}_1\right)\left(\boldsymbol{\beta}_1^*-\boldsymbol{\beta}_1\right)^{\top}\right\}.$$

Theorem 5.2 *Firstly, the asymptotic covariance of* $\widehat{\boldsymbol{\beta}}_1^{\mathrm{FM}}$ *is given by*

$$\begin{aligned}
\mathrm{Cov}\left(\widehat{\boldsymbol{\beta}}_1^{\mathrm{FM}}\right) &= \mathrm{E}\left\{\lim_{n\to\infty}\sqrt{n}\left(\widehat{\boldsymbol{\beta}}_1^{\mathrm{FM}}-\boldsymbol{\beta}_1\right)\sqrt{n}\left(\widehat{\boldsymbol{\beta}}_1^{\mathrm{FM}}-\boldsymbol{\beta}_1\right)^{\top}\right\} \\
&= \mathrm{E}\left(\vartheta_1\vartheta_1^{\top}\right) \\
&= \mathrm{Cov}\left(\vartheta_1\vartheta_1^{\top}\right)+\mathrm{E}\left(\vartheta_1\right)\mathrm{E}\left(\vartheta_1^{\top}\right) \\
&= \sigma^2\tilde{\boldsymbol{Q}}_{11.2}^{-1}+\boldsymbol{\eta}_{11.2}\boldsymbol{\eta}_{11.2}^{\top}.
\end{aligned}$$

The asymptotic covariance of $\widehat{\boldsymbol{\beta}}_1^{\mathrm{SM}}$ *is given by*

$$\begin{aligned}
\mathrm{Cov}\left(\widehat{\boldsymbol{\beta}}_1^{\mathrm{SM}}\right) &= \mathrm{E}\left\{\lim_{n\to\infty}\sqrt{n}\left(\widehat{\boldsymbol{\beta}}_1^{\mathrm{SM}}-\boldsymbol{\beta}_1\right)\sqrt{n}\left(\widehat{\boldsymbol{\beta}}_1^{\mathrm{SM}}-\boldsymbol{\beta}_1\right)^{\top}\right\} \\
&= \mathrm{E}\left(\vartheta_2\vartheta_2^{\top}\right) \\
&= \mathrm{Cov}\left(\vartheta_2\vartheta_2^{\top}\right)+\mathrm{E}\left(\vartheta_2\right)\mathrm{E}\left(\vartheta_2^{\top}\right) \\
&= \sigma^2\tilde{\boldsymbol{Q}}_{11}^{-1}+\boldsymbol{\xi}\boldsymbol{\xi}^{\top},
\end{aligned}$$

The asymptotic covariance of $\widehat{\boldsymbol{\beta}}_1^{\mathrm{S}}$ is given by

$$\begin{aligned}
\operatorname{Cov}\left(\widehat{\boldsymbol{\beta}}_1^{\mathrm{S}}\right) &= \mathrm{E}\left\{\lim_{n\to\infty}\sqrt{n}\left(\widehat{\boldsymbol{\beta}}_1^{\mathrm{S}}-\boldsymbol{\beta}_1\right)\sqrt{n}\left(\widehat{\boldsymbol{\beta}}_1^{\mathrm{S}}-\boldsymbol{\beta}_1\right)^{\top}\right\}\\
&= \mathrm{E}\left\{\lim_{n\to\infty} n\left[\left(\widehat{\boldsymbol{\beta}}_1^{\mathrm{FM}}-\boldsymbol{\beta}_1\right)-\left(\widehat{\boldsymbol{\beta}}_1^{\mathrm{FM}}-\widehat{\boldsymbol{\beta}}_1^{\mathrm{SM}}\right)(p_2-2)\mathcal{T}_n^{-1}\right]\right.\\
&\quad \left.\left[\left(\widehat{\boldsymbol{\beta}}_1^{\mathrm{FM}}-\boldsymbol{\beta}_1\right)-\left(\widehat{\boldsymbol{\beta}}_1^{\mathrm{FM}}-\widehat{\boldsymbol{\beta}}_1^{\mathrm{SM}}\right)(p_2-2)\mathcal{T}_n^{-1}\right]^{\top}\right\}\\
&= \mathrm{E}\left\{\vartheta_1\vartheta_1^{\top}-2(p_2-2)\vartheta_3\vartheta_1^{\top}\mathcal{T}_n^{-1}+(p_2-2)^2\vartheta_3\vartheta_3^{\top}\mathcal{T}_n^{-2}\right\}.
\end{aligned}$$

Note that, by using Lemma 3.2 and the formula for a conditional mean of a bivariate normal, we have

$$\begin{aligned}
\mathrm{E}\left\{\vartheta_3\vartheta_1^{\top}\mathcal{T}_n^{-1}\right\} &= \mathrm{E}\left\{\mathrm{E}\left(\vartheta_3\vartheta_1^{\top}\mathcal{T}_n^{-1}|\vartheta_3\right)\right\}\\
&= \mathrm{E}\left\{\vartheta_3\mathrm{E}\left(\vartheta_1^{\top}\mathcal{T}_n^{-1}|\vartheta_3\right)\right\}\\
&= \mathrm{E}\left\{\vartheta_3\left[-\boldsymbol{\eta}_{11.2}+(\vartheta_3-\boldsymbol{\delta})\right]^{\top}\mathcal{T}_n^{-1}\right\}\\
&= -\mathrm{E}\left\{\vartheta_3\boldsymbol{\eta}_{11.2}^{\top}\mathcal{T}_n^{-1}\right\}+\mathrm{E}\left\{\vartheta_3(\vartheta_3-\boldsymbol{\delta})^{\top}\mathcal{T}_n^{-1}\right\}\\
&= -\boldsymbol{\eta}_{11.2}^{\top}\mathrm{E}\left\{\vartheta_3\mathcal{T}_n^{-1}\right\}+\mathrm{E}\left\{\vartheta_3\vartheta_3^{\top}\mathcal{T}_n^{-1}\right\}\\
&\quad -\mathrm{E}\left\{\vartheta_3\boldsymbol{\delta}^{\top}\mathcal{T}_n^{-1}\right\}\\
&= -\boldsymbol{\eta}_{11.2}^{\top}\boldsymbol{\delta}\mathrm{E}\left(\chi_{p_2+2}^{-2}(\Delta)\right)+\left\{\operatorname{Cov}(\vartheta_3\vartheta_3^{\top})\mathrm{E}\left(\chi_{p_2+2}^{-2}(\Delta)\right)\right.\\
&\quad \left.+\mathrm{E}(\vartheta_3)\mathrm{E}\left(\vartheta_3^{\top}\right)\mathrm{E}\left(\chi_{p_2+4}^{-2}(\Delta)\right)-\boldsymbol{\delta}\boldsymbol{\delta}^{\top}\mathrm{H}_{p_2+2}\left(\chi_{p_2,\alpha}^{2};\Delta\right)\right\}\\
&= -\boldsymbol{\eta}_{11.2}^{\top}\boldsymbol{\delta}\mathrm{E}\left(\chi_{p_2+2}^{-2}(\Delta)\right)+\boldsymbol{\Phi}_*\mathrm{E}\left(\chi_{p_2+2}^{-2}(\Delta)\right)\\
&\quad +\boldsymbol{\delta}\boldsymbol{\delta}^{\top}\mathrm{E}\left(\chi_{p_2+4}^{-2}(\Delta)\right)-\boldsymbol{\delta}\boldsymbol{\delta}^{\top}\mathrm{E}\left(\chi_{p_2+2}^{-2}(\Delta)\right),
\end{aligned}$$

$$\begin{aligned}
\operatorname{Cov}\left(\widehat{\boldsymbol{\beta}}_1^{\mathrm{S}}\right) &= \sigma^2\tilde{\boldsymbol{Q}}_{11.2}^{-1}+\boldsymbol{\eta}_{11.2}\boldsymbol{\eta}_{11.2}^{\top}+2(p_2-2)\boldsymbol{\eta}_{11.2}^{\top}\boldsymbol{\delta}\mathrm{E}\left(\chi_{p_2+2,\alpha}^{-2}(\Delta)\right)\\
&\quad -(p_2-2)\boldsymbol{\Phi}_*\left\{2\mathrm{E}\left(\chi_{p_2+2}^{-2}(\Delta)\right)-(p_2-2)\mathrm{E}\left(\chi_{p_2+2}^{-4}(\Delta)\right)\right\}\\
&\quad +(p_2-2)\boldsymbol{\delta}\boldsymbol{\delta}^{\top}\left\{-2\mathrm{E}\left(\chi_{p_2+4}^{-2}(\Delta)\right)+2\mathrm{E}\left(\chi_{p_2+2}^{-2}(\Delta)\right)\right.\\
&\quad \left.+(p_2-2)\mathrm{E}\left(\chi_{p_2+4}^{-4}(\Delta)\right)\right\}.
\end{aligned}$$

Finally, the asymptotic covariance matrix of positive shrinkage ridge regression estimator is derived as follows:

$$\begin{aligned}
\operatorname{Cov}\left(\widehat{\boldsymbol{\beta}}_1^{\mathrm{PS}}\right) &= \mathrm{E}\left\{\lim_{n\to\infty} n\left(\widehat{\boldsymbol{\beta}}_1^{\mathrm{PS}}-\boldsymbol{\beta}_1\right)\left(\widehat{\boldsymbol{\beta}}_1^{\mathrm{PS}}-\boldsymbol{\beta}_1\right)^{\top}\right\}\\
&= \operatorname{Cov}\left(\widehat{\boldsymbol{\beta}}_1^{\mathrm{S}}\right)-2\mathrm{E}\left\{\lim_{n\to\infty}\sqrt{n}\left[\left(\widehat{\boldsymbol{\beta}}_1^{\mathrm{FM}}-\widehat{\boldsymbol{\beta}}_1^{\mathrm{SM}}\right)\left(\widehat{\boldsymbol{\beta}}_1^{\mathrm{S}}-\boldsymbol{\beta}_1\right)^{\top}\right.\right.\\
&\quad \left.\left.\times\left\{1-(p_2-2)\mathcal{T}_n^{-1}\right\}\mathrm{I}(\mathcal{T}_n\le p_2-2)\right]\right\}\\
&\quad +\mathrm{E}\left\{\lim_{n\to\infty}\sqrt{n}\left[\left(\widehat{\boldsymbol{\beta}}_1^{\mathrm{FM}}-\widehat{\boldsymbol{\beta}}_1^{\mathrm{SM}}\right)\left(\widehat{\boldsymbol{\beta}}_1^{\mathrm{FM}}-\widehat{\boldsymbol{\beta}}_1^{\mathrm{SM}}\right)^{\top}\right.\right.\\
&\quad \left.\left.\times\left\{1-(p_2-2)\mathcal{T}_n^{-1}\right\}^2\mathrm{I}(\mathcal{T}_n\le p_2-2)\right]\right\}\\
&= \operatorname{Cov}\left(\widehat{\boldsymbol{\beta}}_1^{\mathrm{S}}\right)-2\mathrm{E}\left\{\vartheta_3\vartheta_1^{\top}\left\{1-(p_2-2)\mathcal{T}_n^{-1}\right\}\mathrm{I}(\mathcal{T}_n\le p_2-2)\right\}\\
&\quad +2\mathrm{E}\left\{\vartheta_3\vartheta_3^{\top}(p_2-2)\mathcal{T}_n^{-1}\mathrm{I}(\mathcal{T}_n\le p_2-2)\right\}\\
&\quad -2\mathrm{E}\left\{\vartheta_3\vartheta_3^{\top}(p_2-2)^2\mathcal{T}_n^{-2}\mathrm{I}(\mathcal{T}_n\le p_2-2)\right\}\\
&\quad +\mathrm{E}\left\{\vartheta_3\vartheta_3^{\top}\mathrm{I}(\mathcal{T}_n\le p_2-2)\right\}
\end{aligned}$$

$$
\begin{aligned}
& -2\mathrm{E}\left\{\vartheta_3\vartheta_3^\top (p_2-2)\mathcal{T}_n^{-1}\mathrm{I}(\mathcal{T}_n \le p_2-2)\right\} \\
& +\mathrm{E}\left\{\vartheta_3\vartheta_3^\top (p_2-2)^2\mathcal{T}_n^{-2}\mathrm{I}(\mathcal{T}_n \le p_2-2)\right\} \\
= \; & \mathrm{Cov}\left(\widehat{\boldsymbol{\beta}}_1^{\mathrm{S}}\right) - 2\mathrm{E}\left\{\vartheta_3\vartheta_1^\top\left\{1-(p_2-2)\mathcal{T}_n^{-1}\right\}\mathrm{I}(\mathcal{T}_n \le p_2-2)\right\} \\
& -\mathrm{E}\left\{\vartheta_3\vartheta_3^\top (p_2-2)^2\mathcal{T}_n^{-2}\mathrm{I}(\mathcal{T}_n \le p_2-2)\right\} \\
& +\mathrm{E}\left\{\vartheta_3\vartheta_3^\top\mathrm{I}(\mathcal{T}_n \le p_2-2)\right\}.
\end{aligned}
$$

Based on Lemma 3.2 and the formula for a conditional mean of a bivariate normal, we have

$$
\begin{aligned}
& \mathrm{E}\left\{\vartheta_3\vartheta_1^\top\left\{1-(p_2-2)\mathcal{T}_n^{-1}\right\}\mathrm{I}(\mathcal{T}_n \le p_2-2)\right\} \\
= \; & \mathrm{E}\left\{\mathrm{E}\left(\vartheta_3\vartheta_1^\top\left\{1-(p_2-2)\mathcal{T}_n^{-1}\right\}\mathrm{I}(\mathcal{T}_n \le p_2-2)\,|\vartheta_3\right)\right\} \\
= \; & \mathrm{E}\left\{\vartheta_3\mathrm{E}\left(\vartheta_1^\top\left\{1-(p_2-2)\mathcal{T}_n^{-1}\right\}\mathrm{I}(\mathcal{T}_n \le p_2-2)\,|\vartheta_3\right)\right\} \\
= \; & \mathrm{E}\left\{\vartheta_3\left[-\boldsymbol{\eta}_{11.2}+(\vartheta_3-\boldsymbol{\delta})\right]^\top\left\{1-(p_2-2)\mathcal{T}_n^{-1}\right\}\mathrm{I}(\mathcal{T}_n \le p_2-2)\right\} \\
= \; & -\boldsymbol{\eta}_{11.2}\mathrm{E}\left(\vartheta_3\left\{1-(p_2-2)\mathcal{T}_n^{-1}\right\}\mathrm{I}(\mathcal{T}_n \le p_2-2)\right) \\
& +\mathrm{E}\left(\vartheta_3\vartheta_3^\top\left\{1-(p_2-2)\mathcal{T}_n^{-1}\right\}\mathrm{I}(\mathcal{T}_n \le p_2-2)\right) \\
& -\mathrm{E}\left(\vartheta_3\boldsymbol{\delta}^\top\left\{1-(p_2-2)\mathcal{T}_n^{-1}\right\}\mathrm{I}(\mathcal{T}_n \le p_2-2)\right) \\
= \; & -\boldsymbol{\delta}\boldsymbol{\eta}_{11.2}^\top\boldsymbol{E}\left(\left\{1-(p_2-2)\chi_{p_2+2}^{-2}(\Delta)\right\}\mathrm{I}\left(\chi_{p_2+2}^{2}(\Delta) \le p_2-2\right)\right) \\
& +\boldsymbol{\Phi}_*\mathrm{E}\left(\left\{1-(p_2-2)\chi_{p_2+2}^{-2}(\Delta)\right\}\mathrm{I}\left(\chi_{p_2+2}^{2}(\Delta) \le p_2-2\right)\right) \\
& +\boldsymbol{\delta}\boldsymbol{\delta}^\top\mathrm{E}\left(\left\{1-(p_2-2)\chi_{p_{2+4}}^{-2}(\Delta)\right\}\mathrm{I}\left(\chi_{p_{2+4}}^{2}(\Delta) \le p_2-2\right)\right) \\
& -\boldsymbol{\delta}\boldsymbol{\delta}^\top\mathrm{E}\left(\left\{1-(p_2-2)\chi_{p_2+2}^{-2}(\Delta)\right\}\mathrm{I}\left(\chi_{p_2+2}^{2}(\Delta) \le p_2-2\right)\right),
\end{aligned}
$$

$$
\begin{aligned}
\mathrm{Cov}\left(\widehat{\boldsymbol{\beta}}_1^{\mathrm{PS}}\right) = \; & \mathrm{Cov}\left(\widehat{\boldsymbol{\beta}}_1^{\mathrm{S}}\right) \\
& +2\boldsymbol{\delta}\boldsymbol{\eta}_{11.2}^\top\mathrm{E}\left(\left\{1-(p_2-2)\chi_{p_2+2}^{-2}(\Delta)\right\}\mathrm{I}\left(\chi_{p_2+2}^{2}(\Delta) \le p_2-2\right)\right) \\
& -2\boldsymbol{\Phi}_*\mathrm{E}\left(\left\{1-(p_2-2)\chi_{p_2+2}^{-2}(\Delta)\right\}\mathrm{I}\left(\chi_{p_2+2}^{-2}(\Delta) \le p_2-2\right)\right) \\
& -2\boldsymbol{\delta}\boldsymbol{\delta}^\top\mathrm{E}\left(\left\{1-(p_2-2)\chi_{p_2+4}^{-2}(\Delta)\right\}\mathrm{I}\left(\chi_{p_2+4}^{2}(\Delta) \le p_2-2\right)\right) \\
& +2\boldsymbol{\delta}\boldsymbol{\delta}^\top\mathrm{E}\left(\left\{1-(p_2-2)\chi_{p_2+2}^{-2}(\Delta)\right\}\mathrm{I}\left(\chi_{p_2+2}^{2}(\Delta) \le p_2-2\right)\right) \\
& -(p_2-2)^2\boldsymbol{\Phi}_*\mathrm{E}\left(\chi_{p_2+2,\alpha}^{-4}(\Delta)\mathrm{I}\left(\chi_{p_2+2,\alpha}^{2}(\Delta) \le p_2-2\right)\right) \\
& -(p_2-2)^2\boldsymbol{\delta}\boldsymbol{\delta}^\top\mathrm{E}\left(\chi_{p_2+4}^{-4}(\Delta)\mathrm{I}\left(\chi_{p_2+2}^{2}(\Delta) \le p_2-2\right)\right) \\
& +\boldsymbol{\Phi}_*\mathrm{H}_{p_2+2}(p_2-2;\Delta)+\boldsymbol{\delta}\boldsymbol{\delta}^\top\mathrm{H}_{p_2+4}(p_2-2;\Delta) \\
= \; & \mathrm{Cov}\left(\widehat{\boldsymbol{\beta}}_1^{\mathrm{S}}\right) \\
& +2\boldsymbol{\delta}\boldsymbol{\eta}_{11.2}^\top\mathrm{E}\left(\left\{1-(p_2-2)\chi_{p_2+2}^{-2}(\Delta)\right\}\mathrm{I}\left(\chi_{p_2+2}^{2}(\Delta) \le p_2-2\right)\right) \\
& +(p_2-2)\sigma^2\tilde{\boldsymbol{Q}}_{11}^{-1}\tilde{\boldsymbol{Q}}_{12}\tilde{\boldsymbol{Q}}_{22.1}^{-1}\tilde{\boldsymbol{Q}}_{21}\tilde{\boldsymbol{Q}}_{11}^{-1} \\
& \times\left[2\mathrm{E}\left(\chi_{p_2+2}^{-2}(\Delta)\mathrm{I}\left(\chi_{p_2+2}^{2}(\Delta) \le p_2-2\right)\right)\right. \\
& \left.-(p_2-2)\mathrm{E}\left(\chi_{p_2+2}^{-4}(\Delta)\mathrm{I}\left(\chi_{p_2+2}^{2}(\Delta) \le p_2-2\right)\right)\right] \\
& -\sigma^2\tilde{\boldsymbol{Q}}_{11}^{-1}\tilde{\boldsymbol{Q}}_{12}\tilde{\boldsymbol{Q}}_{22.1}^{-1}\tilde{\boldsymbol{Q}}_{21}\tilde{\boldsymbol{Q}}_{11}^{-1}\mathrm{H}_{p_2+2}(p_2-2;\Delta) \\
& +\boldsymbol{\delta}\boldsymbol{\delta}^\top\left[2\mathrm{H}_{p_2+2}(p_2-2;\Delta)-\mathrm{H}_{p_2+4}(p_2-2;\Delta)\right] \\
& -(p_2-2)\boldsymbol{\delta}\boldsymbol{\delta}^\top\left[2\mathrm{E}\left(\chi_{p_2+2}^{-2}(\Delta)\mathrm{I}\left(\chi_{p_2+2}^{2}(\Delta) \le p_2-2\right)\right)\right. \\
& -2\mathrm{E}\left(\chi_{p_2+4}^{-2}(\Delta)\mathrm{I}\left(\chi_{p_2+4}^{2}(\Delta) \le p_2-2\right)\right) \\
& \left.+(p_2-2)\mathrm{E}\left(\chi_{p_2+2}^{-4}(\Delta)\mathrm{I}\left(\chi_{p_2+2}^{2}(\Delta) \le p_2-2\right)\right)\right].
\end{aligned}
$$

Theorem 5.3 *The asymptotic risks of the estimators can be derived by following the definition of ADR*

$$\begin{aligned}
\mathrm{ADR}\left(\boldsymbol{\beta}_1^*\right) &= n\mathrm{E}\left[\left(\boldsymbol{\beta}_1^* - \boldsymbol{\beta}_1\right)^\top \boldsymbol{W}\left(\boldsymbol{\beta}_1^* - \boldsymbol{\beta}_1\right)\right] \\
&= n\mathrm{tr}\left[\boldsymbol{W}\mathrm{E}\left(\boldsymbol{\beta}_1^* - \boldsymbol{\beta}_1\right)\left(\boldsymbol{\beta}_1^* - \boldsymbol{\beta}_1\right)^\top\right] \\
&= \mathrm{tr}\left(\boldsymbol{W}\mathrm{Cov}\left(\boldsymbol{\beta}_1^*\right)\right).
\end{aligned}$$

6

Shrinkage Strategies : Generalized Linear Models

6.1 Introduction

Generalized linear models (GLMs) are the natural extensions of classical linear models that allow for greater flexibility. GLMs are useful in modeling data in he social sciences, biology, medicine, and survival analysis, to mention a few. GLMs are based on an assumed relationship between the mean of the response variable and a linear combination of explanatory variables. Data may be assumed to be from a host of probability distribution functions, including Bernoulli, normal, binomial, Poisson, negative binomial, and gamma distributions, many of which generally provide good fits to non-normal error structures. Technically, this strategy models the conditional distribution of a random variable Y given a set of predictors to follow a distribution in the exponential family using a linear combination $\boldsymbol{x}^\top\boldsymbol{\beta}$, where $\boldsymbol{\beta}$ is a vector of regression coefficients. The parameter vector $\boldsymbol{\beta}$ is unknown and to estimate it or to test hypotheses about it is achieved by using the maximum likelihood estimation method and the likelihood ratio test.

Generally, the observations pertaining to a given statistical model can usually be summarized in terms of a random component and a systematic component. In the GLM, the random component is inherent in the exponential family distribution of the observation, and the systematic component assumes a linear structure in the predictor variables for a function of the conditional mean. We refer to Dobson and Barnett (2018)for some insights. This function is known as the *link function.* When the parameter θ_i is modeled as a linear function of the predictors, the link function is known as a *canonical link.* Thus, for a given vector of observations $\boldsymbol{Y} = (y_1, y_2, \ldots, y_n)^\top$, where y_i is assumed to have a distribution in the exponential family of distributions with predictor values $\boldsymbol{x}_i = (x_{i1}, x_{i2}, \ldots, x_{in})^\top$. Then a probability density/mass function has the form

$$f_Y(y_i; \theta_i, \phi) = \exp((y_i\theta_i - b(\theta_i))/a_i(\phi) + c(y_i, \phi)), \tag{6.1}$$

where $a(\cdot)$, $b(\cdot)$, and $c(\cdot)$ are known functions, and ϕ is the *dispersion parameter.* If this parameter is unknown, then it is treated as a nuisance parameter in the inferential process. However, if ϕ is known, then this is an exponential-family model with canonical parameter θ_i. In practice, researchers are interested in applying GLM procedures where the dispersion parameter ϕ is known i.e., when the responses are binary or count data. In this case, the above density function can be written as

$$f_Y(y_i; \theta_i) = c(y_i)exp(y_i\theta_i - b(\theta_i)). \tag{6.2}$$

GLMs have the following key features, see McCullagh and Nelder (1989).

(i) The random component of a GLM specifies the distribution of the response variable Y_i. The distribution has the form (6.2) and for any distribution of this form, the mean and variance of Y_i are given by

$$E(Y_i) = \mu_i = \frac{db(\theta_i)}{d\theta_i} = g^{-1}(\boldsymbol{x}_i^\top\boldsymbol{\beta}) \text{ and } Var(Y_i) = V(\mu_i) = \frac{D_n^2 b(\theta_i)}{d\theta_i^2}.$$

DOI: 10.1201/9781003170259-6

(ii) The systematic component of a GLM is a linear combination of predictors or regressor variables, termed the linear predictor $\theta_i = \boldsymbol{x}_i^\top \boldsymbol{\beta}$, where $\boldsymbol{x}_i^\top = (x_{i1}, x_{i2}, \cdots, x_{in})$ is the predictors and $\boldsymbol{\beta}$ is the vector of model parameters. The linear form of the systematic component places the predictors on an additive scale, making the interpretation of their respective effects simple. Further, the significance of each predictor can be tested with linear restrictions $H_0 : \boldsymbol{H}\boldsymbol{\beta} = \boldsymbol{h}$ versus $H_a : \boldsymbol{H}\boldsymbol{\beta} \neq \boldsymbol{h}$, where $\boldsymbol{H}$ is an restriction matrix, and $\boldsymbol{h}$ is an vector of constants.

(iii) The *link* function of a GLM specifies a monotonic differentiable function, which connects the random and systematic components. This connection has been done by equating the mean response μ_i to the linear predictor θ_i by $\theta_i = g(\mu_i)$, that is

$$g(\mu_i) \overset{link}{=} \theta_i = \boldsymbol{x}_i^\top \boldsymbol{\beta}.$$

The link function that equates the linear predictor to the canonical parameter is called the canonical link, $\theta_i = \boldsymbol{x}_i^\top \boldsymbol{\beta} = g(\mu_i)$. The link function $g(\mu_i) = \mu_i$ is the identity link function which equates the conditional mean response to the linear predictor. Therefore the link function for the regression model with normally distributed response variable Y_i is the identity link.

In application, a given data set may be distributed according to some unknown member of the exponential family and therefore, different link functions have to be examined. The link is a linearizing transformation of the mean. In other words, a function that maps the response mean onto a scale on which linearity is assumed. One purpose of the link is to allow θ_i to range freely while restricting the range of μ_i. For example, the inverse logit link $\mu_i = 1/1 + e^{-\theta_i}$ maps $(-\infty, \infty)$ onto $(0, 1)$, which is an appropriate range if μ_i is a probability. The monotonicity of the link function guarantees that this mapping is one-to-one. Consequentially, we can express the GLM in terms of the inverse link function

$$E(Y_i) = \mu_i = g^{-1}(\boldsymbol{x}_i^\top \boldsymbol{\beta}).$$

In summary, the canonical link is a useful link function in many cases and is a reasonable function to try, unless the subject matter suggests otherwise. Indeed, the canonical link does simplify the estimation method slightly. Having said that, there is no need to restrict generalized linear models to canonical link functions.

Finally, the generalized linear models form a general class of probabilistic regression models with the assumptions that:

(i) the response probability distribution is a member of the exponential family of distributions;

(ii) the responses Y_i, $i = 1, 2, \cdots, n$ form a set of independent response random variables;

(iii) the predicting variables are linearly combined to explain systematic variation in a function of the mean.

In practice, generalized linear model fitting involves the following:

- choosing a relevant error distribution.
- identifying the predicting variables to be included in the systematic components.
- specifying the link function.

One important task is to estimate the parameters involved in a given model. In the following section we describe the maximum likelihood method for estimating the regression parameters under the usual assumptions specified earlier.

6.2 Maximum Likelihood Estimation

If the data follows an exponential family model, the maximum likelihood method is the best way to estimate the parameters. By maximum likelihood method; we refer to Green and Silverman (1993). The maximum likelihood estimators (MLE) possess rich properties including consistency, efficiency, and asymptotic normality. Thus, it is natural to consider the maximum likelihood procedure for estimating model parameters.

Let the responses $y_1, y_2, \cdots, y_n$ be generated from a member of the exponential family (6.2). The log-likelihood is given by:

$$\partial l(\boldsymbol{\beta}) = \sum_{i=1}^{n} ((y_i\theta_i - b(\theta_i)) + \ln c(y_i)) = \sum_{i=1}^{n} \ell_i, \tag{6.3}$$

where ℓ_i is the ith component of the log-likelihood. The likelihood implicitly depends on the parameters $\beta_j,\ \ j = 1, 2, \cdots, k$, firstly through the link function $g(\mu_i)$ and secondly through the linearity that it encompasses with respect to β_j values. The derivatives of the log-likelihood with respect to β_j are evaluated by the chain rule:

$$U_j(\boldsymbol{\beta}) = \frac{\partial l}{\partial \beta_j} = \sum_{i=1}^{n} \frac{\partial \ell_i}{\partial \theta_i}\frac{\partial \theta_i}{\partial \mu_i}\frac{\partial \mu_i}{\partial \theta_i}\frac{\partial \theta_i}{\partial \beta_j} = 0; \quad j = 1, 2, \cdots, k \tag{6.4}$$

it reduces to

$$\frac{\partial l}{\partial \beta_j} = \sum_{i=1}^{n} \frac{y_i - \mu_i}{V(\mu_i)}\frac{d\mu_i}{d\theta_i} x_{ij}; \quad j = 1, 2, \cdots, k. \tag{6.5}$$

Alternatively, the equations can be represented in a vector form for convenience:

$$(\boldsymbol{Y} - \boldsymbol{\mu})^\top \boldsymbol{D_n}(\boldsymbol{\mu})\boldsymbol{X} = \boldsymbol{0}, \tag{6.6}$$

where $\boldsymbol{X} = (\boldsymbol{x}_1, \boldsymbol{x}_2, \cdots, \boldsymbol{x}_n)^\top$, $\boldsymbol{D_n}(\boldsymbol{\mu}) = \text{diag}(d_{ii})$ and $d_{ii} = 1/V(\mu_i)g'(\mu_i)$ with $g'(\mu_i) = \partial g/\partial \theta_i$. The maximum likelihood estimator of $\boldsymbol{\beta}$ is obtained by solving the these equations (6.6) for $\widehat{\boldsymbol{\beta}}^{\text{FM}}$. The numerical methods to solve (6.6) are essentially iterative in nature. The following technique is employed using the approximate linearized form of $g(y_i)$, where

$$g(y_i) \approx g(\mu_i) + (y_i - \mu_i)g'(\mu_i) = \theta_i + (y_i - \mu_i)\frac{d\theta_i}{d\mu_i} = z_i,$$

and z_i is the adjusted dependent variable that depends on both y_i and μ_i. The variance of z_i is $(g'(\mu_i))^2 V(\mu_i)$, an initial estimate of $\boldsymbol{\beta}$ may be obtained by weighted least squares of $\boldsymbol{z}$ on $\boldsymbol{X}$, with variance-covariance matrix given by a diagonal matrix $\boldsymbol{W}$ whose components are, $w_{ii} = 1/V(\mu_i)(g'(\mu_i))^2$. The equations (6.6) can be written as

$$\sum_{i=1}^{n}(y_i - \mu_i)g'(\mu_i)w_{ii}x_{ij} = \sum_{i=1}^{n}(z_i - g(\mu_i))w_{ii}x_{ij} = 0. \tag{6.7}$$

Both $\boldsymbol{z}$ and $\boldsymbol{W}$ are used for maximum likelihood estimation through weighted least squares regression. This process is iterative, since both $\boldsymbol{z}$ and $\boldsymbol{W}$ depend on the fitted values of current estimates. Let $(\widehat{\boldsymbol{\beta}}^{\text{FM}})^{(r)}$ be an approximation to the maximum likelihood estimator $\boldsymbol{\beta}$, the Fisher's scoring algorithm for computing the MLE of $\widehat{\boldsymbol{\beta}}^{\text{FM}}$ yields

$$(\widehat{\boldsymbol{\beta}}^{\text{FM}})^{(r+1)} = (\widehat{\boldsymbol{\beta}}^{\text{FM}})^{(r)} + (\boldsymbol{X}^\top \boldsymbol{W}\boldsymbol{X})^{-1}\boldsymbol{X}^\top \boldsymbol{W}\boldsymbol{z}^*,$$

where $\boldsymbol{W} = \text{diag}(w_{ii})$ with $w_{ii} = \left(g'(\mu_i)^2 b''(\theta_i)\right)^{-1}$ and $\boldsymbol{z}^*$ the n-vector with $z_i^* = (y_i - \mu_i)g'(\mu_i)$. Fahrmeir and Kaufmann (1985) proved that under usual regularity conditions, $\widehat{\boldsymbol{\beta}}^{\text{FM}}$ is asymptotically normal $N(\boldsymbol{0}, (\boldsymbol{X}^\top \boldsymbol{W}\boldsymbol{X})^{-1})$.

In this Chapter, our focus is on the logit model, a very important member of the generalized linear model family. Many researchers prefer to work directly with this model instead of the GLM, therefore it is treated as an independent model in its own right. We are primarily interested in logistic regression model that is widely used to model independent binary response data in medical and epidemiologic studies, among others. Essentially, the model assumes that the logit of the response variable can be modeled by a linear combination of unknown parameters. For detailed information on logistic regression we refer to Hilbe (2009) and Hosmer Jr et al. (2013), and among others.

6.3 A Genle Introduction of Logistic Regression Model

Suppose that $y_1, y_2, \cdots, y_n$ are independent binary response variables that take a value of 0 or 1, and that $\boldsymbol{x}_i = (x_{i1}, x_{i2}, \cdots, x_{ip})^\top$ is a $p \times 1$ vector of predictors for the ith subject. Let us define $\pi(z) = e^z/1 + e^z$. The logistic regression model assumes that

$$\text{P}(y = 1|\boldsymbol{x}_i) = \pi(\boldsymbol{x}_i^\top \boldsymbol{\beta}) = \frac{\exp(\boldsymbol{x}_i^\top \boldsymbol{\beta})}{1 + \exp(\boldsymbol{x}_i^\top \boldsymbol{\beta})}, \quad 1 \leq i \leq n,$$

where $\boldsymbol{\beta}$ is a $p \times 1$ vector of regression parameters. The log-likelihood is given as follows:

$$\partial l(\boldsymbol{\beta}) = \sum_{i=1}^{n} \left[y_i \ln \pi(\boldsymbol{x}_i^\top \boldsymbol{\beta}) + (1 - y_i)\ln(1 - \pi(\boldsymbol{x}_i^\top \boldsymbol{\beta}))\right]. \tag{6.8}$$

Naturally, the log-likelihood function depends on unknown regression parameter vectors $\boldsymbol{\beta}$, which need to be estimated based on sample information. The derivative of the log-likelihood with respect to $\boldsymbol{\beta}$ is obtained by using the chain rule:

$$\frac{\partial l}{\partial \boldsymbol{\beta}} = \sum_{i=1}^{n} \left[y_i - \pi(\boldsymbol{x}_i^\top \boldsymbol{\beta})\right] \boldsymbol{x}_i = 0. \tag{6.9}$$

The maximum likelihood estimator $\widehat{\boldsymbol{\beta}}^{\text{FM}}$ of $\boldsymbol{\beta}$ is obtained by solving the score equation (6.9). Clearly, the equation is non-linear in parameter vector $\boldsymbol{\beta}$. In an effort to solve the likelihood equation we take recourse in using an iterative procedure, such as Newton-Raphson, to determine the value of $\widehat{\boldsymbol{\beta}}^{\text{FM}}$ of $\boldsymbol{\beta}$ that maximizes the log-likelihood function $l(\boldsymbol{\beta})$. It has been shown that under the assumed usual regularity conditions, $\widehat{\boldsymbol{\beta}}^{\text{FM}}$ is consistent and asymptotically normal with variance-covariance matrix $(\boldsymbol{I}(\boldsymbol{\beta}))^{-1}$, where

$$\boldsymbol{I}(\boldsymbol{\beta}) = \sum_{i=1}^{n} \pi(\boldsymbol{x}_i^\top \boldsymbol{\beta})(1 - \pi(\boldsymbol{x}_i^\top \boldsymbol{\beta}))\boldsymbol{x}_i \boldsymbol{x}_i^\top.$$

6.3.1 Statement of the Problem

In this chapter, we focus on the problem of estimating the regression coefficients of a logistic regression model when many predictors are included in the model and some of them may be less relevant or less influential on the response of interest. In other words, some predictors are active (influential) while many others are inactive. This information leads to two choices for practitioners: a full model with all predictors or a candidate submodel that contains active predictors only. In this situation, we consider the information from inactive predictors and either use the full model or the submodel. Model selection is a fundamental task in statistical data analysis and is one of the most pervasive problems in statistical applications. Two aspects of modeling are accurately interpret the selected predictors relationship with the response variable, and for prediction of the fitted model. Classical methods for developing these aspects have not had much success when there are a moderate-to-large number of predictors in the model. In this chapter, shrinkage and penalized likelihood approaches are proposed to deal with these kinds of problems in the case of binary data. The penalized methods select variables and estimate coefficients simultaneously. The shrinkage method that combines the full and submodel estimators is inspired by Stein's result, where efficient estimates can be obtained by shrinking a full model estimator in the direction of a submodel estimator when there are more than two dimensions. Existing literature (see, Ahmed et al. (2012) and Ahmed et al. (2007)) show that the shrinkage estimators improve upon the penalty Tibshirani (1996) over other classical estimators. Several authors developed the shrinkage estimation strategy for parametric, semi-parametric, and non-parametric linear models for censored and uncensored data, (see Ahmed et al. (2012), Ahmed et al. (2007), Ahmed et al. (2006) and others). Here we present the shrinkage estimation method for modeling binary data by amalgamating the ideas from recent literature on sparsity patterns and comparing the resulting estimator to the full and submodel estimators as well as to penalty estimators.

We now present a motivating example to illustrate the above situation. There are many such examples available in the reviewed scientific literature.

A motivating example: Hosmer Jr et al. (2013) considered low birth weight data. The data was collected at Baystate Medical Center in Springfield, Massachusetts in 1996 as low birth weight has been a concern for physicians. The goal of this study was to identify risk factors associated with giving birth to a low birth weight baby (weighing less than 2500 grams). Data was collected from 189 women, of whom 59 had low birth weight babies and 130 had the normal birth weight babies. The predictor variables are age, weight of the mother at the last menstrual period, race, smoking status, history of premature labor, history of hypertension, presence of uterine irritability, and the number of physician visits during the first trimester of pregnancy. The shrinkage method uses a two-step approach to estimate the coefficients of active predictors. In the first step, an AIC or BIC criterion is used to form a subset of the total set of predictors. This criterion shows that the history of premature labor, history of hypertension, weight of the mother at the last menstrual period, smoking status, and race of the mother are the active predictors, and the effects of the other inactive predictors may be ignored. In this situation, we can partition the full parameter vector $\boldsymbol{\beta}$ into active and inactive parameter sub-vectors as $\boldsymbol{\beta} = (\boldsymbol{\beta}_1^\top, \boldsymbol{\beta}_2^\top)^\top$, where $\boldsymbol{\beta}_1$ and $\boldsymbol{\beta}_2$ are assumed to have dimensions $p_1 \times 1$ and $p_2 \times 1$, respectively, such that $p = p_1 + p_2$. Our interest lies in the estimation of the parameter sub-vector $\boldsymbol{\beta}_1$ when the information on $\boldsymbol{\beta}_2$ is available. The information about the inactive parameters may be used to estimate $\boldsymbol{\beta}_1$ when their values are near some specified value which, without loss of generality, may be set to a null vector,

$$\boldsymbol{\beta}_2 = \boldsymbol{\beta}_2^0 = \mathbf{0}. \tag{6.10}$$

In the second step, we combine the submodel and full model estimators using the shrinkage strategy to achieve an improved estimator for the remaining active predictors. This approach has been implemented for low-dimensional scenarios, where n is relatively larger than p. For some insights on the application of the shrinkage strategy in GLMs; we refer to Hossain et al. (2015); Lisawadi et al. (2016); Hossain and Ahmed (2014); Hossain et al. (2014); Reangsephet et al. (2021); Hossain and Ahmed (2012); Lisawadi et al. (2021).

However, when p is large and n is small, a common goal is to find genetic explanations that are responsible for observable traits in biomedical studies. Understanding the genetic associations of diseases helps medical researchers to further investigate and develop corresponding treatment methods. Suppose a medical researcher measured about 600 microRNA (miRNA) expressions in serum samples from two groups of participants. One group consisted of 30 oral cancer patients and the other group consisted of 26 individuals without cancer. The question is whether these miRNA readings can be used to distinguish cancer patients from others. If the treatment method is successful, this genetic information might be further used to predict whether an oral-cancer patient will progress from a minor tumor to a serious one. Using all 600 miRNAs for classification leads to a poor predictive value because of the high level of noise. Consequently, it is important to select those that make a major contribution to identifying oral-cancer patients. A logistic regression of the tumor type on the miRNA readings can be used to identify the relevant miRNAs by selecting the most important predictors. However, the number of predictors p is 600, and the number of participants n is just 56. This large-p-small-n situation places this problem outside the domain of classical model statistical methods. However, penalization methods such as LASSO, ALASSO, and SCAD are available to deal with the high dimensionality.

Meier et al. (2008) studied group LASSO for logistic regression. They showed that the group LASSO is consistent under certain conditions and proposed a block coordinate descent algorithm that can handle high-dimensional data. Zou (2006) studied a one-step approach in non-concave penalized likelihood methods in models with a fixed p. This approach is closely related to the ALASSO. Park and Hastie (2007) proposed an algorithm for computing the entire solution path of the L_1 regularized maximum likelihood estimates, which facilitates the choice of a tuning parameter. This algorithm does both shrinkage and variable selection due to the nature of the constraint region, which leads to exact zeros for some of the coefficients. However, it does not satisfy oracle properties, meaning it does not yield unbiased estimates Fan and Li (2001). Zhu and Hastie (2004) used L_2-penalized method for logistic regression to pursue classification in the context of microarray cancer studies with categorical outcomes. These methods have been extensively studied in the literature; for example, Radchenko and James (2011), Wang et al. (2010), Huang et al. (2008), Wang and Leng (2007), Yuan and Lin (2006), Efron et al. (2004), and others.

The rest of the chapter is organized as follows. In Section 6.4, we present the estimation strategies for the logistic regression model. Section 6.5 is devoted to developing the asymptotic properties of the non-penalty estimators and their asymptotic distributional biases and risks. In Section 6.6, a simulation study is conducted to assess the relative performance of the listed estimators. Several real data sets are used to illustrate the listed strategies in Section 6.7, and high-dimensional simulations and real data example are given in Section 6.8. In Section 6.9, we give brief information about the negative binomial regression model. Followed by the estimation strategies for negative binomial regression model in Section 6.10. Section 6.11 is devoted to developing the asymptotic properties of the non-penalty estimators and Monte Carlo simulation examples are given in Section 6.12. Two real data examples are given in Section 6.13. Section 6.14 demonstrates uses for a high-dimensional model. The R codes can be found in Section 6.15. Finally, we give our concluding remarks in Section 6.16.

6.4 Estimation Strategies

Suppose that $y_1, y_2, \cdots, y_n$ are independent binary response variables that take a value of 0 or 1, and that $\boldsymbol{x}_i = (x_{i1}, x_{i2}, \cdots, x_{ip})'$ is a $p \times 1$ vector of predictors for the ith subject. Define $\pi(z) = e^z/1 + e^z$. The logistic regression model assumes that

$$\mathrm{P}(y = 1|\boldsymbol{x}_i) = \pi(\boldsymbol{x}_i^\top \boldsymbol{\beta}) = \frac{\exp(\boldsymbol{x}_i^\top \boldsymbol{\beta})}{1 + \exp(\boldsymbol{x}_i^\top \boldsymbol{\beta})}, \quad 1 \le i \le n,$$

where $\boldsymbol{\beta}$ is a $p \times 1$ vector of regression parameters. The log-likelihood is given by

$$l(\boldsymbol{\beta}) = \sum_{i=1}^{n} \left[y_i \ln \pi(\boldsymbol{x}_i^\top \boldsymbol{\beta}) + (1 - y_i) \ln(1 - \pi(\boldsymbol{x}_i^\top \boldsymbol{\beta})) \right]. \tag{6.11}$$

In low-dimensional models, the full model estimator that includes all the available predictors, the maximum likelihood estimator $\widehat{\boldsymbol{\beta}}^{\mathrm{FM}}$ of $\boldsymbol{\beta}$ is obtained by solving the score equation (6.9).

Conversely, under the sparsity condition $\boldsymbol{\beta}_2 = \mathbf{0}$ theoretically the restricted maximum likelihood estimator of the submodel estimator, $\widehat{\boldsymbol{\beta}}^{\mathrm{SM}}$ of $\boldsymbol{\beta}$ is obtained by maximizing the log-likelihood function (6.11).

Let us define a distance measure, D_n to construct the shrinkage estimators. In fact, the likelihood ratio statistic for testing $H_0 : \boldsymbol{\beta}_2 = \mathbf{0}$ can be used as a standardized distance measure between full model and submodel estimators. If $l(\widehat{\boldsymbol{\beta}}^{\mathrm{FM}})$ and $l(\widehat{\boldsymbol{\beta}}^{\mathrm{SM}})$ are the values of the log-likelihood at the full model estimate and submodel estimates respectively, then

$$\begin{aligned} D_n &= 2[l(\widehat{\boldsymbol{\beta}}^{\mathrm{FM}}; y_1, \cdots, y_n) - l(\widehat{\boldsymbol{\beta}}^{\mathrm{SM}}; y_1, \cdots, y_n)], \\ &= n(\widehat{\boldsymbol{\beta}}_2^{\mathrm{MLE}})^\top \left(\boldsymbol{I}_{22} - \boldsymbol{I}_{21}\boldsymbol{I}_{11}^{-1}\boldsymbol{I}_{12}\right) \widehat{\boldsymbol{\beta}}_2^{\mathrm{MLE}} + o_p(1), \end{aligned} \tag{6.12}$$

where $\boldsymbol{I}_{..}$ are the partition matrices of matrix $\boldsymbol{I}(\widehat{\boldsymbol{\beta}}^{\mathrm{FM}})$

$$\begin{pmatrix} \boldsymbol{I}_{11} & \boldsymbol{I}_{12} \\ \boldsymbol{I}_{21} & \boldsymbol{I}_{22} \end{pmatrix},$$

when $\boldsymbol{\beta} = (\boldsymbol{\beta}_1^\top, \boldsymbol{\beta}_2^\top)^\top$. Under the restriction, the distribution of D_n converges to a χ^2 distribution with p_2 degrees of freedom as $n \to \infty$.

6.4.1 The Shrinkage Estimation Strategies

The shrinkage estimator (SE) that shrinks the full model estimator $\widehat{\boldsymbol{\beta}}^{\mathrm{FM}}$ toward $\widehat{\boldsymbol{\beta}}^{\mathrm{SM}}$ is defined on soft thresholding as:

$$\widehat{\boldsymbol{\beta}}^{\mathrm{S}} = \widehat{\boldsymbol{\beta}}^{\mathrm{SM}} + \left(1 - (p_2 - 2)D_n^{-1}\right)(\widehat{\boldsymbol{\beta}}^{\mathrm{FM}} - \widehat{\boldsymbol{\beta}}^{\mathrm{SM}}), \quad p_2 \ge 3.$$

The shrinkage estimator is based on the optimal soft threshold Ahmed (2014). It is a weighted average of the submodel and full model estimators, the weight being a function of D_n.

To adjust for possible over-shrinking, we suggest using a truncated estimator called a positive-part shrinkage (PS) estimator, defined as

$$\widehat{\boldsymbol{\beta}}^{\mathrm{PS}} = \widehat{\boldsymbol{\beta}}^{\mathrm{SM}} + \left(1 - (p_2 - 2)D_n^{-1}\right)^+ (\widehat{\boldsymbol{\beta}}^{\mathrm{FM}} - \widehat{\boldsymbol{\beta}}^{\mathrm{SM}}),$$

where $z^+ = max(0, z)$.

In the next section, we present the asymptotic properties and theoretical comparison of the listed estimators under the following regularity conditions:

(i) The parameter $\boldsymbol{\beta}$ is defined in an open subset $\mathcal{B}$ of $\Re^p$. The function $g(\mu_i)$ has continuous partial derivative with respect to $\boldsymbol{\beta}$ and μ_i up to the third order.

(ii) There exists a unique MLE $\widehat{\boldsymbol{\beta}}^{\text{FM}}$ in $\mathcal{B}$ for $\ell(\boldsymbol{\beta})$. The moments of $\partial\ell/\partial\boldsymbol{\beta}$ exist at least up to the third order.

(iii) The information matrix $E\left(-\frac{\partial^2 \ell}{\partial\boldsymbol{\beta}\partial\boldsymbol{\beta}^\top}\right)$ is finite and positive definite at $\boldsymbol{\beta} = \boldsymbol{\beta}_0$, where $\boldsymbol{\beta}_0$ is the true vector of coefficients.

6.5 Asymptotic Properties

We consider the properties of the shrinkage estimators for a logistic regression model, where the subspace $\boldsymbol{\beta}_2 = \mathbf{0}$ may not hold, $\boldsymbol{\beta}_2 = \mathbf{0} + \boldsymbol{D_n}$. Since the statistic D_n converges to ∞ for fixed $\boldsymbol{D_n} \neq \mathbf{0}$, the SE and PSE will be asymptotically equivalent in probability to $\widehat{\boldsymbol{\beta}}^{\text{FM}}$ for such $\boldsymbol{D_n}$. This leads us to consider the following local alternatives:

$$K_{(n)} : \boldsymbol{\beta}_2 = \frac{\boldsymbol{\delta}}{\sqrt{n}}, \tag{6.13}$$

where $\boldsymbol{\delta} = (\delta_1, \cdots, \delta_{p_2}) \in \Re^{p_2}$ is a fixed vector. The vector $\boldsymbol{\delta}/n^{1/2}$ is a measure of how much the local alternatives $K_{(n)}$ differs from the subspace (6.10). The shrinkage estimators are generally (asymptotically) biased estimators of $\boldsymbol{\beta}$. We consider the following loss function:

$$L(\widehat{\boldsymbol{\beta}}^*, \boldsymbol{\beta}; \boldsymbol{W}) = \left[\sqrt{n}(\widehat{\boldsymbol{\beta}}^* - \boldsymbol{\beta})\right]^\top \boldsymbol{W} \left[\sqrt{n}(\widehat{\boldsymbol{\beta}}^* - \boldsymbol{\beta})\right], \tag{6.14}$$

where $\boldsymbol{W}$ is a positive semi-definite weight matrix and $\widehat{\boldsymbol{\beta}}^*$ is any one of $\widehat{\boldsymbol{\beta}}^{\text{FM}}$, $\widehat{\boldsymbol{\beta}}^{\text{SM}}$, $\widehat{\boldsymbol{\beta}}^{\text{S}}$, or $\widehat{\boldsymbol{\beta}}^{\text{PS}}$. A common choice of $\boldsymbol{W}$ is the identity matrix $\boldsymbol{I}$, which will be used in the numerical study. The asymptotic distribution function of $\boldsymbol{\beta}^*$ under $K_{(n)}$ by

$$G(\boldsymbol{y}) = \lim_{n\to\infty} P\left[\sqrt{n}(\boldsymbol{\beta}^* - \boldsymbol{\beta}) \leq \boldsymbol{y} | K_{(n)}\right],$$

where $G(\boldsymbol{y})$ is a non-degenerate distribution function.

The *asymptotic distributional bias* (ADB) of an estimator $\boldsymbol{\beta}^*$ is defined as

$$\text{ADB}(\boldsymbol{\beta}^*) = \lim_{n\to\infty} E\left\{n^{\frac{1}{2}}(\boldsymbol{\beta}^* - \boldsymbol{\beta})\right\} = \int \cdots \int \boldsymbol{y} dG(\boldsymbol{y}).$$

The *asymptotic distributional risk* (ADR) is defended by using the distribution of $G(\boldsymbol{y})$ and taking the expected value in both sides of (6.14)

$$\begin{aligned} R(\boldsymbol{\beta}^*; \boldsymbol{W}) &= \int \cdots \int \boldsymbol{y}^\top \boldsymbol{W} \boldsymbol{y} dG(\boldsymbol{y}), \\ &= \text{tr}(\boldsymbol{W}\boldsymbol{\Sigma}^*), \end{aligned} \tag{6.15}$$

where $\boldsymbol{\Sigma}^* = \int \cdots \int \boldsymbol{y}\boldsymbol{y}^\top dG(\boldsymbol{y})$ is the dispersion matrix for the distribution $G(\boldsymbol{y})$.

The expressions for ADR and ADB of the shrinkage estimators can be established using the following theorem.

Theorem 6.1 *Under the local alternatives $K_{(n)}$ in (6.13) and the usual regularity conditions, as $n \to \infty$,*

1. $\sqrt{n}\widehat{\boldsymbol{\beta}}_2^{\mathrm{MLE}} \xrightarrow{L} N(\boldsymbol{\delta}, \boldsymbol{I}_{22.1})$

2. *The test statistic* D_n *converges to a non-central chi-squared distribution* $\chi^2_{p_2}(\Delta)$ *with* p_2 *degrees of freedom and non-centrality parameter* $\Delta = \boldsymbol{\delta}^\top \boldsymbol{I}_{22.1}\boldsymbol{\delta}$*, where* $\boldsymbol{I}_{22.1} = \boldsymbol{I}_{22} - \boldsymbol{I}_{21}\boldsymbol{I}_{11}^{-1}\boldsymbol{I}_{12}$ *is a positive definite matrix.*

Using the above theorem, the respective bias expressions are displayed in the following theorem.

Theorem 6.2 *Under the local alternatives* $K_{(n)}$ *and the condition of Theorem 6.1, the ADBs of the estimators are*

$$\begin{aligned}
ADB(\widehat{\boldsymbol{\beta}}^{\mathrm{FM}}) &= \boldsymbol{0}, \\
ADB(\widehat{\boldsymbol{\beta}}^{\mathrm{SM}}) &= \boldsymbol{M\delta}, \quad \boldsymbol{M} = \boldsymbol{I}_{11}^{-1}\boldsymbol{I}_{12}, \\
ADB(\widehat{\boldsymbol{\beta}}^{\mathrm{S}}) &= (p_2-2)\boldsymbol{M\delta}E(\chi^{-2}_{p_2+2}(\Delta)), \\
ADB(\widehat{\boldsymbol{\beta}}^{\mathrm{PS}}) &= ADB(\widehat{\boldsymbol{\beta}}^{\mathrm{S}}) + \boldsymbol{M\delta}\Psi_{p_2+2}(p_2-2,\Delta) \\
&- (p_2-2)\boldsymbol{M\delta}E(\chi^{-2}_{p_2+2}(\Delta)I(\chi^{2}_{p_2+2}(\Delta) < (p_2-2))),
\end{aligned}$$

where the notation $\Psi_\nu(x,\Delta)$ is a non-central chi-square distribution function with ν degrees of freedom and non-centrality parameter Δ, and

$$E\left(x_\nu^{-2j}(\Delta)\right) = \int_0^\infty x^{-2j}d\Psi_g(x,\Delta)$$

Proof *See the Appendix.*

Since the components of $\boldsymbol{M\delta}$ are common for the ADB of $\widehat{\boldsymbol{\beta}}^{\mathrm{SM}}$, $\widehat{\boldsymbol{\beta}}^{\mathrm{S}}$, and $\widehat{\boldsymbol{\beta}}^{\mathrm{PS}}$, they differ by a scalar factor only. So it suffices to compare the scalar factors. For fixed p_2, we notice that the ADB of both shrinkage estimators are bounded in Δ. Note that, $E(\chi^{-2}_{p_2+2}(\Delta))$ is a decreasing log-convex function of Δ and the ADB of $\widehat{\boldsymbol{\beta}}^{\mathrm{S}}$ starts from the origin at $\Delta = 0$, increases to a maximum, and then decreases toward 0. The properties of $\widehat{\boldsymbol{\beta}}^{\mathrm{PS}}$ are similar to those of $\widehat{\boldsymbol{\beta}}^{\mathrm{S}}$. Interestingly, the bias curve of $\widehat{\boldsymbol{\beta}}^{\mathrm{PS}}$ remains below the bias curve of $\widehat{\boldsymbol{\beta}}^{\mathrm{S}}$ for all values of Δ. We now consider the ADRs of the estimators.

Theorem 6.3 *Under the local alternatives* $K_{(n)}$ *and the assumptions of Theorem 6.1, the ADRs of the estimators are*

$$\begin{aligned}
ADR(\widehat{\boldsymbol{\beta}}^{\mathrm{FM}};\boldsymbol{W}) &= tr[\boldsymbol{W}\boldsymbol{I}_{11.2}^{-1}], \\
ADR(\widehat{\boldsymbol{\beta}}^{\mathrm{SM}};\boldsymbol{W}) &= ADR(\widehat{\boldsymbol{\beta}}^{\mathrm{FM}};\boldsymbol{W}) - tr[\boldsymbol{I}_{22.1}^{-1}\boldsymbol{M}'\boldsymbol{W}\boldsymbol{M}] + \boldsymbol{\delta}^\top(\boldsymbol{M}'\boldsymbol{W}\boldsymbol{M})\boldsymbol{\delta}, \\
ADR(\widehat{\boldsymbol{\beta}}^{\mathrm{S}};\boldsymbol{W}) &= ADR(\widehat{\boldsymbol{\beta}}^{\mathrm{FM}};\boldsymbol{W}) + (p_2-2)tr[\boldsymbol{I}_{22.1}^{-1}\boldsymbol{M}^\top\boldsymbol{W}\boldsymbol{M}] \\
&\times ((p_2-2)E(Z_1^2) - 2E(Z_1)) \\
&- (p_2-2)\boldsymbol{\delta}^\top(\boldsymbol{M}^\top\boldsymbol{W}\boldsymbol{M})\boldsymbol{\delta} \\
&\times [2E(Z_1) - (p_2-2)E(Z_2^2) + 2E(Z_2)], \\
ADR(\widehat{\boldsymbol{\beta}}^{\mathrm{PS}};\boldsymbol{W}) &= ADR(\widehat{\boldsymbol{\beta}}^{\mathrm{S}};\boldsymbol{W}) + 2\boldsymbol{\delta}^\top(\boldsymbol{M}^\top\boldsymbol{W}\boldsymbol{M})\boldsymbol{\delta} \\
&\times E\left[(1-(p_2-2)Z_1)I((p_2-2)Z_1 > 1)\right] \\
&- tr[\boldsymbol{I}_{22.1}^{-1}\boldsymbol{M}^\top\boldsymbol{W}\boldsymbol{M}]E\left[(1-(p_2-2)Z_1)^2 I((p_2-2)Z_1 > 1)\right] \\
&- \boldsymbol{\delta}^\top(\boldsymbol{M}^\top\boldsymbol{W}\boldsymbol{M})\boldsymbol{\delta}E\left[(1-(p_2-2)Z_2)^2 I((p_2-2)Z_2 > 1)\right],
\end{aligned}$$

where $Z_1 = \chi^{-2}_{p_2+2}(\Delta)$, $Z_2 = \chi^{-2}_{p_2+4}(\Delta)$, and $\boldsymbol{I}_{11.2} = \boldsymbol{I}_{11} - \boldsymbol{I}_{12}\boldsymbol{I}_{22}^{-1}\boldsymbol{I}_{21}$.

Proof *See the Appendix.*

The expressions for risk can be simplified by restricting and imposing certain conditions on the matrices involved. We are interested in assessing the relative performances of the estimator in term of the sparsity parameter $\Delta \in (0, \infty)$.

The full model estimator is independent of sparsity assumption and thus has a constant risk in the entire parameter space induced by Δ. However, the submodel estimator is highly efficient for small values of Δ. Alternatively, when Δ moves away from the origin, the submodel estimator becomes inconsistent and inefficient as its ADR increases and becomes unbounded. It can be shown that under certain conditions, the ADR of the shrinkage estimators is smaller than or equal to the ADR of the full model estimator in the entire parameter space and the upper limit is attained when $\Delta \to \infty$. Finally, it can be established that

$$\mathrm{ADR}(\widehat{\boldsymbol{\beta}}^{\mathrm{PS}}) \leq \mathrm{ADR}(\widehat{\boldsymbol{\beta}}^{\mathrm{S}})$$

with a strict inequality holding for some Δ. Therefore, the risk of $\widehat{\boldsymbol{\beta}}^{\mathrm{PS}}$ is smaller than the risk of $\widehat{\boldsymbol{\beta}}^{\mathrm{S}}$ in the entire parameter space and the upper limit is attained when $\Delta \to \infty$.

We provide an extensive simulation study in the next section, which compares the performance of the listed estimators. We also investigate the relative properties of non-penalty estimators with five penalty estimators.

6.6 Simulation Experiment

In an effort to assess the relative performance of the listed estimators numerically, we conducted a Monte Carlo simulation study using the relative MSE metric as an efficiency measure. A binary response is generated using the following model:

$$ln\left(\frac{\pi_i}{1-\pi_i}\right) = \eta_i = \boldsymbol{x}_i^\top \boldsymbol{\beta}, \quad i = 1, \cdots, n,$$

where $\pi_i = P(Y = 1 |\ x_i)$ and the predictor values $\boldsymbol{x}_i{}^\top = (x_{i1}, x_{i2}, \cdots, x_{in})$ have been drawn from a standard multivariate normal distribution.

We consider the regression coefficients $\boldsymbol{\beta} = \left(\boldsymbol{\beta}_1^\top, \boldsymbol{\beta}_2^\top, \boldsymbol{\beta}_3^\top\right)^\top$ are set with dimensions p_1, p_2 and p_3, respectively. $\boldsymbol{\beta}_1$ represent strong signals, i.e. $\boldsymbol{\beta}_1$ is a vector of 1 value or higher, $\boldsymbol{\beta}_2$ represent the weak signals and $\boldsymbol{\beta}_3$ denote no signals, that is, $\boldsymbol{\beta}_3 = \mathbf{0}$. We simulated 1000 data sets consisting of $n = 250, 500$, with $p_1 = 3, 6$, $p_2 = 0, 3, 6, 9$ and $p_3 = 4, 8, 12, 16$.

We examine the characteristics of the estimators by using the MSE of the listed respective estimator relative to the MSE of the full model estimator, $\widehat{\boldsymbol{\beta}}^{\mathrm{FM}}$. Thus, the simulated relative MSE (RMSE) of an estimator $\widehat{\boldsymbol{\beta}}^*$ to $\widehat{\boldsymbol{\beta}}^{\mathrm{FM}}$ is defined by

$$\mathrm{RMSE}(\widehat{\boldsymbol{\beta}}^{\mathrm{FM}} : \widehat{\boldsymbol{\beta}}^*) = \frac{\text{Simulated MSE}(\widehat{\boldsymbol{\beta}}^{\mathrm{FM}})}{\text{Simulated MSE}(\widehat{\boldsymbol{\beta}}^*)}.$$

If the RMSE is larger than 1 then it means that $\widehat{\boldsymbol{\beta}}^*$ is relatively better than $\widehat{\boldsymbol{\beta}}^{\mathrm{FM}}$. For each of the different simulation set-ups, the results are given in Tables 6.1–6.5.

In Table 6.1, we report the RMSE of submodel and shrinkage estimators relative to full model estimators at the selected values of Δ. In this study, we only include strong and sparse signals, that is, $p_2 = 0$. The results are as expected, when Δ is close to zero, the submodel estimator outperforms the full model and shrinkage estimators. However, for larger values of Δ it becomes inconsistent as the RMSE converges to 0. Hence a submodel estimator may not be desirable. However, the performance of shrinkage estimators is robust, as for small

TABLE 6.1: RMSE of the Estimators for $p_2 = 0$.

		$n = 250$						$n = 500$					
		$p_1 = 3$			$p_1 = 6$			$p_1 = 3$			$p_1 = 6$		
p_3	Δ	SM	S	PS	SM	S	PS	SM	S	PS	SM	S	PS
4	0.0	2.03	1.36	1.45	1.78	1.26	1.33	1.99	1.35	1.45	1.58	1.22	1.27
	0.4	0.89	1.13	1.14	1.23	1.16	1.17	0.53	1.08	1.08	0.72	1.07	1.07
	0.8	0.32	1.08	1.08	0.55	1.10	1.10	0.16	1.04	1.04	0.26	1.04	1.04
	1.2	0.17	1.06	1.06	0.28	1.08	1.08	0.08	1.03	1.03	0.13	1.04	1.04
	1.8	0.10	1.05	1.05	0.15	1.06	1.06	0.04	1.03	1.03	0.07	1.03	1.03
	2.4	0.07	1.05	1.05	0.11	1.06	1.06	0.03	1.02	1.02	0.04	1.03	1.03
	3.0	0.06	1.05	1.05	0.09	1.06	1.06	0.03	1.03	1.03	0.04	1.03	1.03
8	0.0	3.32	2.15	2.33	2.64	1.87	1.98	3.05	2.09	2.25	2.25	1.71	1.80
	0.4	1.42	1.52	1.56	1.73	1.57	1.59	0.79	1.30	1.30	0.98	1.29	1.30
	0.8	0.50	1.28	1.28	0.76	1.35	1.35	0.23	1.13	1.13	0.34	1.15	1.15
	1.2	0.26	1.21	1.21	0.39	1.27	1.27	0.11	1.10	1.10	0.17	1.12	1.12
	1.8	0.15	1.16	1.16	0.21	1.21	1.21	0.06	1.08	1.08	0.09	1.11	1.11
	2.4	0.12	1.16	1.16	0.16	1.19	1.19	0.04	1.08	1.08	0.06	1.09	1.09
	3.0	0.09	1.16	1.16	0.14	1.17	1.17	0.04	1.08	1.08	0.05	1.10	1.10
12	0.0	4.74	3.05	3.27	3.71	2.58	2.73	4.18	2.86	3.12	3.05	2.25	2.39
	0.4	1.93	1.98	2.02	2.34	2.07	2.11	1.07	1.57	1.58	1.27	1.56	1.57
	0.8	0.68	1.49	1.49	1.01	1.65	1.65	0.32	1.25	1.25	0.44	1.28	1.28
	1.2	0.36	1.36	1.36	0.53	1.48	1.48	0.16	1.17	1.17	0.22	1.21	1.21
	1.8	0.21	1.27	1.27	0.28	1.36	1.36	0.08	1.14	1.14	0.12	1.18	1.18
	2.4	0.16	1.27	1.27	0.23	1.33	1.33	0.06	1.13	1.13	0.08	1.16	1.16
	3.0	0.13	1.26	1.26	0.20	1.29	1.29	0.05	1.14	1.14	0.06	1.17	1.17
16	0.0	6.24	3.90	4.20	4.89	3.42	3.58	5.39	3.61	3.96	3.82	2.84	3.02
	0.4	2.53	2.48	2.55	2.99	2.65	2.69	1.35	1.85	1.86	1.54	1.85	1.87
	0.8	0.88	1.73	1.73	1.28	1.97	1.98	0.41	1.36	1.36	0.54	1.43	1.43
	1.2	0.47	1.52	1.52	0.67	1.71	1.71	0.20	1.25	1.25	0.27	1.31	1.31
	1.8	0.28	1.41	1.41	0.38	1.52	1.52	0.11	1.20	1.20	0.14	1.26	1.26
	2.4	0.22	1.39	1.39	0.32	1.46	1.46	0.08	1.19	1.19	0.10	1.23	1.23
	3.0	0.18	1.38	1.38	0.28	1.42	1.42	0.06	1.19	1.19	0.08	1.24	1.24

values of Δ shrinkage estimators behave much better than the full model estimator. More importantly, they dominate the full model estimator in the entire parameter space induced by the sparsity parameter Δ. The graphical analysis, based on Figure 6.1 and 6.2 provides a similar analysis.

In Table 6.2, we showcase the RMSE of the submodel and shrinkage estimators relative to the full model estimators at selected values of Δ in the presence of weak signals. We use $p_2 = 3, 6, 9$. Again, the performance of the submodel estimator in terms of Δ is similar to that previously reported. However, the amount of the improvement is much larger than the in presence of weak coefficients. Similarly, both shrinkage estimators outshine the full model estimator for all the values of Δ. As both p_1 and p_2 increase and Δ moves away from the origin, the shrinkage estimators perform better than the submodel estimator in the remaining parameter space. As expected, the positive-part shrinkage estimator is uniformly better than the shrinkage estimator. The graphical analysis, based on Figures 6.3 and 6.4 provide a similar analysis. The amount of RMSE of the estimators is better for larger sample sizes, as reported in Tables 6.4 and 6.5. Figures 6.5 and 6.6 display the similar characteristics of the estimators observed in the tables for different configurations of simulation parameters.

In passing, we would like to remark here that we have calculated the MSE based on only strong signals in Table 6.2 while the RMSE of the estimators in Table 6.3 includes both strong and weak signals in the MSE calculation to provide a fair comparison. The behavior of estimators did not change. However, the magnitude of gain in RMSE is smaller for shrinkage estimators when weak signals are included in the calculation. Clearly, when the number of parameters due to weak signals increases, the MSE of the submodel estimator increases accordingly. Consequently, the MSE of the shrinkage estimators increases, which is consistent with theory.

6.6.1 Penalized Strategies

In this section, we provide a numerical analysis of the submodel, shrinkage, and five penalty estimators with respect to the full model estimator when a submodel is assumed to be correct with $n = 200$ and 400, respectively.

In this simulation, we consider the regression coefficients are set: $\boldsymbol{\beta} = \left(\boldsymbol{\beta}_1^\top, \boldsymbol{\beta}_2^\top, \boldsymbol{\beta}_3^\top\right)^\top = \left(\mathbf{1}_{p_1}^\top, \mathbf{0.1}_{p_2}^\top, \mathbf{0}_{p_3}^\top\right)^\top$, with dimensions p_1, p_2 and p_3, respectively.

In this setting, we simulated 50 data sets consisting of $n = 200, 400$, with $p_1 = 4, 8$, $p_2 = 0, 3, 6, 9$ and $p_3 = 4, 8, 12, 16$. The tuning parameters for penalized methods are selected by 5-fold cross-validation. In Table 6.6, we present the simulated relative MSE of the listed estimators with respect to the full model estimator for and $n = 200$. First, we note that the relative RMSE of all the estimators with respect to the full model estimator increases as the number of inactive predictors (p_3) increases. As one would expect, the RMSE of the submodel estimator is the best, and all the other estimators perform better than the full model estimator. Table 6.6 reveals that some penalty methods perform better than the shrinkage strategy when the number of inactive predictors p_3 in the model is small. On the other hand, the shrinkage estimators outshine the penalty estimators for larger values of p_3. Generally speaking, in the presence of a relatively large number of inactive predictors in the model, the shrinkage strategy does well relative to the penalty estimators.

Interestingly, the performance of ALASSO and SCAD is comparable to shrinkage estimates when the number of weak signals increases in the model if weak signals are included with sparse signals in the MSE calculation of the estimators. The MSE of the submodel estimator is larger in this case, which negatively impacted the MSE of the shrinkage estimators. This is consistent with the theory in the sense that the submodel estimator no longer holds the Oracle property. For this reason, we calculate the MSE of the submodel based on strong signals only. Tables 6.8 and 6.9 showcase the RMSE of the listed estimators. It is evident

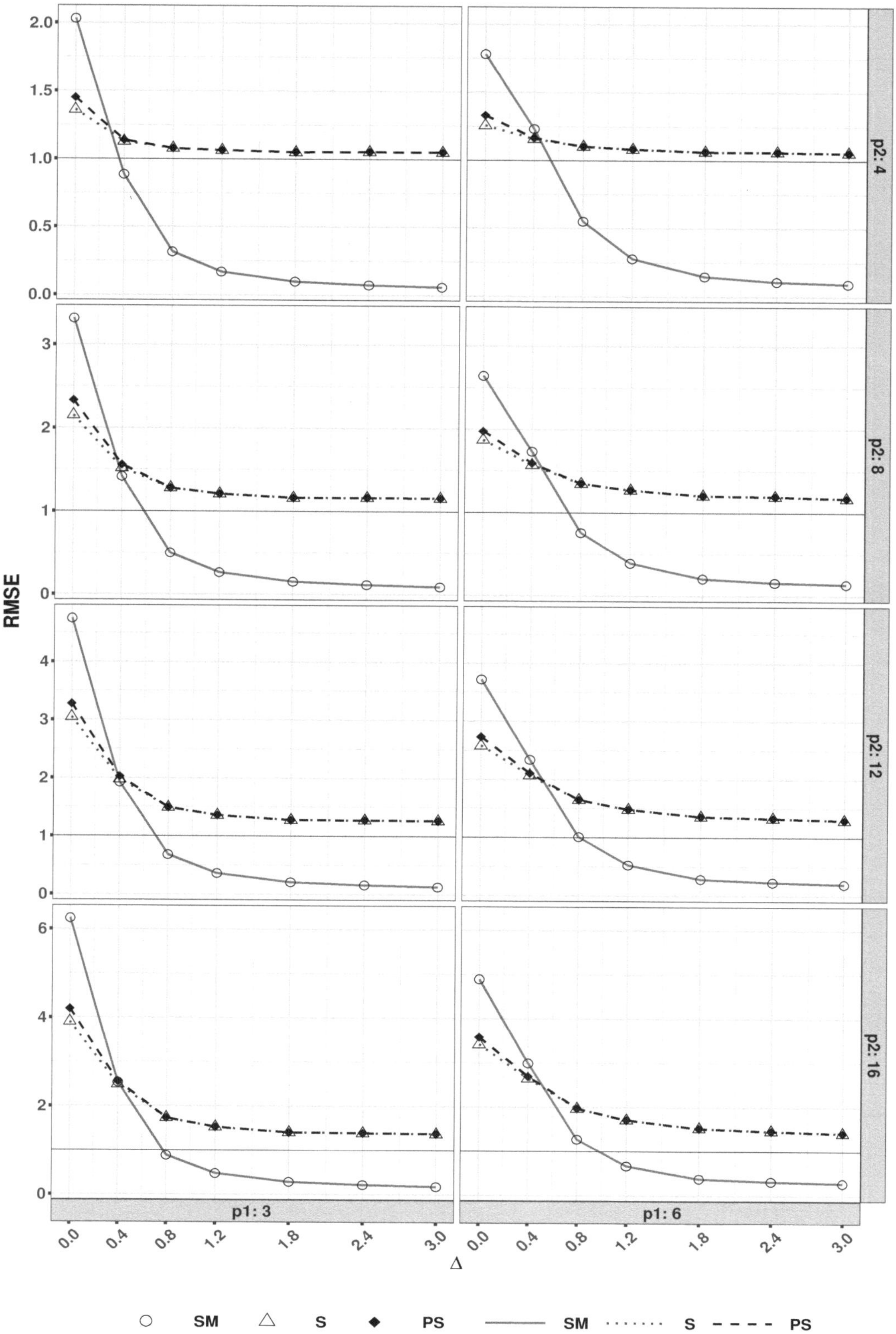

FIGURE 6.1: RMSE of the Estimators for $n = 250$ and $p_2 = 0$.

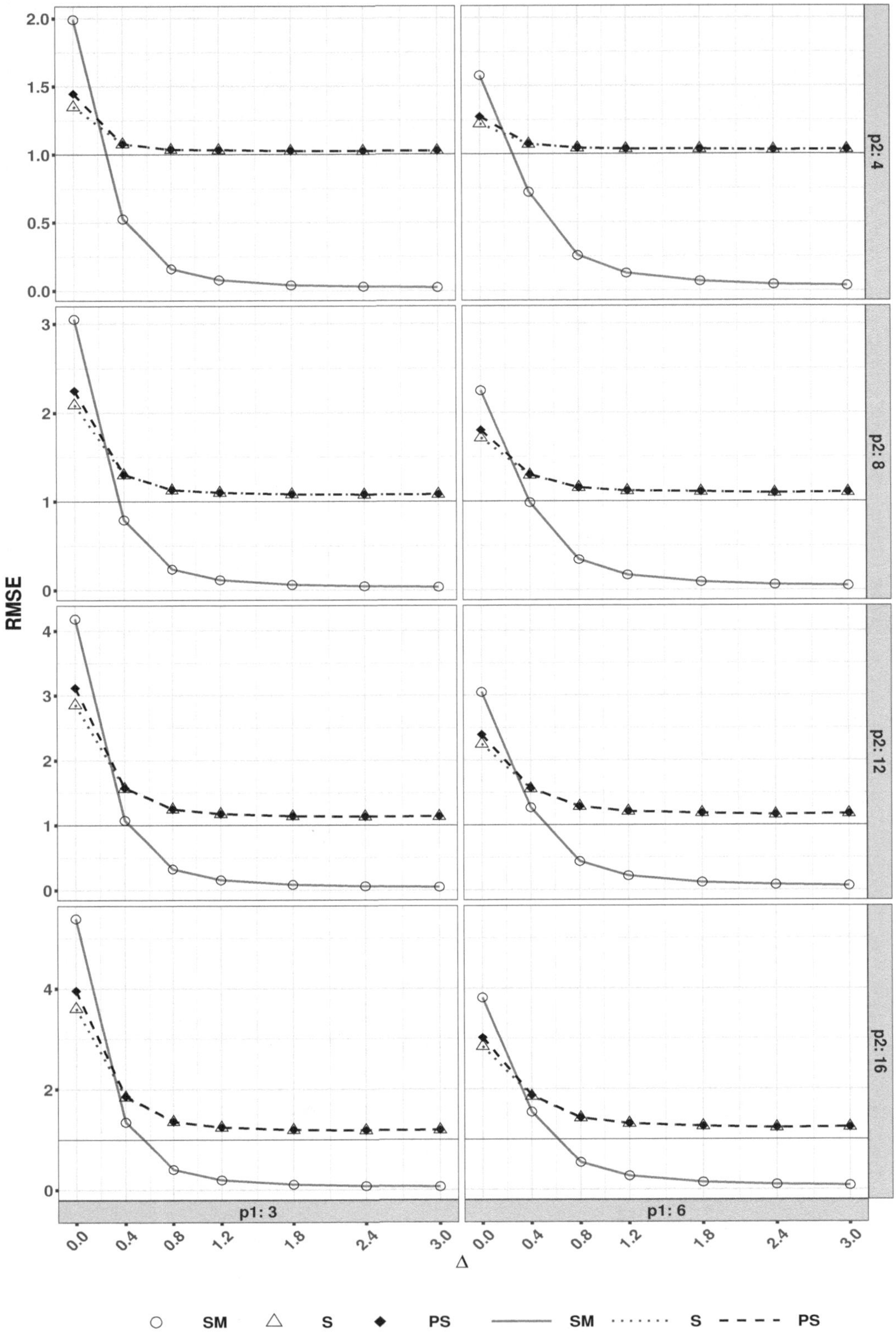

FIGURE 6.2: RMSE of the Estimators for $n = 500$ and $p_2 = 0$.

TABLE 6.2: RMSE of the Estimators for $n = 250$ – Submodel Contains Strong Signals.

			$p_2 = 3$			$p_2 = 6$			$p_2 = 9$		
p_1	p_3	Δ	SM	S	PS	SM	S	PS	SM	S	PS
3	4	0.0	2.93	1.98	2.07	3.94	2.48	2.64	5.37	3.01	3.23
		0.3	0.90	1.28	1.30	0.77	1.43	1.43	0.68	1.52	1.52
		0.6	0.28	1.16	1.16	0.21	1.22	1.22	0.21	1.29	1.29
		0.9	0.17	1.15	1.15	0.14	1.21	1.21	0.18	1.28	1.28
	8	0.0	4.25	2.64	2.85	5.62	3.17	3.47	6.61	3.81	4.19
		0.3	1.32	1.65	1.67	0.99	1.68	1.68	0.87	1.72	1.72
		0.6	0.39	1.32	1.32	0.29	1.33	1.33	0.29	1.44	1.44
		0.9	0.23	1.27	1.27	0.20	1.34	1.34	0.24	1.40	1.40
	12	0.0	6.09	3.42	3.70	6.90	4.13	4.57	7.99	5.18	5.59
		0.3	1.71	2.01	2.04	1.27	1.93	1.93	1.06	1.94	1.94
		0.6	0.50	1.50	1.50	0.39	1.48	1.48	0.39	1.58	1.58
		0.9	0.30	1.41	1.41	0.27	1.45	1.45	0.33	1.50	1.50
	16	0.0	7.39	4.45	4.83	8.68	5.51	5.79	9.64	6.05	6.46
		0.3	2.19	2.44	2.50	1.53	2.23	2.23	1.36	2.23	2.23
		0.6	0.64	1.67	1.67	0.52	1.64	1.64	0.52	1.72	1.72
		0.9	0.39	1.53	1.53	0.33	1.57	1.57	0.45	1.61	1.61
6	4	0.0	2.45	1.76	1.81	3.36	2.35	2.40	4.17	2.95	3.03
		0.3	1.40	1.39	1.40	1.21	1.58	1.59	1.04	1.61	1.62
		0.6	0.45	1.23	1.23	0.38	1.34	1.34	0.33	1.41	1.41
		0.9	0.26	1.19	1.19	0.21	1.29	1.29	0.27	1.35	1.35
	8	0.0	3.61	2.52	2.58	4.49	3.12	3.23	5.48	3.73	3.87
		0.3	1.95	1.79	1.82	1.47	1.93	1.94	1.32	1.94	1.95
		0.6	0.60	1.42	1.42	0.47	1.55	1.55	0.45	1.60	1.60
		0.9	0.35	1.37	1.37	0.27	1.46	1.46	0.36	1.47	1.47
	12	0.0	4.87	3.40	3.47	5.94	4.01	4.21	6.57	4.54	4.74
		0.3	2.34	2.21	2.23	1.96	2.33	2.33	1.71	2.33	2.34
		0.6	0.79	1.69	1.69	0.63	1.76	1.76	0.54	1.80	1.80
		0.9	0.48	1.57	1.57	0.38	1.63	1.63	0.51	1.60	1.60
	16	0.0	6.36	4.20	4.44	6.83	4.85	5.08	7.42	5.43	5.78
		0.3	2.83	2.63	2.65	2.48	2.71	2.73	2.17	2.80	2.80
		0.6	1.06	1.97	1.97	0.91	1.98	1.98	0.72	2.04	2.04
		0.9	0.58	1.77	1.77	0.53	1.78	1.78	0.67	1.76	1.76

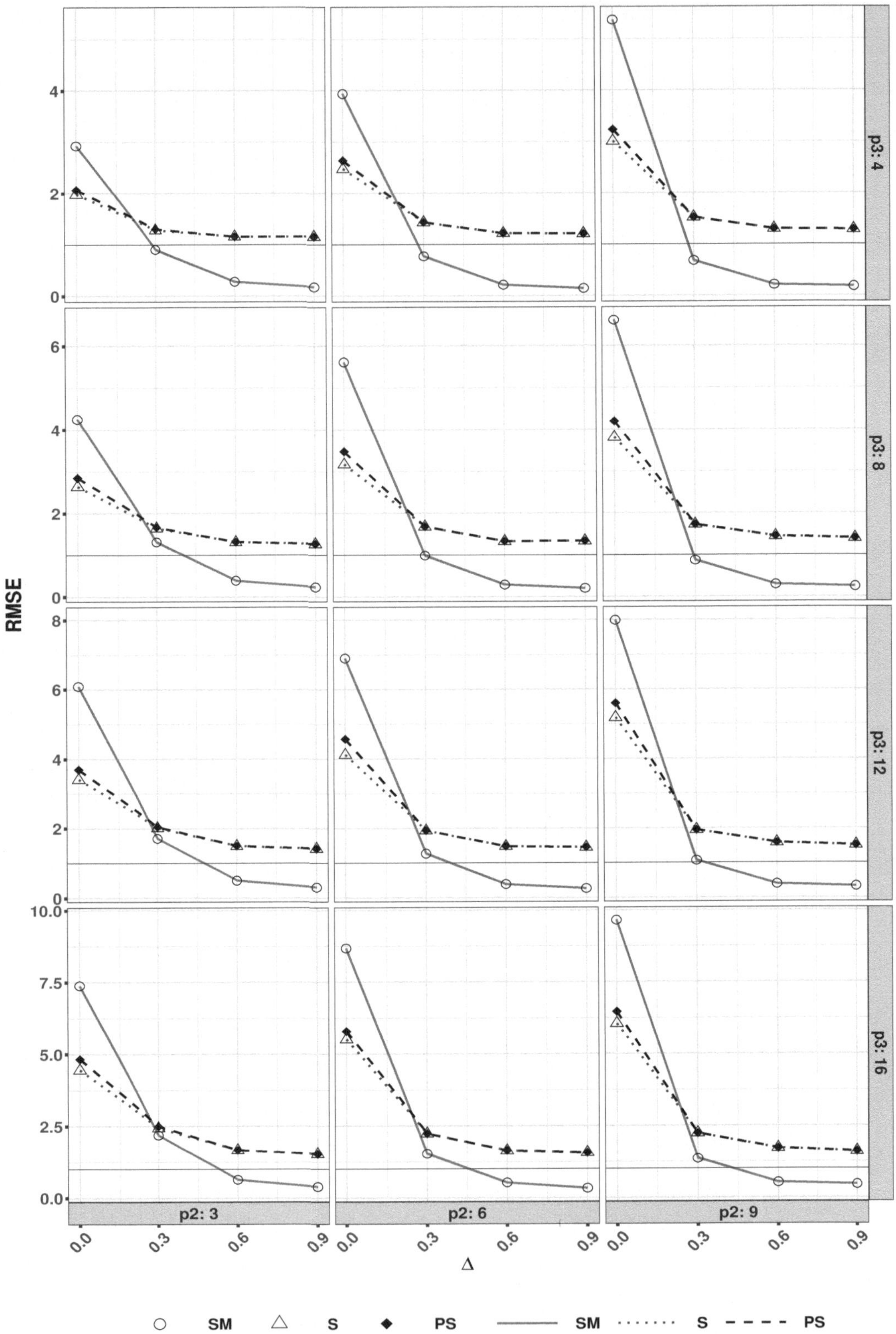

FIGURE 6.3: RMSE of the Estimators for $n = 250$ and $p_1 = 3$, – Submodel Contains Strong Signals.

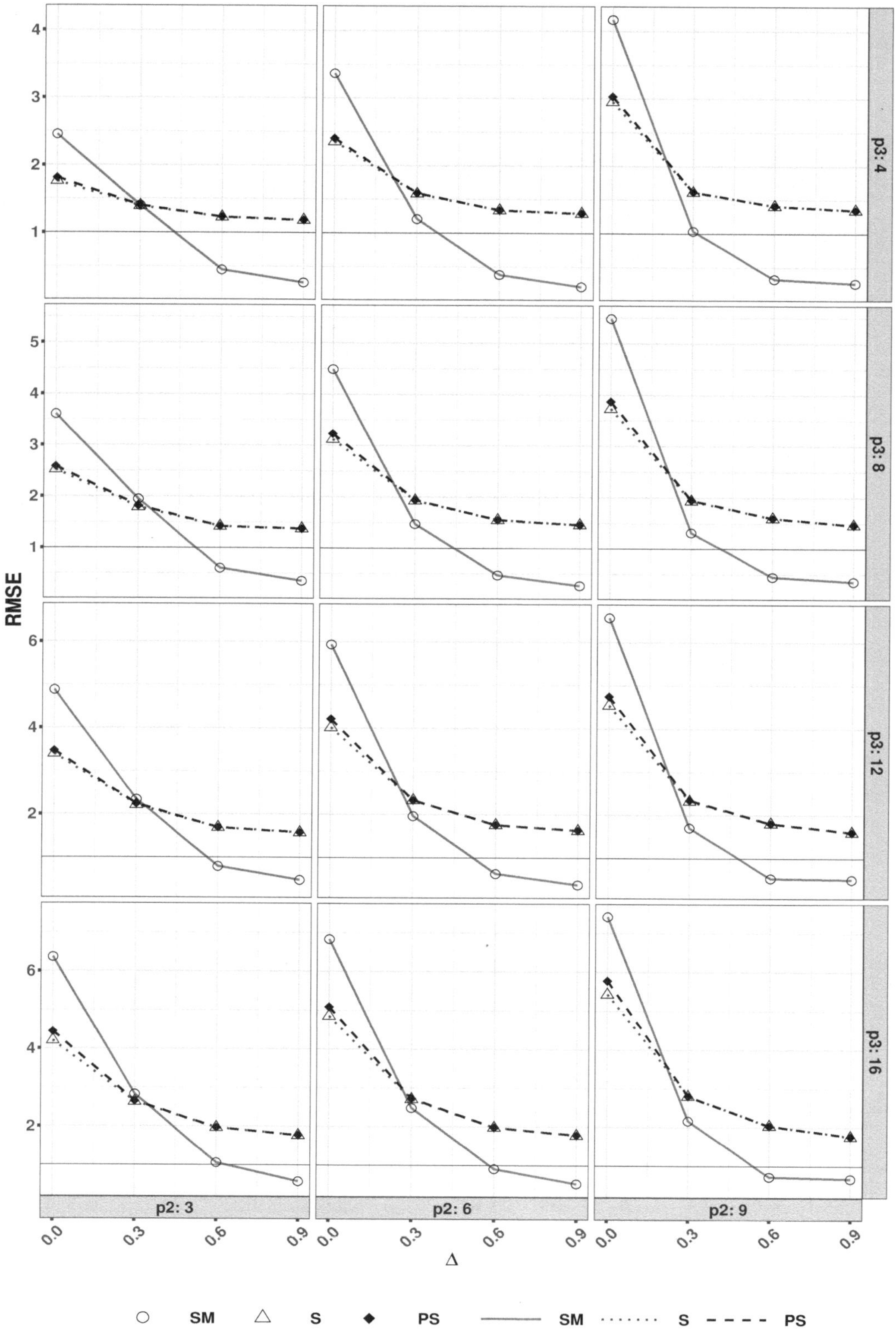

FIGURE 6.4: RMSE of the Estimators for $n = 250$ and $p_1 = 6$, – Submodel Contains Strong Signals.

TABLE 6.3: RMSE of the Estimators for $n = 250$ – Submodel Contains both Strong and Weak Signals.

			$p_2 = 3$			$p_2 = 6$			$p_2 = 9$		
p_1	p_3	Δ	SM	S	PS	SM	S	PS	SM	S	PS
3	4	0.0	1.72	1.33	1.34	1.56	1.23	1.26	1.49	1.20	1.22
		0.4	1.02	1.09	1.11	1.15	1.13	1.14	1.20	1.11	1.11
		0.8	0.46	1.08	1.08	0.52	1.07	1.07	0.62	1.07	1.07
		1.2	0.23	1.06	1.06	0.31	1.06	1.06	0.38	1.06	1.06
	8	0.0	2.52	1.95	2.03	2.22	1.72	1.77	2.01	1.61	1.66
		0.4	1.58	1.47	1.50	1.60	1.46	1.46	1.59	1.39	1.39
		0.8	0.63	1.26	1.26	0.70	1.24	1.24	0.84	1.24	1.24
		1.2	0.34	1.20	1.20	0.40	1.21	1.21	0.49	1.20	1.20
	12	0.0	3.73	2.63	2.77	3.18	2.31	2.36	2.73	2.15	2.20
		0.4	2.05	1.90	1.94	2.02	1.78	1.81	2.08	1.68	1.70
		0.8	0.82	1.46	1.46	0.92	1.44	1.45	1.13	1.45	1.45
		1.2	0.44	1.37	1.37	0.51	1.37	1.37	0.59	1.34	1.34
	16	0.0	4.80	3.37	3.55	4.43	3.03	3.12	3.69	2.75	2.83
		0.4	2.63	2.40	2.44	2.55	2.20	2.23	2.68	2.07	2.08
		0.8	1.00	1.70	1.70	1.21	1.70	1.70	1.43	1.68	1.68
		1.2	0.55	1.52	1.52	0.68	1.55	1.55	0.73	1.49	1.49
6	4	0.0	1.58	1.26	1.27	1.48	1.20	1.21	1.49	1.19	1.21
		0.4	1.29	1.16	1.16	1.32	1.13	1.13	1.47	1.14	1.14
		0.8	0.61	1.11	1.11	0.71	1.09	1.09	0.82	1.09	1.09
		1.2	0.29	1.09	1.09	0.45	1.08	1.08	0.58	1.08	1.08
	8	0.0	2.33	1.77	1.79	2.15	1.70	1.69	2.23	1.68	1.69
		0.4	1.80	1.52	1.55	1.85	1.47	1.48	2.03	1.49	1.49
		0.8	0.74	1.35	1.35	0.97	1.33	1.33	1.10	1.32	1.32
		1.2	0.41	1.30	1.30	0.60	1.26	1.26	0.71	1.24	1.24
	12	0.0	3.28	2.44	2.46	2.87	2.31	2.27	2.95	2.19	2.21
		0.4	2.33	1.94	1.98	2.32	1.88	1.89	2.77	1.91	1.92
		0.8	0.98	1.66	1.67	1.25	1.60	1.61	1.44	1.56	1.56
		1.2	0.51	1.48	1.48	0.79	1.45	1.45	0.90	1.43	1.43
	16	0.0	4.36	3.16	3.21	3.78	2.97	2.96	3.67	2.84	2.89
		0.4	2.87	2.37	2.46	3.25	2.37	2.39	3.76	2.38	2.38
		0.8	1.22	1.93	1.93	1.60	1.92	1.93	1.82	1.89	1.89
		1.2	0.66	1.70	1.70	1.01	1.70	1.70	1.19	1.64	1.64

TABLE 6.4: RMSE of the Estimators for $n = 500$ – Submodel Contains both Strong and Weak Signals.

			$p_2 = 3$			$p_2 = 6$			$p_2 = 9$		
p_1	p_3	Δ	SM	S	PS	SM	S	PS	SM	S	PS
3	10	0.0	4.47	3.06	3.34	5.39	3.61	3.96	6.21	4.15	4.60
		0.3	0.78	1.46	1.46	0.49	1.40	1.40	0.40	1.40	1.40
		0.6	0.22	1.23	1.23	0.14	1.23	1.23	0.13	1.26	1.26
		0.9	0.12	1.17	1.17	0.08	1.19	1.19	0.08	1.22	1.22
	20	0.0	7.51	4.96	5.50	8.38	5.52	6.13	9.17	6.11	6.79
		0.3	1.26	1.98	1.99	0.76	1.75	1.75	0.61	1.70	1.70
		0.6	0.38	1.47	1.47	0.23	1.41	1.41	0.20	1.42	1.42
		0.9	0.19	1.34	1.34	0.14	1.35	1.35	0.13	1.37	1.37
	40	0.0	12.37	8.58	9.58	12.87	9.04	10.05	13.36	9.61	10.66
		0.3	2.38	3.25	3.26	1.44	2.63	2.63	1.14	2.44	2.44
		0.6	0.75	2.03	2.03	0.47	1.86	1.86	0.42	1.83	1.83
		0.9	0.41	1.75	1.75	0.29	1.71	1.71	0.30	1.69	1.69
	60	0.0	14.93	11.63	12.67	15.18	11.98	13.07	15.43	12.30	13.35
		0.3	3.82	4.79	4.81	2.36	3.76	3.76	1.88	3.33	3.33
		0.6	1.32	2.76	2.76	0.84	2.41	2.41	0.77	2.26	2.26
		0.9	0.76	2.23	2.23	0.56	2.10	2.10	0.64	1.99	1.99
6	10	0.0	3.24	2.42	2.55	3.82	2.84	3.02	4.43	3.30	3.52
		0.3	0.86	1.47	1.48	0.60	1.47	1.47	0.52	1.50	1.50
		0.6	0.26	1.27	1.27	0.18	1.29	1.29	0.17	1.32	1.32
		0.9	0.13	1.20	1.20	0.10	1.23	1.23	0.11	1.27	1.27
	20	0.0	5.25	3.91	4.18	5.79	4.32	4.68	6.30	4.73	5.14
		0.3	1.32	2.06	2.06	0.91	1.92	1.92	0.78	1.91	1.91
		0.6	0.41	1.55	1.55	0.28	1.53	1.53	0.27	1.54	1.54
		0.9	0.22	1.42	1.42	0.17	1.42	1.42	0.17	1.44	1.44
	40	0.0	8.20	6.70	7.42	8.49	7.06	7.84	8.80	7.53	8.30
		0.3	2.36	3.51	3.52	1.68	3.10	3.10	1.46	2.93	2.93
		0.6	0.80	2.30	2.30	0.59	2.13	2.13	0.57	2.03	2.03
		0.9	0.45	1.95	1.95	0.38	1.84	1.84	0.43	1.77	1.77
	60	0.0	9.79	9.40	10.13	9.96	9.80	10.50	10.10	10.21	10.88
		0.3	3.53	5.30	5.33	2.69	4.61	4.61	2.47	4.18	4.18
		0.6	1.44	3.28	3.28	1.12	2.83	2.83	1.20	2.51	2.51
		0.9	0.85	2.56	2.56	0.80	2.25	2.25	1.13	1.99	1.99

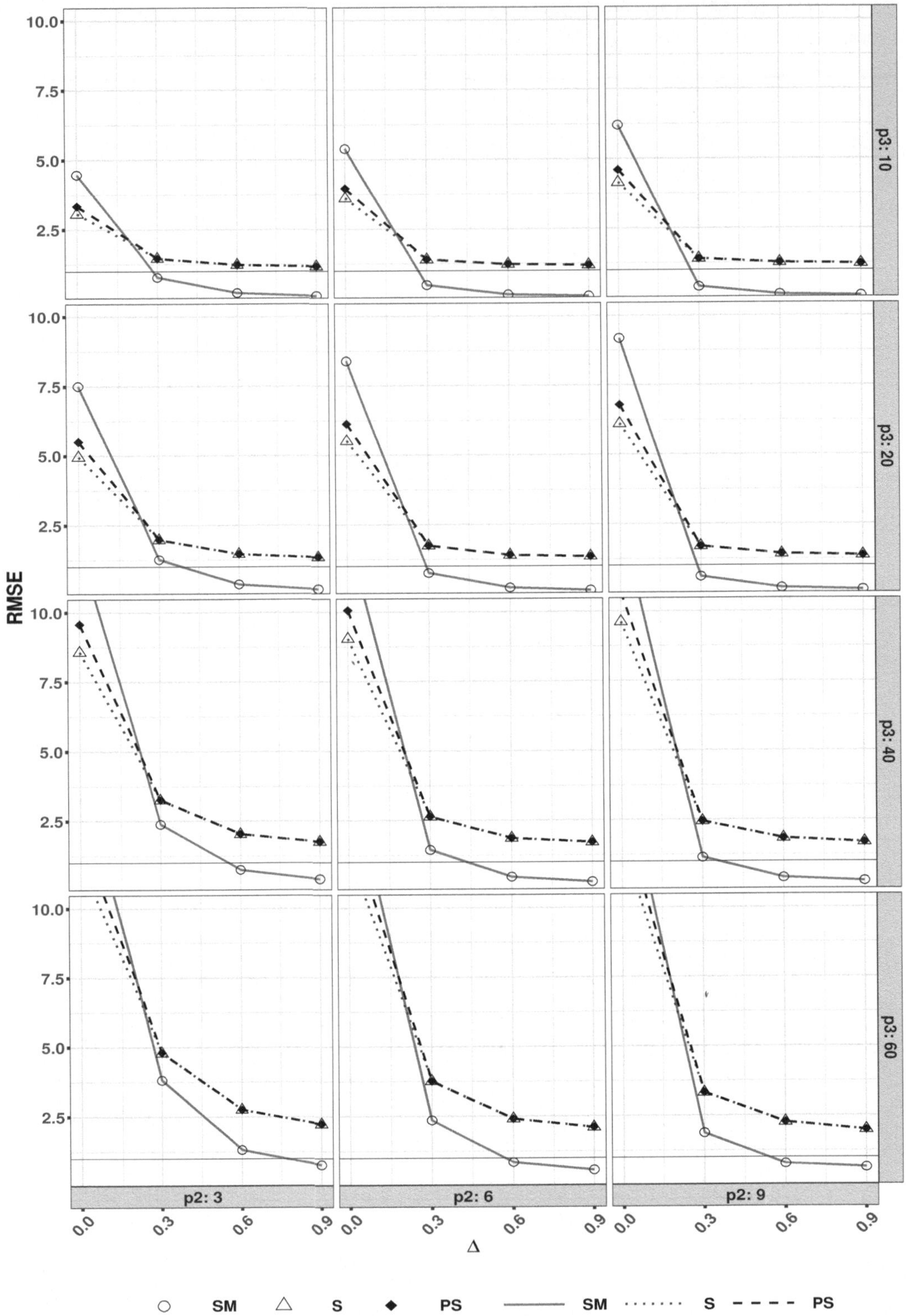

FIGURE 6.5: RMSE of the Estimators for the Submodel is Based on Signals for $n = 500$ and $p_1 = 3$.

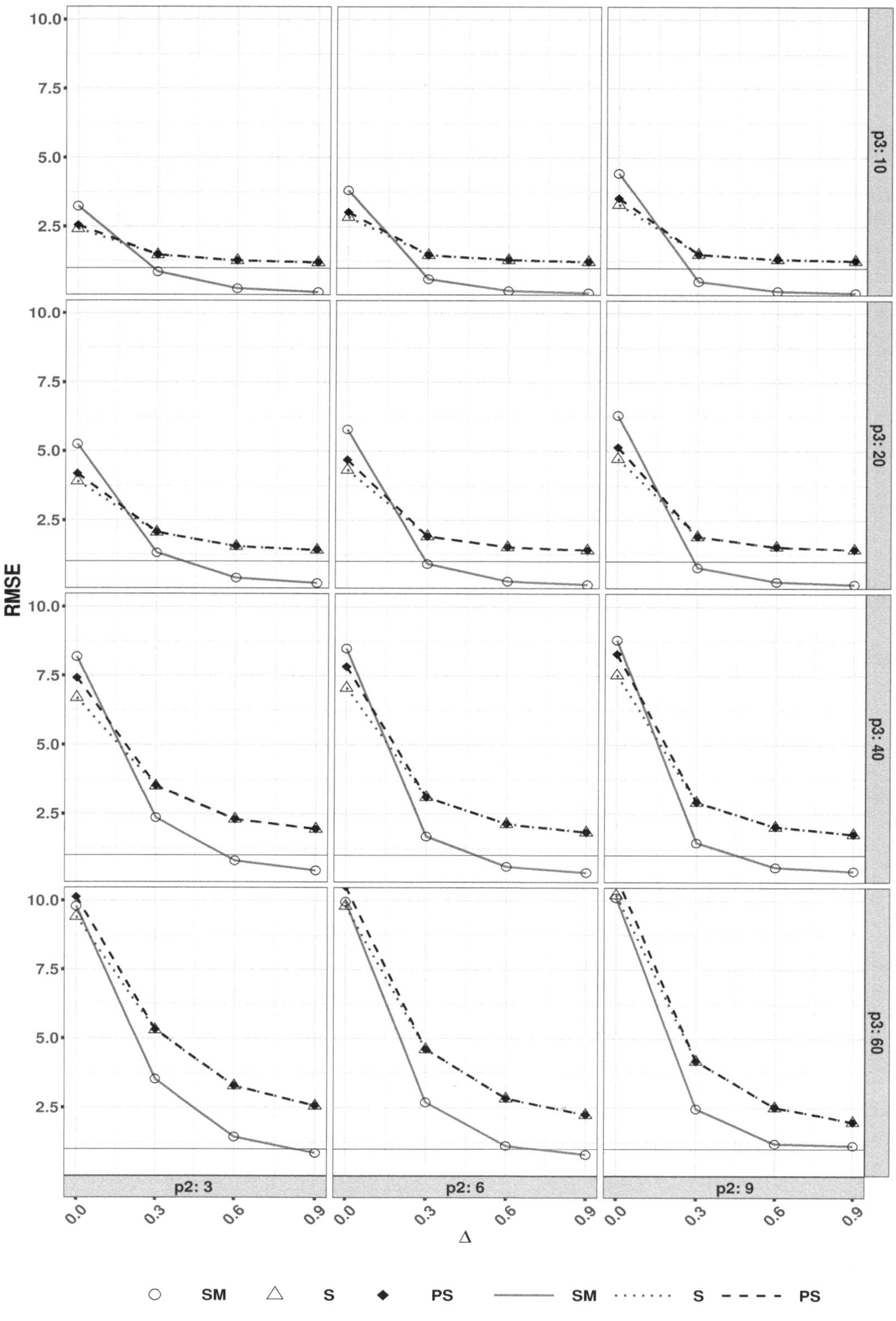

FIGURE 6.6: RMSE of the Estimators for the Submodel is Based on Signals for $n = 500$ and $p_1 = 6$.

TABLE 6.5: RMSE of the Estimators for $n = 250$ – Submodel Contains Strong Signals.

			$p_2 = 3$			$p_2 = 6$			$p_2 = 9$		
p_1	p_3	Δ	SM	S	PS	SM	S	PS	SM	S	PS
3	20	0.0	8.99	5.60	5.93	10.14	6.15	6.70	11.89	7.20	7.74
		0.3	2.69	2.95	3.03	1.91	2.58	2.59	1.71	2.56	2.56
		0.6	0.82	1.90	1.90	0.66	1.81	1.81	0.69	1.84	1.84
		0.9	0.48	1.66	1.66	0.40	1.68	1.68	0.66	1.69	1.69
	40	0.0	17.08	9.62	9.90	17.43	10.18	10.41	18.62	10.23	10.45
		0.3	6.24	5.45	5.49	4.76	4.48	4.48	4.16	4.23	4.23
		0.6	2.44	3.43	3.43	2.32	2.68	2.68	3.47	2.23	2.23
		0.9	1.84	2.53	2.53	2.55	2.11	2.11	3.13	0.95	0.95
	80	0.0	2.47	0.96	0.96	2.46	0.96	0.96	2.75	0.96	0.96
		0.3	2.63	0.96	0.96	2.65	0.95	0.95	2.78	0.94	0.94
		0.6	3.09	0.95	0.95	2.98	0.94	0.94	3.90	0.93	0.93
		0.9	3.54	0.94	0.94	8934	1.94	1.94	2257	2.01	2.01
	100	0.0	3.71	0.94	0.94	3.33	0.95	0.95	4.16	0.94	0.94
		0.3	3.38	0.95	0.95	4.49	0.95	0.95	3.21	0.96	0.96
		0.6	4.65	0.94	0.94	1625	2.53	2.53	773	2.59	2.59
		0.9	4.24	0.93	0.93	758	2.41	2.41	395	2.45	2.45
6	20	0.0	7.25	5.18	5.45	7.55	5.59	5.94	9.69	6.31	6.73
		0.3	3.54	3.26	3.32	2.93	3.23	3.24	2.52	3.25	3.25
		0.6	1.31	2.29	2.29	1.15	2.21	2.21	1.08	2.27	2.27
		0.9	0.71	1.96	1.96	0.76	1.94	1.94	0.90	1.89	1.89
	40	0.0	13.67	8.29	8.63	16.34	8.37	8.57	19.84	6.74	6.81
		0.3	8.48	6.31	6.32	12.05	3.80	3.80	2484	1.72	1.72
		0.6	6.51	3.36	3.36	1754	1.60	1.60	2.26	0.96	0.96
		0.9	4.31	2.88	2.88	15465	1.50	1.50	2.77	0.94	0.94
	80	0.0	2.23	0.96	0.96	1.89	0.96	0.96	2.22	0.96	0.96
		0.3	2.24	0.95	0.95	3.02	0.94	0.94	2.51	0.95	0.95
		0.6	2.20	0.96	0.96	3.04	0.95	0.95	652	2.51	2.51
		0.9	2.33	0.96	0.96	976	2.28	2.28	636	2.28	2.28
	100	0.0	573	3.56	3.56	391	3.81	3.81	285	4.05	4.05
		0.3	308	3.63	3.63	236	3.72	3.72	137	3.75	3.75
		0.6	276	3.29	3.29	169	3.27	3.27	167	3.27	3.27
		0.9	246	3.00	3.00	193	2.90	2.90	176	2.92	2.92

TABLE 6.6: RMSE of the Estimators for $n = 200$.

p_1	p_2	p_3	SM	S	PS	ENET	LASSO	ALASSO	Ridge	SCAD
4	0	4	2.025	1.323	1.422	1.231	1.355	1.609	1.313	1.518
		8	2.896	2.078	2.213	1.301	1.371	1.925	1.243	2.040
		12	5.113	3.170	3.180	1.739	2.113	2.387	1.667	2.538
		16	7.119	3.518	3.733	1.900	2.329	2.824	1.737	3.024
	3	4	1.690	1.287	1.342	1.355	1.627	1.889	1.490	1.708
		8	2.966	1.820	1.868	1.494	1.676	1.981	1.434	2.140
		12	3.722	2.704	2.766	1.700	2.013	2.365	1.581	2.612
		16	6.256	3.733	3.866	2.081	2.563	3.158	1.844	3.753
	6	4	1.597	1.254	1.260	1.425	1.598	1.807	1.486	1.795
		8	2.656	1.643	1.824	1.744	1.978	2.552	1.623	2.699
		12	3.051	2.530	2.529	1.446	1.700	2.470	1.324	2.413
		16	4.426	3.500	3.473	2.078	2.934	3.286	2.008	4.079
	9	4	1.387	1.180	1.216	1.163	1.248	1.661	1.043	1.862
		8	2.074	1.619	1.647	1.816	2.278	2.710	1.724	2.986
		12	2.883	2.216	2.199	1.794	2.288	2.748	1.678	3.227
		16	4.412	3.005	3.001	2.313	3.091	3.330	2.157	4.368
8	0	4	1.937	1.259	1.309	1.030	1.088	1.267	0.973	1.005
		8	2.690	2.081	2.116	1.439	1.505	1.855	1.406	1.708
		12	3.982	2.877	2.937	1.444	1.548	2.434	1.404	1.881
		16	5.479	3.636	4.178	1.680	1.894	3.338	1.687	2.446
	3	4	1.753	1.299	1.298	1.117	1.207	1.510	1.089	1.118
		8	2.493	1.869	1.893	1.393	1.498	2.304	1.372	1.705
		12	5.251	2.749	2.778	2.226	2.308	3.490	2.143	2.764
		16	5.463	3.428	3.571	1.977	2.232	3.667	1.967	2.688
	6	4	1.976	1.273	1.267	1.880	1.910	2.667	1.893	1.736
		8	2.350	1.835	1.830	1.478	1.545	2.224	1.474	1.791
		12	4.226	2.529	2.500	2.638	2.819	3.684	2.380	3.009
		16	5.871	3.459	3.517	2.492	2.762	4.494	2.310	3.246
	9	4	1.456	1.215	1.216	1.775	1.770	2.590	1.644	1.874
		8	2.645	1.796	1.784	1.902	2.068	3.073	1.851	2.277
		12	6.024	2.441	2.437	3.564	4.285	6.881	3.560	5.438
		16	7.553	3.255	3.282	3.940	4.718	6.633	3.762	5.776

TABLE 6.7: RMSE of the Estimators for $n = 400$.

p_1	p_2	p_3	SM	S	PS	ENET	LASSO	ALASSO	Ridge	SCAD
4	0	4	1.864	1.300	1.373	1.018	1.054	1.476	0.809	1.394
		8	2.601	1.974	2.091	1.217	1.424	1.901	1.085	1.976
		12	3.167	2.336	2.599	0.949	1.123	1.904	0.936	2.531
		16	4.477	3.155	3.549	1.148	1.419	2.345	1.130	3.283
	3	4	1.403	1.159	1.244	0.873	0.889	1.395	0.789	1.637
		8	2.020	1.642	1.727	1.098	1.247	1.602	1.027	2.054
		12	3.161	2.380	2.424	1.098	1.317	2.019	1.064	2.824
		16	4.040	2.953	3.022	1.226	1.493	2.465	1.153	3.380
	6	4	1.601	1.176	1.241	1.096	1.147	1.453	1.021	1.616
		8	2.096	1.693	1.697	1.147	1.342	1.830	1.131	2.187
		12	2.615	2.134	2.179	1.357	1.634	2.045	1.234	3.015
		16	2.974	2.325	2.455	1.177	1.428	2.176	1.123	3.102
	9	4	1.467	1.184	1.188	1.244	1.328	1.597	1.198	1.837
		8	1.863	1.468	1.537	1.050	1.133	1.790	1.015	2.190
		12	2.362	1.911	1.932	1.220	1.438	1.858	1.186	2.439
		16	3.019	2.333	2.372	1.283	1.626	2.241	1.166	3.224
8	0	4	1.480	1.189	1.244	1.170	1.126	1.339	0.671	1.136
		8	2.024	1.790	1.782	1.176	1.206	1.638	0.792	1.438
		12	2.872	2.293	2.348	1.172	1.219	1.934	0.937	2.030
		16	3.849	2.693	2.903	1.223	1.340	2.371	1.055	2.401
	3	4	1.548	1.229	1.234	1.130	1.173	1.491	0.731	1.359
		8	2.157	1.604	1.678	1.169	1.204	1.779	0.890	1.797
		12	3.008	2.155	2.216	1.251	1.356	2.122	1.101	2.284
		16	3.541	2.453	2.541	1.436	1.504	2.437	1.358	2.817
	6	4	1.327	1.149	1.168	1.145	1.145	1.366	0.857	1.360
		8	1.939	1.545	1.565	1.148	1.173	1.825	1.017	1.697
		12	2.231	1.880	1.884	1.368	1.418	2.103	1.226	2.075
		16	2.914	2.371	2.455	1.190	1.301	2.263	1.136	2.446
	9	4	1.314	1.162	1.167	1.224	1.282	1.764	1.004	1.671
		8	1.746	1.499	1.494	1.181	1.251	1.890	1.060	1.809
		12	2.356	1.853	1.889	1.317	1.476	2.213	1.243	2.305
		16	3.199	2.391	2.378	1.618	1.814	2.667	1.542	3.146

TABLE 6.8: RMSE of the Estimators for $n = 200$ – Submodel Contains Strong Signals.

p_1	p_2	p_3	SM	S	PS	Ridge	ENET	LASSO	ALASSO	SCAD
4	3	4	2.526	1.885	1.906	1.490	1.355	1.627	1.889	1.708
		8	3.928	2.313	2.406	1.434	1.494	1.676	1.981	2.140
		12	5.284	3.430	3.627	1.581	1.700	2.013	2.365	2.612
		16	7.594	4.598	4.857	1.844	2.081	2.563	3.158	3.753
		20	7.256	5.088	5.610	1.752	2.011	2.534	3.355	3.941
	6	4	3.169	2.135	2.241	1.486	1.425	1.598	1.807	1.795
		8	4.953	3.008	3.265	1.623	1.744	1.978	2.552	2.699
		12	4.313	3.808	3.947	1.324	1.446	1.700	2.470	2.413
		16	6.584	5.401	5.589	2.008	2.078	2.934	3.286	4.079
		20	7.372	5.368	5.790	2.080	2.332	3.163	3.850	4.861
	9	4	2.606	2.407	2.579	1.043	1.163	1.248	1.661	1.862
		8	4.552	3.602	3.654	1.724	1.816	2.278	2.710	2.986
		12	5.103	4.002	4.191	1.678	1.794	2.288	2.748	3.227
		16	6.719	4.898	5.143	2.157	2.313	3.091	3.330	4.368
		20	10.362	5.859	6.119	3.010	3.420	4.647	4.543	7.091
8	3	4	2.284	1.829	1.824	1.089	1.117	1.207	1.510	1.118
		8	3.382	2.539	2.655	1.372	1.393	1.498	2.304	1.705
		12	6.121	3.608	3.677	2.143	2.226	2.308	3.490	2.764
		16	5.502	3.797	4.127	1.967	1.977	2.232	3.667	2.688
		20	9.236	5.795	6.055	3.302	3.627	3.955	5.551	4.563
	6	4	3.659	2.303	2.325	1.893	1.880	1.910	2.667	1.736
		8	3.573	2.888	3.064	1.474	1.478	1.545	2.224	1.791
		12	6.201	4.292	4.386	2.380	2.638	2.819	3.684	3.009
		16	6.242	4.718	5.003	2.310	2.492	2.762	4.494	3.246
		20	10.451	5.826	6.075	3.772	4.106	4.877	7.925	6.690
	9	4	3.695	2.804	2.859	1.644	1.775	1.770	2.590	1.874
		8	4.508	3.580	3.657	1.851	1.902	2.068	3.073	2.277
		12	9.373	4.197	4.293	3.560	3.564	4.285	6.881	5.438
		16	9.069	5.195	5.364	3.762	3.940	4.718	6.633	5.776
		20	8.046	6.851	7.006	3.227	3.427	4.086	6.524	5.040

TABLE 6.9: RMSE of the Estimators for $n = 400$ – Submodel Contains Strong Signals.

p_1	p_2	p_3	SM	S	PS	Ridge	ENET	LASSO	ALASSO	SCAD
4	3	4	1.683	1.430	1.563	0.789	0.873	0.889	1.395	1.637
		8	2.582	2.134	2.219	1.027	1.098	1.247	1.602	2.054
		12	3.782	2.810	2.952	1.064	1.098	1.317	2.019	2.824
		16	4.474	3.436	3.579	1.153	1.226	1.493	2.465	3.380
		20	5.717	4.170	4.553	1.322	1.466	1.899	2.545	4.369
	6	4	2.202	1.821	1.919	1.021	1.096	1.147	1.453	1.616
		8	2.634	2.451	2.507	1.131	1.147	1.342	1.830	2.187
		12	3.645	3.198	3.267	1.234	1.357	1.634	2.045	3.015
		16	3.333	2.986	3.124	1.123	1.177	1.428	2.176	3.102
		20	4.380	3.850	3.931	1.277	1.410	1.793	2.466	3.905
	9	4	2.274	2.054	2.110	1.198	1.244	1.328	1.597	1.837
		8	2.167	2.247	2.280	1.015	1.050	1.133	1.790	2.190
		12	2.809	2.845	2.890	1.186	1.220	1.438	1.858	2.439
		16	3.531	3.491	3.576	1.166	1.283	1.626	2.241	3.224
		20	3.854	3.759	3.921	1.268	1.413	1.784	2.336	3.531
8	3	4	1.935	1.480	1.546	0.731	1.130	1.173	1.491	1.359
		8	2.495	1.916	2.025	0.890	1.169	1.204	1.779	1.797
		12	3.440	2.477	2.600	1.101	1.251	1.356	2.122	2.284
		16	4.120	2.879	2.965	1.358	1.436	1.504	2.437	2.817
		20	4.562	3.626	3.686	1.506	1.634	1.860	2.855	2.912
	6	4	2.023	1.715	1.751	0.857	1.145	1.145	1.366	1.360
		8	2.526	2.187	2.194	1.017	1.148	1.173	1.825	1.697
		12	2.797	2.627	2.618	1.226	1.368	1.418	2.103	2.075
		16	3.296	3.094	3.286	1.136	1.190	1.301	2.263	2.446
		20	4.127	3.670	3.909	1.373	1.396	1.585	2.873	2.822
	9	4	2.025	2.080	2.058	1.004	1.224	1.282	1.764	1.671
		8	2.551	2.361	2.389	1.060	1.181	1.251	1.890	1.809
		12	3.160	2.867	3.035	1.243	1.317	1.476	2.213	2.305
		16	3.943	3.462	3.504	1.542	1.618	1.814	2.667	3.146
		20	3.931	3.710	3.807	1.477	1.505	1.676	2.755	3.102

from Table 6.8 that shrinkage estimators outshine all the penalized estimators in all scenarios. Further, keeping p_1 and p_2 constant as p_3, the weak signals, increases the performance of the PS estimator is remarkable. Thus, reestablishing the beauty and power of the shrinkage strategy.

6.7 Real Data Examples

We analyze four data sets to examine the performance of the listed estimators. More importantly, we are interested in illustrating the proposed shrinkage methodology's characteristics for real applications.

6.7.1 Pima Indians Diabetes (PID) Data

First, we consider diabetes data. This data set is freely available at *mlbench* package in R. The description of the data is given in Table 6.10. Scientists are interested in predicting diabetes, specifically if a given patient tests positive or negative for the chronic disease. We can safely model the data using a logistic regression. The study is based on 768 patients with eight predictors, thus the data matrix consists of $p = 8$ and $n = 768$.

TABLE 6.10: Description of Diabetes Data.

Variable	Description
pregnant	Number of times pregnant
glucose	Plasma glucose concentration (glucose tolerance test)
pressure	Diastolic blood pressure (mm Hg)
triceps	Triceps skin fold thickness (mm)
insulin	2-Hour serum insulin (mu U/ml)
mass	Body mass index (weight in kg/(height in m)^2)
pedigree	Diabetes pedigree function
age	Age (years)
diabetes (Response)	Class variable (test for diabetes)

The full model can be written as follows:

$$\begin{aligned} diabetes_i &= \beta_0 + \beta_1 \text{pregnant}_i + \beta_2 \text{glucose}_i + \beta_3 \text{pressure}_i \\ &+ \beta_4 \text{triceps}_i + \beta_5 \text{insulin}_i \\ &+ \beta_6 \text{mass}_i + \beta_7 \text{pedigree}_i + \beta_8 \text{age}_i + \epsilon_i. \end{aligned}$$

In an effort to select a suitable submodel, we employ a BIC variable selection method. This method indicates that the variables pregnant, pressure, triceps, and insulin are not significantly important, so these predictors can be deleted from the initial model. The submodel is now given by

$$diabetes_i = \beta_0 + \beta_2 \text{glucose}_i + \beta_6 \text{mass}_i + \beta_7 \text{pedigree}_i + \beta_8 \text{age}_i + \epsilon_i.$$

We estimate the parameters of the model and asses the performance of the estimators by considering the components of the confusion matrix and by using the RMSE criterion.

The confusion matrix or contingency table is useful for classifier evaluation. The matrix contains information regarding predicted and actual values. The four components of the matrix are:

TABLE 6.11: Confusion Matrix.

	Actual positive	Actual negative
Predicted positive	True positive (TP)	False positive (FP)
Predicted negative	False negatives (FN)	True negatives (TN)

In Table 6.11, the confusion matrix displays the accuracy of classification tasks. It is an overall proportion of correct classification and describes what percentage of test data is classified correctly. This measure is widely used to examine the data based on logistic and other models. However, it may not be a good measure for imbalanced data sets. There are other metrics that are suitable to examine the performance of the sparse model in the presence of weak signals.

The commonly used ratios are accuracy, recall(sensitivity), precision (specificity), and the F-measure:

i) Accuracy $= (TP + TN)/\text{total}$

ii) Recall (Sensitivity) $= TP/(TP + FN)$

iii) Precision (Specificity) $= TP/(TP + FP)$

iv) F_1 Score $= \frac{2(Recall)(Precision)}{(Recall+Precision)}$

In Table 6.12, we report the relative accuracy (RA), relative precision (RP), relative recall (RR) and relative F_1 score (RFS) of the listed estimators with respect to full model estimator. In this data analysis, we have included a machine learning method, namely support vector machine (SVM) based on the linear and radial kernel functions. Looking at the results, there is not much gain in using shrinkage or penalized strategies. However, ratios based on SVM have an edge over other methods.

We further investigate the relative performance of the estimators using the MSE criteria. Table 6.13 reports the estimated values of the regression coefficients, estimated standard

TABLE 6.12: RA, RP, RR, and RFS of the PID Data.

	Accuracy	RA	Precision	RP	Recall	RR	F_1 Score	RFS
FM	0.746	1.000	0.879	1.000	0.718	1.000	0.789	1.000
SM	0.760	1.019	0.888	1.010	0.732	1.021	0.802	1.016
S	0.753	1.010	0.885	1.006	0.725	1.010	0.796	1.008
PS	0.753	1.010	0.885	1.007	0.724	1.009	0.796	1.008
ENET	0.753	1.010	0.887	1.008	0.723	1.007	0.795	1.008
LASSO	0.751	1.006	0.884	1.005	0.721	1.005	0.793	1.005
Ridge	0.753	1.010	0.888	1.009	0.722	1.006	0.795	1.007
SCAD	0.741	0.994	0.875	0.995	0.714	0.995	0.786	0.995
SVM_1	0.779	1.044	0.802	0.912	0.888	1.238	0.843	1.067
SVM_2	0.757	1.015	0.788	0.896	0.872	1.215	0.827	1.047

The kernel functions are Linear and Radial in SVM_1 and SVM_2, respectively.

error, and estimated bias of the estimators. Finally, the last column of the table gives the RMSE of the listed estimators to their respective full model estimator for comparison purposes. Interestingly, in terms of RMSE, the ridge estimator is outperforming all the estimators. One reason may be due to the multicollinearity presented in the model that the ridge model handles well, and consequently ENET also performs well. The shrinkage estimator's performance is reasonable. However, keep in mind that ridge estimator does well with respect to MLE and shrinkage estimator based on MSE. In this case, one may replace the MLE by the ridge estimator to improve the performance of shrinkage estimators. Another reason shrinkage estimators are not doing their best could be that the selected submodel by BIC may not be the optimal one. Keep in mind that the MSEs of shrinkage estimators are bound and the RMSEs will not go below one if the assumption of sparsity is violated, while the other estimators do not enjoy such an important property.

6.7.2 South Africa Heart-Attack Data

This data set is freely available in *bestglm* package in R. The description of the data is given in Table 6.14 gives a retrospective sample of males in a heart-disease high-risk region of the Western Cape, South Africa. The scientists are interested in predicting coronary heart disease based on nine predictors. We model the data using logistic regression with $p = 9$ and $n = 462$.

The full model can be written as follows:

$$\begin{aligned} chd_i &= \beta_0 + \beta_1 \text{sbp}_i + \beta_2 \text{tobacco}_i + \beta_3 \text{ldl}_i \\ &+ \beta_4 \text{adiposity}_i + \beta_5 \text{famhist}_i \\ &+ \beta_6 \text{typea}_i + \beta_7 \text{obesity}_i + \beta_8 \text{alcohol}_i + +\beta_9 \text{age}_i + \epsilon_i. \end{aligned}$$

We use the BIC variable selection method to select a submodel, which suggests tobacco, ldl, famhist, typea and age are relatively more important predictors for building a submodel. Thus, the submodel is given as follows:

$$chd_i = \beta_0 + \beta_2 \text{tobacco}_i + \beta_3 \text{ldl}_i + \beta_5 \text{famhist}_i + \beta_6 \text{typea}_i + \beta_9 \text{age}_i + \epsilon_i.$$

Now, we build the shrinkage strategy by combining the above submodel with the full model and the weight measure as defined in (6.12). Table 6.15 provides the results in terms of coefficient estimates and their associated standard error and bias. The RMSE for each method is also recorded in the last column of the table. To select the best strategy, we look at the respective value of the RMSE. As expected, the selected submodel will mostly perform better than others by design. For this data example, the RMSE for the submodel is the highest with a value of 2.893. Interestingly, the RMSE of the ridge is 1.284 higher than the positive shrinkage strategy, which is 1.216.

A possible explanation is that the data may be subject to multicollinearity, so the MLE is not behaving well. In this situation, it would be fruitful to construct shrinkage estimators using the ridge estimator. The positive shrinkage strategy outperforms all four remaining penalized estimators. Perhaps multicollinearity is playing a role in the poor performance of penalized methods.

6.7.3 Orinda Longitudinal Study of Myopia (OLSM) Data

The data set in this study is a subset of data from the Orinda longitudinal study of Myopia (OLSM), a cohort study of ocular component development and risk factors for the onset of myopia in children. The data collection began in the 1989-1990 school year and continued

TABLE 6.13: Estimates (First Row), Standard Errors (Second Row) and Bias (Third Row). The RMSE Column Gives the Relative Mean Squared Error of the Estimators with Respect to the FM for PID Data.

	β_2	β_6	β_7	β_8	RMSE
FM	1.133	0.426	0.408	0.328	1.000
	0.039	0.016	0.013	0.043	
	0.209	0.207	0.188	0.288	
SM	1.040	0.416	0.384	0.493	3.212
	0.012	0.010	0.016	0.021	
	0.160	0.147	0.186	0.179	
S	1.059	0.418	0.389	0.460	1.149
	0.093	0.024	0.029	-0.068	
	0.199	0.194	0.178	0.265	
PS	1.059	0.418	0.389	0.460	**1.215**
	0.076	0.015	0.032	-0.070	
	0.198	0.174	0.188	0.266	
Ridge	0.922	0.364	0.318	0.315	**2.650**
	-0.008	0.001	0.007	0.017	
	0.125	0.136	0.134	0.150	
ENET	0.974	0.368	0.287	0.314	**1.810**
	0.047	0.016	0.038	0.008	
	0.171	0.170	0.166	0.184	
LASSO	1.046	0.403	0.310	0.320	1.729
	0.045	-0.010	0.011	-0.001	
	0.171	0.177	0.181	0.196	
ALASSO	1.106	0.495	0.333	0.369	1.152
	0.063	-0.064	0.006	-0.030	
	0.183	0.221	0.216	0.253	
SCAD	1.117	0.535	0.366	0.448	0.907
	0.107	-0.055	-0.001	-0.087	
	0.198	0.246	0.224	0.269	

TABLE 6.14: Description of South Africa Heart-Attack Data.

Predictors	Description
sbp	systolic blood pressure
tobacco	cumulative tobacco (kg)
ldl	low density lipoprotein cholesterol
adiposity	a numeric vector
famhist	family history of heart disease, a factor with levels Absent Present
typea	type-A behavior
alcohol	current alcohol consumption
obesity	a numeric vector
age	age at onset
chd (Response variable)	coronary heart disease

annually through the 2000-2001 school year. The study is concerned with eye health. Information about the parts that make up the eye (the ocular components) was collected during an examination during the school day. The data on family history and visual activities were also collected on a yearly basis in a survey completed by a parent or guardian. The data set used in this example is from 618 children who had at least five years of follow-up and were not myopic when they entered the study.

This data set is freely available at *aplore3* package in R. The description of data is given in Table 6.16. The logistic regression model is built with $p = 12$ and $n = 618$.

The full model can be written as follows:

$$\begin{aligned} myopic_i &= \beta_0 + \beta_1 \text{age}_i + \beta_2 \text{gender}_i + \beta_3 \text{acd}_i \\ &+ \beta_4 \text{lt}_i + \beta_5 \text{vcd}_i \\ &+ \beta_6 \text{sporthr}_i + \beta_7 \text{readhr}_i + \beta_8 \text{comphr}_i \\ &+ \beta_9 \text{studyhr}_i + \beta_{10} \text{tvhr}_i \\ &+ \beta_{11} \text{mommy}_i + \beta_{12} \text{dadmy}_i + \epsilon_i. \end{aligned}$$

In an effort to identify the most influential predictors, we apply the AIC variable selection method. As a result, gender, acd, sporthr, readhr, mommy and dadmy are identified as relative important predictors to build a submodel:

$$\begin{aligned} myopic_i &= \beta_0 + \beta_2 \text{gender}_i + \beta_3 \text{acd}_i + \beta_6 \text{sporthr}_i \\ &+ \beta_7 \text{readhr}_i + \beta_{11} \text{mommy}_i + \beta_{12} \text{dadmy}_i + \epsilon_i. \end{aligned}$$

Table 6.17 showcases the RMSE for the listed estimation methods. It is evident from the respective values of the RMSE that the PS strategy is the clear winner with highest value of RMSE 1.274, excluding submodel. The performance of the penalized methods is questionable for this particular data set.

6.8 High-Dimensional Data

In this section, we compare the relative performance of the listed estimators in high-dimensional cases, that is, when $n < p$. Mainly, we are interested in assessing the performance of the shrinkage estimators with the penalized methods. To construct the shrinkage

TABLE 6.15: Estimates (First Row), Standard Errors (Second Row) and Bias (Third Row). The RMSE Column Gives the Relative Mean Squared Error of the Estimators with Respect to the FM for South Africa Heart-Attack Data.

	β_2	β_3	β_5	β_6	β_9	RMSE
FM	0.334	0.410	-0.421	0.378	0.711	1.000
	0.018	0.011	-0.010	0.007	0.014	
	0.140	0.139	0.075	0.115	0.169	
SM	0.343	0.380	-0.417	0.351	0.760	2.893
	0.011	-0.004	-0.001	-0.001	0.003	
	0.127	0.132	0.072	0.108	0.138	
S	0.340	0.390	-0.418	0.361	0.742	1.152
	0.010	0.028	-0.012	0.021	-0.010	
	0.134	0.140	0.076	0.118	0.156	
PS	0.340	0.390	-0.418	0.361	0.742	**1.216**
	0.010	0.016	-0.014	0.026	-0.004	
	0.130	0.138	0.075	0.121	0.151	
Ridge	0.340	0.331	0.661	0.329	0.582	**1.284**
	-0.003	0.035	0.004	-0.019	0.001	
	0.098	0.115	0.228	0.089	0.130	
ENET	0.338	0.329	0.698	0.323	0.615	1.101
	-0.002	-0.010	0.041	0.030	0.015	
	0.105	0.123	0.220	0.127	0.137	
LASSO	0.336	0.333	0.779	0.324	0.672	0.885
	0.013	-0.011	-0.014	-0.002	-0.033	
	0.101	0.106	0.201	0.125	0.152	
ALASSO	0.317	0.289	0.816	0.293	0.710	0.616
	0.006	0.050	0.074	0.041	-0.021	
	0.157	0.126	0.283	0.104	0.183	
SCAD	0.368	0.351	0.912	0.370	0.733	0.697
	0.003	0.003	-0.004	-0.008	-0.024	
	0.153	0.162	0.197	0.122	0.172	

TABLE 6.16: Description of OSLM Data.

Variable	Description
age	Age at first visit (Years)
gender	Gender (1: Male, 2: Female)
acd	Anterior Chamber Depth (mm)
lt	Lens Thickness (mm)
vcd	Vitreous Chamber Depth (mm)
sporthr	How many hours per week outside of school the child spent engaging in sports/outdoor activities (Hours per week)
readhr	How many hours per week outside of school the child spent reading for pleasure (Hours per week)
comphr	How many hours per week outside of school the child spent playing video/computer games or working on the computer (Hours per week)
studyhr	How many hours per week outside of school the child spent reading or studying for school assignments (Hours per week)
tvhr	How many hours per week outside of school the child spent watching television (Hours per week)
mommy	Was the subject's mother myopic? (1: No, 2: Yes)
dadmy	Was the subject's father myopic? (1: No, 2: Yes)
myopic (Dependent)	Myopia within the first five years of follow up (1: No, 2: Yes)

estimator, we need estimators based on two different models: a model with a larger number of predictors and another model based on a relatively small number of predictors. We then combine these two model-based estimators using a distance measure. The construction of shrinkage estimators for high-dimensional cases follows the same principle as in the low-dimensional case. In passing, we would like to remark that it is up to the data analyst to select two penalized methods and then construct a shrinkage estimator. We suggest that ridge, LASSO, or ENET can be used to obtain a model with a larger number of predictors, and subsequently either SCAD, ALASSO, or any other aggressive penalized procedures can be implemented to obtain a submodel with relatively smaller number of predictors. The idea is to reduce the inherited selection bias and improve the prediction by combining estimators from two models by adapting the shrinkage strategy.

6.8.1 Simulation Experiments

In our simulation experiment, we use ENET to select a model with relatively large predictors. We then choose ALASSO to produce the submodel estimator as it selects fewer predictors than ENET.

A binary response is generated using the following model:

$$ln\left(\frac{\pi_i}{1-\pi_i}\right) = \eta_i = \boldsymbol{x}_i^\top \boldsymbol{\beta}, \quad i = 1, \cdots, n,$$

where $\pi_i = P(Y = 1| \, x_i)$ and the predictor values $\boldsymbol{x}_i{}^\top = (x_{i1}, x_{i2}, \cdots, x_{in})$ have been drawn from a standard multivariate normal distribution. We consider the regression coefficients are set $\boldsymbol{\beta} = \left(\boldsymbol{\beta}_1^\top, \boldsymbol{\beta}_2^\top, \boldsymbol{\beta}_3^\top\right)^\top$ with dimensions p_1, p_2, and p_3, respectively. Further, $\boldsymbol{\beta}_1$ represent

TABLE 6.17: The RMSE of the Estimators for OLSM Data.

	β_2	β_3	β_6	β_7	β_{11}	β_{12}	RMSE
FM	-0.686	0.267	-0.414	0.301	-0.361	-0.202	1.000
	-0.029	0.004	-0.021	0.008	0.001	-0.008	
	0.219	0.130	0.165	0.113	0.191	0.182	
SM	-0.651	0.265	-0.419	0.262	-0.367	-0.216	1.852
	-0.009	0.005	-0.017	-0.006	-0.001	0.002	
	0.194	0.114	0.156	0.105	0.181	0.180	
S	-0.631	0.264	-0.421	0.240	-0.370	-0.224	1.009
	-0.050	0.008	-0.013	0.042	0.003	0.010	
	0.207	0.120	0.158	0.107	0.190	0.186	
PS	-0.651	0.265	-0.419	0.262	-0.367	-0.216	1.274
	-0.047	-0.001	-0.014	0.020	0.008	0.009	
	0.200	0.117	0.157	0.111	0.187	0.190	
Ridge	0.436	0.313	-0.333	0.260	0.599	0.751	0.915
	-0.008	-0.001	-0.019	-0.016	-0.016	0.003	
	0.216	0.111	0.148	0.107	0.232	0.259	
ENET	0.374	0.300	-0.316	0.243	0.569	0.716	0.789
	0.023	-0.002	-0.013	-0.001	0.029	0.029	
	0.266	0.132	0.158	0.128	0.248	0.275	
LASSO	0.380	0.301	-0.314	0.241	0.632	0.808	0.750
	0.005	-0.018	-0.010	-0.012	-0.009	0.002	
	0.279	0.125	0.152	0.126	0.274	0.275	
ALASSO	0.202	0.241	-0.263	0.169	0.638	0.813	0.494
	0.191	0.041	-0.056	0.046	0.065	0.085	
	0.341	0.163	0.201	0.144	0.313	0.309	
SCAD	0.284	0.286	-0.299	0.234	0.797	0.946	0.412
	0.220	0.022	-0.045	0.025	-0.034	0.026	
	0.364	0.171	0.200	0.155	0.344	0.343	

strong signals, where $\boldsymbol{\beta}_1$ is a vector of 1 values or higher, $\boldsymbol{\beta}_2$ stands for the weak signals with the signal strength κ values, and $\boldsymbol{\beta}_3$ represents no signals where $\boldsymbol{\beta}_3 = \mathbf{0}$.

The result of the simulation study is reported in Tables 6.18–6.20 for different configurations of n, p_1, p_2, p_3 and weak signals strength. It is evident from the Tables that performance of the estimators are similar as in the low-dimensional case. The positive shrinkage estimator outperforms the penalty estimators in many cases. The performance of SCAD is also competitive. However, positive shrinkage estimator has an edge over when the number and strength of weak coefficients increase.

6.8.2 Gene Expression Data

Alon et al. (1999) considered gene expression data from microarray experiments of colon tissue samples. In this experiment, the data matrix is of dimension (62×2000), giving the expression levels of 2000 genes for the 62 colon tissue samples. The response contains 40 tumor tissues and 22 normal tissues. We calculate the accuracy and relative accuracy of each method. We construct positive shrinkage in four ways and report it in Table 6.21. For example, notion "PS(ENET-ALASSO)" means that ENET and ALASSO are used to obtain model with larger and smaller number of predictors, respectively. The results indicate the supremacy of the positive shrinkage estimators over the penalized methods. In all cases, the shrinkage estimator enjoys higher accuracy and reliability. In the case when models are selected by LASSO and ALASSO, the values are a little lower, but not significantly lower.

6.9 A Gentle Introduction of Negative Binomial Models

The negative binomial (NB) model is commonly used to model count data with over-dispersion, where the variance is larger than the mean. In this chapter, we interested in estimation problems when a number of predictors are available using NB regression models; we refer to Hilbe (2011) and Cameron and Trivedi (2005) for a detailed information on the host NB regression model.

For brevity, we introduce a NB regression model for predicting a count response y_i (for $i = 1, 2, ..., n$):

$$f(y_i \mid \boldsymbol{x}_i) = \frac{\Gamma\left(y_i + \frac{1}{\theta}\right)}{y_i!\Gamma\left(\frac{1}{\theta}\right)} \left(\frac{\theta\mu_i}{1+\theta\mu_i}\right)^{y_i} \left(\frac{1}{1+\theta\mu_i}\right)^{\frac{1}{\theta}}, y_i = 0, 1, 2, \ldots. \tag{6.16}$$

Here, θ is the dispersion index, $\Gamma(.)$ is the usual gamma function, and $\mu_i = \exp(\boldsymbol{x}_i^\top \boldsymbol{\beta})$ is the mean parameter, where $\boldsymbol{\beta} = (\beta_1, \beta_2, ..., \beta_p)^\top \in \mathcal{R}^p$ is a vector of regression coefficients and $\boldsymbol{x}_i = (x_{1i}, x_{2i}, ..., x_{pi})^\top \in \mathcal{R}^p$ is a vector of p are the predictors. The conditional mean and variance of y_i are given by $E[y_i|\boldsymbol{x}_i] = \mu_i$ and $\mathbb{V}[y_i|\boldsymbol{x}_i] = \mu_i + \theta\mu_i^2$, respectively. The NB regression model can be viewed as an extension of Poisson regression model. The Poisson regression model can be considered as a special case of NB regression model in which θ is zero and it can be seen that $E[y_i|\boldsymbol{x}_i] = \mathbb{V}[y_i|\boldsymbol{x}_i] = \mu_i$. In the framework of the zero-inflated Poisson regression model, which is used to model count data that has an excess of zero counts, Asar et al. (2018) proposed a more accurate estimation. The maximum likelihood method is widely used in a host statistical models for estimating model parameters. Here our main focus is on the estimation of the parameter vector of $\boldsymbol{\beta}$. The estimation the dispersion index θ can be found in Hauer et al. (2002), Zou et al. (2015), and Wu and Lord (2017).

TABLE 6.18: RMSE of the Estimators for $p_1 = 4$ and $p_3 = 1000$.

n	p_2	κ	SM	PS	LASSO	SCAD	Ridge
300	150	0.00	1.811	2.575	2.041	3.561	0.725
		0.05	1.414	1.948	1.691	2.387	0.751
		0.10	1.088	1.423	1.351	1.600	0.826
		0.20	0.938	1.111	1.045	1.054	0.927
		0.40	0.948	1.011	0.967	0.965	0.996
	200	0.00	1.605	2.439	2.116	3.473	0.713
		0.05	1.446	1.956	1.694	2.369	0.753
		0.10	1.223	1.443	1.298	1.483	0.837
		0.20	0.979	1.097	1.039	1.055	0.940
		0.40	0.990	1.006	0.976	0.978	1.013
	250	0.00	1.914	2.294	1.942	3.224	0.726
		0.05	1.414	1.671	1.643	2.284	0.765
		0.10	1.155	1.292	1.258	1.431	0.842
		0.20	0.985	1.064	1.032	1.043	0.963
		0.40	0.971	1.001	0.975	0.975	0.999
600	150	0.00	4.654	5.142	2.695	9.029	0.633
		0.05	3.232	2.776	1.979	3.837	0.698
		0.10	1.664	1.484	1.402	1.815	0.787
		0.20	1.128	1.093	1.050	1.106	0.891
		0.40	1.037	1.036	0.960	0.935	0.946
	200	0.00	5.901	3.640	2.727	9.875	0.649
		0.05	2.759	1.621	1.782	3.010	0.700
		0.10	1.576	1.190	1.290	1.604	0.811
		0.20	1.059	1.068	1.011	1.042	0.920
		0.40	1.006	1.027	0.947	0.936	0.961
	250	0.00	4.878	2.362	2.708	8.739	0.647
		0.05	2.432	1.675	1.726	2.818	0.720
		0.10	1.373	1.241	1.227	1.449	0.848
		0.20	0.993	1.050	0.984	0.985	0.956
		0.40	0.962	1.009	0.947	0.940	0.978

TABLE 6.19: RMSE of the Estimators for $n = 200$ and $p_1 = 3$.

p_2	p_3	κ	SM	PS	LASSO	SCAD	Ridge
10	500	0.0	1.047	1.559	1.690	2.020	0.749
		0.2	0.803	1.475	1.469	1.658	0.774
		0.4	0.789	1.173	1.212	1.277	0.825
		0.6	0.904	1.098	1.123	1.132	0.855
		0.8	1.047	1.086	1.118	1.122	0.845
	1000	0.0	1.042	1.378	1.561	1.765	0.751
		0.2	0.923	1.479	1.386	1.486	0.822
		0.4	0.705	1.154	1.186	1.206	0.863
		0.6	0.814	1.076	1.084	1.128	0.894
		0.8	0.897	1.064	1.087	1.097	0.883
30	500	0.0	0.880	1.300	1.629	1.962	0.763
		0.2	0.876	1.230	1.176	1.220	0.852
		0.4	0.876	1.066	1.008	1.028	0.926
		0.6	0.861	1.019	0.966	0.966	0.958
		0.8	0.917	1.017	0.955	0.943	0.947
	1000	0.0	1.159	1.534	1.637	1.870	0.792
		0.2	0.871	1.243	1.118	1.147	0.839
		0.4	0.880	1.059	1.024	1.021	0.956
		0.6	0.819	1.022	0.998	0.993	0.977
		0.8	0.902	1.013	0.978	0.976	0.975
50	500	0.0	0.898	1.293	1.689	2.012	0.741
		0.2	0.879	1.140	1.099	1.134	0.880
		0.4	0.791	1.033	0.987	0.976	0.948
		0.6	0.886	1.025	0.973	0.963	0.959
		0.8	0.958	1.018	0.959	0.948	0.954
	1000	0.0	0.759	1.168	1.567	1.712	0.739
		0.2	0.859	1.143	1.102	1.120	0.909
		0.4	0.836	1.021	0.990	0.994	0.973
		0.6	0.907	1.016	0.974	0.976	0.977
		0.8	0.938	1.014	0.973	0.973	0.976

TABLE 6.20: RMSE of the Estimators for $n = 200$ and $p_1 = 9$.

p_2	p_3	κ	SM	PS	LASSO	SCAD	Ridge
10	500	0.0	1.082	1.471	1.368	1.731	0.791
		0.2	1.120	1.576	1.309	1.557	0.817
		0.4	1.058	1.271	1.206	1.256	0.864
		0.6	1.027	1.079	1.021	1.028	0.896
		0.8	1.106	1.044	0.994	0.997	0.900
	1000	0.0	1.102	1.346	1.240	1.343	0.870
		0.2	1.054	1.426	1.260	1.368	0.876
		0.4	1.040	1.179	1.096	1.102	0.903
		0.6	0.975	1.064	1.012	1.016	0.922
		0.8	0.990	1.031	1.005	0.999	0.952
30	500	0.0	1.164	1.700	1.398	1.661	0.809
		0.2	1.184	1.483	1.160	1.245	0.863
		0.4	0.934	1.107	1.036	1.043	0.928
		0.6	0.910	1.033	0.973	0.951	0.940
		0.8	0.980	1.020	0.962	0.956	0.954
	1000	0.0	1.104	1.516	1.276	1.466	0.852
		0.2	1.067	1.349	1.131	1.204	0.894
		0.4	0.933	1.071	1.015	1.015	0.948
		0.6	0.944	1.019	0.973	0.970	0.968
		0.8	0.946	1.008	0.974	0.977	0.983
50	500	0.0	1.225	1.596	1.336	1.663	0.804
		0.2	1.169	1.245	1.090	1.112	0.891
		0.4	0.895	1.014	0.967	0.968	0.977
		0.6	0.911	1.004	0.955	0.956	0.976
		0.8	0.932	1.006	0.954	0.951	0.965
	1000	0.0	1.051	1.469	1.287	1.470	0.864
		0.2	1.006	1.169	1.053	1.068	0.918
		0.4	0.881	1.009	0.984	0.981	0.986
		0.6	0.901	1.003	0.977	0.974	0.995
		0.8	0.939	1.005	0.975	0.971	0.987

TABLE 6.21: Colon Data Accuracy and Relative Accuracy.

	Accuracy	RA
ENET	0.773	1.000
LASSO	0.775	1.002
ALASSO	0.793	1.025
SCAD	0.786	1.017
Ridge	0.798	1.033
PS(ENET-ALASSO)	0.819	1.060
PS(LASSO-ALASSO)	0.792	1.025
PS(ENET-SCAD)	0.818	1.058
PS(LASSO-SCAD)	0.812	1.050

Let us consider the classical maximum likelihood estimation for the regression parameters in NB regression model. The log-likelihood function is defined as:

$$\begin{aligned}\ell(\boldsymbol{\beta},\theta) &= \sum_{i=1}^{n}\left\{y_i\ln\left(\frac{\theta\mu_i}{1+\mu_i}\right)+\frac{1}{\theta}\ln\left(\frac{1}{1+\mu_i}\right)\right. \\ &+ \left.\ln\Gamma\left(y_i+\frac{1}{\theta}\right)-\ln\Gamma(y_i+1)-\ln\Gamma\left(\frac{1}{\theta}\right)\right\}.\end{aligned} \tag{6.17}$$

The maximum likelihood estimator of $\boldsymbol{\beta}$ and θ are obtained by solving the following score equations:

$$\frac{\partial\ell(\boldsymbol{\beta},\theta)}{\partial\boldsymbol{\beta}}=\sum_{i=1}^{n}\frac{\boldsymbol{x}_i(y_i-\mu_i)}{(1+\theta\mu_i)}, \tag{6.18}$$

and

$$\begin{aligned}\frac{\partial\ell(\boldsymbol{\beta},\theta)}{\partial\theta} &= \sum_{i=1}^{n}\left\{\frac{1}{\theta^2}\left(\ln(1+\theta\mu_i)+\frac{\theta(y_i-\mu_i)}{1+\theta\mu_i}\right)\right. \\ &+ \left.\frac{\partial}{\partial\theta}\ln\Gamma\left(y_i+\frac{1}{\theta}\right)-\frac{\partial}{\partial\theta}\ln\Gamma\left(\frac{1}{\theta}\right)\right\}=0.\end{aligned} \tag{6.19}$$

The likelihood equations can be solved using the Newton-Raphson method. Let $\widehat{\boldsymbol{\beta}}^{\mathrm{FM}}$ be MLE of $\boldsymbol{\beta}$ based on the full model, where all the predictors in the model are to be estimated, and $\widehat{\theta}$ is the MLE for dispersion parameter, θ. For $\theta>0$, Lawless (1987) showed that, under the assumed regularity condition and for $n\to\infty$,

$$\begin{bmatrix} n^{1/2}(\widehat{\boldsymbol{\beta}}^{\mathrm{FM}}-\boldsymbol{\beta}) \\ n^{1/2}(\widehat{\theta}-\theta)\end{bmatrix} \xrightarrow{D} \mathcal{N}_{p+1}\left(\begin{bmatrix}\mathbf{0}_p \\ 0\end{bmatrix},\begin{bmatrix}\boldsymbol{I}_1^*(\boldsymbol{\beta},\theta)^{-1} & \mathbf{0}_p \\ \mathbf{0}_p^\top & I_2^*(\boldsymbol{\beta},\theta)^{-1}\end{bmatrix}\right). \tag{6.20}$$

Here, $\boldsymbol{I}_1^*(\boldsymbol{\beta},\theta) = \lim\limits_{n\to\infty}\frac{1}{n}\sum_{i=1}^{n}\frac{\mu_i\boldsymbol{x}_i\boldsymbol{x}_i^\top}{1+\theta\mu_i}$, and $I_2^*(\boldsymbol{\beta},\theta) = \lim\limits_{n\to\infty}\frac{1}{n}\sum_{i=1}^{n}$ $\left\{\frac{1}{\theta^4}E\left[\sum_{j=0}^{y_j-1}(\frac{1}{\theta}+j)^{-2}-\frac{\theta\mu_i}{\mu_i+\frac{1}{\theta}}\right]\right\}$.

However, MLEs are not stable if there are too many predictors in the model. Estimation based on a full model with numerous predictors may lead to an array of disadvantages, including mis-specification on the scale of estimation and high variability. In the following section, we consider the estimation problem when a model is sparse.

6.9.1 Sparse NB Regression Model

In many practical situations, an initial or full model may have a large number of predictors, but it may have only a few important predictors, while the rest of the parameters can be ignored. In other words, a submodel with a few influential predictors may be available. Suppose the parameter vector $\boldsymbol{\beta}$ can be partitioned as $\boldsymbol{\beta} = (\boldsymbol{\beta}_1^\top, \boldsymbol{\beta}_2^\top)^\top$, where $\boldsymbol{\beta}_1 = (\beta_1, \beta_2, ..., \beta_{p_1})^\top \in \mathcal{R}^{p_1}$ is a vector of coefficients related with p_1 relevant predictors and $\boldsymbol{\beta}_2 = (\beta_{p_1+1}, \beta_{p_1+2}, ..., \beta_{p_1+p_2})^\top \in \mathcal{R}^{p_2}$ is a vector of coefficients related with p_2 irrelevant predictors. If $\boldsymbol{\beta}_2$ can be set as a zero vector $\mathbf{0}_{p_2}$, then we say the model is sparse. However, sparsity is a stringent assumption and may not be judiciously justified. The submodel estimators based on p_1 predictors are highly efficient when the assumption of sparsity holds. However, if too many important predictors are ignored, in which $\boldsymbol{\beta}_2$ is significantly different from $\mathbf{0}_{p_2}$, the performance of the submodel will be poorer than that of full model. The MLE performance based on both competing models (FM and SM) suffer from the uncertainty of the subspace information, $\boldsymbol{\beta}_2 = \mathbf{0}_{p_2}$. We consider some alternative estimation strategies for estimating the regression parameter vector for NB regression models in the presence of uncertain subspace information. As such, we include shrinkage (based on Stein-rule) and some penalized maximum likelihood estimation strategies and we also consider two penalized methods, LASSO and ALASSO.

6.10 Shrinkage and Penalized Strategies

We are interested in estimating $\boldsymbol{\beta}_1$ when $\boldsymbol{\beta}_2$ may or may not equal to $\mathbf{0}_{p_2}$. Thus, we are considering full and submodels. The MLE of $\boldsymbol{\beta}_1$, be denoted as $\widehat{\boldsymbol{\beta}}_1^{\mathrm{FM}}$, is obtained by solving equations (6.18) and (6.19). The SM-based ML estimator of $\boldsymbol{\beta}_1$, denoted as $\widehat{\boldsymbol{\beta}}_1^{\mathrm{SM}}$, is obtained by solving Eqs. (6.18) and (6.19) under the restriction $\boldsymbol{\beta}_2 = \mathbf{0}_{p_2}$. We now propose the improved shrinkage estimators based on effective strategies, designed to outperform both $\widehat{\boldsymbol{\beta}}_1^{\mathrm{FM}}$ and $\widehat{\boldsymbol{\beta}}_1^{\mathrm{SM}}$ under mild conditions. These estimators optimally integrate $\widehat{\boldsymbol{\beta}}_1^{\mathrm{FM}}$ and $\widehat{\boldsymbol{\beta}}_1^{\mathrm{SM}}$, and can be generally formulated as

$$\widehat{\boldsymbol{\beta}}_1^{\mathrm{S}} = \widehat{\boldsymbol{\beta}}_1^{\mathrm{FM}} - c(.)(\widehat{\boldsymbol{\beta}}_1^{\mathrm{FM}} - \widehat{\boldsymbol{\beta}}_1^{\mathrm{SM}}),$$

where $c(.)$ is a suitably function to reflect our estimators. To allow a suitable choice to be made between $\widehat{\boldsymbol{\beta}}_1^{\mathrm{FM}}$ and $\widehat{\boldsymbol{\beta}}_1^{\mathrm{SM}}$, we examine the accuracy of the subspace information using the following normalize distance:

$$T_n = n(\widehat{\boldsymbol{\beta}}_2^{\mathrm{MLE}} - \boldsymbol{\beta}_2)^\top \boldsymbol{C}_{22.1}(\widehat{\boldsymbol{\beta}}_2^{\mathrm{MLE}} - \boldsymbol{\beta}_2) + o_{p_2}(1). \quad (6.21)$$

Here, $\boldsymbol{C}_{22.1}^{-1}$ is the asymptotic covariance matrix of $n^{1/2}(\widehat{\boldsymbol{\beta}}_2^{\mathrm{MLE}} - \boldsymbol{\beta}_2)$, see Appendix for details. Aitchison and Silvey (1958) showed that under the restriction $\boldsymbol{\beta}_2 = \mathbf{0}_{p_2}$, as $n \to \infty$, the distribution of T_n converges to a χ^2 distribution with p_2 degrees of freedom and $o_{p_2}(1)$ tends to zero.

A suitable function for combining $\widehat{\boldsymbol{\beta}}_1^{\mathrm{FM}}$ and $\widehat{\boldsymbol{\beta}}_1^{\mathrm{SM}}$ can be imposed by setting $c_1(.) = (p_2 - 2)T_n^{-1}$, which gives shrinkage estimator of $\boldsymbol{\beta}_1$ and is given by Eq. (6.22).

$$\widehat{\boldsymbol{\beta}}_1^{\mathrm{S}} = \widehat{\boldsymbol{\beta}}_1^{\mathrm{FM}} - (p_2 - 2)T_n^{-1}(\widehat{\boldsymbol{\beta}}_1^{\mathrm{FM}} - \widehat{\boldsymbol{\beta}}_1^{\mathrm{SM}}), \qquad p_2 \geq 3. \quad (6.22)$$

Alternatively, we may write this as

$$\widehat{\boldsymbol{\beta}}_1^{\mathrm{S}} = \widehat{\boldsymbol{\beta}}_1^{\mathrm{SM}} + (1 - (p_2 - 2)T_n^{-1})(\widehat{\boldsymbol{\beta}}_1^{\mathrm{FM}} - \widehat{\boldsymbol{\beta}}_1^{\mathrm{SM}}), \qquad p_2 \geq 3. \quad (6.23)$$

A truncated version, called the positive-part shrinkage (PS) estimator, is defined as follows:

$$\widehat{\boldsymbol{\beta}}_1^{\mathrm{PS}} = \widehat{\boldsymbol{\beta}}_1^{\mathrm{S}} - (1 - (p_2 - 2)T_n^{-1})(\widehat{\boldsymbol{\beta}}_1^{\mathrm{FM}} - \widehat{\boldsymbol{\beta}}_1^{\mathrm{SM}})I(T_n \leq (p_2 - 2)), \qquad p_2 \geq 3. \tag{6.24}$$

Further, we consider LASSO, ALASSO, and SCAD penalty strategies of variable selection and parameter estimation. We compare the relative performance of the penalty strategies with shrinkage strategies through simulation.

6.11 Asymptotic Analysis

The asymptotic properties of estimators are studied under the sequence of local alternatives $K_{(n)}$:

$$K_{(n)} : \boldsymbol{\beta}_2 = n^{-1/2}\boldsymbol{\delta}. \tag{6.25}$$

Here, $\boldsymbol{\delta}^\top = (\delta_1, \delta_2, ..., \delta_{p_2}) \in \mathcal{R}^{p_2}$ is a p_2-dimensional vector of fixed values.

Suppose that $\widehat{\boldsymbol{\beta}}_1^*$ is any estimator of $\boldsymbol{\beta}_1$. Under the sequence $K_{(n)}$, the asymptotic distribution function of $\widehat{\boldsymbol{\beta}}_1^*$ can be written as $F(\boldsymbol{y}) = \lim_{n\to\infty} \mathbb{P}[n^{1/2}(\widehat{\boldsymbol{\beta}}_1^* - \boldsymbol{\beta}_1) \leq \boldsymbol{y}|K_{(n)}]$, where $F(\boldsymbol{y})$ is a non-degenerate distribution function. Then, the asymptotic distributional bias (ADB) of $\widehat{\boldsymbol{\beta}}_1^*$ is given by

$$\mathrm{ADB}\widehat{\boldsymbol{\beta}}_1^*) = \int ... \int \boldsymbol{y}dF(\boldsymbol{y}) = \lim_{n\to\infty} E[n^{1/2}(\widehat{\boldsymbol{\beta}}_1^* - \boldsymbol{\beta}_1)]. \tag{6.26}$$

Let $\boldsymbol{M}$ be a positive-definite matrix and consider the following weighted quadratic loss function: $L(\widehat{\boldsymbol{\beta}}_1^*; \boldsymbol{M}) = n^{1/2}(\widehat{\boldsymbol{\beta}}_1^* - \boldsymbol{\beta}_1)^\top \boldsymbol{M} n^{1/2}(\widehat{\boldsymbol{\beta}}_1^* - \boldsymbol{\beta}_1)$. Then, we define the asymptotic distributional risk (ADR) of $\widehat{\boldsymbol{\beta}}_1^*$ as

$$ADR(\widehat{\boldsymbol{\beta}}_1^*; \boldsymbol{M}) = tr[\boldsymbol{M}\boldsymbol{\Sigma}(\widehat{\boldsymbol{\beta}}_1^*)], \tag{6.27}$$

where $tr(.)$ is the trace of the matrix, and $\boldsymbol{\Sigma}(\widehat{\boldsymbol{\beta}}_1^*)$ is the asymptotic mean square error matrix for the distribution $F(\boldsymbol{y})$ of $\widehat{\boldsymbol{\beta}}_1^*$, given by

$$\boldsymbol{\Sigma}(\widehat{\boldsymbol{\beta}}_1^*) = \int ... \int \boldsymbol{y}\boldsymbol{y}^\top dF(\boldsymbol{y}) = \lim_{n\to\infty} E[n^{1/2}(\widehat{\boldsymbol{\beta}}_1^* - \boldsymbol{\beta}_1) n^{1/2}(\widehat{\boldsymbol{\beta}}_1^* - \boldsymbol{\beta}_1)^\top]. \tag{6.28}$$

We say that $\widehat{\boldsymbol{\beta}}_1^*$ strictly dominates $\widehat{\boldsymbol{\beta}}_1^{**}$ if $ADR(\widehat{\boldsymbol{\beta}}_1^*; \boldsymbol{M}) < ADR(\widehat{\boldsymbol{\beta}}_1^{**}; \boldsymbol{M})$ for some $(\boldsymbol{\beta}_1, \boldsymbol{M})$.

Lemma 6.4 *Under the usual regularity condition of ML estimation and the sequence of local alternatives $K_{(n)}$, as $n \to \infty$, the distribution of the statistic T_n converges to a non-central χ^2 distribution with p_2 degrees of freedom and non-centrality parameter $\triangle = \boldsymbol{\delta}^\top \boldsymbol{C}_{22.1}\boldsymbol{\delta}$.*

We refer to Davidson and Lever (1970) for a proof of Lemma 6.4. By virtue of Lemmas 6.4 and 3.2, we obtain the ADB and ADR results in the following theorems.

Theorem 6.5 *Under $K_{(n)}$ and the usual regularity conditions of ML estimation, as $n \to \infty$, $\widehat{\boldsymbol{\beta}}_1^{\mathrm{FM}}$ is an asymptotically unbiased estimator of $\boldsymbol{\beta}_1$. The ADB results of the other* *estimators are*

$$\begin{aligned}
\mathrm{ADB}(\widehat{\boldsymbol{\beta}}_1^{\mathrm{SM}}) &= \boldsymbol{B}, \\
\mathrm{ADB}(\widehat{\boldsymbol{\beta}}_1^{\mathrm{S}}) &= \omega_S \boldsymbol{B}, \\
\mathrm{ADB}(\widehat{\boldsymbol{\beta}}_1^{\mathrm{PS}}) &= \omega_{PS} \boldsymbol{B}.
\end{aligned}$$

Here, $\omega_{PS} = \left\{\psi_{p_2+2}((p_2-2);\triangle) + (p_2-2)E\left[\chi^{-2}_{p_2+2}(\triangle)I\left(\chi^2_{p_2+2}(\triangle) > (p_2-2)\right)\right]\right\}$, $\boldsymbol{B} = \boldsymbol{C}_{11}^{-1}\boldsymbol{C}_{12}\boldsymbol{\delta}$, $\omega_S = (p_2-2)E[\chi^{-2}_{p_2+2}(\triangle)]$, *and* $\psi_\nu(z;\triangle)$ *is the cumulative distribution function of a noncentral* χ^2 *distribution with non-centrality parameter* $\triangle$ *and* ν *degrees of freedom, where* $\triangle = \boldsymbol{\delta}^\top \boldsymbol{C}_{22.1}\boldsymbol{\delta}$.

Proof *See Appendix.*

To simplify comparison of the bias results, we present the scalar quantity of ADBs by use of the quadratic formula, as follows $QADB(\widehat{\boldsymbol{\beta}}_1^*) = \mathrm{ADB}(\widehat{\boldsymbol{\beta}}_1^*)^\top \boldsymbol{C}_{11.2}\mathrm{ADB}(\widehat{\boldsymbol{\beta}}_1^*)$. Here, $QADB(\widehat{\boldsymbol{\beta}}_1^*)$ is called the quadratic asymptotic distributional bias (QADB) of $\widehat{\boldsymbol{\beta}}_1^*$.

Theorem 6.6 *Under* $K_{(n)}$ *and the usual regularity conditions of ML estimation, as* $n \to \infty$*, the QADBs of the estimators are given as* $QADB(\widehat{\boldsymbol{\beta}}_1^{\mathrm{SM}}) = \delta_B$, $QADB(\widehat{\boldsymbol{\beta}}_1^{\mathrm{S}}) = \omega_S^2\delta_B$, $QADB(\widehat{\boldsymbol{\beta}}_1^{\mathrm{PS}}) = \omega_{PS}^2\delta_B$*, where* $\delta_B = \boldsymbol{\delta}^\top \boldsymbol{C}_{21}\boldsymbol{C}_{11}^{-1}\boldsymbol{C}_{11.2}\boldsymbol{C}_{11}^{-1}\boldsymbol{C}_{12}\boldsymbol{\delta}$.

Proof *See Appendix.*

Assuming that $\delta_{\mathrm{B}} \neq 0$, then all estimators, except $\widehat{\boldsymbol{\beta}}_1^{\mathrm{FM}}$, are biased estimators. The bias functions of $\widehat{\boldsymbol{\beta}}_1^{\mathrm{SM}}$ is the unbounded functions of δ_{B}. The behavioral biases of $\widehat{\boldsymbol{\beta}}_1^{\mathrm{S}}$ and $\widehat{\boldsymbol{\beta}}_1^{\mathrm{PS}}$ from 0 to their maximum points, and then gradually decrease to zero as $\triangle \to \infty$, because $E[\chi^{-2}_{p_2+2}(\triangle)]$ is a decreasing log-convex function of $\triangle$. Lastly, the bias of $\widehat{\boldsymbol{\beta}}_1^{\mathrm{S}}$ is smeller than or equal to $\widehat{\boldsymbol{\beta}}_1^{\mathrm{PS}}$ in all scenarios.

Theorem 6.7 *Under* $K_{(n)}$ *and the usual regularity conditions, as* $n \to \infty$*, the ADRs of the estimators are given as*

$$\begin{aligned}
ADR(\widehat{\boldsymbol{\beta}}_1^{\mathrm{FM}};\boldsymbol{M}) &= tr[\boldsymbol{M}\boldsymbol{C}_{11.2}^{-1}]. \\
ADR(\widehat{\boldsymbol{\beta}}_1^{\mathrm{SM}};\boldsymbol{M}) &= ADR(\widehat{\boldsymbol{\beta}}_1^{\mathrm{FM}};\boldsymbol{M}) - c^* + \delta_R. \\
ADR(\widehat{\boldsymbol{\beta}}_1^{\mathrm{S}};\boldsymbol{M}) &= ADR(\widehat{\boldsymbol{\beta}}_1^{\mathrm{FM}};\boldsymbol{M}) + (p_2-2) \\
&\times \left\{(p_2-2)E[\chi^{-4}_{p_2+2}(\triangle)] - 2E[\chi^{-2}_{p_2+2}(\triangle)]\right\}c^* \\
&+ (p_2-2) \\
&\times \left\{(p_2-2)E[\chi^{-4}_{p_2+4}(\triangle)] - 2E[\chi^{-2}_{p_2+4}(\triangle)]\right. \\
&+ \left.2E[\chi^{-2}_{p_2+2}(\triangle)]\right\}\delta_R. \\
ADR(\widehat{\boldsymbol{\beta}}_1^{\mathrm{PS}};\boldsymbol{M}) &= ADR(\widehat{\boldsymbol{\beta}}_1^{\mathrm{S}};\boldsymbol{M}) \\
&- E\left[(1-(p_2-2)\chi^{-2}_{p_2+2}(\triangle))^2 I\left(\chi^2_{p_2+2}(\triangle) \le (p_2-2)\right)\right]c^* \\
&+ \left\{\begin{array}{l} 2E\left[(1-(p_2-2)\chi^{-2}_{p_2+2}(\triangle))I\left(\chi^2_{p_2+2}(\triangle) \le (p_2-2)\right)\right] \\ -E\left[(1-(p_2-2)\chi^{-2}_{p_2+4}(\triangle))^2 I\left(\chi^2_{p_2+4}(\triangle) \le (p_2-2)\right)\right] \end{array}\right\}\delta_R.
\end{aligned}$$

Here,

$$\begin{aligned} c^* &= tr[\boldsymbol{M}\boldsymbol{C}_{11}^{-1}\boldsymbol{C}_{12}\boldsymbol{C}_{22.1}^{-1}\boldsymbol{C}_{21}\boldsymbol{C}_{11}^{-1}], \text{ and} \\ \delta_R &= (\boldsymbol{C}_{11}^{-1}\boldsymbol{C}_{12}\boldsymbol{\delta})^\top \boldsymbol{M}(\boldsymbol{C}_{11}^{-1}\boldsymbol{C}_{12}\boldsymbol{\delta}). \end{aligned}$$

Proof *See Appendix B.*

Assuming that $\boldsymbol{C}_{12} \neq \boldsymbol{0}$, then the risk of $\widehat{\boldsymbol{\beta}}_1^{\text{FM}}$ is a constant with $tr[\boldsymbol{M}\boldsymbol{C}_{11.2}^{-1}]$, whereas that of all other estimators are impacted by the magnitude of the vector $\boldsymbol{\delta}$. The risk function of $\widehat{\boldsymbol{\beta}}_1^{\text{SM}}$ is unbound since $\triangle \in [0, \infty]$. When $\triangle$ is zero or is in the neighborhood of zero, $ADR(\widehat{\boldsymbol{\beta}}_1^{\text{SM}}; \boldsymbol{M}) \leq ADR(\widehat{\boldsymbol{\beta}}_1^{\text{FM}}; \boldsymbol{M})$, but $ADR(\widehat{\boldsymbol{\beta}}_1^{\text{SM}}; \boldsymbol{M}) \geq ADR(\widehat{\boldsymbol{\beta}}_1^{\text{FM}}; \boldsymbol{M})$ otherwise. For $p_2 \geq 3$, $ADR(\widehat{\boldsymbol{\beta}}_1^{\text{PS}}; \boldsymbol{M}) \leq ADR(\widehat{\boldsymbol{\beta}}_1^{\text{S}}; \boldsymbol{M}) \leq ADR(\widehat{\boldsymbol{\beta}}_1^{\text{FM}}; \boldsymbol{M})$ in all scenarios.

6.12 Simulation Experiments

In this section, we examined the performance and utility of the estimation strategies by using a Monte Carlo simulation and application to a real data set. Monte Carlo simulations were conducted to investigate and compare the performance of the estimators under different scenarios. We generated the overdispersed count data with the mean $\mu_i = \exp(\boldsymbol{x}_i^\top \boldsymbol{\beta})$, where $\boldsymbol{x}_i \in \mathcal{R}^p$ has a standard multivariate normal distribution with $n = 100, 500$, and the overdispersion θ=1.5, 2.5, and 3.5. We consider the regression coefficients are set $\boldsymbol{\beta} = \left(\boldsymbol{\beta}_1^\top, \boldsymbol{\beta}_2^\top, \boldsymbol{\beta}_3^\top\right)^\top$ with dimensions p_1, p_2 and p_3, respectively. We simulated 500 data sets with different configurations of (n, p_1, p_2, p_3). In simulation, we use value 1 for strong signals and value 0.1 for weak signals.

The simulated relative MSE (RMSE) of $\widehat{\boldsymbol{\beta}}_1^*$ to $\widehat{\boldsymbol{\beta}}_1^{\text{FM}}$ is given as follows:

$$\text{RMSE}(\widehat{\boldsymbol{\beta}}_1^{\text{FM}}, \widehat{\boldsymbol{\beta}}_1^*) = \frac{E[\| \widehat{\boldsymbol{\beta}}_1^{\text{FM}} - \boldsymbol{\beta}_1 \|_2]}{E[\| \widehat{\boldsymbol{\beta}}_1^* - \boldsymbol{\beta}_1 \|_2]}. \tag{6.29}$$

Here, $\widehat{\boldsymbol{\beta}}_1^*$ is one of proposed estimators and $\| \cdot \|_2$ is the Euclidean distance. Under this criterion, $\widehat{\boldsymbol{\beta}}_1^*$ is superior to $\widehat{\boldsymbol{\beta}}_1^{\text{FM}}$ if $\text{RMSE}(\widehat{\boldsymbol{\beta}}_1^{\text{FM}}, \widehat{\boldsymbol{\beta}}_1^*) > 1$, and less efficient than $\widehat{\boldsymbol{\beta}}_1^{\text{FM}}$ otherwise.

Following Ahmed and Yüzbaşı (2016) and Gao et al. (2017a), we designed the true parameter vector for p predictors with p_1 strong effect, p_2 weak effect, and p_3 no effect (noise), such that $p = p_1 + p_2 + p_3$, as in the following examples: $\boldsymbol{\beta} = (\underbrace{1, 1, ..., 1}_{p_1}, \underbrace{0.1, 0.1, ..., 0.1}_{p_2}, \underbrace{0, 0, ..., 0}_{p_3})^\top$. We considered $n = 100, 500$, $p_1 = 4$, $p_2 = 0, 3, 6, 9$ and $p_3 = 4, 8, 12, 16$.

Tables 6.22 and 6.23 report the values of RMSE when there are no weak signals in the model for given values of Δ. The performances of the estimators are similar to those reported for the logistic model. Figures 6.7 and 6.9 also provide similar characteristics of the estimators. Tables 6.24 and 6.25 give the RMSE of the listed estimators when weak signals are entered in the models, the performance of shrinkage estimators is not much affected by weak signals and stays stable. Figures 6.8 and 6.10 also provide similar characteristics of the estimators. Tables 6.26 and 6.27 display the RMSE values of submodel, PS and penalty estimators for different values of $(\theta, p_1, p_2, p_3, n)$. The relative performance of shrinkage estimators is better than penalty estimation almost in all cases.

TABLE 6.22: RMSE of the Estimators for $n = 100$, $p_1 = 4$, and $p_2 = 0$.

		$\theta = 1.5$			$\theta = 2.5$			$\theta = 3.5$		
p_3	Δ	SM	S	PS	SM	S	PS	SM	S	PS
4	0.0	2.63	1.44	1.54	2.86	1.51	1.65	3.31	1.52	1.69
	0.2	1.56	1.24	1.28	1.38	1.25	1.29	1.20	1.20	1.25
	0.4	0.64	1.08	1.09	0.51	1.07	1.07	0.43	1.06	1.07
	0.6	0.32	1.03	1.03	0.24	1.03	1.03	0.19	1.01	1.01
	0.8	0.19	1.01	1.01	0.14	1.01	1.01	0.12	1.01	1.01
	1.6	0.05	1.00	1.00	0.03	1.00	1.00	0.03	1.00	1.00
	3.2	0.01	1.00	1.00	0.01	1.00	1.00	0.01	1.00	1.00
8	0.0	4.23	2.11	2.29	4.60	2.18	2.64	5.75	2.34	2.70
	0.2	2.56	1.67	1.83	2.30	1.76	1.86	2.09	1.68	1.81
	0.4	1.09	1.34	1.37	0.88	1.30	1.31	0.73	1.26	1.27
	0.6	0.55	1.15	1.15	0.42	1.14	1.14	0.35	1.11	1.11
	0.8	0.33	1.09	1.09	0.25	1.07	1.07	0.21	1.06	1.06
	1.6	0.08	1.01	1.01	0.06	1.01	1.01	0.05	1.01	1.01
	3.2	0.02	0.99	0.99	0.01	1.00	1.00	0.01	1.00	1.00
12	0.0	5.90	2.57	2.87	6.82	2.90	3.50	8.43	3.01	3.58
	0.2	3.60	2.10	2.30	3.45	2.29	2.45	3.26	2.17	2.41
	0.4	1.62	1.64	1.67	1.31	1.58	1.59	1.12	1.50	1.53
	0.6	0.80	1.31	1.31	0.64	1.28	1.28	0.54	1.25	1.25
	0.8	0.49	1.18	1.18	0.38	1.15	1.15	0.31	1.12	1.12
	1.6	0.12	1.03	1.03	0.09	1.02	1.02	0.07	1.02	1.02
	3.2	0.03	0.99	0.99	0.02	1.00	1.00	0.02	1.00	1.00
16	0.0	7.82	3.13	3.33	9.36	3.48	4.15	11.59	3.74	4.28
	0.2	4.76	2.42	2.63	4.67	2.78	2.98	4.48	2.52	2.88
	0.4	2.17	1.89	1.91	1.80	1.83	1.84	1.58	1.74	1.77
	0.6	1.10	1.47	1.47	0.87	1.41	1.41	0.76	1.37	1.37
	0.8	0.68	1.28	1.28	0.51	1.23	1.23	0.43	1.19	1.19
	1.6	0.17	1.04	1.04	0.12	1.04	1.04	0.10	1.03	1.03
	3.2	0.04	0.99	0.99	0.03	1.00	1.00	0.02	1.00	1.00

TABLE 6.23: RMSE of the Estimators for $n = 500$, $p_1 = 4$, and $p_2 = 0$.

		$\theta = 1.5$			$\theta = 2.5$			$\theta = 3.5$		
p_3	Δ	SM	S	PS	SM	S	PS	SM	S	PS
4	0.0	2.62	1.43	1.62	2.85	1.56	1.69	2.95	1.44	1.70
	0.2	0.46	1.06	1.06	0.35	1.04	1.04	0.28	1.04	1.04
	0.4	0.14	1.01	1.01	0.09	1.01	1.01	0.08	1.01	1.01
	0.6	0.06	1.01	1.01	0.04	1.00	1.00	0.03	1.00	1.00
	0.8	0.04	1.00	1.00	0.02	1.00	1.00	0.02	1.00	1.00
	1.6	0.01	1.00	1.00	0.01	1.00	1.00	0.00	1.00	1.00
	3.2	0.00	1.00	1.00	0.00	1.00	1.00	0.00	1.00	1.00
8	0.0	4.09	2.30	2.64	4.69	2.16	2.70	4.82	2.42	2.86
	0.2	0.75	1.25	1.26	0.56	1.19	1.19	0.46	1.17	1.17
	0.4	0.22	1.06	1.06	0.15	1.04	1.04	0.12	1.04	1.04
	0.6	0.10	1.03	1.03	0.07	1.02	1.02	0.05	1.02	1.02
	0.8	0.06	1.02	1.02	0.04	1.01	1.01	0.03	1.01	1.01
	1.6	0.01	1.00	1.00	0.01	1.00	1.00	0.01	1.00	1.00
	3.2	0.00	1.00	1.00	0.00	1.00	1.00	0.00	1.00	1.00
12	0.0	5.56	3.10	3.53	6.24	3.33	3.81	6.66	3.30	3.89
	0.2	1.02	1.46	1.46	0.77	1.34	1.35	0.64	1.32	1.32
	0.4	0.29	1.12	1.12	0.21	1.08	1.08	0.18	1.07	1.07
	0.6	0.13	1.05	1.05	0.09	1.04	1.04	0.08	1.03	1.03
	0.8	0.08	1.03	1.03	0.05	1.02	1.02	0.04	1.01	1.01
	1.6	0.02	1.00	1.00	0.01	1.00	1.00	0.01	1.00	1.00
	3.2	0.00	1.00	1.00	0.00	1.00	1.00	0.00	1.00	1.00
16	0.0	7.07	3.78	4.40	8.07	4.23	4.88	8.59	4.08	5.04
	0.2	1.33	1.68	1.69	1.01	1.53	1.53	0.85	1.48	1.48
	0.4	0.38	1.18	1.18	0.28	1.13	1.13	0.23	1.11	1.11
	0.6	0.17	1.09	1.09	0.12	1.06	1.06	0.10	1.05	1.05
	0.8	0.10	1.04	1.04	0.07	1.03	1.03	0.06	1.02	1.02
	1.6	0.03	1.01	1.01	0.02	1.00	1.00	0.01	1.00	1.00
	3.2	0.01	1.00	1.00	0.00	1.00	1.00	0.00	1.00	1.00

TABLE 6.24: RMSE of the Estimators for $n = 100$, $p_1 = 4$, and $p_2 = 6$.

		$\theta = 1.5$			$\theta = 2.5$			$\theta = 3.5$		
p_3	Δ	SM	S	PS	SM	S	PS	SM	S	PS
4	0.0	1.53	1.20	1.23	1.59	1.20	1.26	1.59	1.18	1.28
	0.2	1.17	1.10	1.12	1.15	1.10	1.12	1.09	1.11	1.12
	0.4	0.72	1.01	1.02	0.62	1.02	1.02	0.56	1.01	1.02
	0.6	0.43	1.00	1.00	0.35	0.99	0.99	0.30	1.00	1.00
	0.8	0.28	0.99	0.99	0.22	0.99	0.99	0.18	1.00	1.00
	1.6	0.08	0.99	0.99	0.06	1.00	1.00	0.04	1.00	1.00
	3.2	0.02	1.00	1.00	0.01	1.00	1.00	0.01	1.00	1.00
8	0.0	2.13	1.55	1.61	2.27	1.53	1.71	2.27	1.63	1.78
	0.2	1.62	1.35	1.40	1.70	1.42	1.46	1.58	1.40	1.44
	0.4	1.02	1.18	1.19	0.89	1.17	1.18	0.82	1.17	1.17
	0.6	0.61	1.07	1.07	0.50	1.05	1.05	0.45	1.06	1.06
	0.8	0.40	1.03	1.03	0.31	1.02	1.02	0.26	1.02	1.02
	1.6	0.11	0.99	0.99	0.08	1.00	1.00	0.07	1.00	1.00
	3.2	0.03	0.99	0.99	0.02	1.00	1.00	0.02	1.00	1.00
12	0.0	2.74	1.85	1.92	2.99	1.97	2.12	3.02	2.00	2.22
	0.2	2.09	1.62	1.66	2.26	1.73	1.78	2.10	1.70	1.80
	0.4	1.32	1.35	1.37	1.19	1.38	1.39	1.12	1.35	1.36
	0.6	0.81	1.19	1.19	0.67	1.15	1.15	0.60	1.15	1.15
	0.8	0.53	1.09	1.09	0.42	1.07	1.07	0.36	1.07	1.07
	1.6	0.15	1.00	1.00	0.11	1.00	1.00	0.09	1.00	1.00
	3.2	0.04	0.99	0.99	0.03	1.00	1.00	0.02	1.00	1.00
16	0.0	3.32	2.10	2.17	3.69	2.28	2.44	3.85	2.44	2.62
	0.2	2.53	1.85	1.88	2.77	1.98	2.06	2.61	2.01	2.13
	0.4	1.60	1.51	1.54	1.49	1.55	1.56	1.43	1.53	1.53
	0.6	1.01	1.30	1.30	0.84	1.25	1.25	0.76	1.25	1.25
	0.8	0.66	1.16	1.16	0.53	1.13	1.13	0.45	1.13	1.13
	1.6	0.19	1.01	1.01	0.13	1.01	1.01	0.11	1.01	1.01
	3.2	0.06	0.99	0.99	0.04	1.00	1.00	0.03	1.00	1.00

TABLE 6.25: RMSE of the Estimators for $n = 500$, $p_1 = 4$, and $p_2 = 6$.

		$\theta = 1.5$			$\theta = 2.5$			$\theta = 3.5$		
p_3	Δ	SM	S	PS	SM	S	PS	SM	S	PS
4	0.0	1.46	1.20	1.24	1.47	1.18	1.24	1.47	1.18	1.25
	0.2	0.60	1.02	1.02	0.50	1.01	1.01	0.43	1.01	1.01
	0.4	0.21	1.00	1.00	0.16	1.00	1.00	0.13	1.00	1.00
	0.6	0.10	1.00	1.00	0.08	1.00	1.00	0.06	1.00	1.00
	0.8	0.06	1.00	1.00	0.04	1.00	1.00	0.04	1.00	1.00
	1.6	0.01	1.00	1.00	0.01	1.00	1.00	0.01	1.00	1.00
	3.2	0.00	1.00	1.00	0.00	1.00	1.00	0.00	1.00	1.00
8	0.0	1.96	1.55	1.64	1.97	1.55	1.67	1.94	1.55	1.67
	0.2	0.80	1.14	1.14	0.67	1.10	1.10	0.59	1.10	1.10
	0.4	0.29	1.04	1.04	0.22	1.03	1.03	0.18	1.02	1.02
	0.6	0.14	1.01	1.01	0.10	1.01	1.01	0.09	1.01	1.01
	0.8	0.08	1.00	1.00	0.06	1.01	1.01	0.05	1.00	1.00
	1.6	0.02	1.00	1.00	0.01	1.00	1.00	0.01	1.00	1.00
	3.2	0.00	1.00	1.00	0.00	1.00	1.00	0.00	1.00	1.00
12	0.0	2.48	1.92	2.05	2.53	1.98	2.11	2.46	2.00	2.14
	0.2	1.02	1.30	1.30	0.85	1.23	1.23	0.76	1.22	1.22
	0.4	0.37	1.09	1.09	0.28	1.07	1.07	0.23	1.05	1.05
	0.6	0.18	1.04	1.04	0.14	1.03	1.03	0.11	1.02	1.02
	0.8	0.10	1.02	1.02	0.07	1.02	1.02	0.06	1.01	1.01
	1.6	0.03	1.00	1.00	0.02	1.00	1.00	0.01	1.00	1.00
	3.2	0.01	1.00	1.00	0.00	1.00	1.00	0.00	1.00	1.00
16	0.0	2.93	2.23	2.38	3.00	2.33	2.47	3.00	2.42	2.60
	0.2	1.22	1.45	1.46	1.04	1.39	1.39	0.92	1.35	1.35
	0.4	0.44	1.14	1.14	0.34	1.11	1.11	0.28	1.09	1.09
	0.6	0.22	1.06	1.06	0.16	1.05	1.05	0.13	1.04	1.04
	0.8	0.13	1.03	1.03	0.09	1.03	1.03	0.08	1.02	1.02
	1.6	0.03	1.00	1.00	0.02	1.00	1.00	0.02	1.00	1.00
	3.2	0.01	1.00	1.00	0.00	1.00	1.00	0.00	1.00	1.00

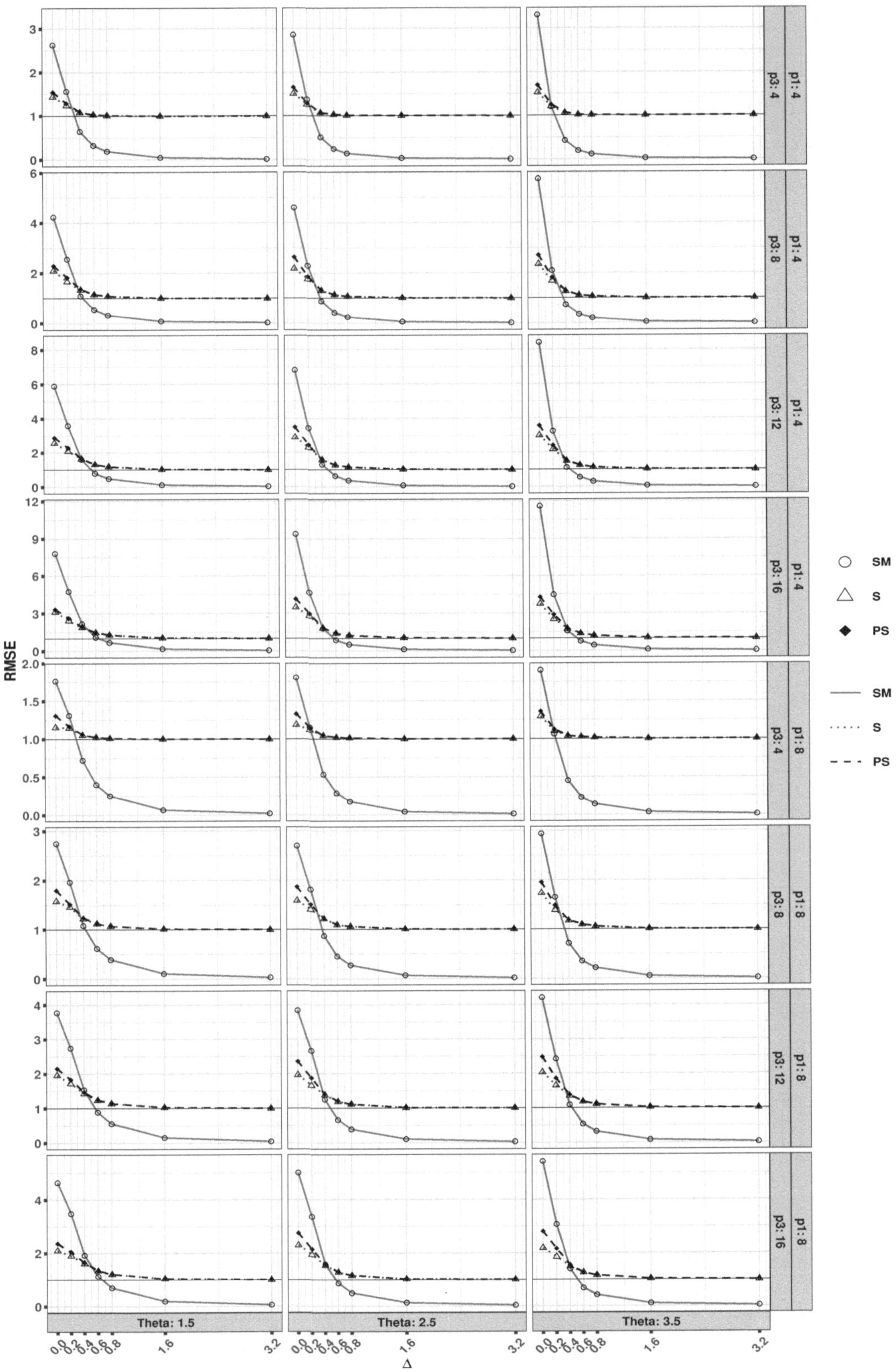

FIGURE 6.7: RMSE of the Estimators for $n = 100$ and $p_2 = 0$.

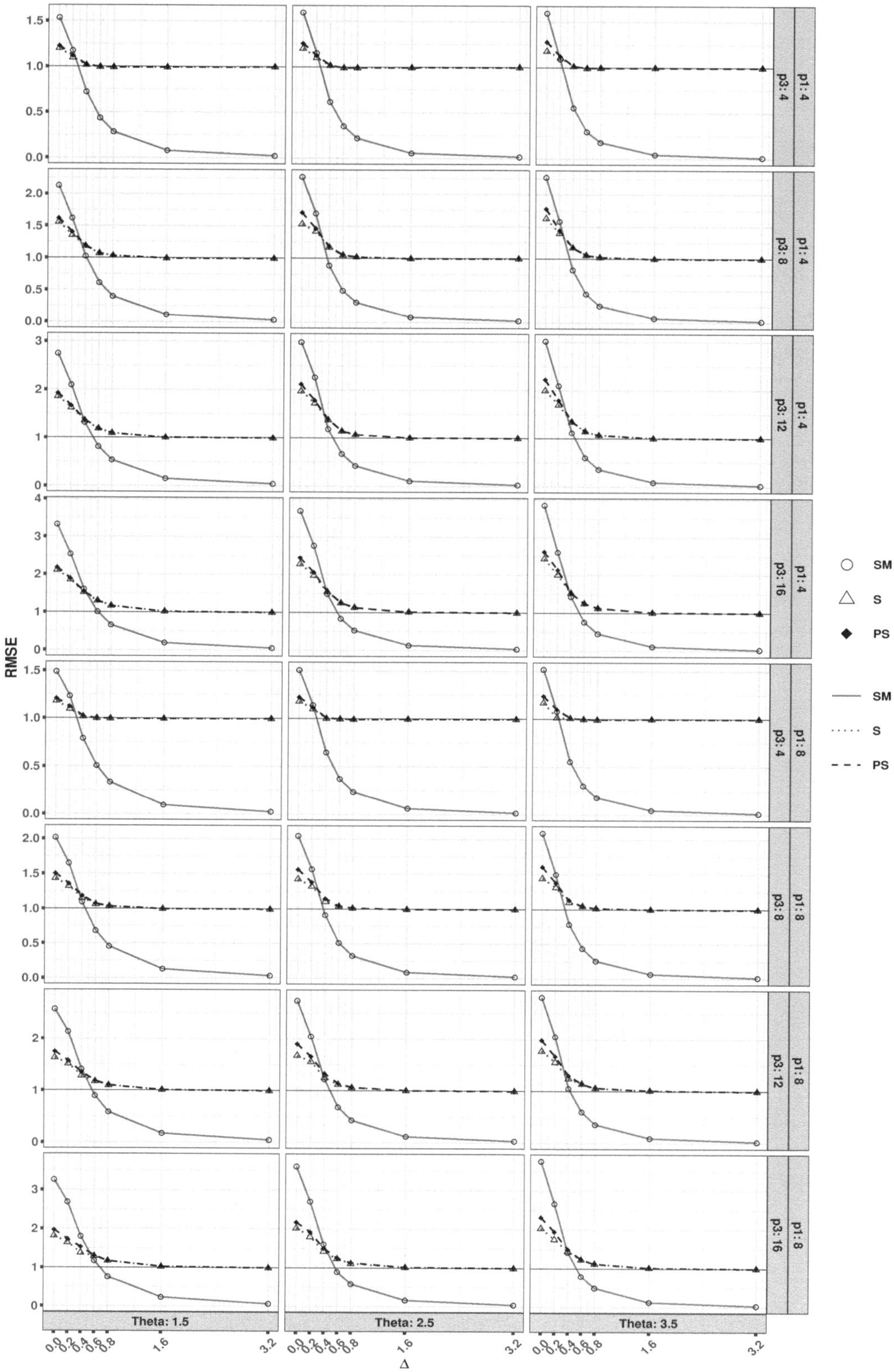

FIGURE 6.8: RMSE of the Estimators for $n = 100$ and $p_2 = 6$.

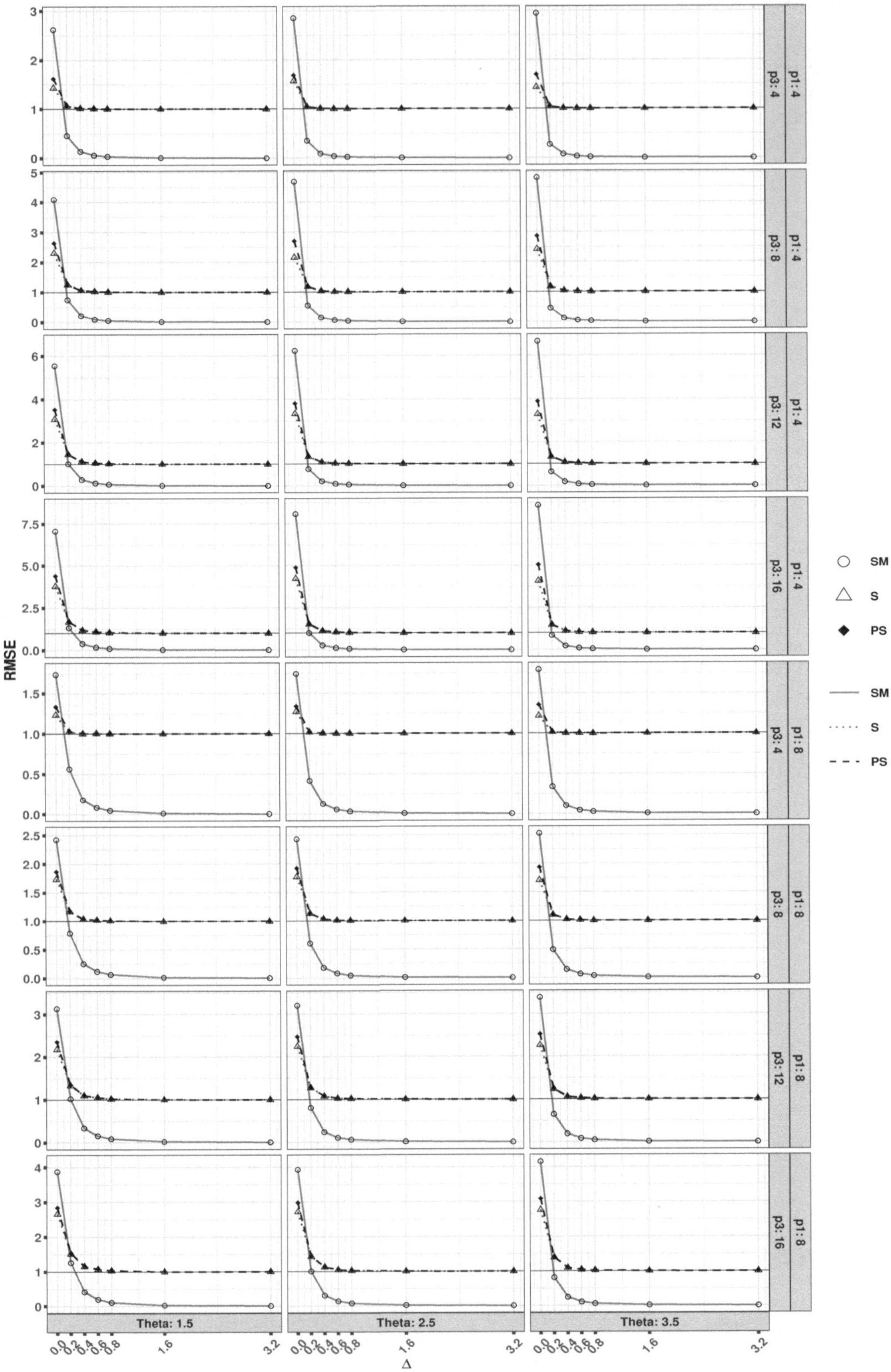

FIGURE 6.9: RMSE of the Estimators for $n = 500$ and $p_2 = 0$.

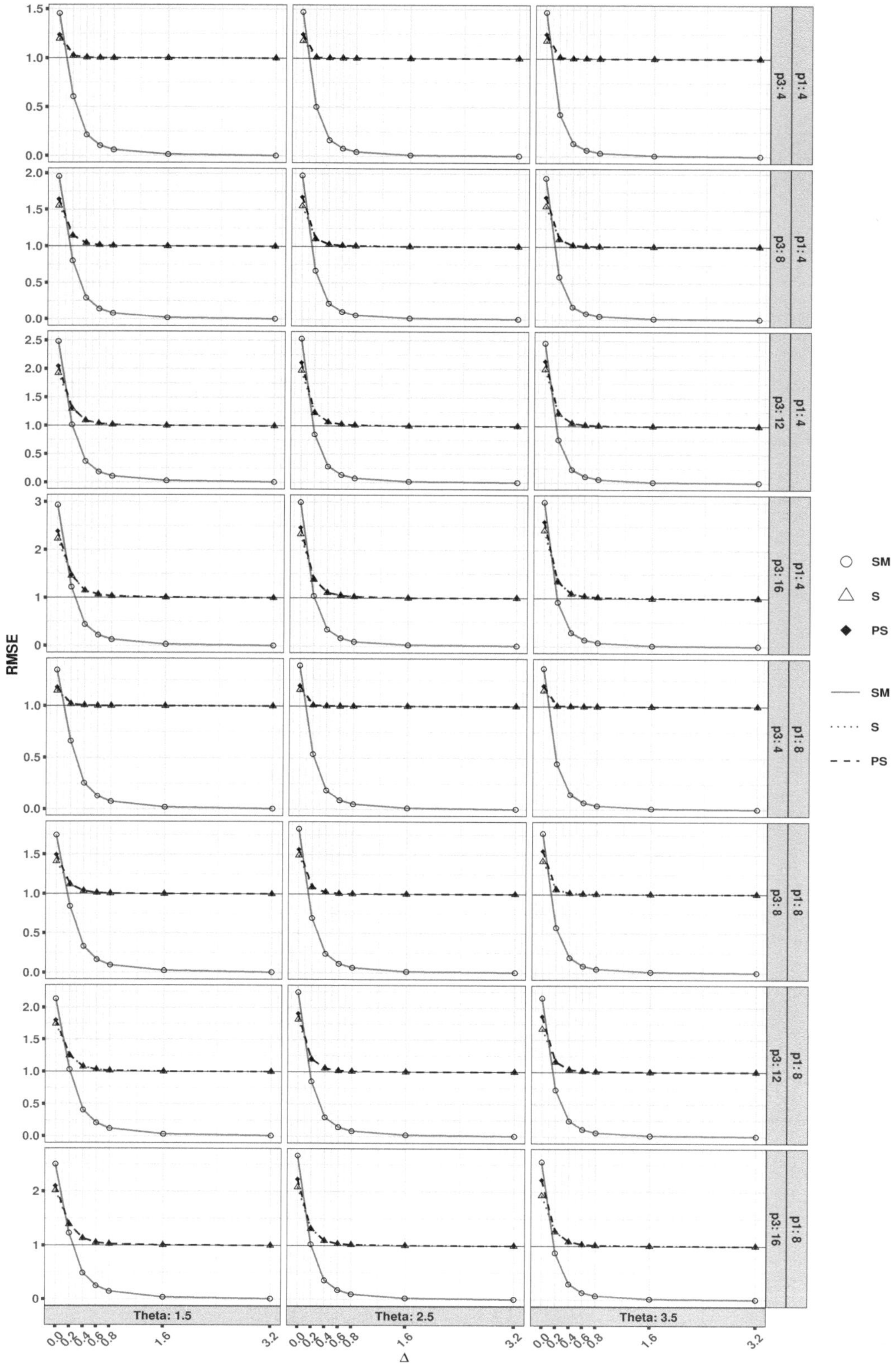

FIGURE 6.10: RMSE of the Estimators for $n = 500$ and $p_2 = 6$.

TABLE 6.26: RMSE of the Estimators for $n = 150$.

θ	p_1	p_2	p_3	SM	PS	LASSO	ALASSO
2.5	4	3	4	1.515	1.467	1.166	1.419
			8	2.369	2.068	1.369	2.207
			12	2.971	2.537	1.522	2.573
			16	3.966	3.115	1.943	3.299
		6	4	1.524	1.495	1.281	1.433
			8	2.005	1.873	1.547	1.920
			12	2.550	2.290	1.682	2.464
			16	2.828	2.677	1.523	2.203
	8	3	4	1.441	1.374	1.028	1.280
			8	1.847	1.673	1.261	1.737
			12	2.397	2.080	1.285	2.043
			16	3.164	2.594	1.397	2.331
		6	4	1.479	1.448	1.141	1.306
			8	1.781	1.711	1.313	1.586
			12	2.255	1.987	1.254	1.812
			16	2.799	2.432	1.485	2.357
3.5	4	3	4	1.742	1.626	1.146	1.440
			8	2.060	1.980	1.229	1.613
			12	3.375	2.609	1.483	2.553
			16	4.080	3.246	1.857	3.230
		6	4	1.403	1.392	1.193	1.333
			8	2.086	1.973	1.516	1.729
			12	2.707	2.533	1.681	2.451
			16	3.246	2.846	1.579	2.422
	8	3	4	1.531	1.399	1.039	1.269
			8	1.988	1.755	1.143	1.515
			12	2.405	2.037	1.309	2.154
			16	3.049	2.363	1.398	2.179
		6	4	1.390	1.380	1.052	1.122
			8	1.871	1.778	1.239	1.454
			12	2.294	2.045	1.187	1.718
			16	2.717	2.426	1.324	1.969

TABLE 6.27: RMSE of the Estimators for $n = 300$.

θ	p_1	p_2	p_3	SM	PS	LASSO	ALASSO
2.5	4	3	4	1.641	1.510	1.178	1.245
			8	2.091	1.947	1.175	1.493
			12	2.703	2.440	1.393	2.058
			16	3.881	3.378	1.663	2.485
		6	4	1.501	1.446	1.050	1.033
			8	1.802	1.753	1.247	1.361
			12	2.419	2.325	1.365	1.567
			16	2.960	2.863	1.345	1.736
	8	3	4	1.391	1.342	1.047	1.102
			8	1.827	1.669	1.098	1.340
			12	2.239	2.018	1.104	1.506
			16	2.828	2.374	1.288	2.002
		6	4	1.363	1.353	1.029	0.946
			8	1.753	1.703	1.073	1.227
			12	2.038	1.964	1.145	1.362
			16	2.358	2.147	1.221	1.613
3.5	4	3	4	1.599	1.517	1.104	1.165
			8	2.353	2.229	1.365	1.568
			12	2.751	2.538	1.301	1.834
			16	4.058	3.396	1.633	2.548
		6	4	1.364	1.363	1.058	0.936
			8	1.876	1.829	1.212	1.234
			12	2.443	2.333	1.381	1.487
			16	2.834	2.678	1.466	1.945
	8	3	4	1.480	1.364	1.051	1.073
			8	1.795	1.641	1.056	1.146
			12	2.345	2.079	1.120	1.543
			16	2.692	2.341	1.218	1.937
		6	4	1.369	1.357	0.979	0.803
			8	1.666	1.598	1.052	1.039
			12	2.028	1.957	1.158	1.276
			16	2.303	2.227	1.183	1.556

6.13 Real Data Examples

In this section, we consider two real data examples.

6.13.1 Resume Data

Here, we take into account a data set from the study of Bertrand and Mullainathan (2004). The data originally has $n = 4870$ observations of $p = 63$ variables. Since has a number of levels for categorical variables, we omitted NA values and eliminated certain predictors while cleaning the data. Finally, the cleaned data has $n = 447$ observations of $p = 56$ variables including response variable. The data is related to fake resumes that is submitted to Boston and Chicago help-wanted advertising to examine race in the workplace. The data is freely available from *openintro* package in R software. In this study, our goal is to model the number of years of work experience on the resume, using ($p = 55$) covariates.

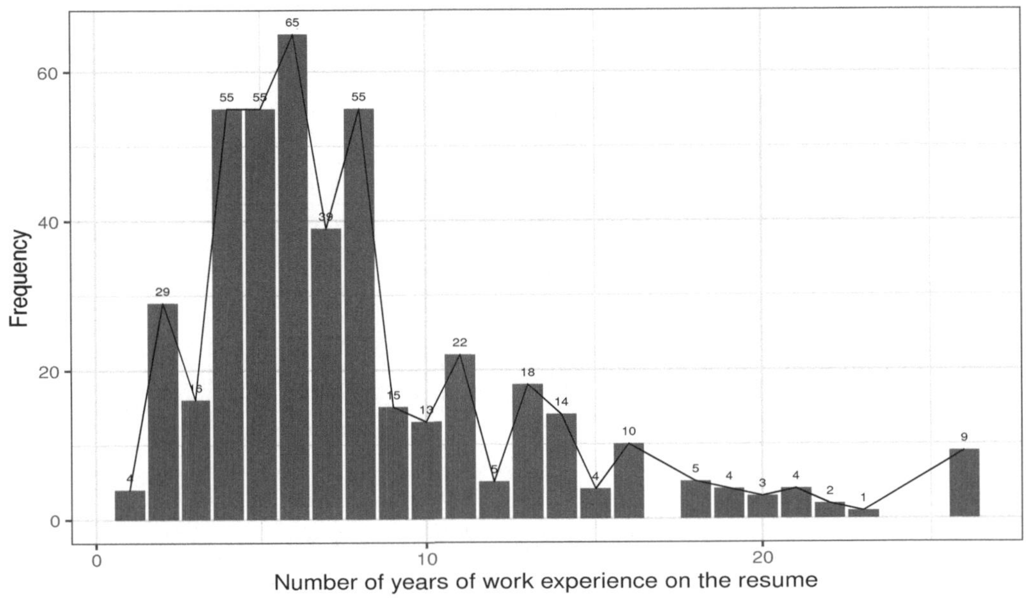

FIGURE 6.11: Frequency Distribution for Number of Years of Work Experience.

To obtain a preliminary overview of the dependent variable, a histogram of the observed count frequencies is utilized (see Figure 6.11). The variance (24.821) of dependent variable is greater than its mean value (7.910). This indicates that there is over dispersion and that a Negative Binomial model is appropriate.

In order to apply the proposed methods, we implement a two-step approach since prior information is not available here. In the first step, we apply the usual variable selection methods to select the best possible submodel. We use likelihood-ratio test (LRT) via *drop1* function of *stats* package in R. It is observed that 16 variables seem to be significantly important. The selected variable descriptions are only given in Table 6.28.

To evaluate the prediction accuracy of the stated estimators, we randomly divided the data into two sets of observations: a training set of 70% of the observations and a test set of 30% of the observations. We fitted the model on the training set only. We consider the

TABLE 6.28: Lists and Descriptions of Variables of Resume Data.

Response	
years_exp	Number of years of work experience on the resume.
Predictors	
n_jobs	Number of jobs listed on resume.
honors	1 = resume mentions some honors.
military	1 = resume mentions some military experience.
occup_specific	1990 Census Occupation Code
occup_broad	Occupation broad with levels 1 = executives and managerial occupations, 2 = administrative supervisors, 3 = sales representatives, 4 = sales workers, 5 = secretaries and legal assistants, 6 = clerical occupations
work_in_school	1 = resume mentions some work experience while at school
special_skills	1 = resume mentions some special skills.
city	City, with levels of "c" = chicago; "b" = boston.
ad_id	Employment ad identifier.
col	1 = applicant has college degree or more.
school_req	Specific education requirement, if any. "hsg" = high school graduate, "somcol" = some college, "colp" = four year degree or higher
eoe	1 = ad mentions employer is "Equal Opportunity Employer."
manager	1 = executives or managers wanted.
supervisor	1 = administrative supervisors wanted.
secretary	1 = secretaries or legal assistants wanted.
off_support	1 = clerical workers wanted.

TABLE 6.29: RPEs of Estimators for Resume Data.

	SM	PS	LASSO	ALASSO
RPE	1.066	1.061	1.050	1.029

following measure to assess the performance of the estimators.

$$\mathrm{PE}(\widehat{\boldsymbol{\beta}}^*) = \left\| \boldsymbol{y}_{\mathrm{test}} - exp(\boldsymbol{X}_{\mathrm{test}}\widehat{\boldsymbol{\beta}}^*) \right\|^2, \tag{6.30}$$

where $\widehat{\boldsymbol{\beta}}^*$ is the one of the listed estimators. This process is repeated 200 times, and the mean values are reported. For the ease of comparison, we calculate the $\mathrm{RPE}(\widehat{\boldsymbol{\beta}}^*) = \frac{\mathrm{PE}(\widehat{\boldsymbol{\beta}}^{\mathrm{FM}})}{\mathrm{PE}(\widehat{\boldsymbol{\beta}}^*)}$. If the RPE is larger than one, then this indicates the superiority of that method over the full model estimator. The results are given in Table 6.29.

Looking at the RPE values in Table 6.29, it is clear that PS has an edge over on all other estimators. While SM has the highest RPE, its effectiveness depends on choosing the right submodel. Otherwise, its RPE may converge to zero.

6.13.2 Labor Supply Data

Here, we consider a data set from the study of Mroz (1987). The data originally has $n = 753$ observations of labor supply behavior of married women. The data is freely available from

TABLE 6.30: Lists and Descriptions of Variables of Labor Supply Data.

Response	
experience	Actual years of wife's previous labor market experience
Predictors	
work	work at home in 1975?
hoursw	wife's hours of work in 1975
child6	number of children less than 6 years old in household
child618	number of children between ages 6 and 18 in household
agew	wife's age
educw	wife's educational attainment, in years
hearnw	wife's average hourly earnings, in 1975 dollars
wagew	wife's wage reported at the time of the 1976 interview
hoursh	husband's hours worked in 1975
ageh	husband's age
educh	husband's educational attainment, in years
wageh	husband's wage, in 1975 dollars
income	family income, in 1975 dollars
educwm	wife's mother's educational attainment, in years
educwf	wife's father's educational attainment, in years
unemprate	unemployment rate in county of residence, in percentage points
city	lives in large city (SMSA)?

Ecdat package in R software. In this study, our goal is to model the actual years of wife's previous labor market experience, using ($p = 17$) covariates that are given in Table 6.30.

To obtain a preliminary overview of the dependent variable, a histogram of the observed count frequencies is utilized (see Figure 6.12). The variance (65.110) of dependent variable is greater than its mean value (10.630). This indicates that there is over dispersion and that a Negative Binomial model is appropriate.

For $i = 1, 2, \ldots, 753$, we consider the full model is given by

$$\begin{aligned}\text{experience}_i &= \beta_0 + \beta_1\text{work}_i + \beta_2\text{hoursw}_i + \beta_3\text{child6}_i + \beta_4\text{child618}_i \\ &+ \beta_5\text{agew}_i + \beta_6\text{educw}_i + \beta_7\text{hearnw}_i + \beta_8\text{wagew}_i + \beta_9\text{hoursh}_i \\ &+ \beta_{10}\text{ageh}_i + \beta_{11}\text{educh}_i + \beta_{12}\text{wageh}_i + \beta_{13}\text{income}_i \\ &+ \beta_{14}\text{educwm}_i + \beta_{15}\text{educwf}_i + \beta_{16}\text{unemprate}_i + \beta_{17}\text{city}_i + \epsilon_i.\end{aligned}$$

Following same way the example of resume data, it is observed that work, hoursw, child618, agew, wagew and income variables seem to be significantly important. Hence, the submodel is then given by

$$\begin{aligned}\text{experience}_i &= \beta_0 + \beta_1\text{work}_i + \beta_2\text{hoursw}_i + \beta_4\text{child618}_i + \beta_5\text{agew}_i \\ &+ \beta_8\text{wagew}_i + \beta_{13}\text{income}_i + \epsilon_i.\end{aligned}$$

To evaluate the prediction accuracy of the specified estimators, we randomly separated the data into two sets of observations: a training set consisting of 70% of the observations and a test set consisting of 30% of the observations. Again, the prediction error in equation (6.30) is computed using 1000 bootstrap samples. The average RPE values are displayed in Table 6.31.

Looking at the RPE values in Table 6.29, it is clear that PS performs better than penalized methods.

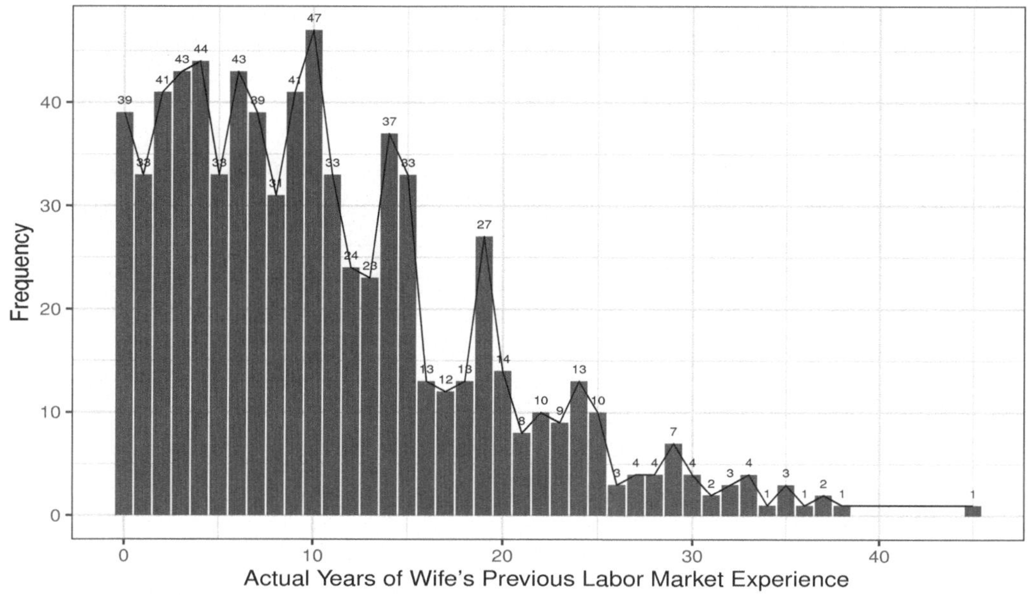

FIGURE 6.12: Frequency Distribution for Number of Years of Labor Market Experience.

TABLE 6.31: RPEs of Estimators for Labor Supply Data.

	SM	PS	LASSO	ALASSO
RPE	1.023	1.018	1.013	0.990

6.14 High-Dimensional Data

Through Monte Carlo simulations, we extended the proposed estimators to address the parameter estimation problem for the high-dimensional sparse NB regression model ($n < p$). Generally, the p existing predictors have different effect sizes containing strong, weak, or none. Following Ahmed and Yüzbaşı (2016) and Gao et al. (2017a), we designed the true parameter vector for p predictors with p_1 strong effect, p_2 weak effect, and p_3 no effect (noise), such that $p = p_1 + p_2 + p_3$, as in the following examples: $\boldsymbol{\beta} = (\underbrace{3, 3, ..., 3}_{p_1}, \underbrace{\kappa, \kappa, \ldots, \kappa}_{p_2}, \underbrace{0, 0, ..., 0}_{p_3})^\top$, where the magnitude of weak effect k was set as 0.05, 0.50, and 1.00 to study whether the performance of methods was affected by changing the very weak effect to moderate. The signs (+ or $-$) of the coefficients were randomly assigned. Assuming that $p_1 + p_2 \leq n$ and $p_3 > n$, we considered $(n, p_1, p_2, p_3) = (75, 5, 40, 150)$ and $(75, 5, 60, 150)$. Under high-dimensional settings, two-stage procedures are generally used: 1. Variable selection to provide the subset of the significant predictors for dimensionality reduction of the predictor vector, and 2. Post-selection parameter estimation based on the resulting parsimonious model obtained from Stage 1. As we know, the MLE is widely used for effectively eliminating irrelevant predictors in the high-dimensional regime, making the model parsimonious. However, the different variable selection methods may produce different subsets of relevant predictors Ahmed and Yüzbaşı (2016). Since

ALASSO performs close to SCAD Gao et al. (2017a), we used both LASSO and ALASSO for detecting the presence of significant predictors in the initial stage. Based on 100 Monte Carlo runs, the selection percentage of predictors for each effect level is presented in Table (6.32).

TABLE 6.32: Percentage of Selection of Predictors for Each Effect Level for $(n, p_1, p_3) = (75, 5, 150)$.

		Strong Effect		Weak Effect		No Effect	
p_2	κ	LASSO	ALASSO	LASSO	ALASSO	LASSO	ALASSO
40	0.05	99.6	99.0	36.825	24.125	31.813	19.547
	0.10	99.2	99.2	38.300	24.750	31.973	19.573
	0.20	93.4	92.0	38.450	26.975	28.933	18.180
60	0.05	99.2	98.4	33.400	20.300	29.033	16.933
	0.10	99.2	99.0	34.733	21.133	29.853	17.040
	0.20	99.0	97.6	39.267	25.550	29.467	16.273

For both LASSO and ALASSO, as κ increased, the performance in selecting predictors with strong effects decreased, the performance in selecting predictors with weak effects increased, and the performance in eliminating noise decreased. LASSO selected more predictors than ALASSO in addition to being more effective than ALASSO in choosing strong and weak predictors. Unfortunately, more predictors with no effect were retained in the LASSO-based parsimonious model.

For small κ, predictors with weak effects may be considered irrelevant for predicting the response and should be eliminated from the resulting parsimonious model. In contrast, they are relevant and should be selected for the model when κ becomes large. For this reason, either LASSO-based or ALASSO-based subsets of selected predictors may not be the best for constructing an optimally parsimonious model in all situations. LASSO resulted in an overfitted model with too many selected predictors for small κ, whereas the ALASSO-based model was considered underfitted with fewer relevant predictors when κ was large.

Assuming that a LASSO-based subset of variables contained $\widehat{p}$ selected predictors, while ALASSO selected only $\widehat{p}_1$ predictors as relevant, where $\widehat{p}_1 < \widehat{p} < p$. To provide post-selection estimation given in Section 2, we constructed an FM based on a LASSO-based subset of selected predictors: $\mathcal{S}_1 = \{x_{i1}, x_{i2}, ..., x_{i\widehat{p}}\}$. The SM contained only selected predictors in the ALASSO-based subset: $\mathcal{S}_2 = \{x_{i1}, x_{i2}, ..., x_{i\widehat{p}_1}\}$. Here, $\boldsymbol{\beta}_2 = (\beta_1, \beta_2, ..., \beta_{\widehat{p}-\widehat{p}_1})^\top$ is the coefficient vector of $\widehat{p} - \widehat{p}_1$ predictors selected as relevant by LASSO but not by ALASSO, that is $\mathcal{S}_1 \cap \mathcal{S}_2^c$. Under H_0, as $n \to \infty$, the distribution of T_n in Eq. (6.21) converges to a chi-square distribution with $\widehat{p} - \widehat{p}_1$ degrees of freedom. The RMSE results are reported in Table (6.33).

We found that the submodel had the best performance for only $\kappa = 0.05$. It confirmed that ALASSO provided a proper subset of relevant predictors, whereas LASSO produced an overfitted model when the weak effect size was small. However, as κ increased, ALASSO produced underfitting and consequently the performance of submodel decreased. The RMSEs of the proposed post-selection estimators were strongly consistent with the results in low-dimensional data. Overall, the estimators based on James-Stein rule strategy performed well in both low- and high-dimensional data.

TABLE 6.33: RMSE of the Estimators for $(n, p_1, p_3) = (75, 5, 150)$.

p_2	κ	SM	S	PS
40	0.05	1.144	1.158	1.153
	0.10	1.072	1.103	1.102
	0.20	0.961	1.104	1.104
60	0.05	1.090	1.057	1.093
	0.10	1.034	1.063	1.062
	0.20	0.958	1.013	1.013

6.15 R-Codes

```
> library('MASS')    # It is for 'mvrnorm' function
> library('lmtest') # It is for 'lrtest' function
> library('caret')  # It is for 'split' function
> library('ncvreg') # It is for 'cv.ncvreg' function
> library('glmnet') # It is for 'cv.glmnet' function
> library('yardstick') # require for prediction accuracy
> set.seed(2023)
> ####
> For Low Dimensional Logistic Regression
> n  <- 200  # The number of sample
> p1 <- 5    # The number of strong signals
> p2 <- 30    # The number of weak signals
> p3 <- 55   # The number of zeros, no signal
> p <- p1+p2+p3
> beta1_true <- rep(1  ,p1)
> beta2_true <- rep(0.1,p2)
> beta3_true <- rep(0  ,p3)
> # The ture value of covariates
> beta_true <- c(beta1_true,beta2_true,beta3_true)
> # The design matrix
> X <- matrix(0, n, p)
> for (i in 1:p){
+   X[,i]= rnorm(n)
+ }
> ## assigning colnames of X to "X1","X2",....
> v<-NULL
> for(i in 1:p){
+   v[i] <- paste("X",i,sep="")
+   assign(v[i],X[,i])
+ }
> epsilon <- rnorm(n) # The errors
> sigma <- 1
> pr = 1/(1+exp(-(X%*%beta_true)))
> y = rbinom(n,1,pr)# The response
> y = factor(y, labels = c(0, 1))
> # Split data into train and test set
```

```
> all.folds <- split (sample (1 : n), rep (1 : 2, length = n))
> train_ind <- all.folds$'1'
> test_ind  <- all.folds$'2'
> y_train <- y[train_ind]
> X_train <- X[train_ind,]
> ### Centering train data of y and X
> X_train_mean  <- apply(X_train,2,mean)
> X_train_scale <- scale(X_train,X_train_mean,F)
> train_scale_df <- data.frame(y_train,X_train_scale)
> #test data
> y_test <- y[test_ind]
> X_test <- X[test_ind,]
> ### Centering test data of X based on train means
> X_test_scale <- scale(X_test,X_train_mean,F)
> # Formula of the Full model
> xcount.FM <- c(0,paste("X", 1:(p1+p2+p3), sep=""))
> Formula_FM<- as.formula(paste("y_train~",
+                               paste(xcount.FM, collapse= "+")))
> # Formula of the Sub model
> xcount.SM <- c(0,paste("X", 1:(p1), sep=""))
> Formula_SM<- as.formula(paste("y_train~",
+                               paste(xcount.SM, collapse= "+")))
>
>
> Full_model  <- glm(Formula_FM, family = binomial(link = "logit"),
+                    data = train_scale_df)
Warning messages:
1: glm.fit: algorithm did not converge
2: glm.fit: fitted probabilities numerically 0 or 1 occurred
> beta_FM <- array(c(Full_model$coefficients), dim = c(p,1))
>
> Sub_model   <- glm(Formula_SM, family = binomial(link = "logit"),
+                    data = train_scale_df)
>
> # Estimation of the sub model estimator
> Cinv <- vcov(Full_model) # inverse of the information matrix
> H <- diag(1,p,p)[-c(1:(p-p2-p3)),]
> h <- array(0, dim = c(p2+p3,1))
> beta_SM <- beta_FM-Cinv%*%t(H)%*%
+   solve(H%*%Cinv%*%t(H))%*%(H%*%beta_FM-h)
>
>
> #test stat
> tn <- -2*(logLik(Sub_model)-logLik(Full_model))[1]
>
> #Shrinkage Estimation
> beta.S <-beta_FM-(beta_FM-beta_SM)*(p2+p3-2)/tn
> beta.PS<-beta_SM+ max(0,(1-(p2+p3-2)/tn))*(beta_FM-beta_SM)
>
>
>
> ##PENALIZED Methods
> #ENET
> alphas=seq(0.1,0.9,0.2)
> fits.enet <- list()
```

```
> for (ind in 1:length(alphas)){
+   fits.enet[[ind]] <- cv.glmnet(X_train_scale, y_train,
+                               alpha = alphas[ind],nfolds = 5,
+                               family = binomial(link = "logit"))
+ }
> cvs.enet <- sapply(fits.enet, function(x) {
+   pred.model  <- predict(x, X_test_scale,
+                        s = "lambda.min",type = "response")
+   pred.model  <- factor(ifelse(pred.model > .5, 1, 0),
+                        levels = c(0, 1) )
+   act_pred <- data.frame(observed = y_test, predicted=pred.model)
+   mse.model  <- accuracy(act_pred, observed, predicted)$.estimate
+   MSEs       <- mse.model
+   return(MSEs)})
> ind_opt   <- which.max(cvs.enet)
> beta.enet <- coef(fits.enet[[ind]],s='lambda.min')[-1]
>
>
>
> #LASSO
> lasso.fit <- cv.glmnet(X_train_scale, y_train, alpha = 1,
+                      nfolds = 5, family = binomial(link =
  "logit"))
> beta.lasso = coef(lasso.fit,s='lambda.min')[-1]
>
> #ALASSO
> weight <- beta.lasso
> weight <- ifelse(weight == 0, 0.00001, weight)
> alasso.fit <- cv.glmnet(X_train_scale, y_train, alpha = 1,
+                       penalty.factor=1/abs(weight),nfolds = 5,
+                       family = binomial(link = "logit"))
> beta.alasso = coef(alasso.fit,s='lambda.min')[-1]
>
>
> #RIDGE
> ridge.fit <- cv.glmnet(X_train_scale, y_train, alpha = 0,
+                      nfolds = 5,
+                      family = binomial(link = "logit"))
> beta.ridge = coef(ridge.fit,s='lambda.min')[-1]
>
> #SCAD
> scad.fit <- cv.ncvreg(X_train_scale, y_train, penalty=c("SCAD"),
+                     nfolds = 5, family = 'binomial')
Warning message:
In ncvreg(X = X, y = y, ...) : Maximum number of iterations reached
> beta.scad = coef(scad.fit,s='lambda.min')[-1]
>
>
>
>
> act_pred_FM  <- data.frame(observed = factor(y_test),
+ predicted = factor(round(1/(1+exp(-(X_test_scale%*%beta_FM)))),
+ levels = c(0,1)))
> act_pred_SM  <- data.frame(observed = factor(y_test),
+ predicted = factor(round(1/(1+exp(-(X_test_scale%*%beta_SM)))),
```

```
+ levels = c(0,1)))
> act_pred_S  <- data.frame(observed = factor(y_test),
+ predicted = factor(round(1/(1+exp(-(X_test_scale%*%beta.S)))),
+ levels = c(0,1)))
> act_pred_PS  <- data.frame(observed = factor(y_test),
+ predicted = factor(round(1/(1+exp(-(X_test_scale%*%beta.PS)))),
+ levels = c(0,1)))
> act_pred_enet  <- data.frame(observed = factor(y_test),
+ predicted = factor(round(1/(1+exp(-(X_test_scale%*%beta.enet)))),
+ levels = c(0,1)))
> act_pred_lasso  <- data.frame(observed = factor(y_test),
+ predicted = factor(round(1/(1+exp(-(X_test_scale%*%beta.lasso)))),
+ levels = c(0,1)))
> act_pred_alasso  <- data.frame(observed = factor(y_test),
+ predicted = factor(round(1/(1+exp(-(X_test_scale%*%beta.alasso)))),
+ levels = c(0,1)))
> act_pred_ridge  <- data.frame(observed = factor(y_test),
+ predicted = factor(round(1/(1+exp(-(X_test_scale%*%beta.ridge)))),
+ levels = c(0,1)))
> act_pred_scad  <- data.frame(observed = factor(y_test),
+ predicted = factor(round(1/(1+exp(-(X_test_scale%*%beta.scad)))),
+ levels = c(0,1)))
>
> # Calculte prediction accuracy of estimators based on test data
> PEs<- c(FM=accuracy(act_pred_FM, observed, predicted)$.estimate,
+         SM=accuracy(act_pred_SM, observed, predicted)$.estimate,
+         PS=accuracy(act_pred_PS, observed, predicted)$.estimate,
+         ENET=accuracy(act_pred_enet, observed,
   predicted)$.estimate,
+         LASSO=accuracy(act_pred_lasso, observed,
   predicted)$.estimate,
+         ALASSO=accuracy(act_pred_alasso, observed,
   predicted)$.estimate,
+         RIDGE=accuracy(act_pred_ridge, observed,
   predicted)$.estimate,
+         SCAD=accuracy(act_pred_scad, observed,
   predicted)$.estimate)
> # print and sort the results
> cbind(Accuracy=PEs,Best_Ranking= 9-rank(PEs))
       Accuracy Best_Ranking
FM         0.57          8.0
SM         0.82          1.0
PS         0.69          3.0
ENET       0.68          4.5
LASSO      0.67          6.5
ALASSO     0.68          4.5
RIDGE      0.67          6.5
SCAD       0.70          2.0
### High Dimensional Case for Logistic Regression
> ####
> n  <- 300  # The number of sample
> p1 <- 5    # The number of strong signals
> p2 <- 100   # The number of weak signals
> p3 <- 5000  # The number of zeros, no signal
> p <- p1+p2+p3
```

```
> beta1_true <- rep(1  ,p1)
> beta2_true <- rep(0.1,p2)
> beta3_true <- rep(0  ,p3)
> # The ture value of covariates
> beta_true <- c(beta1_true,beta2_true,beta3_true)
> # The design matrix
> X <- matrix(0, n, p)
> for (i in 1:p){
+   X[,i]= rnorm(n)
+ }
> ## assigning colnames of X to "X1","X2",...,"X10".
> v<-NULL
> for(i in 1:p){
+   v[i] <- paste("X",i,sep="")
+   assign(v[i],X[,i])
+ }
> epsilon <- rnorm(n) # The errors
> sigma <- 1
> pr = 1/(1+exp(-(X%*%beta_true)))
> y = rbinom(n,1,pr)# The response
> y = factor(y, labels = c(0, 1))
> # Split data into train and test set
> all.folds <- split (sample (1 : n), rep (1 : 2, length = n))
> train_ind <- all.folds$'1'
> test_ind  <- all.folds$'2'
> y_train <- y[train_ind]
> X_train <- X[train_ind,]
> ### Centering train data of y and X
> X_train_mean  <- apply(X_train,2,mean)
> X_train_scale <- scale(X_train,X_train_mean,F)
> train_scale_df <- data.frame(y_train,X_train_scale)
> #test data
> y_test <- y[test_ind]
> X_test <- X[test_ind,]
> ### Centering test data of X based on train means
> X_test_scale <- scale(X_test,X_train_mean,F)
> # Formula of the Full model
> xcount.FM <- c(0,paste("X", 1:(p1+p2), sep=""))
> Formula_FM<- as.formula(paste("y_train~",
+                               paste(xcount.FM, collapse= "+")))
> # Formula of the Sub model
> xcount.SM <- c(0,paste("X", 1:(p1), sep=""))
> Formula_SM<- as.formula(paste("y_train~",
+                               paste(xcount.SM, collapse= "+")))
>
> #test stat
> Full_model  <- glm(Formula_FM, family = binomial(link = "logit"),
+                    data = train_scale_df)
Warning messages:
1: glm.fit: algorithm did not converge
2: glm.fit: fitted probabilities numerically 0 or 1 occurred
> Sub_model   <- glm(Formula_SM, family = binomial(link = "logit"),
+                    data = train_scale_df)
> tn <- -2*(logLik(Sub_model)-logLik(Full_model))[1]
>
```

```
>
> ##PENALIZED Methods
> #ENET
> alphas=seq(0.1,0.9,0.2)
> fits.enet <- list()
> for (ind in 1:length(alphas)){
+   fits.enet[[ind]] <- cv.glmnet(X_train_scale, y_train,
+                                alpha = alphas[ind],nfolds = 5,
+                                family = binomial(link = "logit"))
+ }
> cvs.enet <- sapply(fits.enet, function(x) {
+   pred.model  <- predict(x, X_test_scale,
+                         s = "lambda.min",type = "response")
+   pred.model  <- factor(ifelse(pred.model > .5, 1, 0),
+                        levels = c(0, 1) )
+   act_pred <- data.frame(observed = y_test, predicted=pred.model)
+   mse.model  <- accuracy(act_pred, observed, predicted)$.estimate
+   MSEs       <- mse.model
+   return(MSEs)})
> ind_opt  <- which.max(cvs.enet)
> beta.enet <- coef(fits.enet[[ind]],s='lambda.min')[-1]
>
>
>
> #LASSO
> lasso.fit <- cv.glmnet(X_train_scale, y_train, alpha = 1,
+                       nfolds = 5, family = binomial(link =
   "logit"))
> beta.lasso = coef(lasso.fit,s='lambda.min')[-1]
>
> #ALASSO
> weight <- beta.lasso
> weight <- ifelse(weight == 0, 0.00001, weight)
> alasso.fit <- cv.glmnet(X_train_scale, y_train, alpha = 1,
+                        penalty.factor=1/abs(weight),nfolds = 5,
+                        family = binomial(link = "logit"))
> beta.alasso = coef(alasso.fit,s='lambda.min')[-1]
>
>
> #RIDGE
> ridge.fit <- cv.glmnet(X_train_scale, y_train, alpha = 0,
+                       nfolds = 5,
+                       family = binomial(link = "logit"))
> beta.ridge = coef(ridge.fit,s='lambda.min')[-1]
>
> #SCAD
> scad.fit <- cv.ncvreg(X_train_scale, y_train, penalty=c("SCAD"),
+                      nfolds = 5, family = 'binomial')
> beta.scad = coef(scad.fit,s='lambda.min')[-1]
>
>
> #Shrinkage Estimation
> beta.S <-beta.enet-(beta.enet-beta.alasso)*(p2-2)/tn
> beta.PS<-beta.alasso+ max(0,(1-(p2-2)/tn))*(beta.enet-beta.alasso)
> ###
```

```
> act_pred_FM   <- data.frame(observed = factor(y_test),
+                           predicted =
   factor(round(1/(1+exp(-(X_test_scale%*%beta.enet)))),
+                                               levels = c(0,1)))
> act_pred_SM   <- data.frame(observed = factor(y_test),
+                           predicted =
   factor(round(1/(1+exp(-(X_test_scale%*%beta.alasso)))),
+                                               levels = c(0,1)))
> act_pred_S   <- data.frame(observed = factor(y_test),
+                          predicted =
   factor(round(1/(1+exp(-(X_test_scale%*%beta.S)))),
+                                              levels = c(0,1)))
> act_pred_PS   <- data.frame(observed = factor(y_test),
+                           predicted =
   factor(round(1/(1+exp(-(X_test_scale%*%beta.PS)))),
+                                               levels = c(0,1)))
> act_pred_lasso   <- data.frame(observed = factor(y_test),
+                              predicted =
   factor(round(1/(1+exp(-(X_test_scale%*%beta.lasso)))),
+                                                  levels = c(0,1)))
> act_pred_ridge   <- data.frame(observed = factor(y_test),
+                              predicted =
   factor(round(1/(1+exp(-(X_test_scale%*%beta.ridge)))),
+                                                  levels = c(0,1)))
> act_pred_scad   <- data.frame(observed = factor(y_test),
+                             predicted =
   factor(round(1/(1+exp(-(X_test_scale%*%beta.scad)))),
+                                                 levels = c(0,1)))
>
> # Calculte prediction accuracy of estimators based on test data
> PEs<- c(FM=accuracy(act_pred_FM, observed, predicted)$.estimate,
+         SM=accuracy(act_pred_SM, observed, predicted)$.estimate,
+         PS=accuracy(act_pred_PS, observed, predicted)$.estimate,
+         LASSO=accuracy(act_pred_lasso, observed,
   predicted)$.estimate,
+         RIDGE=accuracy(act_pred_ridge, observed,
   predicted)$.estimate,
+         SCAD=accuracy(act_pred_scad, observed,
   predicted)$.estimate)
> # print and sort the results
> cbind(Accuracy=PEs,Best_Ranking= 7-rank(PEs))
       Accuracy Best_Ranking
FM    0.5866667            4
SM    0.6000000            3
PS    0.6066667            2
LASSO 0.6133333            1
RIDGE 0.5400000            6
SCAD  0.5600000            5
> # Negative Binomial Case
> library('MASS')   # It is for 'mvrnorm' function
> library('lmtest') # It is for 'lrtest' function
> library('mpath') # It is for 'cv.glmreg' function
> set.seed(2023)
> ####
> MSE <- function(beta,beta_hat){mean((beta-beta_hat)^2)}
```

```
> ####
> n  <- 200  # The number of sample
> p1 <- 5   # The number of strong signals
> p2 <- 30   # The number of weak signals
> p3 <- 55  # The number of zeros, no signal
> theta <- 2.5
> p <- p1+p2+p3
> beta1_true <- rep(1  ,p1)
> beta2_true <- rep(0.1,p2)
> beta3_true <- rep(0  ,p3)
> # The ture value of covariates
> beta_true <- c(beta1_true,beta2_true,beta3_true)
> # The design matrix
> X <- matrix(0, n, p)
> for (i in 1:p){
+   X[,i]= rnorm(n)
+ }
> ## assigning colnames of X to "X1","X2",...,"X10".
> v<-NULL
> for(i in 1:p){
+   v[i] <- paste("X",i,sep="")
+   assign(v[i],X[,i])
+ }
> epsilon <- rnorm(n) # The errors
> sigma <- 1
> eta = X %*% beta_true
> # The response
> y = rnbinom(length(eta), mu = exp(eta), size = theta)
> # Split data into train and test set
> y_train <- y
> X_train <- X
> ### Centering train data of y and X
> X_train_mean  <- apply(X_train,2,mean)
> X_train_scale <- scale(X_train,X_train_mean,F)
> train_scale_df <- data.frame(y_train,X_train_scale)
> # Formula of the Full model
> xcount.FM <- c(0,paste("X", 1:(p1+p2+p3), sep=""))
> Formula_FM<- as.formula(paste("y_train~",
+                              paste(xcount.FM, collapse= "+")))
> # Formula of the Sub model
> xcount.SM <- c(0,paste("X", 1:(p1), sep=""))
> Formula_SM<- as.formula(paste("y_train~",
+                              paste(xcount.SM, collapse= "+")))
> Full_model  <- glm(Formula_FM, family = 'negative.binomial'(theta
    = theta),
+                    data = train_scale_df)
> beta_FM <- array(c(Full_model$coefficients), dim = c(p,1))
> Sub_model   <- glm(Formula_SM, family = 'negative.binomial'(theta
    = theta),
+                    data = train_scale_df)
> # Estimation of the sub model estimator
> Cinv <- vcov(Full_model) # inverse of the information matrix
> H <- diag(1,p,p)[-c(1:(p-p2-p3)),]
> h <- array(0, dim = c(p2+p3,1))
> beta_SM <- beta_FM-Cinv%*%t(H)%*%
```

```
+    solve(H%*%Cinv%*%t(H))%*%(H%*%beta_FM-h)
> #test stat
> tn <- -2*(logLik(Sub_model)-logLik(Full_model))[1]
> #Shrinkage Estimation
> beta.S <-beta_FM-(beta_FM-beta_SM)*(p2+p3-2)/tn
> beta.PS<-beta_SM+ max(0,(1-(p2+p3-2)/tn))*(beta_FM-beta_SM)
> ##PENALIZED Methods
> #LASSO
> lasso.fit <- cv.glmreg(X_train_scale, y_train,
+                        alpha = 1, nfolds = 5,
+                        family = 'negbin',
+                        theta=theta,plot.it=F)
> beta.lasso = coef(lasso.fit,s='lambda.min')[-1]
> #ALASSO
> weight <- beta.lasso
> weight <- ifelse(weight == 0, 0.00001, weight)
> alasso.fit <- cv.glmreg(X_train_scale, y_train, alpha = 1,
+                         penalty.factor=1/abs(weight),nfolds = 5,
+                         family = 'negbin',
+                         theta=theta,plot.it=F)
> beta.alasso = coef(alasso.fit,s='lambda.min')[-1]
> # Calculte prediction accuracy of estimators based on test data
> MSEs<-c(FM=MSE(beta_true,beta_FM),
+         SM=MSE(beta_true,beta_SM),
+         PS=MSE(beta_true,beta.PS),
+         LASSO=MSE(beta_true,beta.lasso),
+         ALASSO=MSE(beta_true,beta.alasso))
> # print and rank the results
> cbind(MSE=MSEs,Best_Ranking= rank(PEs))
               MSE Best_Ranking
FM     0.011420070            5
SM     0.004029296            3
PS     0.003884260            1
LASSO  0.005948567            4
ALASSO 0.003909536            2
```

6.16 Concluding Remarks

We critically assessed the relative performance (both analytically and numerically) of the full model estimators based on maximum likelihood estimators, submodel estimators, shrinkage estimators, and some penalty estimators for the logistic regression model. The submodel and shrinkage estimators are constructed when we have subject matter or prior knowledge of a candidate subspace. In other words, when it is suspected that some of the regression parameters of the model are under linear constraint or the model is sparse. Our analytical and simulation results demonstrate that shrinkage estimators provide excellent prediction accuracy and outperform the maximum likelihood estimator uniformly. In fact, the asymptotic risk of the shrinkage estimator is strictly less than that of the maximum likelihood estimator when the sparsity assumption of the model holds. The properties of the penalty estimators were evaluated in terms of a Monte Carlo simulation study, and it was found that it is competitive when there are many inactive predictors in the model. On the other hand, the adaptive shrinkage estimator performs better when the number of inactive predictors is relatively large. Thus, the penalty estimators are preferable when there are few inactive variables; otherwise, in the presence of nuisance variables such as lab effects, Stein-type shrinkage estimators are attractive alternatives to penalty estimators.

Finally, we applied the shrinkage, penalized strategies to several real data sets to assess the relative performance of the listed estimators. These results corroborate with our analytical and simulated findings. The theoretical and numerical results can be extended to the entire class of generalized linear models.

Appendix A – Logistic

To prove the Theorems 6.2 and 6.3, we use $\boldsymbol{\beta} = (\boldsymbol{\beta}_1^\top, \mathbf{0})^\top$. Under the local alternative $K_{(n)}$ and the assumed regularity conditions, the joint distributions are:

$$i).\quad \begin{pmatrix}\boldsymbol{\eta}_1\\ \boldsymbol{\eta}_2\end{pmatrix} \sim N_{p_1+p_2}\left[\begin{pmatrix}0\\ -\boldsymbol{M\delta}\end{pmatrix}, \begin{pmatrix}\boldsymbol{I}_{11.2}^{-1} & \boldsymbol{M}\boldsymbol{I}_{22.1}^{-1}\boldsymbol{M}^\top\\ \boldsymbol{M}^\top\boldsymbol{I}_{22.1}^{-1}\boldsymbol{M} & \boldsymbol{M}\boldsymbol{I}_{22.1}^{-1}\boldsymbol{M}^\top\end{pmatrix}\right]$$

$$ii).\quad \begin{pmatrix}\boldsymbol{\eta}_2\\ \boldsymbol{\eta}_3\end{pmatrix} \sim N_{p_1+p_2}\left[\begin{pmatrix}-\boldsymbol{M\delta}\\ \boldsymbol{M\delta}\end{pmatrix}, \begin{pmatrix}\boldsymbol{M}\boldsymbol{I}_{22.1}^{-1}\boldsymbol{M}^\top & 0\\ 0 & \boldsymbol{\Phi}\end{pmatrix}\right],$$

where $\boldsymbol{\eta}_1 = \sqrt{n}(\widehat{\boldsymbol{\beta}}^{\mathrm{FM}} - \boldsymbol{\beta})$, $\boldsymbol{\eta}_2 = \sqrt{n}(\widehat{\boldsymbol{\beta}}^{\mathrm{FM}} - \widehat{\boldsymbol{\beta}}^{\mathrm{SM}})$, $\boldsymbol{\eta}_3 = \sqrt{n}(\widehat{\boldsymbol{\beta}}^{\mathrm{SM}} - \boldsymbol{\beta})$, $\boldsymbol{M} = \boldsymbol{I}_{11}^{-1}\boldsymbol{I}_{12}$, and $\boldsymbol{\Phi} = \boldsymbol{I}_{11.2}^{-1} - \boldsymbol{I}_{11}^{-1}\boldsymbol{I}_{12}\boldsymbol{I}_{22.1}^{-1}\boldsymbol{I}_{21}\boldsymbol{I}_{11}^{-1}$.

Justification of (i) and (ii):

Since $\boldsymbol{\eta}_2$ and $\boldsymbol{\eta}_3$ are linear functions of $\widehat{\boldsymbol{\beta}}^{\mathrm{FM}}$, they are also asymptotically normally distributed

$$\begin{aligned}
E(\boldsymbol{\eta}_1) &= E[\sqrt{n}(\widehat{\boldsymbol{\beta}}^{\mathrm{FM}} - \boldsymbol{\beta})] = \mathbf{0}.\\
E(\boldsymbol{\eta}_3) &= E[\sqrt{n}(\widehat{\boldsymbol{\beta}}^{\mathrm{SM}} - \boldsymbol{\beta})]\\
&= E[\sqrt{n}(\widehat{\boldsymbol{\beta}}_1^{\mathrm{FM}} - \boldsymbol{\beta}_1 + \boldsymbol{I}_{11}^{-1}\boldsymbol{I}_{12}\widehat{\boldsymbol{\beta}}_2^{\mathrm{MLE}}), \text{ see Lawless and Singhal (1978)}\\
&= \mathbf{0} + \boldsymbol{I}_{11}^{-1}\boldsymbol{I}_{12}\cdot\sqrt{n}\frac{\boldsymbol{\delta}}{\sqrt{n}} = \boldsymbol{M\delta}.\\
E(\boldsymbol{\eta}_2) &= E[\sqrt{n}(\widehat{\boldsymbol{\beta}}^{\mathrm{FM}} - \widehat{\boldsymbol{\beta}}^{\mathrm{SM}})]\\
&= E[\sqrt{n}((\widehat{\boldsymbol{\beta}}^{\mathrm{FM}} - \boldsymbol{\beta}) - (\widehat{\boldsymbol{\beta}}^{\mathrm{SM}} - \boldsymbol{\beta}))]\\
&= \mathbf{0} - (\boldsymbol{M\delta})\\
&= -\boldsymbol{M\delta}.
\end{aligned}$$

$$\begin{aligned}
\mathrm{Var}(\boldsymbol{\eta}_1) &= \boldsymbol{I}_{11.2}^{-1}.\\
\mathrm{Var}(\boldsymbol{\eta}_3) &= \mathrm{Var}[\sqrt{n}(\widehat{\boldsymbol{\beta}}^{\mathrm{SM}} - \boldsymbol{\beta})]\\
&= \mathrm{Var}[\sqrt{n}\left(\widehat{\boldsymbol{\beta}}_1^{\mathrm{FM}} - \boldsymbol{\beta}_1\right) + \boldsymbol{I}_{11}^{-1}\boldsymbol{I}_{12}\widehat{\boldsymbol{\beta}}_2^{\mathrm{MLE}}]\\
&= \mathrm{Var}[\sqrt{n}\left(\widehat{\boldsymbol{\beta}}_1^{\mathrm{FM}} - \boldsymbol{\beta}_1\right)] + \boldsymbol{I}_{11}^{-1}\boldsymbol{I}_{12}\mathrm{Var}[\sqrt{n}\widehat{\boldsymbol{\beta}}_2^{\mathrm{MLE}}]\boldsymbol{I}_{21}\boldsymbol{I}_{11}^{-1}\\
&+ 2\mathrm{Cov}[\sqrt{n}(\widehat{\boldsymbol{\beta}}_1^{\mathrm{FM}} - \boldsymbol{\beta}_1), \boldsymbol{I}_{11}^{-1}\boldsymbol{I}_{12}\sqrt{n}\widehat{\boldsymbol{\beta}}_2^{\mathrm{MLE}}]\\
&= \boldsymbol{I}_{11.2}^{-1} + \boldsymbol{I}_{11}^{-1}\boldsymbol{I}_{12}\boldsymbol{I}_{22.1}^{-1}\boldsymbol{I}_{21}\boldsymbol{I}_{11}^{-1}
\end{aligned}$$

$$
\begin{aligned}
&+ \quad 2\text{Cov}[\sqrt{n}(\widehat{\boldsymbol{\beta}}_1^{\text{FM}} - \boldsymbol{\beta}_1), \sqrt{n}\widehat{\boldsymbol{\beta}}_2^{\text{MLE}}](\boldsymbol{I}_{11}^{-1}\boldsymbol{I}_{12})' \\
&= \boldsymbol{I}_{11.2}^{-1} + \boldsymbol{I}_{11}^{-1}\boldsymbol{I}_{12}\boldsymbol{I}_{22.1}^{-1}\boldsymbol{I}_{21}\boldsymbol{I}_{11}^{-1} - 2\boldsymbol{I}_{11}^{-1}\boldsymbol{I}_{12}\boldsymbol{I}_{22.1}^{-1}\boldsymbol{I}_{21}\boldsymbol{I}_{11}^{-1} \\
&= \boldsymbol{I}_{11.2}^{-1} - \boldsymbol{I}_{11}^{-1}\boldsymbol{I}_{12}\boldsymbol{I}_{22.1}^{-1}\boldsymbol{I}_{21}\boldsymbol{I}_{11}^{-1} \\
&= \boldsymbol{I}_{11.2}^{-1} - \boldsymbol{M}\boldsymbol{I}_{22.1}^{-1}\boldsymbol{M}' = \boldsymbol{\Phi}.
\end{aligned}
$$

$$
\begin{aligned}
\text{Var}(\boldsymbol{\eta}_2) &= \text{Var}[\sqrt{n}(\widehat{\boldsymbol{\beta}}^{\text{FM}} - \widehat{\boldsymbol{\beta}}^{\text{SM}})] \\
&= \text{Var}[\sqrt{n}((\widehat{\boldsymbol{\beta}}^{\text{FM}} - \boldsymbol{\beta}) - (\widehat{\boldsymbol{\beta}}^{\text{SM}} - \boldsymbol{\beta}))] \\
&= \text{Var}[\boldsymbol{\eta}_1 - \boldsymbol{\eta}_3] \\
&= \text{Var}[\boldsymbol{\eta}_1] + \text{Var}[\boldsymbol{\eta}_3] - 2\text{cov}[\boldsymbol{\eta}_1, \boldsymbol{\eta}_3] \\
&= \boldsymbol{I}_{11.2}^{-1} + \boldsymbol{I}_{11.2}^{-1} - \boldsymbol{I}_{11}^{-1}\boldsymbol{I}_{12}\boldsymbol{I}_{22.1}^{-1}\boldsymbol{I}_{21}\boldsymbol{I}_{11}^{-1} - 2\boldsymbol{I}_{11.2}^{-1} \\
&+ \quad 2\boldsymbol{I}_{11}^{-1}\boldsymbol{I}_{12}\boldsymbol{I}_{22.1}^{-1}\boldsymbol{I}_{21}\boldsymbol{I}_{11}^{-1} \\
&= \boldsymbol{I}_{11}^{-1}\boldsymbol{I}_{12}\boldsymbol{I}_{22.1}^{-1}\boldsymbol{I}_{21}\boldsymbol{I}_{11}^{-1} \\
&= \boldsymbol{M}\boldsymbol{I}_{22.1}^{-1}\boldsymbol{M}'.
\end{aligned}
$$

$$
\begin{aligned}
\text{Cov}(\boldsymbol{\eta}_1, \boldsymbol{\eta}_2) &= \text{Cov}[\sqrt{n}(\widehat{\boldsymbol{\beta}}^{\text{FM}} - \boldsymbol{\beta}), \sqrt{n}(\widehat{\boldsymbol{\beta}}^{\text{FM}} - \widehat{\boldsymbol{\beta}}^{\text{SM}})] \\
&= \text{Cov}[\sqrt{n}(\widehat{\boldsymbol{\beta}}^{\text{FM}} - \boldsymbol{\beta}), \sqrt{n}((\widehat{\boldsymbol{\beta}}^{\text{FM}} - \boldsymbol{\beta}) - (\widehat{\boldsymbol{\beta}}^{\text{SM}} - \boldsymbol{\beta}))] \\
&= \text{Var}[\sqrt{n}(\widehat{\boldsymbol{\beta}}^{\text{FM}} - \boldsymbol{\beta})] - \text{Cov}[\sqrt{n}(\widehat{\boldsymbol{\beta}}^{\text{FM}} - \boldsymbol{\beta}), \sqrt{n}(\widehat{\boldsymbol{\beta}}^{\text{SM}} - \boldsymbol{\beta})] \\
&= \boldsymbol{I}_{11.2}^{-1} + \text{Cov}[\sqrt{n}(\widehat{\boldsymbol{\beta}}_1^{\text{FM}} - \boldsymbol{\beta}_1), \boldsymbol{I}_{11}^{-1}\boldsymbol{I}_{12}\sqrt{n}\widehat{\boldsymbol{\beta}}_2^{\text{MLE}}] \\
&= \boldsymbol{I}_{11.2}^{-1} - \boldsymbol{I}_{11.2}^{-1} + \text{Cov}[\sqrt{n}(\widehat{\boldsymbol{\beta}}_1^{\text{FM}} - \boldsymbol{\beta}_1), \sqrt{n}\widehat{\boldsymbol{\beta}}_2^{\text{MLE}}](\boldsymbol{I}_{11}^{-1}\boldsymbol{I}_{12})' \\
&= \boldsymbol{I}_{11}^{-1}\boldsymbol{I}_{12}\boldsymbol{I}_{22.1}^{-1}\boldsymbol{I}_{21}\boldsymbol{I}_{11}^{-1} \\
&= \boldsymbol{M}\boldsymbol{I}_{22.1}^{-1}\boldsymbol{M}'.
\end{aligned}
$$

$$
\begin{aligned}
\text{Cov}(\boldsymbol{\eta}_1, \boldsymbol{\eta}_3) &= \text{Cov}[\sqrt{n}(\widehat{\boldsymbol{\beta}}^{\text{FM}} - \boldsymbol{\beta}), \sqrt{n}(\widehat{\boldsymbol{\beta}}^{\text{SM}} - \boldsymbol{\beta})] \\
&= \boldsymbol{I}_{11.2}^{-1} - \boldsymbol{M}\boldsymbol{I}_{22.1}^{-1}\boldsymbol{M}'.
\end{aligned}
$$

$$
\begin{aligned}
\text{Cov}(\boldsymbol{\eta}_2, \boldsymbol{\eta}_3) &= \text{Cov}[\sqrt{n}(\widehat{\boldsymbol{\beta}}^{\text{FM}} - \widehat{\boldsymbol{\beta}}^{\text{SM}}), \sqrt{n}(\widehat{\boldsymbol{\beta}}^{\text{SM}} - \boldsymbol{\beta})] \\
&= \text{Cov}[\sqrt{n}((\widehat{\boldsymbol{\beta}}^{\text{FM}} - \boldsymbol{\beta}) - (\widehat{\boldsymbol{\beta}}^{\text{SM}} - \boldsymbol{\beta})), \sqrt{n}(\widehat{\boldsymbol{\beta}}^{\text{SM}} - \boldsymbol{\beta})] \\
&= \text{Cov}[\sqrt{n}((\widehat{\boldsymbol{\beta}}^{\text{FM}} - \boldsymbol{\beta}), \sqrt{n}(\widehat{\boldsymbol{\beta}}^{\text{SM}} - \boldsymbol{\beta})] - \text{Var}[\sqrt{n}(\widehat{\boldsymbol{\beta}}^{\text{SM}} - \boldsymbol{\beta})] \\
&= \boldsymbol{I}_{11.2}^{-1} - \boldsymbol{I}_{11}^{-1}\boldsymbol{I}_{12}\boldsymbol{I}_{22.1}^{-1}\boldsymbol{I}_{21}\boldsymbol{I}_{11}^{-1} - \boldsymbol{I}_{11.2}^{-1} + \boldsymbol{I}_{11}^{-1}\boldsymbol{I}_{12}\boldsymbol{I}_{22.1}^{-1}\boldsymbol{I}_{21}\boldsymbol{I}_{11}^{-1} \\
&= \mathbf{0}.
\end{aligned}
$$

We use the Lemma 3.2, which will help us to derive the results of Theorems 6.2 and 6.3.

Proof of Theorem 6.2 *Here, we provide the proof of the bias expressions. It is obvious that* $ADB(\widehat{\boldsymbol{\beta}}^{\text{FM}}) = 0$. *The ADB of the reduced, shrinkage, and positive shrinkage estimators are as follows:*

$$
\begin{aligned}
ADB(\widehat{\boldsymbol{\beta}}^{\text{FM}}) &= E\left[\lim_{n\to\infty} \sqrt{n}(\widehat{\boldsymbol{\beta}}^{\text{FM}} - \boldsymbol{\beta})\right] = \mathbf{0}. \\
ADB(\widehat{\boldsymbol{\beta}}^{\text{SM}}) &= E\left[\lim_{n\to\infty} \sqrt{n}(\widehat{\boldsymbol{\beta}}^{\text{SM}} - \boldsymbol{\beta})\right] \\
&= E\left[\lim_{n\to\infty} \sqrt{n}(\widehat{\boldsymbol{\beta}}_1^{\text{FM}} - \boldsymbol{\beta}_1)\right] + \boldsymbol{I}_{11}^{-1}\boldsymbol{I}_{12}\sqrt{n}\boldsymbol{\delta}/\sqrt{n}) \\
&= \boldsymbol{I}_{11}^{-1}\boldsymbol{I}_{12}\boldsymbol{\delta} = \boldsymbol{M}\boldsymbol{\delta}.
\end{aligned}
$$

$$\begin{aligned}
ADB(\widehat{\boldsymbol{\beta}}^{\rm S}) &= E\left[\lim_{n\to\infty}\sqrt{n}(\widehat{\boldsymbol{\beta}}^{\rm S}-\boldsymbol{\beta})\right]\\
&= E\left[\lim_{n\to\infty}\left(\sqrt{n}(\widehat{\boldsymbol{\beta}}^{\rm FM}-\boldsymbol{\beta})-(\widehat{\boldsymbol{\beta}}^{\rm FM}-\widehat{\boldsymbol{\beta}}^{\rm SM})\right)(p_2-2)D_n^{-1}\right]\\
&= \mathbf{0}-E\left[\lim_{n\to\infty}\left(\sqrt{n}(\widehat{\boldsymbol{\beta}}^{\rm FM}-\boldsymbol{\beta})-\sqrt{n}(\widehat{\boldsymbol{\beta}}^{\rm SM}-\boldsymbol{\beta})\right)(p_2-2)D_n^{-1}\right]\\
&= (p_2-2)E\left[\boldsymbol{\eta}_3 D_n^{-1}\right]\\
&= (p_2-2)\boldsymbol{M\delta}E(Z_1), \quad \textit{where} \quad Z_1=\chi^{-2}_{p_2+2}(\Delta).
\end{aligned}$$

$$\begin{aligned}
ADB(\widehat{\boldsymbol{\beta}}^{\rm PS}) &= E\left[\lim_{n\to\infty}\sqrt{n}(\widehat{\boldsymbol{\beta}}^{\rm PS}-\boldsymbol{\beta})\right]\\
&= E\left[\lim_{n\to\infty}\sqrt{n}(\widehat{\boldsymbol{\beta}}^{\rm S}-\boldsymbol{\beta})\right.\\
&- \left.\lim_{n\to\infty}\sqrt{n}(\widehat{\boldsymbol{\beta}}^{\rm FM}-\widehat{\boldsymbol{\beta}}^{\rm SM})(1-(p_2-2)D_n^{-1})I(D_n<(p_2-2)\right]\\
&= ABD(\widehat{\boldsymbol{\beta}}^{\rm S})-E\left[\boldsymbol{\eta}_2(1-(p_2-2)D_n^{-1})I(D_n<(p_2-2)\right]\\
&= ABD(\widehat{\boldsymbol{\beta}}^{\rm S})+\boldsymbol{M\delta}E\left[(1-(p_2-2)Z_1)I((p_2-2)Z_1>1)\right]\\
&= ABD(\widehat{\boldsymbol{\beta}}^{\rm S})+\boldsymbol{M\delta}\\
&\times \left[\Psi_{(p_2+2)}((p_2-2),\Delta)-(p_2-2)E\left(Z_1 I(Z_1>1)\right)\right].
\end{aligned}$$

Proof of Theorem 6.3 *In order to prove this theorem, we first derive the asymptotic covariance matrices for the four estimators. The covariance matrix* $\boldsymbol{\Sigma}^*(\widehat{\boldsymbol{\beta}}^*)$ *of any estimator* $\widehat{\boldsymbol{\beta}}^*$ *is defined as:*

$$\boldsymbol{\Sigma}^*(\widehat{\boldsymbol{\beta}}^*)=E\left[\lim_{n\to\infty}n(\widehat{\boldsymbol{\beta}}^*-\boldsymbol{\beta})(\widehat{\boldsymbol{\beta}}^*-\boldsymbol{\beta})^\top\right].$$

First, we derive the covariance matrices of FM and SM:

$$\begin{aligned}
\boldsymbol{\Sigma}^*(\widehat{\boldsymbol{\beta}}^{\rm FM}) &= E\left[\lim_{n\to\infty}\sqrt{n}(\widehat{\boldsymbol{\beta}}^{\rm FM}-\boldsymbol{\beta})\sqrt{n}(\widehat{\boldsymbol{\beta}}^{\rm FM}-\boldsymbol{\beta})^\top\right]\\
&= E(\boldsymbol{\eta}_1\boldsymbol{\eta}_1^\top)=Var(\boldsymbol{\eta}_1)+E(\boldsymbol{\eta}_1)E(\boldsymbol{\eta}_1^\top)\\
&= =Var(\boldsymbol{\eta}_1)=\boldsymbol{I}_{11.2}^{-1}.\\
\boldsymbol{\Sigma}^*(\widehat{\boldsymbol{\beta}}^{\rm SM}) &= E\left[\lim_{n\to\infty}\sqrt{n}(\widehat{\boldsymbol{\beta}}^{\rm SM}-\boldsymbol{\beta})\sqrt{n}(\widehat{\boldsymbol{\beta}}^{\rm SM}-\boldsymbol{\beta})^\top\right]\\
&= E(\boldsymbol{\eta}_3\boldsymbol{\eta}_3^\top)=Var(\boldsymbol{\eta}_3)+E(\boldsymbol{\eta}_3)E(\boldsymbol{\eta}_3^\top)\\
&= \boldsymbol{\Phi}+\boldsymbol{M\delta}\delta^\top\boldsymbol{M}^\top.
\end{aligned}$$

Second, we derive the covariance matrices of the shrinkage estimators:

$$\begin{aligned}
\boldsymbol{\Sigma}^*(\widehat{\boldsymbol{\beta}}^{\rm S}) &= E\left[\lim_{n\to\infty}\sqrt{n}(\widehat{\boldsymbol{\beta}}^{\rm S}-\boldsymbol{\beta})\ \sqrt{n}(\widehat{\boldsymbol{\beta}}^{\rm S}-\boldsymbol{\beta})^\top\right]\\
&= E\left[\lim_{n\to\infty}\sqrt{n}\left(\widehat{\boldsymbol{\beta}}^{\rm FM}-\boldsymbol{\beta}+(p_2-2)D_n^{-1}(\widehat{\boldsymbol{\beta}}^{\rm FM}-\widehat{\boldsymbol{\beta}}^{\rm SM})\right)\right.\\
&\quad \left.\sqrt{n}\left(\widehat{\boldsymbol{\beta}}^{\rm FM}-\boldsymbol{\beta}+(p_2-2)D_n^{-1}(\widehat{\boldsymbol{\beta}}^{\rm FM}-\widehat{\boldsymbol{\beta}}^{\rm SM})\right)^\top\right]\\
&= E\left[\lim_{n\to\infty}(\widehat{\boldsymbol{\beta}}^{\rm FM}-\boldsymbol{\beta})(\widehat{\boldsymbol{\beta}}^{\rm FM}-\boldsymbol{\beta})^\top\right.\\
&- 2(p_2-2)D_n^{-1}(\widehat{\boldsymbol{\beta}}^{\rm FM}-\widehat{\boldsymbol{\beta}}^{\rm SM})(\widehat{\boldsymbol{\beta}}^{\rm FM}-\boldsymbol{\beta})^\top\\
&+ \left.(p_2-2)^2D_n^{-2}(\widehat{\boldsymbol{\beta}}^{\rm FM}-\widehat{\boldsymbol{\beta}}^{\rm SM})(\widehat{\boldsymbol{\beta}}^{\rm FM}-\widehat{\boldsymbol{\beta}}^{\rm SM})^\top\right]\\
&= E\left[\boldsymbol{\eta}_1\boldsymbol{\eta}_1^\top-2(p_2-2)D_n^{-1}\boldsymbol{\eta}_2\boldsymbol{\eta}_1^\top+(p_2-2)^2D_n^{-2}\boldsymbol{\eta}_2\boldsymbol{\eta}_2^\top\right].
\end{aligned}$$

Using the conditional mean of bivariate normal, the second term of $\Sigma^*(\widehat{\beta}^{S})$ without $-2(p_2 - 2)$ is equal to

$$\begin{aligned}
E\left[\eta_2\eta_1^\top D_n^{-1}\right] &= E\left[E\left(\eta_2\eta_1^\top D_n^{-1}|\eta_2\right)\right] \\
&= E\left[\eta_2 E\left(\eta_1^\top D_n^{-1}|\eta_2\right)\right] \\
&= E\left[\eta_2\left(E(\eta_1)+(\eta_2-M\delta)\right)^\top D_n^{-1}\right] \\
&= E\left[\eta_2(\eta_2-M\delta)^\top D_n^{-1}\right] \\
&= E\left[\eta_2\eta_2^\top D_n^{-1}\right]-E\left[\eta_2 D_n^{-1}\right]\delta' M' \\
&= Var(\eta_2)E(Z_1)+E(\eta_2)E(\eta_2)^\top E(Z_2)-E(\eta_2)\delta' M' E(Z_1) \\
&= M I_{22.1}^{-1} M^\top E(Z_1)+M\delta\delta^\top M^\top E(Z_2) \\
&+ M\delta\delta^\top M^\top E(Z_1),
\end{aligned}$$

where $Z_2 = \chi^{-2}_{p_2+4}(\Delta)$. Therefore,

$$\begin{aligned}
\Sigma^*(\widehat{\beta}^{S}) &= I_{11.2}^{-1} - 2(p_2-2)[M I_{22.1}^{-1} M^\top E(Z_1) \\
&+ M\delta\delta^\top M^\top E(Z_2) - M\delta\delta^\top M^\top E(Z_1)] \\
&+ (p_2-2)^2[M I_{22.1}^{-1} M^\top E(Z_1^2) + M\delta\delta^\top M^\top E(Z_2^2)] \\
&= I_{11.2}^{-1} + (p_2-2) M I_{22.1}^{-1} M^\top[(p_2-2)E(Z_1^2) - 2E(Z_1)] \\
&- (p_2-2) M\delta\delta^\top M^\top[2E(Z_1) - (p_2-2)E(Z_2^2) + 2E(Z_2)].
\end{aligned}$$

$$\begin{aligned}
\textit{Again, } \Sigma^*(\widehat{\beta}^{PS}) &= E\left[\lim_{n\to\infty}\sqrt{n}(\widehat{\beta}^{PS}-\beta)\,\sqrt{n}(\widehat{\beta}^{PS}-\beta)^\top\right] \\
&= \Sigma^*(\widehat{\beta}^{S}) \\
&- 2E\left[\lim_{n\to\infty} n\left((\widehat{\beta}^{FM}-\widehat{\beta}^{SM})(\widehat{\beta}^{SM}-\beta)^\top\right.\right. \\
&\times \left.\left.(1-(p_2-2)D_n^{-1})I(D_n<(p_2-2))\right)\right] \\
&- E\left[\lim_{n\to\infty} n(\widehat{\beta}^{FM}-\widehat{\beta}^{SM})(\widehat{\beta}^{FM}-\widehat{\beta}^{SM})^\top\right. \\
&\times \left.(1-(p_2-2)D_n^{-1})^2 I(D_n<(p_2-2))\right] \\
&= \Sigma^*(\widehat{\beta}^{S}) - 2E\left[\eta_2\eta_3^\top(1-(p_2-2)D_n^{-1})\right. \\
&\times \left.I(D_n<(p_2-2))\right] \\
&- E\left[\eta_2\eta_2^\top(1-(p_2-2)D_n^{-1})^2 I(D_n<(p_2-2))\right].
\end{aligned}$$

Consider the second term without -2 and use the rule of conditional expectation

$$\begin{aligned}
& E\left[\eta_2\eta_3^\top(1-(p_2-2)D_n^{-1})I(D_n<(p_2-2))\right] \\
=\; & E\left[\eta_2 E\left(\eta_3^\top(1-(p_2-2)D_n^{-1})I(D_n<(p_2-2))|\eta_2\right)\right] \\
=\; & E\left[\eta_2\left((M\delta)^\top + \underbrace{Cov(\eta_2,\eta_3)}_{\mathbf{0}}\cdot\Phi\cdot(\eta_2-M\delta)\right)^\top\right. \\
\times\; & \left.(1-(p_2-2)D_n^{-1})I(D_n<(p_2-2))\right] \\
=\; & E\left[\eta_2(M\delta)^\top(1-(p_2-2)D_n^{-1})I(D_n<(p_2-2))\right] \\
=\; & -M\delta\delta^\top M^\top E\left[(1-(p_2-2)Z_1)I((p_2-2)Z_1>1)\right].
\end{aligned}$$

Therefore,

$$\begin{aligned}\boldsymbol{\Sigma}^*(\widehat{\boldsymbol{\beta}}^{\rm PS}) &= \boldsymbol{\Sigma}^*(\widehat{\boldsymbol{\beta}}^{\rm S}) + 2\boldsymbol{M}\boldsymbol{\delta}\boldsymbol{\delta}^\top\boldsymbol{M}^\top E\left[(1-(p_2-2)Z_1)I((p_2-2)Z_1>1)\right]\\ &- \boldsymbol{M}\boldsymbol{I}_{22.1}^{-1}\boldsymbol{M}^\top\ E\left[(1-(p_2-2)Z_1)^2 I((p_2-2)Z_1>1)\right]\\ &- \boldsymbol{M}\boldsymbol{\delta}\boldsymbol{\delta}^\top\boldsymbol{M}^\top E\left[(1-(p_2-2)Z_2)^2 I((p_2-2)Z_2>1)\right].\end{aligned}$$

The proof of Theorem 6.3 now follows using (6.15) and the above covariance matrices end.

Appendix B – NB

Using the subspace information, we can write $\widehat{\boldsymbol{\beta}}^{\rm FM}$ in the form: $\widehat{\boldsymbol{\beta}}^{\rm FM} = ((\widehat{\boldsymbol{\beta}}_1^{\rm FM})^\top, (\widehat{\boldsymbol{\beta}}_2^{\rm MLE})^\top)^\top$ and $\boldsymbol{I}_1^*(\boldsymbol{\beta},\theta) = \begin{bmatrix} C_{11} & C_{12}\\ C_{21} & C_{22}\end{bmatrix}$. In consequence, under the sequence of local alternatives $K_{(n)}$, we get

$$\begin{bmatrix} n^{1/2}(\widehat{\boldsymbol{\beta}}_1^{\rm FM}-\boldsymbol{\beta}_1)\\ n^{1/2}(\widehat{\boldsymbol{\beta}}_2^{\rm MLE}-\boldsymbol{\beta}_2)\end{bmatrix} \xrightarrow{D} \mathcal{N}_p\left(\begin{bmatrix}\boldsymbol{0}\\ \boldsymbol{\delta}\end{bmatrix}, \begin{bmatrix} \boldsymbol{C}_{11.2}^{-1} & -\boldsymbol{C}_{11}^{-1}\boldsymbol{C}_{12}\boldsymbol{C}_{22.1}^{-1}\\ -(\boldsymbol{C}_{11}^{-1}\boldsymbol{C}_{12}\boldsymbol{C}_{22.1}^{-1})^\top & \boldsymbol{C}_{22.1}^{-1}\end{bmatrix}\right), \tag{6.31}$$

where $\boldsymbol{C}_{11.2} = \boldsymbol{C}_{11} - \boldsymbol{C}_{12}\boldsymbol{C}_{22}^{-1}\boldsymbol{C}_{21}$ and $\boldsymbol{C}_{22.1} = \boldsymbol{C}_{22} - \boldsymbol{C}_{21}\boldsymbol{C}_{11}^{-1}\boldsymbol{C}_{12}$.

Proof of Theorem 6.5 *: The ADB of $\widehat{\boldsymbol{\beta}}_1^{\rm FM}$ is directly obtained from Eq. (6.31), so that $ADB(\widehat{\boldsymbol{\beta}}_1^{\rm FM}) = \boldsymbol{0}$. Since $\widehat{\boldsymbol{\beta}}_1^{\rm SM}$ can be written as $\widehat{\boldsymbol{\beta}}_1^{\rm SM} = \widehat{\boldsymbol{\beta}}_1^{\rm FM} + \boldsymbol{C}_{11}^{-1}\boldsymbol{C}_{12}(\widehat{\boldsymbol{\beta}}_2^{\rm MLE}-\boldsymbol{\beta}_2)$, so the ADB of $\widehat{\boldsymbol{\beta}}_1^{\rm SM}$ is as follows:*

$$\begin{aligned} ADB(\widehat{\boldsymbol{\beta}}_1^{\rm SM}) &= \lim_{n\to\infty} E[n^{1/2}(\widehat{\boldsymbol{\beta}}_1^{\rm SM}-\boldsymbol{\beta}_1)]\\ &= \lim_{n\to\infty} E[n^{1/2}(\widehat{\boldsymbol{\beta}}_1^{\rm FM}+\boldsymbol{C}_{11}^{-1}\boldsymbol{C}_{12}(\widehat{\boldsymbol{\beta}}_2^{\rm MLE}-\boldsymbol{\beta}_2)-\boldsymbol{\beta}_1)]\\ &= \lim_{n\to\infty} E[n^{1/2}(\widehat{\boldsymbol{\beta}}_1^{\rm FM}-\boldsymbol{\beta}_1)] + \boldsymbol{C}_{11}^{-1}\boldsymbol{C}_{12}\lim_{n\to\infty} E[n^{1/2}(\widehat{\boldsymbol{\beta}}_2^{\rm MLE}-\boldsymbol{\beta}_2)]\\ &= ADB(\widehat{\boldsymbol{\beta}}_1^{\rm FM}) + \boldsymbol{C}_{11}^{-1}\boldsymbol{C}_{12}\lim_{n\to\infty} E[n^{1/2}(\widehat{\boldsymbol{\beta}}_2^{\rm MLE}-\boldsymbol{\beta}_2)]\\ &= \boldsymbol{C}_{11}^{-1}\boldsymbol{C}_{12}\boldsymbol{\delta}.\end{aligned}$$

The ADB of $\widehat{\boldsymbol{\beta}}_1^{\rm S}$ is as follows:

$$\begin{aligned} ADB(\widehat{\boldsymbol{\beta}}_1^{\rm S}) &= \lim_{n\to\infty} E[n^{1/2}(\widehat{\boldsymbol{\beta}}_1^{\rm S}-\boldsymbol{\beta}_1)]\\ &= \lim_{n\to\infty} E[n^{1/2}(\widehat{\boldsymbol{\beta}}_1^{\rm FM}-(p_2-2)T_n^{-1}(\widehat{\boldsymbol{\beta}}_1^{\rm FM}-\widehat{\boldsymbol{\beta}}_1^{\rm SM})-\boldsymbol{\beta}_1)]\\ &= \lim_{n\to\infty} E[n^{1/2}(\widehat{\boldsymbol{\beta}}_1^{\rm FM}-\boldsymbol{\beta}_1)] - (p_2-2)\lim_{n\to\infty} E[n^{1/2}(\widehat{\boldsymbol{\beta}}_1^{\rm FM}-\widehat{\boldsymbol{\beta}}_1^{\rm SM})T_n^{-1}]\\ &= ADB(\widehat{\boldsymbol{\beta}}_1^{\rm FM}) - (p_2-2)\\ &\times \lim_{n\to\infty} E[n^{1/2}(\widehat{\boldsymbol{\beta}}_1^{\rm FM}-\widehat{\boldsymbol{\beta}}_1^{\rm FM}-\boldsymbol{C}_{11}^{-1}\boldsymbol{C}_{12}(\widehat{\boldsymbol{\beta}}_2^{\rm MLE}-\boldsymbol{\beta}_2))T_n^{-1}]\\ &= (p_2-2)\boldsymbol{C}_{11}^{-1}\boldsymbol{C}_{12}\lim_{n\to\infty} E[n^{1/2}(\widehat{\boldsymbol{\beta}}_2^{\rm MLE}-\boldsymbol{\beta}_2)T_n^{-1}].\end{aligned}$$

Using Lemma 3.2, we obtain

$$\lim_{n\to\infty} E[n^{1/2}(\widehat{\boldsymbol{\beta}}_2^{\rm MLE}-\boldsymbol{\beta}_2)T_n^{-1}] = \boldsymbol{\delta} E[\chi_{p_2+2}^{-2}(\triangle)];\ \triangle = \boldsymbol{\delta}^\top\boldsymbol{C}_{22.1}\boldsymbol{\delta}.$$

Hence,

$$ADB(\widehat{\boldsymbol{\beta}}_1^{\rm S}) = (p_2-2)E[\chi_{p_2+2}^{-2}(\triangle)]\boldsymbol{C}_{11}^{-1}\boldsymbol{C}_{12}\boldsymbol{\delta}$$

Lastly, we consider the ADB of $\widehat{\boldsymbol{\beta}}_1^{\mathrm{PS}}$, as follows:

$$\begin{aligned}
ADB(\widehat{\boldsymbol{\beta}}_1^{\mathrm{PS}}) &= \lim_{n\to\infty} E[n^{1/2}(\widehat{\boldsymbol{\beta}}_1^{\mathrm{PS}} - \boldsymbol{\beta}_1)] \\
&= \lim_{n\to\infty} E[n^{1/2}(\widehat{\boldsymbol{\beta}}_1^{\mathrm{S}} - (1-(p_2-2)T_n^{-1})(\widehat{\boldsymbol{\beta}}_1^{\mathrm{FM}} - \widehat{\boldsymbol{\beta}}_1^{\mathrm{SM}}) \\
&\quad \times I(T_n \le (p_2-2)) - \boldsymbol{\beta}_1)] \\
&= \lim_{n\to\infty} E[n^{1/2}(\widehat{\boldsymbol{\beta}}_1^{\mathrm{S}} - (1-(p_2-2)T_n^{-1})\boldsymbol{C}_{11}^{-1}\boldsymbol{C}_{12}(\widehat{\boldsymbol{\beta}}_2^{\mathrm{MLE}} - \boldsymbol{\beta}_2) \\
&\quad \times I(T_n \le (p_2-2)) - \boldsymbol{\beta}_1)] \\
&= ADB(\widehat{\boldsymbol{\beta}}_1^{\mathrm{S}}) + \boldsymbol{C}_{11}^{-1}\boldsymbol{C}_{12} \lim_{n\to\infty} E[n^{1/2}(\widehat{\boldsymbol{\beta}}_2^{\mathrm{MLE}} - \boldsymbol{\beta}_2)(1-(p_2-2)T_n^{-1}) \\
&\quad \times I(T_n \le (p_2-2))],
\end{aligned}$$

By using Lemma 3.2, we get

$$\begin{aligned}
&\lim_{n\to\infty} E[n^{1/2}(\widehat{\boldsymbol{\beta}}_2^{\mathrm{MLE}} - \boldsymbol{\beta}_2)(1-(p_2-2)T_n^{-1})I(T_n \le (p_2-2))] \\
=&\boldsymbol{\delta} E[(1-(p_2-2)\chi_{p_2+2}^{-2}(\triangle))I(\chi_{p_2+2}^{2}(\triangle) \le (p_2-2))]
\end{aligned}$$

So,

$$\begin{aligned}
ADB(\widehat{\boldsymbol{\beta}}_1^{\mathrm{PS}}) =& ADB(\widehat{\boldsymbol{\beta}}_1^{\mathrm{S}}) + \boldsymbol{C}_{11}^{-1}\boldsymbol{C}_{12}\boldsymbol{\delta} E\left[(1-(p_2-2)\chi_{p_2+2}^{-2}(\triangle))\right. \\
&\times \left. I(\chi_{p_2+2}^{2}(\triangle) \le (p_2-2))\right] \\
=& ADB(\widehat{\boldsymbol{\beta}}_1^{\mathrm{S}}) + E[I(\chi_{p_2+2}^{2}(\triangle) \le (p_2-2))]\boldsymbol{C}_{11}^{-1}\boldsymbol{C}_{12}\boldsymbol{\delta} \\
&-(p_2-2)E[\chi_{p_2+2}^{-2}(\triangle)I(\chi_{p_2+2}^{2}(\triangle) \le (p_2-2))]\boldsymbol{C}_{11}^{-1}\boldsymbol{C}_{12}\boldsymbol{\delta} \\
=& ADB(\widehat{\boldsymbol{\beta}}_1^{\mathrm{S}}) + \psi_{p_2+2}((p_2-2);\triangle)\boldsymbol{C}_{11}^{-1}\boldsymbol{C}_{12}\boldsymbol{\delta} \\
&- \Big\{ADB(\widehat{\boldsymbol{\beta}}_1^{\mathrm{S}}) - (p_2-2)E[\chi_{p_2+2}^{-2}(\triangle)I(\chi_{p_2+2}^{2}(\triangle) > (p_2-2))] \\
&\times \boldsymbol{C}_{11}^{-1}\boldsymbol{C}_{12}\boldsymbol{\delta}\} \\
=& \{\psi_{p_2+2}((p_2-2);\triangle) + (p_2-2) \\
&\times E\left[\chi_{p_2+2}^{-2}(\triangle)I\left(\chi_{p_2+2}^{2}(\triangle) > (p_2-2)\right)\right]\} \boldsymbol{C}_{11}^{-1}\boldsymbol{C}_{12}\boldsymbol{\delta}.
\end{aligned}$$

Here, $\psi_{p_2+2}(p_2-2;\triangle)$ is the cumulative distribution function of a noncentral χ^2 distribution with non-centrality parameter $\triangle$ and p_2+2 degrees of freedom, where $\triangle = \boldsymbol{\delta}^\top \boldsymbol{C}_{22.1}\boldsymbol{\delta}$.

Proof of Theorem 6.6 :

$$QADB(\widehat{\boldsymbol{\beta}}_1^{\mathrm{FM}}) = \mathbf{0}^\top \boldsymbol{C}_{11.2}\mathbf{0} = 0.$$

$$\begin{aligned}
QADB(\widehat{\boldsymbol{\beta}}_1^{\mathrm{SM}}) &= \boldsymbol{\Delta}_{ADB}^\top \boldsymbol{C}_{11.2}\boldsymbol{\Delta}_{ADB} = \left[\boldsymbol{C}_{11}^{-1}\boldsymbol{C}_{12}\boldsymbol{\delta}\right]^\top \boldsymbol{C}_{11.2}\left[\boldsymbol{C}_{11}^{-1}\boldsymbol{C}_{12}\boldsymbol{\delta}\right] \\
&= \boldsymbol{\delta}^\top \boldsymbol{C}_{21}\boldsymbol{C}_{11}^{-1}\boldsymbol{C}_{11.2}\boldsymbol{C}_{11}^{-1}\boldsymbol{C}_{12}\boldsymbol{\delta} = \delta_B.
\end{aligned}$$

$$\begin{aligned}
QADB(\widehat{\boldsymbol{\beta}}_1^{\mathrm{S}}) &= \left[(p_2-2)E[\chi_{p_2+2}^{-2}(\triangle)]\boldsymbol{\Delta}_{ADB}\right]^\top \\
&\quad \times \boldsymbol{C}_{11.2}\left[(p_2-2)E[\chi_{p_2+2}^{-2}(\triangle)]\boldsymbol{\Delta}_{ADB}\right] \\
&= \left\{(p_2-2)E[\chi_{p_2+2}^{-2}(\triangle)]\right\}^2 \boldsymbol{\Delta}_{ADB}^\top \boldsymbol{C}_{11.2}\boldsymbol{\Delta}_{ADB} \\
&= \left\{(p_2-2)E[\chi_{p_2+2}^{-2}(\triangle)]\right\}^2 \delta_B
\end{aligned}$$

$$\begin{aligned}
QADB(\widehat{\boldsymbol{\beta}}_1^{\mathrm{PS}}) &= [\{\psi_{p_2+2}((p_2-2);\triangle) + (p_2-2) \\
&\quad \times E\left[\chi_{p_2+2}^{-2}(\triangle)I\left(\chi_{p_2+2}^{2}(\triangle) > (p_2-2)\right)\right]\} \boldsymbol{\Delta}_{ADB}]^\top
\end{aligned}$$

$$\begin{aligned}
&\boldsymbol{C}_{11.2}\left[\{\psi_{p_2+2}((p_2-2);\triangle)+(p_2-2)\right.\\
&\left.\times\ E\left[\chi^{-2}_{p_2+2}(\triangle)I\left(\chi^2_{p_2+2}(\triangle)>(p_2-2)\right)\right]\}\boldsymbol{\Delta}_{ADB}\right]\\
&=\left\{\begin{array}{c}\psi_{p_2+2}((p_2-2);\triangle)\\ +(p_2-2)E\left[\chi^{-2}_{p_2+2}(\triangle)I\left(\chi^2_{p_2+2}(\triangle)>(p_2-2)\right)\right]\end{array}\right\}^2\\
&\times\ \boldsymbol{\Delta}^\top_{ADB}\boldsymbol{C}_{11.2}\boldsymbol{\Delta}_{ADB}\\
&=\left\{\begin{array}{c}\psi_{p_2+2}((p_2-2);\triangle)\\ +(p_2-2)E\left[\chi^{-2}_{p_2+2}(\triangle)I\left(\chi^2_{p_2+2}(\triangle)>(p_2-2)\right)\right]\end{array}\right\}^2\delta_B.
\end{aligned}$$

Proof of Theorem 6.7 *We first derive the asymptotic mean square error matrix of the estimators. Using the asymptotic mean square error matrix of $\widehat{\boldsymbol{\beta}}_1^*$ defined in Eq. (6.28) and applying Lemma 3.2, we have*

$$\begin{aligned}
\widehat{\boldsymbol{\Sigma}}(\widehat{\boldsymbol{\beta}}_1^{\mathrm{FM}}) &= \lim_{n\to\infty} E[n^{1/2}(\widehat{\boldsymbol{\beta}}_1^{\mathrm{FM}}-\boldsymbol{\beta}_1)n^{1/2}(\widehat{\boldsymbol{\beta}}_1^{\mathrm{FM}}-\boldsymbol{\beta}_1)^\top]\\
&= \lim_{n\to\infty}\mathbb{V}[n^{1/2}(\widehat{\boldsymbol{\beta}}_1^{\mathrm{FM}}-\boldsymbol{\beta}_1)]\\
&= \boldsymbol{C}^{-1}_{11.2},\\
\widehat{\boldsymbol{\Sigma}}(\widehat{\boldsymbol{\beta}}_1^{\mathrm{SM}}) &= \lim_{n\to\infty} E[n^{1/2}(\widehat{\boldsymbol{\beta}}_1^{\mathrm{SM}}-\boldsymbol{\beta}_1)n^{1/2}(\widehat{\boldsymbol{\beta}}_1^{\mathrm{SM}}-\boldsymbol{\beta}_1)^\top]\\
&= \lim_{n\to\infty}\mathbb{V}[n^{1/2}(\widehat{\boldsymbol{\beta}}_1^{\mathrm{SM}}-\boldsymbol{\beta}_1)]+\lim_{n\to\infty}E[n^{1/2}(\widehat{\boldsymbol{\beta}}_1^{\mathrm{SM}}-\boldsymbol{\beta}_1)]\\
&\times\lim_{n\to\infty}E[n^{1/2}(\widehat{\boldsymbol{\beta}}_1^{\mathrm{SM}}-\boldsymbol{\beta}_1)^\top]\\
&= \lim_{n\to\infty}\mathbb{V}[n^{1/2}(\widehat{\boldsymbol{\beta}}_1^{\mathrm{FM}}-\boldsymbol{\beta}_1)]+\boldsymbol{C}_{11}^{-1}\boldsymbol{C}_{12}\\
&\times\lim_{n\to\infty}\mathbb{V}[n^{1/2}(\widehat{\boldsymbol{\beta}}_2^{\mathrm{MLE}}-\boldsymbol{\beta}_2)](\boldsymbol{C}_{11}^{-1}\boldsymbol{C}_{12})^\top\\
&+2\lim_{n\to\infty}Cov[n^{1/2}(\widehat{\boldsymbol{\beta}}_1^{\mathrm{FM}}-\boldsymbol{\beta}_1),n^{1/2}(\widehat{\boldsymbol{\beta}}_2^{\mathrm{MLE}}-\boldsymbol{\beta}_2)^\top](\boldsymbol{C}_{11}^{-1}\boldsymbol{C}_{12})^\top\\
&+ADB(\widehat{\boldsymbol{\beta}}_1^{\mathrm{SM}})ADB(\widehat{\boldsymbol{\beta}}_1^{\mathrm{SM}})^\top\\
&= \boldsymbol{C}^{-1}_{11.2}-\boldsymbol{C}^{-1}_{11}\boldsymbol{C}_{12}\boldsymbol{C}^{-1}_{22.1}\boldsymbol{C}_{21}\boldsymbol{C}^{-1}_{11}+\Delta_{\widehat{\boldsymbol{\Sigma}}},
\end{aligned}$$

where $\Delta_{\widehat{\boldsymbol{\Sigma}}}=(\boldsymbol{C}_{11}^{-1}\boldsymbol{C}_{12}\boldsymbol{\delta})(\boldsymbol{C}_{11}^{-1}\boldsymbol{C}_{12}\boldsymbol{\delta})^\top$.
Next, we consider

$$\begin{aligned}
\widehat{\boldsymbol{\Sigma}}(\widehat{\boldsymbol{\beta}}_1^{\mathrm{S}}) &= \lim_{n\to\infty} E[n^{1/2}(\widehat{\boldsymbol{\beta}}_1^{\mathrm{S}}-\boldsymbol{\beta}_1)n^{1/2}(\widehat{\boldsymbol{\beta}}_1^{\mathrm{S}}-\boldsymbol{\beta}_1)^\top]\\
&= \lim_{n\to\infty}\mathbb{V}[n^{1/2}(\widehat{\boldsymbol{\beta}}_1^{\mathrm{FM}}-\boldsymbol{\beta}_1)]+2(p_2-2)\\
&\times\underbrace{\lim_{n\to\infty}E\left[n^{1/2}(\widehat{\boldsymbol{\beta}}_1^{\mathrm{FM}}-\boldsymbol{\beta}_1)n^{1/2}(\widehat{\boldsymbol{\beta}}_2^{\mathrm{MLE}}-\boldsymbol{\beta}_2)T_n^{-1}\right]}_{(A_1)}(\boldsymbol{C}_{11}^{-1}\boldsymbol{C}_{12})^\top\\
&+(p_2-2)^2(\boldsymbol{C}_{11}^{-1}\boldsymbol{C}_{12})\\
&\times\underbrace{\lim_{n\to\infty}E\left[n^{1/2}(\widehat{\boldsymbol{\beta}}_2^{\mathrm{MLE}}-\boldsymbol{\beta}_2)n^{1/2}(\widehat{\boldsymbol{\beta}}_2^{\mathrm{MLE}}-\boldsymbol{\beta}_2)^\top T_n^{-2}\right]}_{(A_2)}(\boldsymbol{C}_{11}^{-1}\boldsymbol{C}_{12})^\top.
\end{aligned}$$

Again, by virtue of the conditional expectation of a multivariate normal distribution and Lemma 3.2, we have

$$(A_1)\ =\ \lim_{n\to\infty}E\left[n^{1/2}(\widehat{\boldsymbol{\beta}}_1^{\mathrm{FM}}-\boldsymbol{\beta}_1)n^{1/2}(\widehat{\boldsymbol{\beta}}_2^{\mathrm{MLE}}-\boldsymbol{\beta}_2)T_n^{-1}\right]$$

$$
\begin{aligned}
&= \lim_{n\to\infty} E\left[E\left[n^{1/2}(\widehat{\boldsymbol{\beta}}_1^{\mathrm{FM}} - \boldsymbol{\beta}_1)|n^{1/2}(\widehat{\boldsymbol{\beta}}_2^{\mathrm{MLE}} - \boldsymbol{\beta}_2)\right]\right.\\
&\times \left. n^{1/2}(\widehat{\boldsymbol{\beta}}_2^{\mathrm{MLE}} - \boldsymbol{\beta}_2)T_n^{-1}\right]\\
&= \lim_{n\to\infty} E\left[\left\{E\left[n^{1/2}(\widehat{\boldsymbol{\beta}}_1^{\mathrm{FM}} - \boldsymbol{\beta}_1)\right]\right\} n^{1/2}(\widehat{\boldsymbol{\beta}}_2^{\mathrm{MLE}} - \boldsymbol{\beta}_2)^\top T_n^{-1}\right]\\
&- \boldsymbol{C}_{11}^{-1}\boldsymbol{C}_{12} \lim_{n\to\infty} E\left[n^{1/2}(\widehat{\boldsymbol{\beta}}_2^{\mathrm{MLE}} - \boldsymbol{\beta}_2)n^{1/2}(\widehat{\boldsymbol{\beta}}_2^{\mathrm{MLE}} - \boldsymbol{\beta}_2)^\top T_n^{-1}\right]\\
&+ \boldsymbol{C}_{11}^{-1}\boldsymbol{C}_{12} \lim_{n\to\infty} E\left[E\left[n^{1/2}(\widehat{\boldsymbol{\beta}}_2^{\mathrm{MLE}} - \boldsymbol{\beta}_2)\right] n^{1/2}(\widehat{\boldsymbol{\beta}}_2^{\mathrm{MLE}} - \boldsymbol{\beta}_2)^\top T_n^{-1}\right]\\
&= -\boldsymbol{C}_{11}^{-1}\boldsymbol{C}_{12} \lim_{n\to\infty} \mathbb{V}\left[n^{1/2}(\widehat{\boldsymbol{\beta}}_2^{\mathrm{MLE}} - \boldsymbol{\beta}_2)\right] E\left[\chi_{p_2+2}^{-2}(\triangle)\right]\\
&+ \boldsymbol{C}_{11}^{-1}\boldsymbol{C}_{12}\boldsymbol{\delta} \lim_{n\to\infty} E\left[n^{1/2}(\widehat{\boldsymbol{\beta}}_2^{\mathrm{MLE}} - \boldsymbol{\beta}_2)^\top T_n^{-1}\right]\\
&- \boldsymbol{C}_{11}^{-1}\boldsymbol{C}_{12} \lim_{n\to\infty} E\left[n^{1/2}(\widehat{\boldsymbol{\beta}}_2^{\mathrm{MLE}} - \boldsymbol{\beta}_2)\right] \lim_{n\to\infty}\\
&\times E\left[n^{1/2}(\widehat{\boldsymbol{\beta}}_2^{\mathrm{MLE}} - \boldsymbol{\beta}_2)^\top\right] E\left[\chi_{p_2+4}^{-2}(\triangle)\right]\\
&= -E\left[\chi_{p_2+2}^{-2}(\triangle)\right] \boldsymbol{C}_{11}^{-1}\boldsymbol{C}_{12}\boldsymbol{C}_{22.1}^{-1} - E\left[\chi_{p_2+4}^{-2}(\triangle)\right] \boldsymbol{C}_{11}^{-1}\boldsymbol{C}_{12}\boldsymbol{\delta}\boldsymbol{\delta}^\top\\
&+ \boldsymbol{C}_{11}^{-1}\boldsymbol{C}_{12}\boldsymbol{\delta} \lim_{n\to\infty} E\left[n^{1/2}(\widehat{\boldsymbol{\beta}}_2^{\mathrm{MLE}} - \boldsymbol{\beta}_2)^\top\right] E\left[\chi_{p_2+2}^{-2}(\triangle)\right]\\
&= -E\left[\chi_{p_2+2}^{-2}(\triangle)\right] \boldsymbol{C}_{11}^{-1}\boldsymbol{C}_{12}\boldsymbol{C}_{22.1}^{-1}\\
&- \left\{E\left[\chi_{p_2+4}^{-2}(\triangle)\right] - E\left[\chi_{p_2+2}^{-2}(\triangle)\right]\right\} \boldsymbol{C}_{11}^{-1}\boldsymbol{C}_{12}\boldsymbol{\delta}\boldsymbol{\delta}^\top. \qquad (6.32)
\end{aligned}
$$

$$
\begin{aligned}
(A_2) &= \lim_{n\to\infty} E\left[n^{1/2}(\widehat{\boldsymbol{\beta}}_2^{\mathrm{MLE}} - \boldsymbol{\beta}_2)n^{1/2}(\widehat{\boldsymbol{\beta}}_2^{\mathrm{MLE}} - \boldsymbol{\beta}_2)^\top T_n^{-2}\right]\\
&= \lim_{n\to\infty} \mathbb{V}\left[n^{1/2}(\widehat{\boldsymbol{\beta}}_2^{\mathrm{MLE}} - \boldsymbol{\beta}_2)\right] E\left[\chi_{p_2+2}^{-4}(\triangle)\right]\\
&+ \lim_{n\to\infty} E\left[n^{1/2}(\widehat{\boldsymbol{\beta}}_2^{\mathrm{MLE}} - \boldsymbol{\beta}_2)\right] \lim_{n\to\infty} E\left[n^{1/2}(\widehat{\boldsymbol{\beta}}_2^{\mathrm{MLE}} - \boldsymbol{\beta}_2)^\top\right] E\left[\chi_{p_2+4}^{-4}(\triangle)\right]\\
&= E\left[\chi_{p_2+2}^{-4}(\triangle)\right] \boldsymbol{C}_{22.1}^{-1} + E\left[\chi_{p_2+4}^{-4}(\triangle)\right] \boldsymbol{\delta}\boldsymbol{\delta}^\top. \qquad (6.33)
\end{aligned}
$$

Substitution of (A_1) in (6.32) and (A_2) in (6.33) into $\widehat{\boldsymbol{\Sigma}}(\widehat{\boldsymbol{\beta}}_1^{\mathrm{S}})$ yields

$$
\begin{aligned}
\widehat{\boldsymbol{\Sigma}}(\widehat{\boldsymbol{\beta}}_1^{\mathrm{S}}) &= \boldsymbol{C}_{11.2}^{-1} + (p_2-2)\left\{(p_2-2)E[\chi_{p_2+2}^{-4}(\triangle)] - 2E[\chi_{p_2+2}^{-2}(\triangle)]\right\}\\
&\times \boldsymbol{C}_{11}^{-1}\boldsymbol{C}_{12}\boldsymbol{C}_{22.1}^{-1}\boldsymbol{C}_{21}\boldsymbol{C}_{11}^{-1}\\
&+ (p_2-2)\left\{(p_2-2)E[\chi_{p_2+4}^{-4}(\triangle)] - 2E[\chi_{p_2+4}^{-2}(\triangle)]\right.\\
&+ \left. 2E[\chi_{p_2+2}^{-2}(\triangle)]\right\} \Delta_{\widehat{\boldsymbol{\Sigma}}}.
\end{aligned}
$$

The asymptotic mean square error matrix of $\widehat{\boldsymbol{\Sigma}}(\widehat{\boldsymbol{\beta}}_1^{\mathrm{PS}})$ is

$$
\begin{aligned}
&\widehat{\boldsymbol{\Sigma}}(\widehat{\boldsymbol{\beta}}_1^{\mathrm{PS}}) = \lim_{n\to\infty} E[n^{1/2}(\widehat{\boldsymbol{\beta}}_1^{\mathrm{PS}} - \boldsymbol{\beta}_1)n^{1/2}(\widehat{\boldsymbol{\beta}}_1^{\mathrm{PS}} - \boldsymbol{\beta}_1)^\top]\\
&= \lim_{n\to\infty} E[n^{1/2}(\widehat{\boldsymbol{\beta}}_1^{\mathrm{S}} - \boldsymbol{\beta}_1)n^{1/2}(\widehat{\boldsymbol{\beta}}_1^{\mathrm{S}} - \boldsymbol{\beta}_1)^\top]\\
&\underbrace{-2\lim_{n\to\infty} E\left[n^{1/2}(\widehat{\boldsymbol{\beta}}_1^{\mathrm{S}} - \boldsymbol{\beta}_1)n^{1/2}(\widehat{\boldsymbol{\beta}}_1^{\mathrm{FM}} - \widehat{\boldsymbol{\beta}}_1^{\mathrm{SM}})^\top(1-(p_2-2)T_n^{-1})I(T_n \le (p_2-2))\right]}_{(B_1)}\\
&\underbrace{+\lim_{n\to\infty} E\left[n^{1/2}(\widehat{\boldsymbol{\beta}}_1^{\mathrm{FM}} - \widehat{\boldsymbol{\beta}}_1^{\mathrm{SM}})n^{1/2}(\widehat{\boldsymbol{\beta}}_1^{\mathrm{FM}} - \widehat{\boldsymbol{\beta}}_1^{\mathrm{SM}})^\top(1-(p_2-2)T_n^{-1})I(T_n \le (p_2-2))^2\right]}_{(B_2)}.
\end{aligned}
$$

Direct the application of the conditional expectation of a multivariate normal distribution and Lemma 3.2 to (B_1) *and* (B_2)*, we get*

$$
\begin{aligned}
(B_2) &= E\begin{bmatrix}(1-(p_2-2)\chi^{-2}_{p_2+2}(\triangle))^2\\ I\left(\chi^2_{p_2+2}(\triangle)\le(p_2-2)\right)\end{bmatrix} \boldsymbol{C}_{11}^{-1}\boldsymbol{C}_{12}\boldsymbol{C}_{22.1}^{-1}\boldsymbol{C}_{21}\boldsymbol{C}_{11}^{-1}\\
&\quad + E\begin{bmatrix}(1-(p_2-2)\chi^{-2}_{p_2+4}(\triangle))^2\\ I\left(\chi^2_{p_2+4}(\triangle)\le(p_2-2)\right)\end{bmatrix}\Delta_{\widehat{\boldsymbol{\Sigma}}}.\\
(B_1) &= \underbrace{\lim_{n\to\infty} E\begin{bmatrix} n^{1/2}(\widehat{\boldsymbol{\beta}}_1^{\mathrm{FM}}-\widehat{\boldsymbol{\beta}}_1)n^{1/2}(\widehat{\boldsymbol{\beta}}_1^{\mathrm{FM}}-\widehat{\boldsymbol{\beta}}_1^{\mathrm{SM}})^\top\\ (1-(p_2-2)T_n^{-1})I(T_n\le(p_2-2))\end{bmatrix}}_{C_1}\\
&\quad \underbrace{-\lim_{n\to\infty} E\begin{bmatrix} n^{1/2}(\widehat{\boldsymbol{\beta}}_1^{\mathrm{FM}}-\widehat{\boldsymbol{\beta}}_1^{\mathrm{SM}})n^{1/2}(\widehat{\boldsymbol{\beta}}_1^{\mathrm{FM}}-\widehat{\boldsymbol{\beta}}_1^{\mathrm{SM}})^\top\\ (p_2-2)T_n^{-1}(1-(p_2-2)T_n^{-1})I(T_n\le(p_2-2))\end{bmatrix}}_{C_2}.
\end{aligned}
$$

Here,

$$
\begin{aligned}
C_1 &= E\begin{bmatrix}(1-(p_2-2)\chi^{-2}_{p_2+2}(\triangle))\\ I(\chi^2_{p_2+2}(\triangle)\le(p_2-2))\end{bmatrix}\boldsymbol{C}_{11}^{-1}\boldsymbol{C}_{12}\boldsymbol{C}_{22.1}^{-1}\boldsymbol{C}_{21}\boldsymbol{C}_{11}^{-1}\\
&\quad+\left\{\begin{matrix} E[(1-(p_2-2)\chi^{-2}_{p_2+4}(\triangle))I(\chi^2_{p_2+4}(\triangle)\le(p_2-2))]\\ -E[(1-(p_2-2)\chi^{-2}_{p_2+2}(\triangle))I(\chi^2_{p_2+2}(\triangle)\le(p_2-2))]\end{matrix}\right\}\Delta_{\widehat{\boldsymbol{\Sigma}}},\\
C_2 &= E[(p_2-2)\chi^{-2}_{p_2+2}(\triangle)(1-(p_2-2)\chi^{-2}_{p_2+2}(\triangle))I(\chi^2_{p_2+2}(\triangle)\le(p_2-2))]\\
&\quad\times\boldsymbol{C}_{11}^{-1}\boldsymbol{C}_{12}\boldsymbol{C}_{22.1}^{-1}\boldsymbol{C}_{21}\boldsymbol{C}_{11}^{-1}\\
&\quad+E[(p_2-2)\chi^{-2}_{p_2+4}(\triangle)(1-(p_2-2)\chi^{-2}_{p_2+4}(\triangle))I(\chi^2_{p_2+4}(\triangle)\le(p_2-2))]\\
&\quad\times(\boldsymbol{C}_{11}^{-1}\boldsymbol{C}_{12}\boldsymbol{\delta})(\boldsymbol{C}_{11}^{-1}\boldsymbol{C}_{12}\boldsymbol{\delta})^\top.
\end{aligned}
$$

By substituting (B_1) *and* (B_2) *into* $\widehat{\boldsymbol{\Sigma}}(\widehat{\boldsymbol{\beta}}_1^{\mathrm{PS}})$ *and then collecting like terms, so we obtain*

$$
\begin{aligned}
\widehat{\boldsymbol{\Sigma}}(\widehat{\boldsymbol{\beta}}_1^{\mathrm{PS}}) &= \widehat{\boldsymbol{\Sigma}}(\widehat{\boldsymbol{\beta}}_1^{\mathrm{S}}) - E\left[(1-(p_2-2)\chi^{-2}_{p_2+2}(\triangle))^2 I\left(\chi^2_{p_2+2}(\triangle)\le(p_2-2)\right)\right]\\
&\quad\times\boldsymbol{C}_{11}^{-1}\boldsymbol{C}_{12}\boldsymbol{C}_{22.1}^{-1}\boldsymbol{C}_{21}\boldsymbol{C}_{11}^{-1}\\
&\quad+\left\{\begin{matrix}2E\left[(1-(p_2-2)\chi^{-2}_{p_2+2}(\triangle))I\left(\chi^2_{p_2+2}(\triangle)\le(p_2-2)\right)\right]\\ -E\left[(1-(p_2-2)\chi^{-2}_{p_2+4}(\triangle))^2 I\left(\chi^2_{p_2+4}(\triangle)\le(p_2-2)\right)\right]\end{matrix}\right\}\Delta_{\widehat{\boldsymbol{\Sigma}}}.
\end{aligned}
$$

Consequently, the proof of ADRs of the estimators is easily obtained by using the (6.27) and the above asymptotic mean square error matrix results.

7

Post-Shrinkage Strategy in Sparse Linear Mixed Models

7.1 Introduction

In a host of research fields, such as bioinformatics and epidemiology, the response variable is often described by repeated measures of predictor variables that are collected over a specified period. This type of data is often referred to as "longitudinal data" and is used in medical research, where the responses are subject to various time-dependent and time-constant effects. These effects consist of pre-and post-treatment types, gender, baseline measures, and others. The linear mixed effects model (LMM) Laird and Ware (1982); Longford (1993) is a widely used statistical method in the analysis and modeling of longitudinal and repeated measures data. The LMM model provides an effective and flexible tool to describe the means and covariance structures of a given response variable after accounting for within-subject correlation. The statistical inference procedures for the LMM have been developed over the years for cases, where the number of predictors is less than the number of observations.

In this chapter, our focus is on estimating the fixed-effect parameters of the initial LMM using a ridge estimation technique when it is assumed that the model is sparse. We consider the estimation problem of fixed-effect regression parameters for LMMs when the initial model has many predictors. These predictors may be classified as active or non-active. Naturally, there are two choices to be considered: a full model with all predictors, and a submodel that contains only active predictors. Assuming that the sparsity assumption is true, the submodel provides more efficient statistical inferences than those based on a full model. Conversely, if the submodel is not correct, the estimates could become biased and inefficient. The consequences of incorporating sparse information depend on the quality and/or reliability of the information being incorporated into the estimation process. As in previous chapters, we consider shrinkage estimation, which shrinks the full model estimator to the submodel estimator by incorporating, simultaneously selecting a submodel, and estimates its regression parameters. Several authors have investigated the pretest, shrinkage, and penalized estimating strategies in a host of models; we refer to Ahmed and Opoku (2017); Opoku et al. (2021); Ahmed and Raheem (2012); Ahmed and Nicol (2012) amongst others.

Suppose the fixed-effects parameter $\boldsymbol{\beta}$ in the model can be partitioned into two sub-vectors $\boldsymbol{\beta} = (\boldsymbol{\beta}_1^\top, \boldsymbol{\beta}_2^\top)^\top$, where $\boldsymbol{\beta}_1$ is the regression coefficient vector of active predictors and $\boldsymbol{\beta}_2$ is the regression coefficient vector of inactive predictors. We focus on the estimation of $\boldsymbol{\beta}_1$ when $\boldsymbol{\beta}_2$ may be assumed to be close to a null vector. To deal with this problem, we implement the shrinkage estimation strategy, which combines full model and submodel estimators in an effective way as a trade-off between bias and variance.

There is also the problem of multicollinearity among predictor variables. Various estimation procedures, such as partial least squares estimation Geladi and Kowalski (1986) and Liu estimators Liu (2003) have been implemented to deal with this problem. However,

DOI: 10.1201/9781003170259-7

the widely used technique is ridge estimation Hoerl and Kennard (1970) to deal with the multicollinearity in the data matrix.

Our primary focus is on the estimation and prediction problems for linear mixed effect models when there are many potential predictors that have a weak or no influence on the response of interest. We consider shrinkage estimation strategies using the ridge estimator as a base estimator. The chapter is organized as follows. In Section 7.2, we present the linear mixed effect model along with the full, submodel, and shrinkage estimators based on ridge estimation. Section 7.3 provides the asymptotic bias and risk of the estimators. A Monte Carlo simulation is used to evaluate the performance of the estimators including a comparison with the penalized estimators in both low and high-dimensional cases. The results are reported in Section 7.4. Section 7.5 showcases high-dimensional applications, specifically resting-state effective brain connectivity and genetic data. We also illustrate the proposed estimation methods in low-dimensional cases as we explore Amsterdam's population growth and health study. We conclude the chapter in Section 7.6.

7.2 Estimation Strategies

In this section, we briefly describe the linear mixed model, submodel, and shrinkage estimation strategies.

7.2.1 A Gentle Introduction to Linear Mixed Model

Let us consider a sample of N subjects. For the ith subject, we collect the response variable y_{ij} for the jth time, where $i = 1\ldots,n; j = 1\ldots,n_i$ and $N = \sum_{i=1}^{n} n_i$. Let $\mathbf{Y_i} = (y_{i1},\ldots y_{in_i})^\top$ denotes the $n_i \times 1$ vector of responses from the ith subject. Let $\mathbf{X}_i = (\mathbf{x}_{i1},\ldots,\mathbf{x}_{in_i})^\top$ and $\mathbf{Z}_i = (\mathbf{z}_{i1},\ldots,\mathbf{z}_{in_i})^\top$ be $n_i\times$ p and $n_i\times$q known fixed-effects and random-effect design matrix for the ith subject of full rank p and q, respectively. The linear mixed model Laird and Ware (1982) for a vector of repeated responses $\mathbf{Y_i}$ on the ith subject takes the following the form

$$\mathbf{Y}_i = \mathbf{X}_i\boldsymbol{\beta} + \mathbf{Z}_i\mathbf{a}_i + \boldsymbol{\epsilon}_i, \tag{7.1}$$

where $\boldsymbol{\beta} = (\beta_1,\ldots,\beta_p)^\top$ is the p $\times$ 1 vector of unknown fixed-effect unknown regression coefficients, $\boldsymbol{a_i}$ is the q $\times$ 1 vector of unobservable random effects for the *ith* subject, we assume that $\boldsymbol{a_i}$ has a multivariate normal distribution with zero mean and a covariance matrix $\mathbf{G}$, where $\mathbf{G}$ is an unknown $q \times q$ covariance matrix. Further, $\boldsymbol{\epsilon_i}$ is $n_i\times 1$ vector of error terms, and are normally distributed with zero mean, covariance matrix $\sigma^2\mathrm{I}_{n_i}$. We also assume that $\boldsymbol{\epsilon_i}$ are independent of the random effects $\boldsymbol{a_i}$. The marginal distribution for the response y_i is normal with mean $\mathbf{X}_i\boldsymbol{\beta}$ and covariance matrix $Cov(Y_i) = \boldsymbol{Z}_i\sigma_i^2\boldsymbol{Z}_i^{\boldsymbol{T}} + \sigma^2\boldsymbol{I_n}$. By stacking the vectors, the mixed model can be can be expressed as $\mathbf{Y} = \mathbf{X}\boldsymbol{\beta} + \mathbf{Z}\mathbf{a} + \boldsymbol{\epsilon}$. From Equation (7.1), the distribution of the model follows $\boldsymbol{Y} \sim \mathcal{N}_n(\boldsymbol{X\beta}, \boldsymbol{V})$, where $E(\boldsymbol{Y}) = \boldsymbol{X\beta}$ with covariance, $\boldsymbol{V} = \sum_{i=1}^{n} \boldsymbol{Z}_i\sigma_i^2\boldsymbol{Z}_i^{\boldsymbol{T}} + \sigma^2\mathrm{I}_n$.

7.2.2 Ridge Regression

The generalized least squares estimator (GLS) of $\boldsymbol{\beta}$ is defined as

$$\widehat{\boldsymbol{\beta}}^{\mathrm{GLS}} = (\mathbf{X}^\top\mathbf{V}^{-1}\mathbf{X})^{-1}\mathbf{X}^\top\mathbf{V}^{-1}\mathbf{Y}$$

and the ridge full model estimator can be obtained by introducing a penalized regression so that

$$\widehat{\boldsymbol{\beta}} = \arg\min_{\boldsymbol{\beta}} \left\{ (\mathbf{Y} - \mathbf{X}\boldsymbol{\beta})^\top \mathbf{V}^{-1} (\mathbf{Y} - \mathbf{X}\boldsymbol{\beta}) + k\boldsymbol{\beta}^\top \boldsymbol{\beta} \right\}$$

and

$$\widehat{\boldsymbol{\beta}}^{\text{Ridge}} = (\mathbf{X}^T \boldsymbol{V}^{-1}\mathbf{X} + k\mathbf{I})^{-1}\mathbf{X}^\top \boldsymbol{V}^{-1}\mathbf{Y},$$

where $\widehat{\boldsymbol{\beta}}^{\text{Ridge}}$ is the ridge full model estimator and $k \in [0, \infty)$ is the tuning parameter. If $k = 0$, $\widehat{\boldsymbol{\beta}}^{\text{Ridge}}$ is the GLS estimator and $\widehat{\boldsymbol{\beta}}^{\text{Ridge}} = 0$ for k is sufficiently large. We select the value of k using cross-validation. Thus, the ridge regression strategy can be viewed as a penalized strategy although it was originally introduced to deal with multicollinearity.

To obtain a submodel, we partition $\mathbf{X} = (\mathbf{X}_1, \mathbf{X}_2)$, where $\mathbf{X}_1$ is an $n \times p_1$ sub-matrix containing the active predictors and $\mathbf{X}_2$ is an $n \times p_2$ sub-matrix that contains the inactive predictors. Similarly, $\boldsymbol{\beta} = (\boldsymbol{\beta}_1, \boldsymbol{\beta}_2)$, where $\boldsymbol{\beta}_1$ and $\boldsymbol{\beta}_2$ will have dimensions p_1 and p_2, respectively, with $p_1 + p_2 = p$. Thus, a submodel is defined as

$$\boldsymbol{Y} = \boldsymbol{X}\boldsymbol{\beta} + \boldsymbol{Z}\boldsymbol{a} + \boldsymbol{\epsilon} \text{ subject to } \boldsymbol{\beta}^\top \boldsymbol{\beta} \leq \phi \text{ and } \boldsymbol{\beta}_2 = \mathbf{0}$$

Alternatively, the above submodel can written as:

$$\boldsymbol{Y} = \boldsymbol{X_1}\boldsymbol{\beta_1} + \boldsymbol{Z}\boldsymbol{a} + \boldsymbol{\epsilon} \text{ subject to } \boldsymbol{\beta_1}^\top \boldsymbol{\beta_1} \leq \phi.$$

The submodel estimator $\widehat{\boldsymbol{\beta}}_1^{\text{SM}}$ of $\boldsymbol{\beta}_1$ has the following form:

$$\widehat{\boldsymbol{\beta}}_1^{\text{SM}} = (\mathbf{X}_1^T \mathbf{V}^{-1}\mathbf{X}_1 + k\mathbf{I})^{-1}\mathbf{X}_1^\top \mathbf{V}^{-1}\mathbf{Y}.$$

On the other hand, $\widehat{\boldsymbol{\beta}}_1^{\text{FM}}$ of the full model ridge estimator $\boldsymbol{\beta}_1$ is:

$$\widehat{\boldsymbol{\beta}}_1^{\text{FM}} = (\mathbf{X}_1^T \mathbf{V}^{-1/2}\mathrm{M}_{X_2}\mathbf{V}^{-1/2}\mathbf{X}_1 + k\mathrm{I})^{-1}\mathbf{X}_1^\top \mathbf{V}^{-1/2}\mathrm{M}_{X_2}\mathbf{V}^{-1/2}\mathbf{Y},$$

where

$$\mathbf{M}_{X_2} = \mathbf{I} - \mathbf{P} = \mathbf{I} - \mathbf{V}^{-1/2}\mathbf{X}_2(\mathbf{X}_2\mathbf{V}^{-1}\mathbf{X}_2)^{-1}\mathbf{X}_2^\top \mathbf{V}^{-1/2}.$$

7.2.3 Shrinkage Estimation Strategy

By construction, the submodel estimator will be more efficient than the full model estimator if the model is nearly sparse, that is $\boldsymbol{\beta}_2$ is close to the zero vector. However, if the assumption of sparsity is not valid, the submodel estimator is likely to be more biased and may have a higher risk than the full model estimator. There is some doubt as to whether or not to impose the sparsity condition on the model's parameter.

For this reason, we suggest setting the shrinkage ridge estimator to bed based on soft thresholding. The shrinkage ridge estimator (S) of $\boldsymbol{\beta}_1$, denoted as $\widehat{\boldsymbol{\beta}}_1^{\text{S}}$, is defined as

$$\widehat{\boldsymbol{\beta}}_1^{\text{S}} = \widehat{\boldsymbol{\beta}}_1^{\text{SM}} + (\widehat{\boldsymbol{\beta}}_1^{\text{FM}} - \widehat{\boldsymbol{\beta}}_1^{\text{SM}})(1 - (p_2 - 2)\mathrm{L}_n^{-1}), \quad p_2 \geq 3.$$

Here, $\widehat{\boldsymbol{\beta}}_1^{\text{S}}$ is a combination of the full model $\widehat{\boldsymbol{\beta}}_1^{\text{FM}}$ and submodel $\widehat{\boldsymbol{\beta}}_1^{\text{SM}}$ estimates. Further,

$$L_n = 2\left\{ l^*\left(\widehat{\boldsymbol{\beta}}^{\text{FM}}|\boldsymbol{Y}\right) - l^*\left(\widehat{\boldsymbol{\beta}}^{\text{SM}}|\boldsymbol{Y}\right) \right\}.$$

To counter the inherited over-shrinkage problem in $\widehat{\boldsymbol{\beta}}_1^{\text{S}}$, we prefer the positive-part shrinkage ridge estimator (PS) over $\widehat{\boldsymbol{\beta}}_1^{\text{S}}$, which is defined as:

$$\widehat{\boldsymbol{\beta}}_1^{\text{PS}} = \widehat{\boldsymbol{\beta}}_1^{\text{SM}} + (\widehat{\boldsymbol{\beta}}_1^{\text{FM}} - \widehat{\boldsymbol{\beta}}_1^{\text{SM}})(1 - (p_2 - 2)L_n^{-1})^+, p_2 \geq 3$$

where $(1 - (p_2 - 2)\mathrm{L}_n^{-1})^+ = \max(0, 1 - (p_2 - 2)\mathrm{L}_n^{-1})$. The PS estimator will control possible over-shrinking in the shrinkage estimator.

7.3 Asymptotic Results

Now we provide the asymptotic distributional bias and risk of the estimators. We assess the properties of the estimators for increasing n and as $\boldsymbol{\beta}_2$ approaches the null vector under the sequence of local alternatives defined as

$$K_n : \boldsymbol{\beta}_2 = \boldsymbol{\beta}_{2(n)} = \frac{\boldsymbol{\omega}}{\sqrt{n}}, \tag{7.2}$$

where $\boldsymbol{\omega} = (\omega_1, \omega_2 \ldots, \omega_{p_2})^\top \in \mathbb{R}^{p_2}$ is a fixed vector. The vector $\frac{\boldsymbol{\omega}}{\sqrt{n}}$ can be viewed as a measure of how far local alternatives K_n differ from the sparsity assumption of $\boldsymbol{\beta}_2 = \mathbf{0}$.

The asymptotic distributional bias of the estimator $\widehat{\boldsymbol{\beta}}_1^*$ is defined as:

$$\mathrm{ADB}(\widehat{\boldsymbol{\beta}}_1^*) = \lim_{n\to\infty} E\{\sqrt{n}(\widehat{\boldsymbol{\beta}}_1^* - \boldsymbol{\beta}_1)\},$$

The asymptotic covariance of an estimator $\widehat{\boldsymbol{\beta}}_1^*$ is:

$$\mathrm{Cov}(\widehat{\boldsymbol{\beta}}_1^*) = \lim_{n\to\infty} E\{n(\widehat{\boldsymbol{\beta}}_1^* - \boldsymbol{\beta}_1)(\widehat{\boldsymbol{\beta}}_1^* - \boldsymbol{\beta}_1)^\top\}.$$

Thus, the asymptotic risk of an estimator $\widehat{\boldsymbol{\beta}}_1^*$ is defined as:

$$\mathrm{R}(\widehat{\boldsymbol{\beta}}_1^*) = \mathrm{tr}\Big(\mathbf{Q}, \mathrm{Cov}(\widehat{\boldsymbol{\beta}}_1^*)\Big),$$

where $\mathbf{Q}$ is a positive-definite matrix of weights with dimensions of $p \times p$. For brevity's sake, we set $\mathbf{Q} = \mathbf{I}$ and we present two regularity conditions to establish the asymptotic properties of the estimators as follows:

Assumption 7.1 *(i)* $\frac{1}{n}\max_{1\le i\le n} \boldsymbol{x}_i^\top \left[\boldsymbol{X}^\top \boldsymbol{V}^{-1}\boldsymbol{X}\right]^{-1}\boldsymbol{x}_i \to \boldsymbol{0}$ *as* $n\to\infty$, *where* $\boldsymbol{x}_i^\top$ *is the ith row of* $\boldsymbol{X}$.

(ii) $\boldsymbol{B}_n = n^{-1}\left[\boldsymbol{X}^\top \boldsymbol{V}^{-1}\boldsymbol{X}\right]^{-1} \to \boldsymbol{B}$, *for some finite* $\boldsymbol{B} = \begin{pmatrix} \boldsymbol{B}_{11} & \boldsymbol{B}_{12} \\ \boldsymbol{B}_{21} & \boldsymbol{B}_{22} \end{pmatrix}$.

Theorem 7.1 *For* $k < \infty$, *If* $k/\sqrt{n} \to \lambda_o$ *and* $\boldsymbol{B}$ *is non-singular, the distribution of the full model ridge estimator,* $\widehat{\boldsymbol{\beta}}_n^{\mathrm{FM}}$ *is*

$$\sqrt{n}(\widehat{\boldsymbol{\beta}}_n^{\mathrm{FM}} - \boldsymbol{\beta}) \xrightarrow{D} \mathcal{N}(-\lambda_o \boldsymbol{B}^{-1}\boldsymbol{\beta}, \boldsymbol{B}^{-1}),$$

where $\xrightarrow{D}$ *denotes convergence in distribution.*

for proof we refer to Knight and Fu (2000)

Proposition 7.2 *Dnder the local alternatives* K_n, *7.1, and Theorem 7.1, we have*

$$\begin{pmatrix} \varphi_1 \\ \varphi_3 \end{pmatrix} \xrightarrow{D} \mathcal{N}\left[\begin{pmatrix} -\boldsymbol{\mu}_{11.2} \\ \boldsymbol{\delta} \end{pmatrix}, \begin{pmatrix} \boldsymbol{B}_{11.2}^{-1} & \boldsymbol{\Phi} \\ \boldsymbol{\Phi} & \boldsymbol{\Phi} \end{pmatrix}\right],$$

$$\begin{pmatrix} \varphi_3 \\ \varphi_2 \end{pmatrix} \xrightarrow{D} \mathcal{N}\left[\begin{pmatrix} \boldsymbol{\delta} \\ -\boldsymbol{\gamma} \end{pmatrix}, \begin{pmatrix} \boldsymbol{\Phi} & \mathbf{0} \\ \mathbf{0} & \boldsymbol{B}_{11}^{-1} \end{pmatrix}\right],$$

where $\varphi_1 = \sqrt{n}(\widehat{\boldsymbol{\beta}}_1^{\mathrm{FM}} - \boldsymbol{\beta}_1)$, $\varphi_2 = \sqrt{n}(\widehat{\boldsymbol{\beta}}_1^{\mathrm{SM}} - \boldsymbol{\beta}_1)$, $\varphi_3 = \sqrt{n}(\widehat{\boldsymbol{\beta}}_1^{\mathrm{FM}} - \widehat{\boldsymbol{\beta}}_1^{\mathrm{SM}})$, $\boldsymbol{\gamma} = \boldsymbol{\mu}_{11.2} + \boldsymbol{\delta}$, $\boldsymbol{\delta} = \boldsymbol{B}_{11}^{-1}\boldsymbol{B}_{12}\boldsymbol{\omega}$, $\boldsymbol{\Phi} = \boldsymbol{B}_{11}^{-1}\boldsymbol{B}_{12}\boldsymbol{B}_{22.1}^{-1}\boldsymbol{B}_{21}\boldsymbol{B}_{11}^{-1}$, $\boldsymbol{B}_{22.1} = \boldsymbol{B}_{22} - \boldsymbol{B}_{21}\boldsymbol{B}_{11}^{-1}\boldsymbol{B}_{12}$, $\boldsymbol{\mu} = -\lambda_o \boldsymbol{B}^{-1}\boldsymbol{\beta} = \begin{pmatrix} \boldsymbol{\mu}_1 \\ \boldsymbol{\mu}_2 \end{pmatrix}$ *and* $\boldsymbol{\mu}_{11.2} = \boldsymbol{\mu}_1 - \boldsymbol{B}_{12}\boldsymbol{B}_{22}^{-1}((\boldsymbol{\beta}_2 - \boldsymbol{\omega}) - \boldsymbol{\mu}_2)$.

Proof *See Appendix 7.6.*

Theorem 7.3 *Under the condition of Theorem 7.1 and the local alternatives K_n, the expressions for asymptotic distributional bias for the estimators are given as follows:*

$$\begin{aligned}
ADB(\widehat{\boldsymbol{\beta}}_1^{\rm FM}) &= -\boldsymbol{\mu}_{11.2},\\
ADB(\widehat{\boldsymbol{\beta}}_1^{\rm SM}) &= -\boldsymbol{\mu}_{11.2} - \boldsymbol{B}_{11}^{-1}\boldsymbol{B}_{12}\boldsymbol{\delta} = -\boldsymbol{\gamma},\\
ADB(\widehat{\boldsymbol{\beta}}_1^{\rm S}) &= -\boldsymbol{\mu}_{11.2} - (p_2-2)\boldsymbol{\delta}E(\chi_{p_2+2}^{-2}(\Delta)),\\
ADB(\widehat{\boldsymbol{\beta}}_1^{\rm PS}) &= -\boldsymbol{\mu}_{11.2} - \boldsymbol{\delta}H_{p_2+2}(\chi_{p_2-2}^{2};\Delta)\\
&\quad - (p_2-2)\boldsymbol{\delta}E\{\chi_{p_2+2}^{-2}(\Delta)I(\chi_{p_2+2}^{-2} > p_2-2)\},
\end{aligned}$$

where $\Delta = \boldsymbol{\omega}^\top \boldsymbol{B}_{22.1}^{-1}\boldsymbol{\omega}$, $\boldsymbol{B}_{22.1} = \boldsymbol{B}_{22} - \boldsymbol{B}_{21}\boldsymbol{B}_{11}^{-1}\boldsymbol{B}_{12}$, and $H_v(x;\Delta)$ is the cumulative distribution function of the non-central chi-squared distribution with non-centrality parameter Δ and v degrees of freedom, and $E(\chi_v^{-2j}(\Delta))$ is the expected value of the inverse of a non-central χ^2 distribution with v degrees of freedom and non-centrality parameter Δ,

$$E(\chi_v^{-2j}(\Delta)) = \int_0^\infty x^{-2j}dH_v(x,\Delta).$$

Proof *See Appendix 7.6.*

Since the ADBs of the estimators are in non-scalar form, we define the following quadratic asymptotic distributional bias (QADB) of $\widehat{\boldsymbol{\beta}}_1^*$ by

$$\text{QADB}(\widehat{\boldsymbol{\beta}}_1^*) = \Big(\text{ADB}(\widehat{\boldsymbol{\beta}}_1^*)\Big)^\top \mathbf{B}_{11.2}\Big(\text{ADB}(\widehat{\boldsymbol{\beta}}_1^*)\Big),$$

where $\mathbf{B}_{11.2} = \mathbf{B}_{11} - \mathbf{B}_{12}\mathbf{B}_{22}^{-1}\mathbf{B}_{21}$.

Corollary 7.1 *Suppose Theorem 7.3 holds. Then, under $\{K_n\}$, the QADBs of the estimators are*

$$\begin{aligned}
QADB(\widehat{\boldsymbol{\beta}}_1^{\rm FM}) &= \boldsymbol{\mu}_{11.2}^\top \boldsymbol{B}_{11.2}\boldsymbol{\mu}_{11.2},\\
QADB(\widehat{\boldsymbol{\beta}}_1^{\rm SM}) &= \boldsymbol{\gamma}^\top \boldsymbol{B}_{11.2}\boldsymbol{\gamma},\\
QADB(\widehat{\boldsymbol{\beta}}_1^{\rm S}) &= \boldsymbol{\mu}_{11.2}^\top \boldsymbol{B}_{11.2}\boldsymbol{\mu}_{11.2} + (p_2-2)\boldsymbol{\mu}_{11.2}^\top \boldsymbol{B}_{11.2}\boldsymbol{\delta}E\big(\chi_{p_2+2}^{-2}(\Delta)\big)\\
&\quad + (p_2-2)\boldsymbol{\delta}^\top \boldsymbol{B}_{11.2}\boldsymbol{\mu}_{11.2}E\big(\chi_{p_2+2}^{-2}(\Delta)\big)\\
&\quad + (p_2-2)^2\boldsymbol{\delta}^\top \boldsymbol{B}_{11.2}\boldsymbol{\delta}\Big(E\big(\chi_{p_2+2}^{-2}(\Delta)\big)\Big)^2,\\
QADB(\widehat{\boldsymbol{\beta}}_1^{\rm PS}) &= \boldsymbol{\mu}_{11.2}^\top \boldsymbol{B}_{11.2}\boldsymbol{\mu}_{11.2} + \big(\boldsymbol{\delta}^\top \boldsymbol{B}_{11.2}\boldsymbol{\mu}_{11.2} + \boldsymbol{\mu}_{11.2}^\top \boldsymbol{B}_{11.2}\boldsymbol{\delta}\big)\\
&\quad \times \big[H_{p_2+2}(p_2-2;\Delta)\\
&\quad + (p_2-2)E\{\chi_{p_2+2}^{-2}(\Delta)I(\chi_{p_2+2}^{-2}(\Delta) > p_2-2)\}\big] + \boldsymbol{\delta}^\top \boldsymbol{B}_{11.2}\boldsymbol{\delta}\\
&\quad \times \Bigg[H_{p_2+2}(p_2-2;\Delta)\\
&\quad + (p_2-2)E\{\chi_{p_2+2}^{-2}(\Delta)I(\chi_{p_2+2}^{-2}(\Delta) > p_2-2)\}\Bigg]^2.
\end{aligned}$$

Clearly, if $\boldsymbol{B}_{11.2} = \mathbf{0}$ then the QADB of all estimators will be equivalent and are therefore asymptotically unbiased. However, it is important to assess the bias function's behavior when $\boldsymbol{B}_{11.2} \neq 0$. Under this condition, we summarize the results for the asymptotic bias of the estimators as follows:

1. *The QADB of* $\widehat{\boldsymbol{\beta}}_1^{\mathrm{FM}}$ *is a constant line since it is independent of the sparsity condition*

2. *The QADB of* $\widehat{\boldsymbol{\beta}}_1^{\mathrm{SM}}$ *is an unbounded function of* $\boldsymbol{\gamma}^\top \boldsymbol{B}_{11.2}\boldsymbol{\gamma}$. *Consequently, it is heavily dependent on the sparsity condition. If the model is sparse, then it would be an unbiased estimator. The magnitude of the bias will depend on the correctness of the sparsity condition.*

3. *The QADB of* $\widehat{\boldsymbol{\beta}}_1^{\mathrm{S}}$ *and* $\widehat{\boldsymbol{\beta}}_1^{\mathrm{PS}}$ *start from* $\boldsymbol{\mu}_{11.2}^\top \boldsymbol{B}_{11.2}\boldsymbol{\mu}_{11.2}$ *at* $\Delta = 0$, *where the sparsity condition is justified, and increases to a point then decrease toward zero, since* $E\{\chi_{p_2+2}^{-2}(\Delta)\}$ *is a non-increasing function of* Δ. *Thus, the shrinkage strategy plays an important role in controlling the magnitude of bias inherited in* $\widehat{\boldsymbol{\beta}}_1^{\mathrm{SM}}$. *Thus, combining the submodel estimator with a full model estimator is clearly advantageous.*

In the following theorem we present the expressions for the covariance matrices of the estimators using Theorem 7.1.

Theorem 7.4 *Under the local alternatives* K_n, *the covariance matrices of the estimators are given:*

$$
\begin{aligned}
Cov(\widehat{\boldsymbol{\beta}}_1^{\mathrm{FM}}) &= \boldsymbol{B}_{11.2}^{-1} + \boldsymbol{\mu}_{11.2}\boldsymbol{\mu}_{11.2}^\top, \\
Cov(\widehat{\boldsymbol{\beta}}_1^{\mathrm{SM}}) &= \boldsymbol{B}_{11}^{-1} + \boldsymbol{\gamma}\boldsymbol{\gamma}^\top, \\
Cov(\widehat{\boldsymbol{\beta}}_1^{\mathrm{S}}) &= \boldsymbol{B}_{11.2}^{-1} + \boldsymbol{\mu}_{11.2}\boldsymbol{\mu}_{11.2}^T + 2(p_2-2)\boldsymbol{\mu}_{11.2}^T\boldsymbol{\delta}E\Big(\chi_{p_2+2}^{-2}(\Delta)\Big) \\
&\quad - (p_2-2)\boldsymbol{\Phi}\Big\{2E\Big(\chi_{p_2+2}^{-2}(\Delta)\Big) - (p_2-2)E\Big(\chi_{p_2+2}^{-4}(\Delta)\Big)\Big\} \\
&\quad + (p_2-2)\boldsymbol{\delta}\boldsymbol{\delta}^\top \\
&\quad \times \Big\{-2E(\chi_{p_2+4}^{-2}(\Delta)) + 2E(\chi_{p_2+2}^{-2}(\Delta)) + (p_2-2)E\Big(\chi_{p_2+4}^{-4}(\Delta)\Big)\Big\}, \\
Cov(\widehat{\boldsymbol{\beta}}_1^{\mathrm{PS}}) &= Cov(\widehat{\boldsymbol{\beta}}_1^{\mathrm{S}}) + 2\boldsymbol{\delta}\boldsymbol{\mu}_{11.2}^\top \\
&\quad \times E\Big(\Big\{1-(p_2-2)\chi_{p_2+2}^{-2}(\Delta)\Big\}I\Big(\chi_{p_2+2}^{2}(\Delta) \le p_2-2\Big)\Big) \\
&\quad - 2\boldsymbol{\Phi}E\Big(\Big\{1-(p_2-2)\chi_{p_2+2}^{-2}(\Delta)\Big\}I\Big(\chi_{p_2+2}^{2}(\Delta) \le p_2-2\Big)\Big) \\
&\quad - 2\boldsymbol{\delta}\boldsymbol{\delta}^\top E(\{1-(p_2-2)\chi_{p_2+4}^{-2}(\Delta)\}I(\chi_{p_2+4}^{2}(\Delta) \le p_2-2)) \\
&\quad + 2\boldsymbol{\delta}\boldsymbol{\delta}^\top E\Big(\Big\{1-(p_2-2)\chi_{p_2+2}^{-2}(\Delta)\Big\}I\Big(\chi_{p_2+2}^{2}(\Delta) \le p_2-2\Big)\Big) \\
&\quad - (p_2-2)^2\boldsymbol{\Phi}E\Big(\chi_{p_2+2}^{-4}(\Delta)I\Big(\chi_{p_2+2,\alpha}^{2}(\Delta) \le p_2-2\Big)\Big) \\
&\quad - (p_2-2)^2\boldsymbol{\delta}\boldsymbol{\delta}^\top E\Big(\chi_{p_2+2,\alpha}^{-4}(\Delta)I\Big(\chi_{p_2+2,\alpha}^{2}(\Delta) \le p_2-2\Big)\Big) \\
&\quad + \boldsymbol{\Phi}H_{p_2+2}(p_2-2;\Delta) + \boldsymbol{\delta}\boldsymbol{\delta}^\top H_{p_2+4}(p_2-2;\Delta).
\end{aligned}
$$

Proof *See Appendix 7.6.*

By definition, the asymptotic distributional risk (ADR) of the estimators are given in the following corollary.

Corollary 7.2

$$
\begin{aligned}
ADR(\widehat{\boldsymbol{\beta}}_1^{\mathrm{FM}}) &= tr\Big(\boldsymbol{Q}\boldsymbol{B}_{11.2}^{-1}\Big) + \boldsymbol{\mu}_{11.2}^\top \boldsymbol{Q}\boldsymbol{\mu}_{11.2},\\
ADR(\widehat{\boldsymbol{\beta}}_1^{\mathrm{SM}}) &= tr\Big(\boldsymbol{Q}\boldsymbol{B}_{11}^{-1}\Big) + \boldsymbol{\gamma}^\top \boldsymbol{Q}\boldsymbol{\gamma},\\
ADR(\widehat{\boldsymbol{\beta}}_1^{\mathrm{S}}) &= tr\Big(\boldsymbol{Q}\boldsymbol{B}_{11.2}^{-1}\Big) + \boldsymbol{\mu}_{11.2}^\top \boldsymbol{Q}\boldsymbol{\mu}_{11.2} + 2(p_2-2)\boldsymbol{\mu}_{11.2}^\top \boldsymbol{Q}\boldsymbol{\delta} E\Big(\chi_{p_2+2}^{-2}(\Delta)\Big)\\
&\quad - (p_2-2)\, tr(\boldsymbol{Q}\boldsymbol{\Phi})\Big[E\Big(\chi_{p_2+2}^{-2}(\Delta)\Big) - (p_2-2)E\Big(\chi_{p_2+2}^{-4}(\Delta)\Big)\Big]\\
&\quad + (p_2-2)\boldsymbol{\delta}^\top \boldsymbol{Q}\boldsymbol{\delta}\\
&\quad \times \Big[2E\Big(\chi_{p_2+2}^{-2}(\Delta)\Big) - 2E\big(\chi_{p_2+4}^{-2}(\Delta)\big) - (p_2-2)E\big(\chi_{p_2+4}^{-4}(\Delta)\big)\Big],\\
ADR(\widehat{\boldsymbol{\beta}}_1^{\mathrm{PS}}) &= ADR[\widehat{\boldsymbol{\beta}}_1^{\mathrm{S}}] + 2\boldsymbol{\delta}\boldsymbol{Q}\boldsymbol{\mu}_{11.2}^\top\\
&\quad \times E\Big(\Big\{1-(p_2-2)\chi_{p_2+2}^{-2}(\Delta)\Big\} I\Big(\chi_{p_2+2}^{2}(\Delta) \le p_2-2\Big)\Big)\\
&\quad - 2tr(\boldsymbol{Q}\boldsymbol{\Phi})E\Big(\Big\{1-(p_2-2)\chi_{p_2+2}^{-2}(\Delta)\Big\} I\Big(\chi_{p_2+2}^{2}(\Delta) \le p_2-2\Big)\Big)\\
&\quad - 2\boldsymbol{\delta}^\top \boldsymbol{Q}\boldsymbol{\delta} E(\{1-(p_2-2)\chi_{p_2+4}^{-2}(\Delta)\} I(\chi_{p_2+4}^{2}(\Delta) \le p_2-2))\\
&\quad + 2\boldsymbol{\delta}^\top \boldsymbol{Q}\boldsymbol{\delta} E\Big(\Big\{1-(p_2-2)\chi_{p_2+2}^{-2}(\Delta)\Big\} I\Big(\chi_{p_2+2}^{2}(\Delta) \le p_2-2\Big)\Big)\\
&\quad - (p_2-2)^2 tr(\boldsymbol{Q}\boldsymbol{\Phi})E\Big(\chi_{p_2+2}^{-4}(\Delta) I\Big(\chi_{p_2+2}^{2}(\Delta) \le p_2-2\Big)\Big)\\
&\quad - (p_2-2)^2 \boldsymbol{\delta}^\top \boldsymbol{Q}\boldsymbol{\delta} E\Big(\chi_{p_2+2}^{-4}(\Delta) I\Big(\chi_{p_2+2}^{2}(\Delta) \le p_2-2\Big)\Big)\\
&\quad + tr(\boldsymbol{Q}\boldsymbol{\Phi})H_{p_2+2}(p_2-2;\Delta) + \boldsymbol{\delta}^\top \boldsymbol{Q}\boldsymbol{\delta} H_{p_2+4}(p_2-2;\Delta).
\end{aligned}
$$

We summarized the key findings from above expressions assuming $\mathbf{B}_{12} \neq \mathbf{0}$ since for $\mathbf{B}_{12} = \mathbf{0}$ the risk of the shrinkage estimators will reduce to the risk of $\widehat{\boldsymbol{\beta}}_1^{\mathrm{FM}}$.

1. The risk of $\widehat{\boldsymbol{\beta}}_1^{\mathrm{FM}}$ remains constant since it does not depend on the sparsity condition.

2. However, the risk of the submodel is subject to the sparsity assumption. The parameter Δ is a function of the sparsity condition, if the sparsity holds then $\Delta = 0$. We observe that $\widehat{\boldsymbol{\beta}}_1^{\mathrm{SM}}$ is an unbounded function of Δ since $\Delta \in [0, \infty)$. Thus, the submodel estimator may not be a wise choice if there are doubts surrounding the sparsity assumption.

3. Interestingly, both shrinkage estimators are a bounded function of Δ. Thus, the shrinkage procedure is a powerful strategy in controlling the magnitude of the bias and risk. It also performs better than the submodel estimator in most parts of the parameter space ($\Delta \in [0, \infty)$) induced by the sparsity assumption.

4. It can been seen that the respective risks of $\widehat{\boldsymbol{\beta}}_1^{\mathrm{PS}}$ and $\widehat{\boldsymbol{\beta}}_1^{\mathrm{S}}$ are smaller than the risk $\widehat{\boldsymbol{\beta}}_1^{\mathrm{FM}}$ in the entire parameter space, where $\Delta \in [0, \infty)$. Thus, both shrinkage estimators are uniformly better than $\widehat{\boldsymbol{\beta}}_1^{\mathrm{FM}}$.

5. Further, it can be established that

$$\mathrm{ADR}\big[\widehat{\boldsymbol{\beta}}_1^{\mathrm{PS}}\big] \le \mathrm{ADR}\big[\widehat{\boldsymbol{\beta}}_1^{\mathrm{S}}\big] \le \mathrm{ADR}\big[\widehat{\boldsymbol{\beta}}_1^{\mathrm{FM}}\big]$$

for all $\Delta \ge 0$, where the strict inequality holds for some small values of Δ.

7.4 High-Dimensional Simulation Studies

To examine the validity of the large-sample properties of the estimators in finite samples, we design a simulation experiment to assess the behavior of the estimators. We use the standard mean squared error criterion for comparing the relative performance of the estimators. Based on the simulated data from the linear mixed model, we calculate the MSE of all the estimators as

$$MSE(\widehat{\boldsymbol{\beta}}) = \frac{1}{5000}\sum_{j=1}^{5000}(\widehat{\beta} - \beta)^\top(\widehat{\beta} - \beta),$$

where $\widehat{\boldsymbol{\beta}}$ denotes any one of $\widehat{\boldsymbol{\beta}}^{\mathrm{SM}}, \widehat{\boldsymbol{\beta}}^{\mathrm{S}}$ and $\widehat{\boldsymbol{\beta}}^{\mathrm{PS}}$, in the jth repetition. We use the relative mean squared efficiency (RMSE) or the ratio of MSE for risk performance comparison. The RMSE of an estimator $\widehat{\boldsymbol{\beta}}^*$ with respect to the baseline full model ridge estimator $\widehat{\boldsymbol{\beta}}_1^{\mathrm{FM}}$ is defined as:

$$\mathrm{RMSE}(\widehat{\boldsymbol{\beta}}_1^{\mathrm{FM}} : \widehat{\boldsymbol{\beta}}_1^*) = \frac{\mathrm{MSE}(\widehat{\boldsymbol{\beta}}_1^{\mathrm{FM}})}{\mathrm{MSE}(\widehat{\boldsymbol{\beta}}_1^*)},$$

where $\boldsymbol{\beta}_1^*$ is one of the listed estimators under consideration.

We simulate the response from the following linear mixed model

$$Y_i = \mathbf{X}_i\boldsymbol{\beta} + \mathbf{Z}_i\mathbf{a}_i + \boldsymbol{\epsilon}_i, \tag{7.3}$$

where $\boldsymbol{\epsilon_i} \sim \mathcal{N}(0, \sigma^2\mathrm{I}_{n_i})$ with $\sigma^2 = 1$. We generate random-effect covariate $\mathbf{a}_i$ from a multivariate normal distribution with zero mean and covariance matrix $\mathbf{G} = 0.3\mathrm{I}_{2\times2}$, where $\mathrm{I}_{2\times2}$ is 2×2 identity matrix. The design matrix $\mathbf{X}_i = (\mathbf{x}_{i1}, \ldots, \mathbf{x}_{in_i})^\top$ is generated from a n_i-multivariate normal distribution with mean vector and covariance matrix Σ_x. For simplicity and ease of calculations, we confine to an intra-class coefficient covariance matrix, where we assume that the off-diagonal elements of the covariance matrix Σ_x are equal to ρ. The simulated parameter ρ is the coefficient of correlation between any two predictors, and we select $\rho = 0.2, 0.5, 0.8$ for our simulation study.

We also calculate the ratio of the largest eigenvalue to the smallest eigenvalue of the matrix $\mathbf{X}^\top\mathbf{V}^{-1}\mathbf{X}$ known as the condition number index (CNI) Goldstein (1993). This ratio is a useful index to assess the existence of multicollinearity in the design matrix. As a rule, if the CNI is larger than 30, the model has significant multicollinearity.

The key feature is to incorporate the sparsity assumption in the model. Therefore, we consider the case when the model is assumed to be sparse. To achieve this objective, we partition the fixed-effects regression coefficients as $\boldsymbol{\beta} = (\boldsymbol{\beta}_1^\top, \boldsymbol{\beta}_2^\top)^\top = (\boldsymbol{\beta}_1^\top, \mathbf{0}_{p_2})^\top$. The coefficients $\boldsymbol{\beta}_1$ and $\boldsymbol{\beta}_2$ are p_1 and p_2 dimensional vectors, respectively, with $p = p_1 + p_2$. In this study, we assume sparsity where $\boldsymbol{\beta}_2 = \mathbf{0}$. We are interested in estimating $\boldsymbol{\beta}_1$ when the sparsity assumption may or may not hold.

In order to investigate the behavior of the estimators, we define $\Delta = ||\boldsymbol{\beta} - \boldsymbol{\beta}_o||$, where $\boldsymbol{\beta}_o = (\boldsymbol{\beta}_1^\top, \mathbf{0}_{p_2})^\top$ and $||.||$ is the Euclidean norm. We considered Δ values between 0 and 4. If $\Delta = 0$ means the sparsity assumption holds, then we have $\boldsymbol{\beta} = (1, 1, 1, 1, \underbrace{0, 0, \ldots, 0}_{p_2})^\top$ to generate the response. Conversely, when $\Delta \geq 0$, say $\Delta^* = 4$, we will have $\boldsymbol{\beta} = (1, 1, 1, 1, 4, \underbrace{0, 0, \ldots, 0}_{p_2-1})^\top$ to generate the response. In our simulation study, $(p_1, p_2) \in \{(4, 50), (4, 700), (4, 1500)\}$. Each realization is repeated 5000 times to obtain consistent results, and the MSE of suggested estimators is computed.

Here we report the results of the simulation study for $n = 75, 150$ and $p_1 = 4$ with different correlation coefficient ρ values and are presented in Tables 7.1 and 7.2. We plot the RMSEs against Δ in Figures 7.1 and 7.2 for some other simulated parameter configurations. We consider both low and high-dimensional cases, and the findings are summarized below.

1. When $\Delta = 0$, the submodel outperforms all other estimators. However, as $\Delta = 0$ departs from zero, the RMSE of the submodel estimator decreases and converges to zero. The behavior of the estimator is consistent with the theoretical results.

2. Both shrinkage estimators outperform the full ridge estimator, irrespective of the corrected submodel selected. This is consistent with the asymptotic theory presented in Section 7.3.

3. The positive shrinkage estimator performs better than the shrinkage estimator in the entire parameter space induced by Δ as presented in Tables 7.1 and 7.2 and associated graphs.

4. Δ measures the degree of deviation from the sparsity assumption. It is clear that one cannot go wrong with the use of shrinkage estimators, even if the selected submodel is misspecified. As evident from Tables 7.1 and 7.2, and Figures 7.1 and 7.2, if the selected submodel is correct, meaning $\Delta = 0$, then the shrinkage estimators are relatively efficient compared with the ridge full model estimator. However, if the submodel is misspecified, the gain slowly diminishes, and shrinkage estimators behave like the full model ridge estimator. In terms of risk, the shrinkage estimators are at least as good as the full ridge model estimator. Therefore, the use of shrinkage estimators is appropriate in applications when a submodel cannot be correctly specified.

7.4.1 Comparing with Penalized Estimation Strategies

We compare our listed estimators with two penalized estimators. A 10-fold cross-validation is used for selecting the optimal value of the penalty parameters that minimizes the mean squared errors for the penalized estimators. The results for $\rho = 0.2, 0.4, 0.6$, $n = 75, 150$, $p_1 = 4$ and $p_2 = 50, 700, 1500, 2500$ are presented in Table 7.3. Keeping in mind that the simulation is based on the sparsity condition, we observed the following from Table 7.3.

1. As expected, the submodel estimator performs better than all other estimators.

2. The shrinkage ridge estimators perform better than both penalized estimators for all values of ρ in Table 7.3. Thus, shrinkage estimators are efficient when there is multicollinearity amongst the predictor variables.

3. For a large number of sparse predictors, p_2, the shrinkage ridge estimators perform much better than the LASSO-type estimators for smaller values of p_2. For example, for fixed ρ and CNI, the RMSE of PS is 4.52 for $p_2 = 2500$ and when $p_2 = 50$ it is 1.71. The RMSE of LASSO is 2.43 for $p_2 = 2500$, and when $p_2 = 50$ it is 1.15. This clearly indicates the noticeable superiority of the shrinkage estimators over the penalized estimators for a large number of sparsity parameters in the model.

4. Finally, tabulated values reveal that the shrinkage estimators are preferable when there is multicollinearity amongst the predictor variables and/or there are too many inactive predictors in the model.

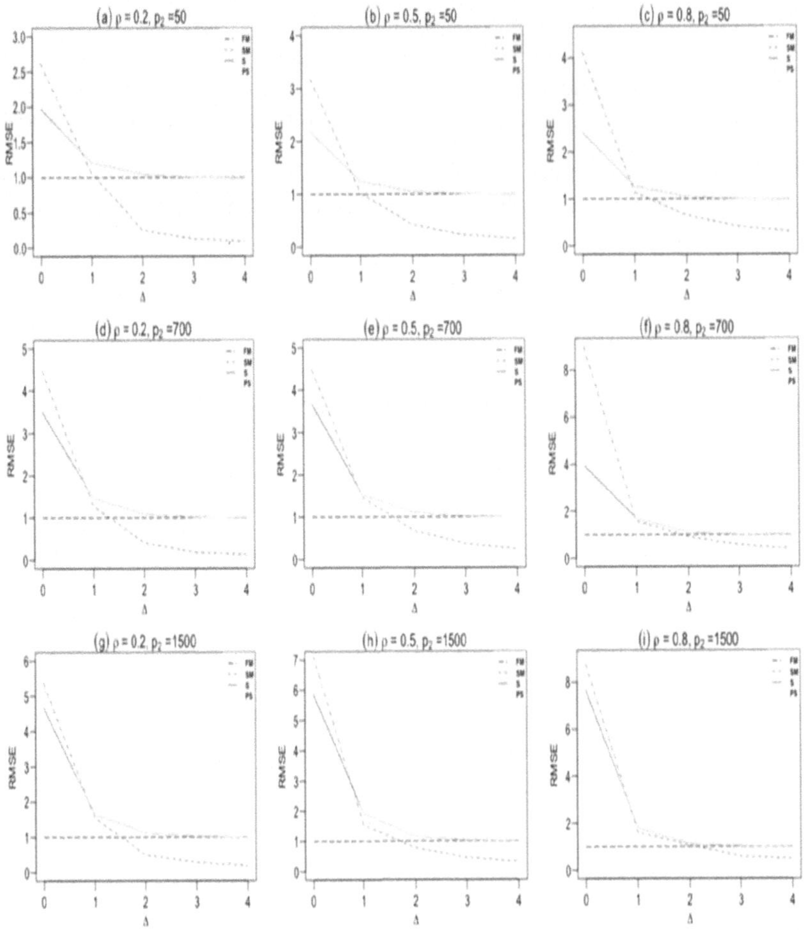

FIGURE 7.1: RMSE of the Estimators as a Function of the Sparsity Parameter Δ for $n = 75$, and $p_1 = 4$.

7.4.2 Weak Signals

The assumption of complete sparsity, where the model contains only strong and no signals is a stringent one. In reality, the model likely contains some weak signals. For this reason, we consider a simulation scenario where we investigate the performance of shrinkage estimators that include weak signals. Specifically, we simply split $p_2 = p_2(\text{weak signals}) + p_3(zeros)$, with no change on p_1 in the estimation of $\boldsymbol{\beta}_1$. We will have $\boldsymbol{\beta} = (1, 1, 1, \underbrace{0.1, 0.1, \ldots, 0.1}_{p_2}, \mathbf{0}_{p_3}^\top)^\top$. In this simulation setting, there are $p_3 = 50$ zero signals and a large amount of weak signals (p_2) that contribute simultaneously, and the number of weak signals is gradually increased.

From Table 7.4 we can observe that as p_2 increases closer to the sample size (n), the submodel estimator performs better than the shrinkage estimators. As the number of weak signals p_2 keeps increasing, the submodel estimator loses superiority and becomes worse than the full model estimator. Similarly, the performance of the penalty estimators becomes inferior in the presence of weak signals and gets worse when the number of weak signals

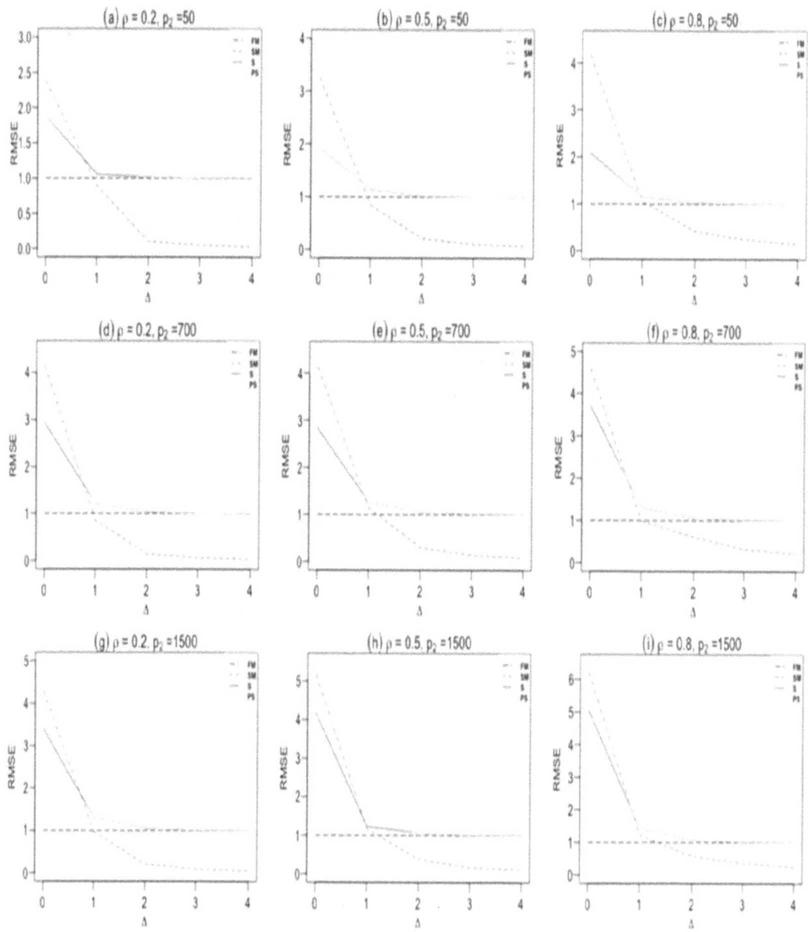

FIGURE 7.2: RMSE of the Estimators as a Function of the Sparsity Parameter Δ for $n = 150$, and $p_1 = 4$.

increases. As a matter of fact, the ridge estimator based on the full model is preferable in the presence of weak signals.

Interestingly, the shrinkage estimators take into account the possible contributions of predictors with weak signals and have dominant performances over LASSO-type methods.

7.5 Real Data Applications

We present two real data analyses to illustrate the performance of the proposed estimators. In the low-dimensional case, we apply the listed estimation strategies to some Amsterdam Growth and Health Data. Next, we consider high-dimensional genetic and brain network connectivity edge weight data. Both data sets were analyzed by Opoku et al. (2021)

TABLE 7.1: RMSEs of the Estimators for $p_1 = 4$ and $n = 75$.

ρ	p_2	Δ	CNI	SM	S	PS
0.4	50	0	403	2.47	1.81	1.83
		1		1.02	1.15	1.17
		2		0.31	1.02	1.03
		3		0.13	0.99	0.99
		4		0.05	1.00	1.00
	700	0	652	3.52	2.71	2.12
		1		1.17	1.21	1.24
		2		0.38	0.98	1.05
		3		0.14	0.99	1.00
		4		0.08	1.00	1.00
	1500	0	753	4.11	3.74	3.77
		1		1.21	1.24	1.26
		2		0.37	1.10	1.11
		3		0.22	1.6	1.04
		4		0.13	1.00	1.00
0.6	50	0	1528	3.05	2.11	2.13
		1		1.01	1.15	1.17
		2		0.63	1.05	1.06
		3		0.30	1.02	1.01
		4		0.12	1.00	1.00
	700	0	1906	4.41	2.91	2.95
		1		1.05	1.41	1.44
		2		0.64	1.04	1.05
		3		0.32	1.00	1.00
		4		0.16	1.00	1.00
	1500	0	2691	4.73	3.31	3.34
		1		1.14	1.39	1.41
		2		0.71	1.08	1.09
		3		0.26	0.99	1.00
		4		0.08	1.00	1.00

7.5.1 Amsterdam Growth and Health Data (AGHD)

The AGHD is obtained from the Amsterdam Growth and Health Study Twisk et al. (1995). The main objective of this study is to understand and reveal the relationship between lifestyle and health from adolescence into young adulthood. The data matrix contains five predictors: X_1 is the baseline fitness level measured as the maximum oxygen uptake on a treadmill, X_2 is the amount of body fat estimated by the sum of the thickness of four skinfolds, X_3 is a smoking indicator ($0 =$ no, $1 =$ yes), X_4 is the gender ($1 =$ female, $2 =$ male), and time measurement as X_5 and subject specific random-effects. The response variable Y is the total serum cholesterol measured over six time points. A total of 147 subjects participated in the study, where all variables were measured at $n_i = 6$ time occasions.

For the AGHD, we fit a linear mixed model with all five covariates for both fixed and subject-specific random effects using a two-stage selection procedure for the purpose of choosing both the random and fixed effects.

TABLE 7.2: RMSEs of the Estimators for $p_1 = 4$ and $n = 150$.

ρ	p_2	Δ	CNI	SM	S	PS
0.40	50	0	172	2.21	1.73	1.76
		1		0.84	1.06	1.08
		2		0.24	1.02	1.01
		3		0.12	0.99	0.99
		4		0.03	1.00	1.00
	700	0	402	3.91	2.74	2.80
		1		1.21	1.06	1.10
		2		0.51	0.99	1.03
		3		0.20	1.00	1.00
		4		0.01	1.00	1.00
	1500	0	563	4.15	2.84	2.86
		1		1.01	1.11	1.12
		2		0.35	0.95	1.01
		3		0.17	1.00	1.00
		4		0.02	1.00	1.00
0.6	50	0	1106	3.24	2.05	2.06
		1		0.81	1.07	1.08
		2		0.32	0.99	1.01
		3		0.05	1.00	0.99
		4		0.01	1.00	1.00
	700	0	1748	3.88	2.21	2.23
		1		1.15	1.17	1.18
		2		0.28	1.02	1.04
		3		0.12	0.99	1.00
		4		0.05	1.00	1.00
	1500	0	1946	4.81	3.21	3.24
		1		1.17	1.32	1.34
		2		0.43	1.07	1.07
		3		0.14	1.01	1.00
		4		0.08	1.00	1.00

To apply the shrinkage methods, we apply a variable selection based on an AIC procedure to select the submodel. The analysis found X_2 and X_5 to be significant variables for prediction of the response variable serum cholesterol, and the other variables were subsequently ignored as they were not significantly important. Therefore, the submodel includes only X_2 and X_5. T he full model includes all the all five predictors. We construct the shrinkage estimators from these two models. The sparsity assumption can be formulated as $\boldsymbol{\beta}_2 = (\beta_1, \beta_3, \beta_4) = (0, 0, 0)$ with $p = 5$, $p_1 = 2$ and $p_2 = 3$.

To assess the performance of the estimators, we implement the mean squared prediction error (MSPE) using bootstrap samples. We draw 1500 bootstrap samples from the data matrix $\{(Y_{ij}, \mathbf{X}_{ij}), i = 1, 2, \ldots, 147; j = 1, 2, \ldots, 6\}$. We calculate the relative prediction error (RPE) of $\boldsymbol{\beta}_1^*$ with respect to $\boldsymbol{\beta}_1^{\text{FM}}$, the full model estimator. The RPE is defined as

$$\text{RPE}(\widehat{\boldsymbol{\beta}}_1^{\text{FM}} : \widehat{\boldsymbol{\beta}}_1^*) = \frac{\text{MSPE}(\widehat{\boldsymbol{\beta}}_1^*)}{\text{MSPE}(\widehat{\boldsymbol{\beta}}_1^{\text{FM}})} = \frac{(\mathbf{Y} - \mathbf{X}_1\widehat{\boldsymbol{\beta}}_1^*)^\top(\mathbf{Y} - \mathbf{X}_1\widehat{\boldsymbol{\beta}}_1^*)}{(\mathbf{Y} - \mathbf{X}_1\widehat{\boldsymbol{\beta}}_1^{\text{FM}})^\top(\mathbf{Y} - \mathbf{X}_1\widehat{\boldsymbol{\beta}}_1^{\text{FM}})},$$

TABLE 7.3: RMSE of Estimators for $p_1 = 4$.

n	ρ	p_2	CNI	SM	S	PS	LASSO	ALASSO
75	0.2	50	30.13	3.25	1.67	1.71	1.15	1.22
		700	381.62	3.78	3.01	3.11	1.34	1.40
		1500	1851.40	4.31	3.94	4.12	1.86	1.93
		2500	4247.13	5.42	4.41	4.50	2.43	2.84
	0.4	50	41.78	3.47	1.97	2.09	1.05	1.11
		700	743.17	4.28	2.26	2.31	1.19	1.24
		1500	2350.89	5.09	3.31	3.52	1.37	1.67
		2500	6908.39	5.54	4.31	4.40	1.57	1.79
	0.6	50	70.88	4.03	2.64	2.66	1.09	1.06
		700	721.96	4.45	2.84	2.87	1.20	1.17
		1500	2781.4	5.12	3.08	3.11	1.35	1.31
		2500	5431.83	6.23	4.13	4.18	1.54	1.51
150	0.2	50	37.12	2.83	2.05	2.10	1.25	1.28
		700	351.79	3.42	2.63	2.71	1.41	1.52
		1500	850.32	4.16	2.77	2.83	1.81	1.91
		2500	3239.09	4.85	3.70	3.90	2.11	2.23
	0.4	50	49.64	3.14	2.28	3.31	1.21	1.34
		700	501.36	3.82	2.79	2.81	1.37	1.85
		1500	1109.64	4.50	3.61	3.70	1.75	2.16
		2500	3589.32	5.71	4.10	4.16	2.09	2.21
	0.6	50	74.11	3.92	3.11	3.17	1.24	1.22
		700	691.25	4.23	3.21	3.34	1.35	1.43
		1500	1389.06	5.41	3.91	4.11	1.64	1.68
		2500	3904.88	5.98	5.11	5.16	1.82	1.74

where $\boldsymbol{\beta}_1^*$ is one of the listed estimators. In this case, if RPE < 1, then $\widehat{\boldsymbol{\beta}}_1^*$ is a better strategy over $\widehat{\boldsymbol{\beta}}_1^{\text{FM}}$.

Table 7.5 reports the estimates, standard error of the non-sparse predictors, and RPEs of the estimators with respect to the ridge estimator including all five predictors. Not surprisingly, the submodel ridge estimator $\widehat{\boldsymbol{\beta}}_1^{\text{SM}}$ has the minimum RPE because it is computed under the assumption that the submodel is correct, i.e. sparsity holds. It is evident from the RPE values in Table 7.5 that the shrinkage estimators are superior to the penalized estimators in terms of RPE. Furthermore, the positive shrinkage is more efficient than the shrinkage ridge estimator. The data result strongly corroborates the theoretical and simulation findings.

7.5.2 Resting-State Effective Brain Connectivity and Genetic Data

This data contains longitudinal resting-state functional magnetic resonance imaging (rs-fMRI) effective brain connectivity network and genetic study data Nie et al. (2020) obtained from a sample of 111 subjects with a total of 319 rs-fMRI scans from the Alzheimer's Disease Neuroimaging Initiative (ADNI) database. The 111 subjects comprise 36 cognitively normal (CN), 63 mild cognitive impairment (MCI) and 12 Alzheimer's Disease (AD) subjects. The response is a network connection between regions of interest estimated from a rs-fMRI scan within the Default Mode Network (DMN).

TABLE 7.4: RMSE of Estimators for $p_1 = 4$, p_3(zero signals) $= 50$ and p_2 is the Number of Weak Signals Gradually Increased.

n	ρ	p_2	SM	S	PS	LASSO	ALASSO
75	0.2	0	3.25	1.67	1.71	1.15	1.22
		100	2.02	1.38	1.42	0.97	0.95
		500	0.84	1.24	1.26	0.88	0.87
		2000	0.75	1.08	1.11	0.79	0.84
	0.4	0	3.47	1.97	2.09	1.05	1.11
		100	2.12	1.43	1.47	1.01	1.03
		500	0.91	1.18	1.23	0.93	0.95
		2000	0.83	1.03	1.12	0.85	0.87
	0.6	0	4.03	2.64	2.66	1.09	1.06
		100	2.31	2.09	2.11	1.02	1.01
		500	1.22	1.41	1.45	0.86	0.90
		2000	0.89	1.10	1.14	0.76	0.82
150	0.2	0	2.83	2.05	2.10	1.25	1.28
		100	1.34	1.20	1.24	0.92	0.94
		500	0.76	1.14	1.17	0.79	0.83
		2000	0.84	1.03	1.05	0.88	0.89
	0.4	0	3.14	2.28	3.31	1.21	1.34
		100	2.05	1.33	1.39	1.12	1.14
		500	0.94	1.10	1.23	0.98	0.99
		2000	0.79	1.04	1.10	0.89	0.79
	0.6	0	3.92	3.11	3.17	1.24	1.22
		100	2.12	1.28	1.31	1.18	1.17
		500	0.98	1.13	1.16	0.93	0.96
		2000	0.82	0.98	1.02	0.89	0.86

TABLE 7.5: Estimate, Standard Error for the Active Predictors and RPE of Estimators for the Amsterdam Growth and Health Study data.

	FM	SM	S	PS	LASSO	ALASSO
Estimate(β_2)	0.294	0.325	0.316	0.320	0.526	0.516
Standard error	0.096	0.093	0.005	0.004	0.072	0.067
Estimate (β_5)	0.237	0.224	0.229	0.231	0.154	0.158
Standard error	0.010	0.009	0.090	0.092	0.010	0.010
RPE	1.000	0.653	0.735	0.730	0.892	0.885

There is a longitudinal sequence of such connections for each subject based on the number of repeated measurements. The DMN consists of a set of brain regions that tend to be active in the resting state, when a subject is mind-wandering with no intended task. For analysis purpose, we consider the network edge weight from the left intraparietal cortex to posterior cingulate cortex (LIPC $\to$ PCC) as the response. The genetic data are single nucleotide polymorphisms (SNPs) from non-sex chromosomes, i.e., chromosome 1 to chromosome 22. SNPs with minor allele frequency less than 5% are removed as are SNPs with a Hardy-Weinberg equilibrium p-value lower than 10^{-6} or a missing rate greater than 5%.

TABLE 7.6: RPEs of Estimators for Resting-State Effective Brain Connectivity and Genetic Data.

	FM	SM	S	PS	LASSO	ALASSO
RPE	1.000	0.825	0.931	0.929	1.143	1.210

After pre-processing, there are 1,220,955 SNPs and the longitudinal rs-fMRI effective connectivity network uses the 111 subjects with rs-fMRI data. The response is network edge weight. Further, there are SNPs which are the fixed-effects and subject specific random-effects.

To obtain a submodel, we use a genome-wide association study (GWAS) to screen the genetic data at 100 SNPs. We implement a second screening by applying a multinomial logistic regression to identify a smaller subset of the 100 SNPs that are potentially associated with the disease (CN/MCI/AD). This gives a subset of the top 10 SNPs. These top 10 SNPs are the most important predictors, and the other 90 SNPs are ignored as not significant. We now have two models, a full model with all 100 SNPs, and a submodel with 10 SNPs. Now, we can construct the shrinkage estimators using these two models.

We draw 1500 bootstrap samples with replacements from the corresponding data matrix. We list the RPE (the smaller the RPE, the better the prediction strategy) of the estimators based on the bootstrap simulation with respect to the full model ridge estimator in Table 7.6. The table values reveal that the RPEs of the shrinkage ridge estimators are smaller than the penalty estimators. The submodel ridge estimator has the smallest RPE since it is computed when the submodel is correct. The positive shrinkage performs better than the shrinkage estimator. Thus, the data analysis is in agreement with the simulation and theoretical findings.

7.6 Concluding Remarks

This chapter presented shrinkage and penalized estimation strategies for low and high-dimensional linear mixed models when multicollinearity exists. We are mainly interested in the estimation of fixed-effects regression parameters in the linear mixed model under the assumption of sparsity. Namely, only the important predictors contribute to prediction, and others can be removed from the model. We considered a more realistic situation when some of the predictors may have weak and very weak influences on the response of interest. We implemented a shrinkage estimation strategy based on ridge estimation as the benchmark estimator. We provided the asymptotic properties of the shrinkage ridge estimators and established that the shrinkage ridge estimation strategy is uniformly better than the ridge estimator that includes all available predictors. The shrinkage strategy also performs better than the submodel ridge estimator in a wide range of the parameter space induced by the sparsity assumption.

A Monte Carlo simulation was conducted to examine the moderate sample behavior of the listed estimators in a broad sense. In other words, we assess the relative performance of the estimators when the sparsity assumption may or may not hold. The simulation results strongly corroborate the large-sample theory. We also investigate the relative performance of the penalized estimators using shrinkage ridge estimators. The simulated results revealed that the performance of shrinkage ridge estimators outshined the penalized estimators, especially when predictors are highly correlated.

Finally, we applied the shrinkage ridge strategy to two real data sets. The data analysis showed the same results, that the shrinkage ridge strategy is superior with the smallest relative prediction error compared to the penalized strategy. The findings of the data analyses strongly confirm the findings of the simulation study and theoretical results.

We suggest the use of the shrinkage ridge estimation strategy when the assumption of sparsity may be in question. The results of our simulation study and real data application are consistent with the available results in Ahmed and Yüzbaşı (2017); Ahmed et al. (2016); Ahmed and Yüzbaşı (2016) and Opoku et al. (2021).

In passing, we would like to remark that we only considered LASSO and ALASSO procedures for brevity and comparison purposes. However, readers may apply other penalty estimators like the Elastic-Net (ENET), the Minimax Concave Penalty (MCP), and the Smoothly Clipped Absolute Deviation method (SCAD) for high-dimensional linear mixed models to compare with the shrinkage ridge estimators. Another interesting extension will be integrating two submodels. The goal is to improve the estimation and prediction accuracy of the non-sparse set of the fixed-effects parameters by combining an over-fitted model with an under-fitted one Ahmed et al. (2016); Ahmed and Yüzbaşı (2016). This approach will include combining two submodels produced by two different variable selection techniques Ahmed and Yüzbaşı (2016).

Appendix

Proof of Proposition 7.2 *Using the argument and equation:* $\widehat{\boldsymbol{Y}} = \boldsymbol{Y} - \boldsymbol{X}_2\widehat{\boldsymbol{\beta}}_2^{\mathrm{FM}}$, *where*

$$\begin{aligned}
\widehat{\boldsymbol{\beta}}_1^{\mathrm{FM}} &= \arg\min_{\boldsymbol{\beta}_1}\left\{(\widehat{\boldsymbol{Y}} - \boldsymbol{X}_1\boldsymbol{\beta}_1)^\top \boldsymbol{V}^{-1}(\widehat{\boldsymbol{Y}} - \boldsymbol{X}_1\boldsymbol{\beta}_1) + \lambda||\boldsymbol{\beta}_1||^2\right\}\\
&= \left[\boldsymbol{X}_1^\top \boldsymbol{V}^{-1}\boldsymbol{X}_1 + \lambda I_{p_1}\right]^{-1}\boldsymbol{X}_1^\top \boldsymbol{V}^{-1}\widehat{\boldsymbol{Y}}\\
&= \left[\boldsymbol{X}_1^\top \boldsymbol{V}^{-1}\boldsymbol{X}_1 + \lambda I_{p_1}\right]^{-1}\boldsymbol{X}_1^\top \boldsymbol{V}^{-1}\boldsymbol{Y} - \left[\boldsymbol{X}_1^\top \boldsymbol{V}^{-1}\boldsymbol{X}_1 + \lambda I_{p_1}\right]^{-1}\\
&\times \boldsymbol{X}_1^\top \boldsymbol{V}^{-1}\boldsymbol{X}_2\widehat{\boldsymbol{\beta}}_2^{\mathrm{FM}}\\
&= \widehat{\boldsymbol{\beta}}_1^{\mathrm{SM}} - \left[\boldsymbol{X}_1 \boldsymbol{V}^{-1}\boldsymbol{X}_1 + \lambda I_{p_1}\right]^{-1}\boldsymbol{X}_1^\top \boldsymbol{V}^{-1}\boldsymbol{X}_2\widehat{\boldsymbol{\beta}}_2^{\mathrm{FM}}\\
&= \widehat{\boldsymbol{\beta}}_1^{\mathrm{SM}} - \boldsymbol{B}_{11}^{-1}\boldsymbol{B}_{12}\widehat{\boldsymbol{\beta}}_2^{\mathrm{FM}}
\end{aligned}$$

From Theorem 7.1, we partition $\sqrt{n}(\widehat{\boldsymbol{\beta}}^{\mathrm{FM}} - \boldsymbol{\beta})$ *as* $\sqrt{n}(\widehat{\boldsymbol{\beta}}^{\mathrm{FM}} - \boldsymbol{\beta}) = \big(\sqrt{n}(\widehat{\boldsymbol{\beta}}_1^{\mathrm{FM}} - \boldsymbol{\beta}_1), \sqrt{n}(\widehat{\boldsymbol{\beta}}_2^{\mathrm{FM}} - \boldsymbol{\beta}_2)\big)$. *We obtain* $\sqrt{n}(\widehat{\boldsymbol{\beta}}_1^{\mathrm{FM}} - \boldsymbol{\beta}_1) \xrightarrow{D} \mathcal{N}_{p_1}(-\boldsymbol{\mu}_{11.2}, \boldsymbol{B}_{11.2}^{-1})$, *where* $\boldsymbol{B}_{11.2}^{-1} = \boldsymbol{B}_{11} - \boldsymbol{B}_{12}\boldsymbol{B}_{22}^{-1}\boldsymbol{B}_{21}$. *We have shown that* $\widehat{\boldsymbol{\beta}}_1^{\mathrm{SM}} = \widehat{\boldsymbol{\beta}}_1^{\mathrm{FM}} + \boldsymbol{B}_{11}^{-1}\boldsymbol{B}_{12}\widehat{\boldsymbol{\beta}}_2^{\mathrm{FM}}$. *Thus, under the local alternative* $\{K_n\}$:

$$\begin{aligned}
\varphi_2 &= \sqrt{n}\big(\widehat{\boldsymbol{\beta}}_1^{\mathrm{SM}} - \boldsymbol{\beta}_1\big)\\
&= \sqrt{n}\big(\widehat{\boldsymbol{\beta}}_1^{\mathrm{FM}} + \boldsymbol{B}_{11}^{-1}\boldsymbol{B}_{12}\widehat{\boldsymbol{\beta}}_2^{\mathrm{FM}} - \boldsymbol{\beta}_1\big)\\
&= \varphi_1 + \boldsymbol{B}_{11}^{-1}\boldsymbol{B}_{12}\sqrt{n}\widehat{\boldsymbol{\beta}}_2^{\mathrm{FM}},\\
\varphi_3 &= \sqrt{n}\big(\widehat{\boldsymbol{\beta}}_1^{\mathrm{FM}} - \widehat{\boldsymbol{\beta}}_1^{\mathrm{SM}}\big)\\
&= \sqrt{n}\big(\widehat{\boldsymbol{\beta}}_1^{\mathrm{FM}} - \boldsymbol{\beta}_1\big) - \sqrt{n}\big(\widehat{\boldsymbol{\beta}}_1^{\mathrm{SM}} - \boldsymbol{\beta}_1\big)\\
&= \varphi_1 - \varphi_2.
\end{aligned}$$

Since φ_2 and φ_3 are linear functions of φ_1, implies that as $n \to \infty$, they are also asymptotically normally distributed with mean vectors and covariance matrices, respectively are:

$$
\begin{aligned}
E(\varphi_1) &= E\Big(\sqrt{n}(\widehat{\boldsymbol{\beta}}_1^{\text{FM}} - \boldsymbol{\beta}_1)\Big) = -\boldsymbol{\mu}_{11.2} \\
E(\varphi_2) &= E\Big(\varphi_1 + \boldsymbol{B}_{11}^{-1}\boldsymbol{B}_{12}\sqrt{n}\widehat{\boldsymbol{\beta}}_2^{\text{FM}}\Big) \\
&= E(\varphi_1) + \boldsymbol{B}_{11}^{-1}\boldsymbol{B}_{12}\sqrt{n}E(\widehat{\boldsymbol{\beta}}_2^{\text{FM}}) \\
&= -\boldsymbol{\mu}_{11.2} + \boldsymbol{B}_{11}^{-1}\boldsymbol{B}_{12}\boldsymbol{\omega} = -(\boldsymbol{\mu}_{11.2} - \boldsymbol{\delta}) = -\boldsymbol{\gamma} \\
E(\varphi_3) &= E(\varphi_1 - \varphi_2) = -\boldsymbol{\mu}_{11.2} - (-(\boldsymbol{\mu}_{11.2} - \boldsymbol{\delta})) = \boldsymbol{\delta} \\
Var(\varphi_1) &= \boldsymbol{B}_{22.1}^{-1} \\
Var(\varphi_2) &= Var\Big(\varphi_1 + \boldsymbol{B}_{11}^{-1}\boldsymbol{B}_{12}\sqrt{n}\widehat{\boldsymbol{\beta}}_2^{\text{FM}}\Big) \\
&= Var(\varphi_1) + \boldsymbol{B}_{11}^{-1}\boldsymbol{B}_{12}\boldsymbol{B}_{22.1}^{-1}\boldsymbol{B}_{21}\boldsymbol{B}_{11}^{-1} \\
&+ 2Cov\Big[\sqrt{n}(\widehat{\boldsymbol{\beta}}_1^{\text{FM}} - \boldsymbol{\beta}_1), \sqrt{n}(\widehat{\boldsymbol{\beta}}_2^{\text{FM}} - \boldsymbol{\beta}_2)\Big](\boldsymbol{B}_{11}^{-1}\boldsymbol{B}_{12})^\top \\
&= \boldsymbol{B}_{22.1}^{-1} - \boldsymbol{B}_{11}^{-1}\boldsymbol{B}_{12}\boldsymbol{B}_{22.1}^{-1}\boldsymbol{B}_{21}\boldsymbol{B}_{11}^{-1} = \boldsymbol{B}_{11}^{-1} \\
Var(\varphi_3) &= Var\Big(\sqrt{n}(\widehat{\boldsymbol{\beta}}_1^{\text{FM}} - \widehat{\boldsymbol{\beta}}_1^{\text{SM}})\Big) \\
&= Var\Big(\sqrt{n}(\widehat{\boldsymbol{\beta}}_1^{\text{FM}} - \widehat{\boldsymbol{\beta}}_1^{\text{FM}} - \boldsymbol{B}_{11}^{-1}\boldsymbol{B}_{12}\widehat{\boldsymbol{\beta}}_2^{\text{FM}})\Big) \\
&= \boldsymbol{B}_{11}^{-1}\boldsymbol{B}_{12}Var\Big[\sqrt{n}\widehat{\boldsymbol{\beta}}_2^{\text{FM}}\Big](\boldsymbol{B}_{11}^{-1}\boldsymbol{B}_{12})^\top \\
&= \boldsymbol{B}_{11}^{-1}\boldsymbol{B}_{12}\boldsymbol{B}_{22.1}^{-1}\boldsymbol{B}_{21}\boldsymbol{B}_{11}^{-1} = \boldsymbol{\Phi} \\
Cov(\varphi_1, \varphi_3) &= Cov\Big[\sqrt{n}(\widehat{\boldsymbol{\beta}}_1^{\text{FM}} - \boldsymbol{\beta}_1), \sqrt{n}(\widehat{\boldsymbol{\beta}}_1^{\text{FM}} - \widehat{\boldsymbol{\beta}}_1^{\text{SM}})\Big] \\
&= Var\Big(\sqrt{n}(\widehat{\boldsymbol{\beta}}_1^{\text{FM}} - \boldsymbol{\beta}_1)\Big) \\
&- Cov\Big[\sqrt{n}(\widehat{\boldsymbol{\beta}}_1^{\text{FM}} - \boldsymbol{\beta}_1), \sqrt{n}(\widehat{\boldsymbol{\beta}}_1^{\text{SM}} - \boldsymbol{\beta}_1)\Big] \\
&= Var(\varphi_1) \\
&- Cov\Big[\sqrt{n}(\widehat{\boldsymbol{\beta}}_1^{\text{FM}} - \boldsymbol{\beta}_1), \sqrt{n}(\widehat{\boldsymbol{\beta}}_1^{\text{FM}} - \boldsymbol{\beta}_1) + \sqrt{n}\boldsymbol{B}_{11}^{-1}\boldsymbol{B}_{12}\widehat{\boldsymbol{\beta}}_2^{\text{FM}}\Big] \\
&= \boldsymbol{B}_{11}^{-1}\boldsymbol{B}_{12}\boldsymbol{B}_{22.1}^{-1}\boldsymbol{B}_{21}\boldsymbol{B}_{11}^{-1} = \boldsymbol{\Phi}
\end{aligned}
$$

$$
\begin{aligned}
Cov(\varphi_2, \varphi_3) &= Cov\Big[\sqrt{n}(\widehat{\boldsymbol{\beta}}_1^{\text{SM}} - \boldsymbol{\beta}_1), \sqrt{n}(\widehat{\boldsymbol{\beta}}_1^{\text{FM}} - \widehat{\boldsymbol{\beta}}_1^{\text{SM}})\Big] \\
&= Cov\Big[\sqrt{n}(\widehat{\boldsymbol{\beta}}_1^{\text{SM}} - \boldsymbol{\beta}_1), \sqrt{n}(\widehat{\boldsymbol{\beta}}_1^{\text{FM}} - \boldsymbol{\beta}_1)\Big] - Var\Big(\sqrt{n}(\widehat{\boldsymbol{\beta}}_1^{\text{SM}} - \boldsymbol{\beta}_1)\Big) \\
&= \boldsymbol{B}_{11.2}^{-1} - \boldsymbol{B}_{11}^{-1}\boldsymbol{B}_{12}\boldsymbol{B}_{22.1}^{-1}\boldsymbol{B}_{21}\boldsymbol{B}_{11}^{-1} - \boldsymbol{B}_{11}^{-1} \\
&= \boldsymbol{B}_{11.2}^{-1} - (\boldsymbol{B}_{11.2}^{-1} - \boldsymbol{B}_{11}^{-1}) - \boldsymbol{B}_{11}^{-1} = \boldsymbol{0}
\end{aligned}
$$

Thus, the asymptotic distributions of φ_2 and φ_3 are:

$$\varphi_2 = \sqrt{n}(\widehat{\boldsymbol{\beta}}_1^{\rm SM} - \boldsymbol{\beta}_1) \xrightarrow{D} \mathcal{N}_{p_1}(-\boldsymbol{\gamma}, \boldsymbol{B}_{11}^{-1})$$
$$\varphi_3 = \sqrt{n}(\widehat{\boldsymbol{\beta}}_1^{\rm FM} - \widehat{\boldsymbol{\beta}}_1^{\rm SM}) \xrightarrow{D} \mathcal{N}_{p_1}(\boldsymbol{\delta}, \boldsymbol{\Phi})$$

The Lemma 3.2 is useful for the proof of the bias and covariance of the estimators.

Proof of Theorem 7.3

$$\begin{aligned} ADB(\widehat{\boldsymbol{\beta}}_1^{\rm FM}) &= E\{\lim_{n\to\infty}\sqrt{n}(\widehat{\boldsymbol{\beta}}_1^{\rm FM} - \boldsymbol{\beta}_1)\} \\ &= -\boldsymbol{\mu}_{11.2}. \end{aligned}$$

$$\begin{aligned} ADB(\widehat{\boldsymbol{\beta}}_1^{\rm SM}) &= E\{\lim_{n\to\infty}\sqrt{n}(\widehat{\boldsymbol{\beta}}_1^{\rm SM} - \boldsymbol{\beta}_1)\} \\ &= E\{\lim_{n\to\infty}\sqrt{n}(\widehat{\boldsymbol{\beta}}_1^{\rm FM} - \boldsymbol{B}_{11}^{-1}\boldsymbol{B}_{12}\widehat{\boldsymbol{\beta}}_2^{\rm FM} - \boldsymbol{\beta}_1)\} \\ &= E\{\lim_{n\to\infty}\sqrt{n}(\widehat{\boldsymbol{\beta}}_{\mathbf{1}}^{\mathbf{FM}} - \boldsymbol{\beta}_{\mathbf{1}})\} - E\{\lim_{n\to\infty}\sqrt{n}(\boldsymbol{B}_{11}^{-1}\boldsymbol{B}_{12}\widehat{\boldsymbol{\beta}}_2^{\rm FM})\} \\ &= -\boldsymbol{\mu}_{11.2} - E\{\lim_{n\to\infty}\sqrt{n}(\boldsymbol{B}_{11}^{-1}\boldsymbol{B}_{12}\widehat{\boldsymbol{\beta}}_2^{\rm FM})\} \\ &= -\boldsymbol{\mu}_{11.2} - \boldsymbol{B}_{11}^{-1}\boldsymbol{B}_{12}\boldsymbol{\omega} = -(\boldsymbol{\mu}_{11.2} + \boldsymbol{\delta}) = -\boldsymbol{\gamma}. \end{aligned}$$

Using Lemma 3.2,

$$\begin{aligned} ADB(\widehat{\boldsymbol{\beta}}_1^{\rm S}) &= E\{\lim_{n\to\infty}\sqrt{n}(\widehat{\boldsymbol{\beta}}_{\mathbf{1}}^{\mathbf{S}} - \boldsymbol{\beta}_{\mathbf{1}})\} \\ &= E\{\lim_{n\to\infty}\sqrt{n}(\widehat{\boldsymbol{\beta}}_{\mathbf{1}}^{\mathbf{FM}} - (\widehat{\boldsymbol{\beta}}_{\mathbf{1}}^{\mathbf{FM}} - \widehat{\boldsymbol{\beta}}_{\mathbf{1}}^{\mathbf{SM}})(p_2-2)L_n^{-1} - \boldsymbol{\beta}_{\mathbf{1}})\} \\ &= E\{\lim_{n\to\infty}\sqrt{n}(\widehat{\boldsymbol{\beta}}_{\mathbf{1}}^{\mathbf{FM}} - \boldsymbol{\beta}_{\mathbf{1}})\} \\ &- E\{\lim_{n\to\infty}\sqrt{n}(\widehat{\boldsymbol{\beta}}_{\mathbf{1}}^{\mathbf{FM}} - \widehat{\boldsymbol{\beta}}_{\mathbf{1}}^{\mathbf{SM}})(p_2-2)L_n^{-1}\} \\ &= -\boldsymbol{\mu}_{11.2} - E\{\lim_{n\to\infty}\sqrt{n}(\widehat{\boldsymbol{\beta}}_{\mathbf{1}}^{\mathbf{FM}} - \widehat{\boldsymbol{\beta}}_{\mathbf{1}}^{\mathbf{SM}})(p_2-2)L_n^{-1}\} \\ &= -\boldsymbol{\mu}_{11.2} - (p_2-2)\boldsymbol{\delta}E(\chi_{p_2+2}^{-2}(\Delta)). \end{aligned}$$

$$\begin{aligned} ADB(\widehat{\boldsymbol{\beta}}_1^{\rm PS}) &= E\{\lim_{n\to\infty}\sqrt{n}(\widehat{\boldsymbol{\beta}}_{\mathbf{1}}^{\mathbf{PS}} - \boldsymbol{\beta}_{\mathbf{1}})\} \\ &= E\left\{\lim_{n\to\infty}\sqrt{n}\left(\widehat{\boldsymbol{\beta}}_{\mathbf{1}}^{\mathbf{SM}} + (\widehat{\boldsymbol{\beta}}_{\mathbf{1}}^{\mathbf{FM}} - \widehat{\boldsymbol{\beta}}_{\mathbf{1}}^{\mathbf{SM}})(1-(p_2-2)L_n^{-1})\right.\right. \\ &\times I(L_n > p_2-2) - \boldsymbol{\beta}_{\mathbf{1}})\} \\ &= E\{\sqrt{n}[\widehat{\boldsymbol{\beta}}_1^{\rm SM} + (\widehat{\beta}_1^{\rm FM} - \widehat{\beta}_1^{\rm SM})(1 - I(L_n \le p_2-2)) \\ &- (\widehat{\beta}_1^{\rm FM} - \widehat{\beta}_1^{\rm SM})(p_2-2)L_n^{-1}I(L_n > p_2-2) - \boldsymbol{\beta}_1]\} \\ &= E\left\{\lim_{n\to\infty}\sqrt{n}(\widehat{\boldsymbol{\beta}}_{\mathbf{1}}^{\mathbf{FM}} - \boldsymbol{\beta}_{\mathbf{1}})\right\} \\ &- E\{\lim_{n\to\infty}\sqrt{n}(\widehat{\boldsymbol{\beta}}_{\mathbf{1}}^{\mathbf{FM}} - \widehat{\boldsymbol{\beta}}_{\mathbf{1}}^{\mathbf{SM}})(p_2-2)I(L_n \le p_2-2)\Big\} \\ &- E\{\lim_{n\to\infty}\sqrt{n}(\widehat{\boldsymbol{\beta}}_{\mathbf{1}}^{\mathbf{FM}} - \widehat{\boldsymbol{\beta}}_{\mathbf{1}}^{\mathbf{SM}})(p_2-2)L_n^{-1}I(L_n > p_2-2) \\ &= -\boldsymbol{\mu}_{11.2} - \boldsymbol{\delta}H_{p_2+2}(\chi_{p_2-2}^2;\Delta)\} \\ &- (p_2-2)\boldsymbol{\delta}E\{\chi_{p_2+2}^{-2}(\Delta)I(\chi_{p_2+2}^{-2} > p_2-2)\}. \end{aligned}$$

By definition

$$\mathrm{Cov}(\widehat{\boldsymbol{\beta}}_1^*) = \lim_{n\to\infty} E\{n(\widehat{\boldsymbol{\beta}}_1^* - \boldsymbol{\beta}_1)(\widehat{\boldsymbol{\beta}}_1^* - \boldsymbol{\beta}_1)^\top\}.$$

Proof of Theorem 7.4

$$\begin{aligned} Cov(\widehat{\beta}_1^{\text{FM}}) &= E\{\lim_{n\to\infty} \sqrt{n}(\widehat{\boldsymbol{\beta}}_\mathbf{1}^{\mathbf{FM}} - \boldsymbol{\beta}_\mathbf{1})\sqrt{n}(\widehat{\boldsymbol{\beta}}_\mathbf{1}^{\mathbf{FM}} - \boldsymbol{\beta}_\mathbf{1})^\top\} \\ &= E(\boldsymbol{\varphi}_1\boldsymbol{\varphi}_1^\top) = Cov(\boldsymbol{\varphi}_1\boldsymbol{\varphi}_1^\top) + E(\boldsymbol{\varphi}_1)E(\boldsymbol{\varphi}_1^\top) \\ &= \boldsymbol{B}_{11.2}^{-1} + \boldsymbol{\mu}_{11.2}\boldsymbol{\mu}_{11.2}^\top. \end{aligned}$$

Similarly,

$$\begin{aligned} Cov(\widehat{\boldsymbol{\beta}}_1^{\text{SM}}) &= E\{\lim_{n\to\infty} \sqrt{n}(\widehat{\boldsymbol{\beta}}_1^{\text{SM}} - \boldsymbol{\beta}_\mathbf{1})\sqrt{n}(\widehat{\boldsymbol{\beta}}_\mathbf{1}^{\mathbf{SM}} - \boldsymbol{\beta}_\mathbf{1})^\top\} \\ &= E(\varphi_2\varphi_2^\top) = Cov(\varphi_2\varphi_2^\top) + E(\varphi_2)E(\varphi_2^\top) \\ &= \boldsymbol{B}_{11}^{-1} + \boldsymbol{\gamma}\boldsymbol{\gamma}^\top. \end{aligned}$$

Now, the asymptotic covariance of $\widehat{\boldsymbol{\beta}}_1^{\text{S}}$ *can be obtained as follows:*

$$\begin{aligned} Cov(\widehat{\boldsymbol{\beta}}_1^{\text{S}}) &= E\{\lim_{n\to\infty} \sqrt{n}(\widehat{\boldsymbol{\beta}}_\mathbf{1}^{\mathbf{S}} - \boldsymbol{\beta}_\mathbf{1})\sqrt{n}(\widehat{\boldsymbol{\beta}}_\mathbf{1}^{\mathbf{S}} - \boldsymbol{\beta}_1)^\top\} \\ &= E\{\lim_{n\to\infty} n\big[(\widehat{\boldsymbol{\beta}}_1^{\text{FM}} - \boldsymbol{\beta}_1) - (\widehat{\boldsymbol{\beta}}_1^{\text{FM}} - \widehat{\boldsymbol{\beta}}_1^{\text{SM}})(p_2-2)L_n^{-1}\big] \\ &\big[(\widehat{\boldsymbol{\beta}}_\mathbf{1}^{\mathbf{FM}} - \boldsymbol{\beta}_1) - (\widehat{\boldsymbol{\beta}}_1^{\text{FM}} - \widehat{\boldsymbol{\beta}}_1^{\text{SM}})(p_2-2)L_n^{-1}\big]^\top\} \\ &= E\Big\{[\boldsymbol{\varphi}_1 - \boldsymbol{\varphi}_3(p_2-2)L_n^{-1}][\boldsymbol{\varphi}_1 - \boldsymbol{\varphi}_3(p_2-2)L_n^{-1}]^\top\Big\} \\ &= E\Big\{\varphi_1\varphi_1^\top - 2(p_2-2)\varphi_3\varphi_1^\top L_n^{-1} + (p_2-2)^2\varphi_3\varphi_3^\top L_n^{-2}\Big\} \end{aligned}$$

We need to compute $E\{\varphi_3\varphi_3^\top L_n^{-2}\}$ *and* $E\{\varphi_3\varphi_1^\top L_n^{-1}\}$*. By using Lemma 3.2, the first term is obtained as follows:*

$$E\{\varphi_3\varphi_3^\top L_n^{-2}\} = \boldsymbol{\Phi} E(\chi_{p_2+2}^{-4}(\Delta)) + \boldsymbol{\delta}\boldsymbol{\delta}^\top E(\chi_{p_2+4}^{-4}(\Delta)).$$

The second term is computed from normal theory

$$\begin{aligned} E\Big\{\varphi_3\varphi_1^\top L_n^{-1}\Big\} &= E\Big\{E(\varphi_3\varphi_1^\top L_n^{-1}|\varphi_3)\Big\} = E\Big\{\varphi_3 E(\varphi_1^\top L_n^{-1}|\varphi_3)\Big\} \\ &= E\{\varphi_3[-\boldsymbol{\mu}_{11.2} + (\varphi_3 - \boldsymbol{\delta})]^\top L_n^{-1}\} \\ &= -E\{\varphi_3\boldsymbol{\mu}_{11.2}L_n^{-1}\} + E\{\varphi_3(\varphi_3 - \boldsymbol{\delta})^\top L_n^{-1}\} \\ &= -\boldsymbol{\mu}_{11.2}^\top E\{\varphi_3 L_n^{-1}\} + E\{\varphi_3\varphi_3^\top L_n^{-1}\} - E\{\varphi_3\boldsymbol{\delta}^\top L_n^{-1}\} \end{aligned}$$

From above, we can find $E\{\varphi_3\boldsymbol{\delta}^\top L_n^{-1}\} = \boldsymbol{\delta}\boldsymbol{\delta}^\top E(\chi_{p_2+2}^{-2}(\Delta))$ *and* $E\{\varphi_3 L_n^{-1}\} = \boldsymbol{\delta} E(\chi_{p_2+2}^{-2}(\Delta))$*. Putting these terms together and simplifying, we obtain*

$$\begin{aligned} Cov(\widehat{\boldsymbol{\beta}}_1^{\text{S}}) &= \boldsymbol{B}_{11.2}^{-1} + \boldsymbol{\mu}_{11.2}\boldsymbol{\mu}_{11.2}^T + 2(p_2-2)\boldsymbol{\mu}_{11.2}^T\boldsymbol{\delta} E\Big(\chi_{p_2+2}^{-2}(\Delta)\Big) \\ &-(p_2-2)\boldsymbol{\Phi}\Big\{2E\Big(\chi_{p_2+2}^{-2}(\Delta)\Big) - (p_2-2)E\Big(\chi_{p_2+2}^{-4}(\Delta)\Big)\Big\} \\ &+(p_2-2)\boldsymbol{\delta}\boldsymbol{\delta}^\top \\ &\times\Big\{-2E(\chi_{p_2+4}^{-2}(\Delta)) + 2E(\chi_{p_2+2}^{-2}(\Delta)) + (p_2-2)E\Big(\chi_{p_2+4}^{-4}(\Delta)\Big)\Big\}. \end{aligned}$$

Since $\widehat{\boldsymbol{\beta}}_1^{\text{PS}} = \widehat{\boldsymbol{\beta}}_1^{\text{S}} - (\widehat{\boldsymbol{\beta}}_1^{\text{FM}} - \widehat{\boldsymbol{\beta}}_1^{\text{SM}})\{1 - (p_2-2)L_n^{-1}\}I(L_n \le p_2 - 2)$.

We derive the covariance of the estimator $\widehat{\boldsymbol{\beta}}_1^{\mathrm{PS}}$ as follows.

$$\begin{aligned}
Cov(\widehat{\boldsymbol{\beta}}_1^{\mathrm{PS}}) &= E\Big\{\lim_{n\to\infty}\sqrt{n}(\widehat{\boldsymbol{\beta}}_1^{\mathrm{PS}}-\boldsymbol{\beta}_1)\sqrt{n}(\widehat{\boldsymbol{\beta}}_1^{\mathrm{PS}}-\boldsymbol{\beta}_1)^\top\Big\} \\
&= E\Bigg\{\lim_{n\to\infty}\sqrt{n}(\widehat{\boldsymbol{\beta}}_1^{\mathrm{S}}-\boldsymbol{\beta}_1)-\sqrt{n}(\widehat{\boldsymbol{\beta}}_1^{\mathrm{FM}}-\widehat{\boldsymbol{\beta}}_1^{\mathrm{SM}})\{1-(p_2-2)L_n^{-1}\} \\
&\quad \times I(L_n\le p_2-2)\Big[\sqrt{n}(\widehat{\boldsymbol{\beta}}_1^{\mathrm{S}}-\boldsymbol{\beta}_1)-\sqrt{n}(\widehat{\boldsymbol{\beta}}_1^{\mathrm{FM}}-\widehat{\boldsymbol{\beta}}_1^{\mathrm{SM}}) \\
&\quad \times\{1-(p_2-2)L_n^{-1}\}I(L_n\le p_2-2)\Big]^\top\Bigg\} \\
&= E\Big\{\lim_{n\to\infty}\sqrt{n}(\widehat{\boldsymbol{\beta}}_1^{\mathrm{S}}-\boldsymbol{\beta}_1)\sqrt{n}(\widehat{\boldsymbol{\beta}}_1^{\mathrm{S}}-\boldsymbol{\beta}_1)^\top-2\boldsymbol{\varphi}_3\sqrt{n}(\widehat{\boldsymbol{\beta}}_1^{\mathrm{S}}-\boldsymbol{\beta}_1)^\top \\
&\quad \times\{1-(p_2-2)L_n^{-1}\}I(L_n\le p_2-2) \\
&\quad +\boldsymbol{\varphi}_3\boldsymbol{\varphi}_3^\top\{1-(p_2-2)L_n^{-1}\}^2 I(L_n\le p_2-2)\Big\} \\
&= Cov(\widehat{\boldsymbol{\beta}}_1^{\mathrm{S}})-2E\Big\{\lim_{n\to\infty}\boldsymbol{\varphi}_3\sqrt{n}(\widehat{\boldsymbol{\beta}}_1^{\mathrm{S}}-\boldsymbol{\beta}_1)^\top \\
&\quad \times\{1-(p_2-2)L_n^{-1}\}^2 I(L_n\le p_2-2)\Big\} \\
&\quad +E\Big\{\lim_{n\to\infty}\boldsymbol{\varphi}_3\boldsymbol{\varphi}_3^\top\{1-(p_2-2)L_n^{-1}\}^2 I(L_n\le p_2-2)\Big\} \\
&= Cov(\widehat{\boldsymbol{\beta}}_1^{\mathrm{S}})-2E\Big\{\lim_{n\to\infty}\boldsymbol{\varphi}_3\boldsymbol{\varphi}_1^\top\{1-(p_2-2)L_n^{-1}\}I(L_n\le p_2-2)\Big\} \\
&\quad +2E\Big\{\lim_{n\to\infty}\boldsymbol{\varphi}_3\boldsymbol{\varphi}_3^\top(p_2-2)L_n^{-1}\{1-(p_2-2)L_n^{-1}\} \\
&\quad \times I(L_n\le p_2-2)\} \\
&\quad +E\Big\{\lim_{n\to\infty}\boldsymbol{\varphi}_3\boldsymbol{\varphi}_3^\top\{1-(p_2-2)L_n^{-1}\}^2 I(L_n\le p_2-2)\Big\} \\
&= Cov(\widehat{\boldsymbol{\beta}}_1^{\mathrm{S}})-2E\Big\{\lim_{n\to\infty}\boldsymbol{\varphi}_3\boldsymbol{\varphi}_1^\top\{1-(p_2-2)L_n^{-1}\}I(L_n\le p_2-2)\Big\} \\
&\quad -E\Big\{\lim_{n\to\infty}\boldsymbol{\varphi}_3\boldsymbol{\varphi}_3^\top(p_2-2)^2L_n^{-2}I(L_n\le p_2-2)\Big\} \\
&\quad +E\Big\{\lim_{n\to\infty}\boldsymbol{\varphi}_3\boldsymbol{\varphi}_3^\top I(L_n\le p_2-2)\Big\}
\end{aligned}$$

We first compute the last term in the equation above $E\Big\{\boldsymbol{\varphi}_3\boldsymbol{\varphi}_3^\top I(L_n \le p_2-2)\Big\}$ *as* $E\Big\{\boldsymbol{\varphi}_3\boldsymbol{\varphi}_3^\top I(L_n\le p_2-2)\Big\}=\boldsymbol{\Phi}\boldsymbol{H}_{p_2+2}(p_2-2;\Delta)+\boldsymbol{\delta}\boldsymbol{\delta}^\top\boldsymbol{H}_{p_2+4}(p_2-2;\Delta)$. *Using Lemma 3.2 and from the normal theory, we find,*

$$\begin{aligned}
& E\Big\{\boldsymbol{\varphi}_3\boldsymbol{\varphi}_1^\top\{1-(p_2-2)L_n^{-1}\}I(L_n\le p_2-2)\Big\} \\
&= E\Big\{E(\boldsymbol{\varphi}_3\boldsymbol{\varphi}_1^\top\{1-(p_2-2)L_n^{-1}\}I(L_n\le p_2-2)|\boldsymbol{\varphi}_3)\Big\} \\
&= E\Big\{\boldsymbol{\varphi}_3E(\boldsymbol{\varphi}_1^\top\{1-(p_2-2)L_n^{-1}\}I(L_n\le p_2-2)|\boldsymbol{\varphi}_3)\Big\} \\
&= E\Big\{\boldsymbol{\varphi}_3[\boldsymbol{\mu}_{11.2}+(\boldsymbol{\varphi}_3-\boldsymbol{\delta})]^\top\{1-(p_2-2)L_n^{-1}\}I(L_n\le p_2-2)\Big\} \\
&= -\boldsymbol{\mu}_{11.2}E\Big(\boldsymbol{\varphi}_3\{1-(p_2-2)L_n^{-1}\}I\big(L_n\le p_2-2\big)\Big) \\
&\quad +E\Big(\boldsymbol{\varphi}_3\boldsymbol{\varphi}_3^\top\{1-(p_2-2)L_n^{-1}\}I\big(L_n\le p_2-2\big)\Big) \\
&\quad -E\Big(\boldsymbol{\varphi}_3\boldsymbol{\delta}^\top\{1-(p_2-2)L_n^{-1}\}I\big(L_n\le p_2-2\big)\Big)
\end{aligned}$$

$$= -\boldsymbol{\delta}\boldsymbol{\mu}_{11.2}^{\top} E\Big(\Big\{1 - (p_2-2)\chi_{p_2+2}^{-2}(\Delta)\Big\} I\Big(\chi_{p_2+2}^{-2}(\Delta) \le p_2 - 2\Big)\Big)$$
$$+\boldsymbol{\Phi} E\Big(\Big\{1 - (p_2-2)\chi_{p_2+2}^{-2}(\Delta)\Big\} I\Big(\chi_{p_2+2}^{-2}(\Delta) \le p_2 - 2\Big)\Big)$$
$$+\boldsymbol{\delta}\boldsymbol{\delta}^{\top} E\Big(\Big\{1 - (p_2-2)\chi_{p_2+4}^{-2}(\Delta)\Big\} I\Big(\chi_{p_2+4}^{-2}(\Delta) \le p_2 - 2\Big)\Big)$$
$$-\boldsymbol{\delta}\boldsymbol{\delta}^{\top} E\Big(\Big\{1 - (p_2-2)\chi_{p_2+4}^{-2}(\Delta)\Big\} I\Big(\chi_{p_2+4}^{-2}(\Delta) \le p_2 - 2\Big)\Big).$$

$$E\Big\{\boldsymbol{\varphi}_3\boldsymbol{\varphi}_3^{\top}(p_2-2)^2 L_n^{-2} I(L_n \le p_2 - 2)\Big\}$$
$$= (p_2-2)^2\boldsymbol{\Phi} E\Big(\chi_{p_2+2}^{-4}(\Delta) I\Big(\chi_{p_2+2}^{2}(\Delta) \le p_2 - 2\Big)\Big)$$
$$+(p_2-2)^2\boldsymbol{\delta}\boldsymbol{\delta}^{\top} E\Big(\chi_{p_2+2}^{-4}(\Delta) I\Big(\chi_{p_2+2}^{2}(\Delta) \le p_2 - 2\Big)\Big)$$

Putting all the terms together, we obtain

$$\begin{aligned}
Cov(\widehat{\boldsymbol{\beta}}_1^{\mathrm{PS}}) \;=\;& Cov(\widehat{\boldsymbol{\beta}}_1^{\mathrm{S}}) + 2\boldsymbol{\delta}\boldsymbol{\mu}_{11.2}^{\top} \\
& \times E\Big(\Big\{1 - (p_2-2)\chi_{p_2+2}^{-2}(\Delta)\Big\} I\Big(\chi_{p_2+2}^{2}(\Delta) \le p_2 - 2\Big)\Big) \\
& -2\boldsymbol{\Phi} E\Big(\Big\{1 - (p_2-2)\chi_{p_2+2}^{-2}(\Delta)\Big\} I\Big(\chi_{p_2+2}^{2}(\Delta) \le p_2 - 2\Big)\Big) \\
& -2\boldsymbol{\delta}\boldsymbol{\delta}^{\top} E(\{1 - (p_2-2)\chi_{p_2+4}^{-2}(\Delta)\} I(\chi_{p_2+4}^{2}(\Delta) \le p_2 - 2)) \\
& +2\boldsymbol{\delta}\boldsymbol{\delta}^{\top} E\Big(\Big\{1 - (p_2-2)\chi_{p_2+2}^{-2}(\Delta)\Big\} I\Big(\chi_{p_2+2}^{2}(\Delta) \le p_2 - 2\Big)\Big) \\
& -(p_2-2)^2\boldsymbol{\Phi} E\Big(\chi_{p_2+2}^{-4}(\Delta) I\Big(\chi_{p_2+2,\alpha}^{2}(\Delta) \le p_2 - 2\Big)\Big) \\
& -(p_2-2)^2\boldsymbol{\delta}\boldsymbol{\delta}^{\top} E\Big(\chi_{p_2+2}^{-4}(\Delta) I\Big(\chi_{p_2+2}^{2}(\Delta) \le p_2 - 2\Big)\Big) \\
& +\boldsymbol{\Phi}\boldsymbol{H}_{p_2+2}(p_2-2;\Delta) + \boldsymbol{\delta}\boldsymbol{\delta}^{\top}\boldsymbol{H}_{p_2+4}(p_2-2;\Delta).
\end{aligned}$$

8

Shrinkage Estimation in Sparse Nonlinear Regression Models

8.1 Introduction

In this chapter we consider the estimation and prediction problem in nonlinear regression models when the model may or may not be sparse.

To formulate the nonlinear regression model, let us consider $\boldsymbol{y} = (y_1, y_2, ..., y_n)^\top$ a $n \times 1$ vector of the response variable and $\boldsymbol{z} = (\boldsymbol{z}_1, \boldsymbol{z}_2, ..., \boldsymbol{z}_n)^\top$ a $n \times p$ data matrix, where $\boldsymbol{z}_i = (z_{i1}, z_{i2}, ..., z_{ip})$ for $i = 1, 2, ..., n$. The nonlinear regression model is given by

$$y_i = f(\boldsymbol{z}_i, \boldsymbol{\theta}) + u_i, \tag{8.1}$$

where $f(\boldsymbol{z}_i, \boldsymbol{\theta})$ is the mean of y_i and nonlinear, $\boldsymbol{\theta} = (\theta_1, \theta_2, ..., \theta_k)^\top$ is a $k \times 1$ vector of the parameter to be estimated, and $\boldsymbol{u} = (u_1, u_2, ..., u_n)^\top$ is a $n \times 1$ vector of random error assumed to be independent and identically distributed with mean zero and variance σ^2.

The least squares method can be applied to estimate the regression parameter θ_j, $j = 1, 2, ..., k$. By definition, the sum of squares error function is given as:

$$S(\boldsymbol{\theta}) = \sum_{i=1}^{n} [y_i - f(\boldsymbol{z}_i, \boldsymbol{\theta})]^2.$$

The nonlinear least squares estimate, $\widehat{\boldsymbol{\theta}}$, is obtained by solving the following normal equation as is done in the linear regression case:

$$\frac{\partial S(\boldsymbol{\theta})}{\partial \theta_j} = \sum_{i=1}^{n} [y_i - f(\boldsymbol{z}_i, \boldsymbol{\theta})] \left[\frac{\partial f(\boldsymbol{z}_i, \boldsymbol{\theta})}{\partial \theta_j}\right]_{\boldsymbol{\theta}=\widehat{\boldsymbol{\theta}}} = 0.$$

Clearly, it is not possible to find a closed form solution. In such situations, iterative methods are applied to minimize $S(\boldsymbol{\theta})$. The commonly used procedure is the Gauss-Newton or linearization method. More detailed information about the nonlinear regression model and the Gauss-Newton procedure can be found in Myers et al. (2010) and in many statistics books.

The objective of this chapter to develop estimation and prediction strategies when the model's sparsity assumption may or may not hold. We consider both low- and high-dimensional regimes using a nonlinear regression model. We consider full model, submodel, shrinkage, and penalty estimations. The mean squared error criterion is used to assess the characteristic of the estimators. For low-dimensional cases we provide some asymptotic properties of non-penalty estimators. We also conduct a simulation study to provide the relative performance of penalty and non-penalty estimators.

The remainder of the chapter is structured as follows. Section 8.2 contains the full, submodel, and shrinkage estimators. Section 8.3 describes the estimators' asymtotic properties.

DOI: 10.1201/9781003170259-8

Section 8.4 presents the results of a Monte Carlo simulation experiment. A real data example is available in Section 8.5 for demonstration purposes. The R codes can be found in Section 8.6. Section 8.7 brings the chapter to a close.

8.2 Model and Estimation Strategies

We are interested in the estimation of the regression parameters when the model is sparse, where there are only a handful number of active predictors in the models and the rest are not useful for estimation and prediction purposes. Suppose the regression parameter vector $\boldsymbol{\theta}$ can be partitioned such that $\boldsymbol{\theta} = (\boldsymbol{\theta}_1^\top, \boldsymbol{\theta}_2^\top)^\top$, the dimensions of $\boldsymbol{\theta}_1$ and $\boldsymbol{\theta}_2$ are $k_1 \times 1$ and $k_2 \times 1$, respectively, and $k = k_1 + k_2$. We set the data matrix $\boldsymbol{P} = (\boldsymbol{P}_1, \boldsymbol{P}_2)$ to have dimensions $n \times k$, where $\boldsymbol{P}_1$ is $n \times k_1$ dimensional and $\boldsymbol{P}_2$ is $n \times k_2$ dimensional. The product matrix decomposition is as follows:

$$\boldsymbol{Q} = \begin{bmatrix} \boldsymbol{Q}_{11} & \boldsymbol{Q}_{12} \\ \boldsymbol{Q}_{21} & \boldsymbol{Q}_{22} \end{bmatrix} = \begin{bmatrix} \boldsymbol{P}_1^\top \boldsymbol{P}_1 & \boldsymbol{P}_1^\top \boldsymbol{P}_2 \\ \boldsymbol{P}_2^\top \boldsymbol{P}_1 & \boldsymbol{P}_2^\top \boldsymbol{P}_2 \end{bmatrix} = \boldsymbol{P}^\top \boldsymbol{P},$$

where $\boldsymbol{Q}$ is a $k \times k$ matrix. Assume $(1/n)\boldsymbol{Q} \to \boldsymbol{G}$ as $n \to \infty$, where $\boldsymbol{G}$ is a $k \times k$ positive-definite matrix decomposed as

$$\boldsymbol{G} = \begin{bmatrix} \boldsymbol{G}_{11} & \boldsymbol{G}_{12} \\ \boldsymbol{G}_{21} & \boldsymbol{G}_{22} \end{bmatrix}, \text{ and } \boldsymbol{G}_{st} = \lim_{n\to\infty} \frac{1}{n}\boldsymbol{Q}_{st},\ s,t = 1,2.$$

The estimator including all predictors $\widehat{\boldsymbol{\theta}}^{\mathrm{FM}}$ of $\boldsymbol{\theta}$ is obtained by solving the Gauss-Newton iterative method in the final iteration of the nonlinear least squares estimator:

$$\widehat{\boldsymbol{\theta}}^{\mathrm{FM}} = \left(\boldsymbol{P}^\top \boldsymbol{P}\right)^{-1} \boldsymbol{P}^\top \boldsymbol{F},$$

where $\boldsymbol{F} = \boldsymbol{P}\widehat{\boldsymbol{\theta}} + \boldsymbol{d}$ and $\boldsymbol{d} = \boldsymbol{y} - f(\boldsymbol{z}, \widehat{\boldsymbol{\theta}})$.

Our main interest is in estimating $\boldsymbol{\theta}_1$, the coefficients of active predictors. In other words, we are interested in estimating the parameter vector $\boldsymbol{\theta}_1$ when it is plausible that $\boldsymbol{\theta}_2$ is close to zero. Thus, the full model estimator of $\boldsymbol{\theta}_1$ is obtained as:

$$\widehat{\boldsymbol{\theta}}_1^{\mathrm{FM}} = \left(\boldsymbol{P}_1^\top \boldsymbol{P}_1\right)^{-1} \boldsymbol{P}_1^\top \left[\boldsymbol{F} - \boldsymbol{P}_2 \widehat{\boldsymbol{\theta}}_2^{\mathrm{FM}}\right].$$

Under the sparsity condition $\boldsymbol{\theta}_2 = \boldsymbol{0}$ we apply the Lagrange multiplier to obtain the submodel estimator of $\boldsymbol{\theta}_1$ which is approximated as

$$\widehat{\boldsymbol{\theta}}_1^{\mathrm{SM}} = \widehat{\boldsymbol{\theta}}_1^{\mathrm{FM}} - \gamma_n \widehat{\boldsymbol{\theta}}_2^{\mathrm{FM}},$$

where $\gamma_n = -\boldsymbol{Q}_{11}^{-1}\boldsymbol{Q}_{12}$. We assume $\gamma_n \xrightarrow{\mathrm{P}} \gamma$ as $n \to \infty$, where $\xrightarrow{\mathrm{P}}$ indicates convergence in probability. In the simulation experiment, the submodel estimator which has $\boldsymbol{\theta}_2 = \boldsymbol{0}$ as a constraint is obtained using the Gauss-Newton iterative method.

8.2.1 Shrinkage Strategy

Let us define a distance measure as follows:

$$W_n = s^{-2}\left[\left(\widehat{\boldsymbol{\theta}}_2^{\mathrm{FM}}\right)^\top \boldsymbol{Q}_{22.1} \widehat{\boldsymbol{\theta}}_2^{\mathrm{FM}}\right], \tag{8.2}$$

where $s^2 = [y - f(\boldsymbol{z}, \widehat{\boldsymbol{\theta}})]^\top [y - f(\boldsymbol{z}, \widehat{\boldsymbol{\theta}})]/(n-k)$ and $\boldsymbol{Q}_{22.1} = \boldsymbol{Q}_{22} - \boldsymbol{Q}_{21}\boldsymbol{Q}_{11}^{-1}\boldsymbol{Q}_{12}$. The positive part shrinkage (PS) estimator is a function of the James-Stein or shrinkage (S) estimator, the general form is given by

$$\widehat{\boldsymbol{\theta}}_1^{\mathrm{PS}} = \widehat{\boldsymbol{\theta}}_1^{\mathrm{S}} - \left[1 - \frac{r \cdot g(W_n)}{W_n}\right] \left(\widehat{\boldsymbol{\theta}}_1^{\mathrm{FM}} - \widehat{\boldsymbol{\theta}}_1^{\mathrm{SM}}\right) I(W_n \le r),$$

where $r = k_2 - 2$ is a shrinkage constant, $k_2 \ge 3$, $g(W_n)$ is a continuous, bounded, and differentiable function of W_n Yüzbaşı et al. (2017b), and $I(\cdot)$ is an indicator function which is one if $W_n \le r$, and zero otherwise. Here, the general form of the shrinkage estimator is

$$\widehat{\boldsymbol{\theta}}_1^{\mathrm{S}} = \widehat{\boldsymbol{\theta}}_1^{\mathrm{SM}} + \left[1 - \frac{r \cdot g(W_n)}{W_n}\right] \left(\widehat{\boldsymbol{\theta}}_1^{\mathrm{FM}} - \widehat{\boldsymbol{\theta}}_1^{\mathrm{SM}}\right).$$

If $r \cdot g(W_n)/W_n > 1$ the sign of the coefficients will reverse. This is an indication of over-shrinkage and the positive-part shrinkage estimator has been used to moderate this effect.

For $g(W_n) = 1$ the widely-used positive-part shrinkage estimator is:

$$\widehat{\boldsymbol{\theta}}_1^{\mathrm{PS1}} = \widehat{\boldsymbol{\theta}}_1^{\mathrm{S1}} - (1 - rW_n^{-1}) \left(\widehat{\boldsymbol{\theta}}_1^{\mathrm{FM}} - \widehat{\boldsymbol{\theta}}_1^{\mathrm{SM}}\right) I(W_n \le r),$$

where $\widehat{\boldsymbol{\theta}}_1^{\mathrm{S1}} = \widehat{\boldsymbol{\theta}}_1^{\mathrm{SM}} + (1 - rW_n^{-1}) \left(\widehat{\boldsymbol{\theta}}_1^{\mathrm{FM}} - \widehat{\boldsymbol{\theta}}_1^{\mathrm{SM}}\right)$. Following Yüzbaşı et al. (2017b) and Reangsephet et al. (2020), we let $g(W_n) = 1/(1 + W_n^{-1})$ and obtain

$$\widehat{\boldsymbol{\theta}}_1^{\mathrm{PS2}} = \widehat{\boldsymbol{\theta}}_1^{\mathrm{S2}} - \left(1 - \frac{r}{W_n + 1}\right) \left(\widehat{\boldsymbol{\theta}}_1^{\mathrm{FM}} - \widehat{\boldsymbol{\theta}}_1^{\mathrm{SM}}\right) I(W_n \le r).$$

Here, $\widehat{\boldsymbol{\theta}}_1^{\mathrm{S2}} = \widehat{\boldsymbol{\theta}}_1^{\mathrm{SM}} + \{1 - [r/(W_n + 1)]\} \left(\widehat{\boldsymbol{\theta}}_1^{\mathrm{FM}} - \widehat{\boldsymbol{\theta}}_1^{\mathrm{SM}}\right)$. Lastly, we propose $g(W_n) = \arctan(W_n)$, similar to the results of Yüzbaşı et al. (2017b) and Reangsephet et al. (2020), yields the following formula

$$\widehat{\boldsymbol{\theta}}_1^{\mathrm{PS3}} = \widehat{\boldsymbol{\theta}}_1^{\mathrm{S2}} - \left\{1 - \frac{r \cdot \arctan(W_n)}{W_n}\right\} \left(\widehat{\boldsymbol{\theta}}_1^{\mathrm{FM}} - \widehat{\boldsymbol{\theta}}_1^{\mathrm{SM}}\right) I(W_n \le r).$$

Note that $\widehat{\boldsymbol{\theta}}_1^{\mathrm{PS3}} = \widehat{\boldsymbol{\theta}}_1^{\mathrm{SM}} + \{1 - [r \cdot \arctan(W_n)/W_n]\} \left(\widehat{\boldsymbol{\theta}}_1^{\mathrm{FM}} - \widehat{\boldsymbol{\theta}}_1^{\mathrm{SM}}\right)$.

where $\boldsymbol{\Gamma}^*(\widehat{\boldsymbol{\theta}}_1^*) = \lim_{n \to \infty} E\left[\sqrt{n}(\widehat{\boldsymbol{\theta}}_1^* - \boldsymbol{\theta}_1)\sqrt{n}(\widehat{\boldsymbol{\theta}}_1^* - \boldsymbol{\theta}_1)^\top\right]$ is the asymptotic covariance matrix of the distribution $\widehat{\boldsymbol{\theta}}_1^*$ and $\boldsymbol{M}$ is a positive semi-definite weighting matrix, see Ahmed (2014) for more information.

8.3 Asymptotic Analysis

We assess the asymptotic performance of the estimators in terms of their respective bias and risk. We define the quadratic asymptotic distributional bias (QADB) of an estimator $\widehat{\boldsymbol{\theta}}_1^*$ of $\widehat{\boldsymbol{\theta}}$ as:

$$\mathrm{QADB}(\widehat{\boldsymbol{\theta}}_1^*) = \left[\mathrm{ADB}(\widehat{\boldsymbol{\theta}}_1^*)\right]^\top \sigma^{-2}\boldsymbol{G}_{11.2} \left[\mathrm{ADB}(\widehat{\boldsymbol{\theta}}_1^*)\right], \tag{8.3}$$

where $\sigma^{-2}\boldsymbol{G}_{11.2} = \sigma^{-2}(\boldsymbol{G}_{11} - \boldsymbol{G}_{12}\boldsymbol{G}_{22}^{-1}\boldsymbol{G}_{21})$ and $\text{ADB}(\widehat{\boldsymbol{\theta}}_1^*) = \lim_{n\to\infty} E\left[\sqrt{n}(\widehat{\boldsymbol{\theta}}_1^* - \boldsymbol{\theta}_1)\right]$ is the asymptotic distributional bias of $\widehat{\boldsymbol{\theta}}_1^*$.

The asymptotic distributional risk (ADR) of an estimator $\widehat{\boldsymbol{\theta}}_1^*$ of of $\widehat{\boldsymbol{\theta}}$ is defined as:

$$\text{ADR}(\widehat{\boldsymbol{\theta}}_1^*; \boldsymbol{M}) = \text{tr}(\boldsymbol{M}\boldsymbol{\Gamma}^*), \tag{8.4}$$

To examine the asymptotic properties of the estimators we consider a sequence $\{K_n\}$ defined as $\{K_n\} : \boldsymbol{\theta}_2 = \boldsymbol{\delta}/\sqrt{n}$, where $\boldsymbol{\delta} = (\delta_1, \delta_2, ..., \delta_{k_2})^\top \in \mathbb{R}^{k_2}$ is a $k_2 \times 1$ fixed vector.

In the following theorem we provide the results for quadratic bias.

Theorem 8.1 *Using the definition of the QADB under the sequence $\{K_n\}$ and usual regularity conditions as $n \to \infty$, the QADBs of the estimators are given:*

1. $QADB(\widehat{\boldsymbol{\theta}}_1^{\text{FM}}) = 0$,
2. $QADB(\widehat{\boldsymbol{\theta}}_1^{\text{SM}}) = \Delta^*$
3. $QADB(\widehat{\boldsymbol{\theta}}_1^{\text{PS}}) = \Delta^* \left\{ rE\left[\frac{g(W_1)}{W_1}\right] + E\left[\left(1 - \frac{r \cdot g(W_1)}{W_1}\right) I\left(\frac{W_1}{g(W_1)} \leq r\right)\right]\right\}^2$

where $r = k_2 - 2$, $k_2 \geq 3$, $\Delta^* = \sigma^{-2}\boldsymbol{\delta}^\top\boldsymbol{G}^*\boldsymbol{\delta}$, $\boldsymbol{G}^* = \boldsymbol{G}_{21}\boldsymbol{G}_{11}^{-1}\boldsymbol{G}_{11.2}\boldsymbol{G}_{11}^{-1}\boldsymbol{G}_{12}$, $W_1 = \chi^2_{k_2+2}(\Delta)$, and $g(W_1) = 1,\ 1/(1 + W_1^{-1})$, or $\arctan(W_1)$.

Proof *See Appendix for the proof.*

The expressions for the asymptotic risk of the estimators are given in the following theorem.

Theorem 8.2 *Under the assumed regularity condition and local alternative $\{K_n\}$ as $n \to \infty$, the ADRs of the estimators are given as follows:*

$$\begin{aligned}
ADR(\widehat{\boldsymbol{\theta}}_1^{\text{FM}}; \boldsymbol{M}) &= \sigma^2 tr(\boldsymbol{M}\boldsymbol{G}_{11.2}^{-1}) \\
ADR(\widehat{\boldsymbol{\theta}}_1^{\text{SM}}; \boldsymbol{M}) &= ADR(\widehat{\boldsymbol{\theta}}_1^{\text{FM}}; \boldsymbol{M}) - \sigma^2 tr(\boldsymbol{G}^\circ\boldsymbol{G}_{22.1}^{-1}) + \boldsymbol{\delta}^\top\mathbf{G}^\circ\boldsymbol{\delta} \\
ADR(\widehat{\boldsymbol{\theta}}_1^{\text{PS}}; \boldsymbol{M}) &= ADR(\widehat{\boldsymbol{\beta}}_1^{\text{S}}; \boldsymbol{M}) - \sigma^2 tr(\boldsymbol{G}^\circ\boldsymbol{G}_{22.1}^{-1}) \\
&\times\ E\left[\left(1 - \frac{r \cdot g(W_1)}{W_1}\right)^2 I\left(\frac{W_1}{g(W_1)} \leq r\right)\right] \\
&-\ \boldsymbol{\delta}^\top\boldsymbol{G}^\circ\boldsymbol{\delta}E\left[\left(1 - \frac{r \cdot g(W_2)}{g(W_2)}\right)^2 I\left(\frac{W_2}{g(W_2)} \leq r\right)\right] \\
&+\ 2\boldsymbol{\delta}^\top\boldsymbol{G}^\circ\boldsymbol{\delta}E\left[\left(1 - \frac{r \cdot g(W_1)}{W_1}\right) I\left(\frac{W_1}{g(W_1)} \leq r\right)\right]
\end{aligned}$$

where $\boldsymbol{G}^\circ = \boldsymbol{G}_{21}\boldsymbol{G}_{11}^{-1}\boldsymbol{M}\boldsymbol{G}_{11}^{-1}\boldsymbol{G}_{12}$, $W_1 = \chi^2_{k_2+2}(\Delta)$, $W_2 = \chi^2_{k_2+4}(\Delta)$, $g(W_1) = 1$, $1/(1 + W_1^{-1})$, or $\arctan(W_1)$, and $g(W_2) = 1,\ 1/(1 + W_2^{-1})$, or $\arctan(W_2)$.

Proof *See Appendix for the proof.*

The asymptotic bias and risk properties of the estimators remain the same as described in earlier chapters. They retain their characteristics for the nonlinear regression model as well. As expected, the bias and risk of a full model estimator is independent of the sparsity parameter. The submodel estimator bias and risk expressions are a function of the sparsity parameter and departure from the sparsity assumption has a serious impact on the efficiency of the estimator. Overall, the shrinkage strategy is an advantageous choice over both full and submodel estimators. The shrinkage estimators dominate the full model estimator and perform better in most of the parameter space induced by the sparsity parameter.

To illustrate the performance of the estimators numerically, we present a simulation study that compares the relative estimator performance. We also include two penalized estimation methods in this study.

8.4 Simulation Experiments

In this study, we generate data from a mono-molecular model, a type of nonlinear model. The mono-molecular model originates from research in physical chemistry. The model represents mono-molecular chemical reactions of the first order and has been used to explain several phenomena, such as cell expansion, crop response to nutrients, and animal growth. In plant growth and nutrient supply, the mono-molecular model is known as the Mitscherlich model and has a long history of applications in agricultural sciences and applied biology. It is an expression of the Law of Diminishing Increments, as it was originally applied to study the effect of fertilization on crop yields. The model has been applied by Khamis et al. (2005), Uckardes (2013), Chukwu and Oyamakin (2015), and Powell et al. (2020), amongst others.

The relationship between crop yield and the amount of fertilizer is expressed through the generating equation

$$\frac{dy}{dx} = K(\alpha - y), \tag{8.5}$$

where y is the yield rate, x is the amount of fertilizer, K is the proportionality constant that Mitscherlich called the effect-factor, and α is a parameter representing a maximum. The integrated form of (8.5) is

$$y = \alpha(1 - be^{-\kappa x}),$$

in which b is a constant of integration Panik (2014). For multiple regression, we apply the mono-molecular model and postulate the model as

$$y_i = \theta_1 \left(1 - \theta_2 e^{-\theta_3 z_{i1} - \theta_4 z_{i2} - \cdots - \theta_k z_{i,p}}\right) + u_i, \tag{8.6}$$

where the assumed errors u_i are uncorrelated with constant variance. We generate the response values from $z_{ij} \sim \mathrm{N}(0, 1)$ and $u_i \sim \mathrm{N}(0, 1)$ for all $i = 1, 2, ..., n$ and $j = 1, 2, ..., p$.

In this simulation study, we are interested in estimating the regression parameter vector when the model is sparse. Specifically, we assume $\boldsymbol{\theta}_2 = \mathbf{0}$ and we aim to estimate $\boldsymbol{\theta}_1$ under this sparsity condition.

To begin the simulation process, we partition regression coefficient as follows:

$$\boldsymbol{\theta} = \left(\boldsymbol{\theta}_1^\top, \boldsymbol{\theta}_2^\top\right)^\top = \left(\boldsymbol{\theta}_1^\top, \mathbf{0}^\top\right)^\top.$$

The deviation between the simulation model and the submodel is defined as $\Delta^{\text{sim}} = \|\boldsymbol{\theta} - \boldsymbol{\theta}^{(0)}\|$, where $\boldsymbol{\theta}$ is the simulated parameter, $\boldsymbol{\theta}^{(0)} = \left(\boldsymbol{\theta}_1^\top, \mathbf{0}^\top\right)^\top$, and $\| \cdot \|$ is the Euclidean norm. This setup allows us to investigate the behavior of the estimators when sparsity is violated, i.e. $\Delta^{\text{sim}} > 0$.

For penalty estimators, tuning parameters of LASSO and ALASSO methods are selected by using BIC criteria. Clearly, when the sparsity condition is true then the submodel at hand is the true model, in this case $\Delta^{\text{sim}} = 0$. We generate the data with values of $\boldsymbol{\theta}_1$ of 2.5, 1.5, 1.5, and 0.5 and the response values were generated from (8.6). We set the number

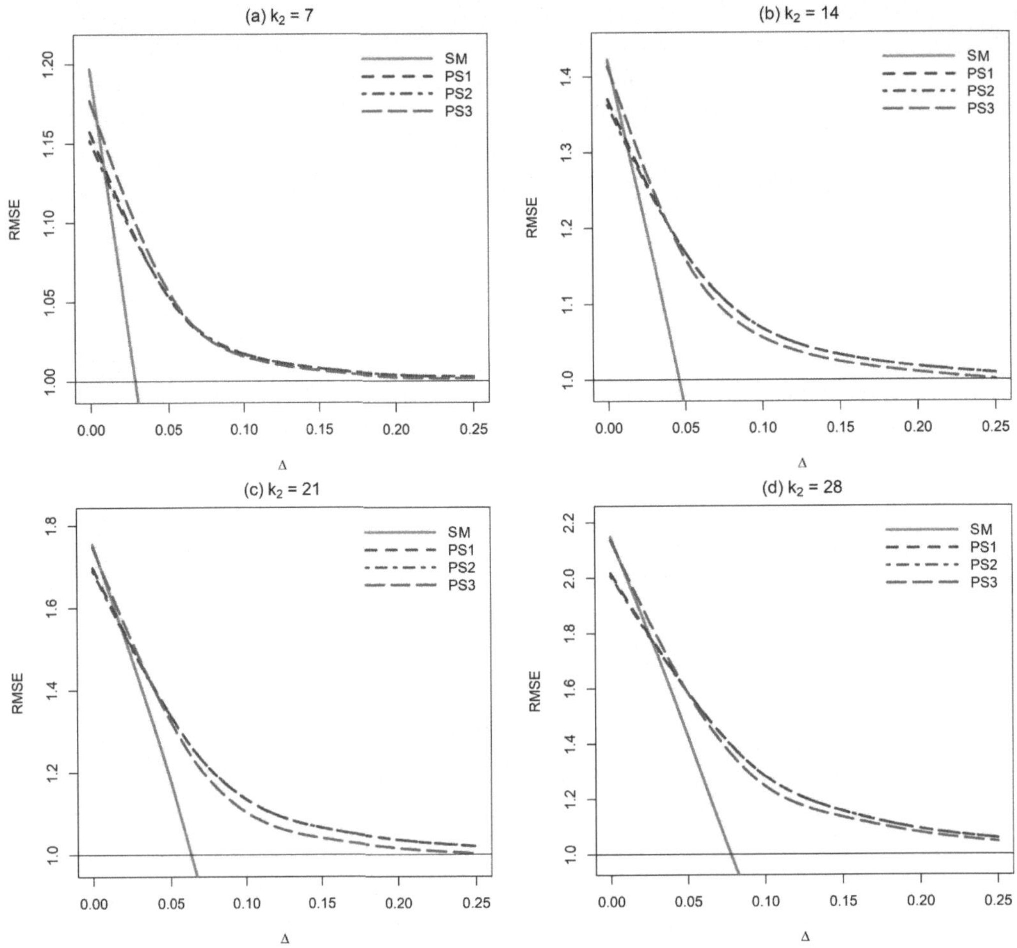

FIGURE 8.1: RMSEs of Estimators for $k_1 = 5$.

of trials $N = 1,000$ to obtain stable results and the MSE of the respective estimators are calculated.

We use the notion of the relative mean squared error (RMSE) of $\widehat{\boldsymbol{\theta}}_1^{\text{FM}}$ of $\widehat{\boldsymbol{\theta}}_1^{*}$ to asses the relative performance of the estimator. By definition,

$$\text{RMSE}(\widehat{\boldsymbol{\theta}}_1^{\text{FM}}, \widehat{\boldsymbol{\theta}}_1^{*}) = \frac{\text{MSE}(\widehat{\boldsymbol{\theta}}_1^{\text{FM}})}{\text{MSE}(\widehat{\boldsymbol{\theta}}_1^{*})}.$$

Clearly, a RMSE larger than one implies that $\widehat{\boldsymbol{\theta}}_1^{*}$ dominates $\widehat{\boldsymbol{\theta}}_1^{\text{FM}}$, the full model estimator.

We first consider the cases where the selected submodel may or may not be true when $\Delta^{\text{sim}} \geq 0$. The vector of active parameters $(\boldsymbol{\theta}_1^{\top})$ is set to $(2, 1, 1, -0.75, -0.75)$. Consider a simple experiment where the simulation model has $\boldsymbol{\theta}_2^{\top} = (\theta_6, \mathbf{0})$, where θ_6 is a scalar and assumes several values. For the value of $\Delta^{\text{sim}} = \theta_6$ we set Δ^{sim} between 0 and 0.25. Here, $k_2 - 1$ is the number of inactive parameters and $k_2 = 7$, 14, 21, and 28 are used in simulating. The sample size is $n = 100$ and the number of iterations is $N = 1000$ times. The curve of estimator RMSEs are displayed in Figure 8.1 for each k_2.

The curves shows the respective pattern for each of estimators and the pattern analysis are summarized as follows. The submodel estimator performs better than the shrinkage estimators when $\Delta^{\text{sim}} = 0$. However, when $\Delta^{\text{sim}} > 0$ the RMSE of the submodel estimator is an increasing function of Δ^{sim} and converges to zero when Δ^{sim} increases. Moreover, the performance of the PS3 estimator is almost equivalent to that of the SM estimator, where k_2 is large at $\Delta^{\text{sim}} = 0$.

The shrinkage estimators perform better than the submodel estimator in most of the parameter range induced by the sparsity parameter Δ^{sim}. All of the three shrinkage estimators perform uniformly better than the full model estimator. The performance of the three shrinkage estimators are comparable with one another as there is no clear winner in the given range.

To include the penalty estimators in the study, we consider the case when the sparsity assumption is correct where $\Delta^{\text{sim}} = 0$ since these estimators are not defined for all possible values of Δ^{sim}. Again, we compute the MSEs of the respective estimators relative to the full model estimator for $k_1 = 4$ and $k_2 = 8, 12, 16, 20$. The RMSE results of various configurations of simulated parameters are reported in Table 8.1.

TABLE 8.1: RMSEs of Estimators when $\Delta^{\text{sim}} = 0$ for $k_1 = 4$, $n = 75$, and $N = 1{,}000$.

Estimator	Number of inactive parameters			
	8	12	16	20
SM	1.2673	1.3935	1.4759	1.8635
PS1	1.2085	1.3089	1.4082	1.6885
PS2	1.1963	1.3005	1.4017	1.6748
PS3	1.2429	1.3615	1.4560	1.8230
LASSO	0.9730	1.0587	1.1156	1.4022
ALASSO	1.0366	1.1174	1.2106	1.5239

Table 8.1 demonstrates that the RMSEs of all estimators is an increasing function of k_2. When keeping all other simulated parameters constant the submodel estimator has the highest RMSE when $\Delta^{\text{sim}} = 0$. For all k_2 the performance of PS1 and PS2 are comparable. However, PS3 dominates both PS1 and PS2. Interestingly, the penalized estimators are dominated by the shrinkage estimators and the performance of the ALASSO estimator is better than that of the LASSO estimator for selected values of k_2.

8.4.1 High-Dimensional Data

We move our attention from the low-dimensional case and design a simulation study with high-dimensional data, where number of predictor variables are larger than the observations. We follow the strategies developed by Ahmed and Yüzbaşı (2017), Gao et al. (2017a) and later Epprecht et al. (2021) followed the same idea.

The study can be classified into the following three groups:

1. Predictors have a strong influence (strong signal) on the response variable.

2. Predictor variables have a weak-to-moderate influence (weak-to-moderate signal), which may or may not contribute to explaining the response variable.

3. Predictor variables have no influence (sparse or no signal) on the response variable, thus related regression coefficients are exactly zero.

We design the simulation study to incorporate the parameter estimation problem for the high-dimensional mono-molecular nonlinear regression model.

TABLE 8.2: Percentage of Selection of Predictors for each Signal Level for $(n, p_1, p_2, p_3) = (75, 5, 50, 200)$.

κ	Strong Signal		Weak Signal		No Signal	
	LASSO	ALASSO	LASSO	ALASSO	LASSO	ALASSO
0.001	100.00	100.00	9.48	3.16	9.49	3.27
0.025	99.44	98.80	9.90	4.91	9.53	4.82
0.050	99.20	97.36	12.67	6.35	10.61	4.99
0.075	99.20	96.96	14.44	7.22	10.77	5.15
0.100	97.92	92.72	16.04	8.74	11.59	5.56

To include all of the above three cases, we partition $\boldsymbol{z} = (\boldsymbol{z}_1, \boldsymbol{z}_2, \boldsymbol{z}_3)$ and $\boldsymbol{\theta} = (\boldsymbol{\theta}_1^\top, \boldsymbol{\theta}_2^\top, \boldsymbol{\theta}_3^\top)^\top$, where $\boldsymbol{z}_1$, $\boldsymbol{z}_2$, and $\boldsymbol{z}_3$ are $n \times p_1$, $n \times p_2$, and $n \times p_3$ submatrices of predictors, respectively, such that $p = p_1 + p_2 + p_3$. Similarly, $\boldsymbol{\theta}_1$, $\boldsymbol{\theta}_2$, and $\boldsymbol{\theta}_3$ are sub-vectors of regression parameters with k_1, k_2, and k_3 dimensions, respectively, and $k = k_1 + k_2 + k_3$. We also assume that total number of strong and weak signals are less than the sample size, i.e. $p_1 + p_2 < n$ and $p_3 > n$.

We generate the response variable for the mono-molecular model from (8.6) with a regression coefficient vector:

$$\boldsymbol{\theta} = (\underbrace{\boldsymbol{\theta}_1^\top}_{k_1}, \underbrace{\boldsymbol{\theta}_2^\top}_{k_2}, \underbrace{\boldsymbol{\theta}_3^\top}_{k_3})^\top = (\underbrace{3, 3, 2, 2, 0.7, 0.7, 0.7}_{k_1}, \underbrace{\kappa, \kappa, ..., \kappa}_{k_2}, \underbrace{0, 0, ..., 0}_{k_3})^\top,$$

having strong, weak-to-moderate, and no signals, respectively. To gain some insight, we set the weak-to-moderate signal values (κ) to 0.001, 0.025, 0.050, 0.075, and 0.100 and randomly assign κ to have either positive or negative signs.

In our simulation design, we consider the sample sizes (n) with a number of strong predictors (p_1), weak-to-moderate predictors (p_2), and no influences (p_3) as 75, 5, 50, and 200, respectively, that satisfy $p_1 + p_2 < n$ and $p_3 > n$. We select the tuning parameters of LASSO and ALASSO by using BIC criteria. The number of simulations run $N = 250$ times for each configuration to obtain a stable result.

Our high-dimensional simulation experiment involves the following two steps:

1. A variable selection step to detect predictors with strong and weak-to-moderate signals to reduce the dimension to a low-dimensional model.

2. A post-selection parameter estimation step using the resulting models obtained from step 1.

LASSO and ALASSO are implemented to obtain two submodels with a different set of predictors.

In the submodel selection step, the performance of the selecting variable methods is examined only for $(n, p_1, p_2, p_3) = (75, 5, 50, 200)$. Based on 250 simulation iterations, the percentage of predictor variables selected based on LASSO and ALASSO procedures for each signal level are reported in Table 8.2. The percentage of selection of each predictor using the LASSO and ALASSO procedures are also graphically represented in Figures 8.2 and 8.3, respectively.

The results in Table 8.2 reveal that the LASSO strategy selects predictors with strong signals for all κ, while the ALASSO strategy selects strong signal predictors decreases when κ changes from 0.025 to 0.050 and then does not change as κ increases. As κ increases, the

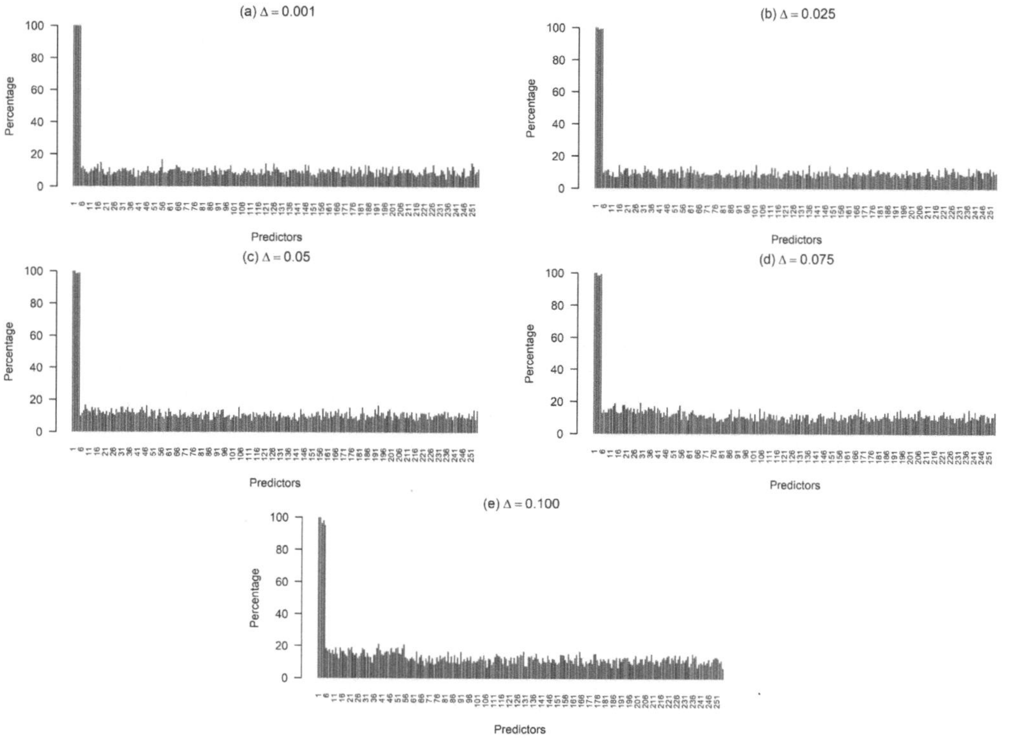

FIGURE 8.2: Percentage of Selection of each Predictor Variable for $(n, p_1, p_2, p_3) = (75, 5, 50, 200)$ using LASSO strategy.

performance of selecting predictors with weak signals of both penalty methods increases. However, the performance of eliminating predictor variables with no signals decreases.

As shown in Figure 8.2, the LASSO strategy selects too many predictors when κ is very small, which may yield an over-fitted model, whereas ALASSO selects fewer substantial predictors for large κ, which may produce an under-fitted model shown in Figure 8.3.

We can see that either the LASSO or ALASSO submodel selection procedure may not be the best to describe an optimal model in all situations. Therefore, we apply the shrinkage strategies in Section 8.2.1 for post-selection parameter estimation to address this concern and remove the deficiency at model selection stage.

8.4.1.1 Post-Selection Estimation Strategy

We set the LASSO submodel selection strategy to contain $\tilde{p}$ selected predictors while the ALASSO strategy chooses $\tilde{p}_1$ predictors, where $\tilde{p}_1 < \tilde{p} < p$.

For post-selection parameter estimation, we suggest the shrinkage estimation strategy given in Section 8.2.1. The full model (with large number of predictors) is constructed using predictors that are selected from the LASSO strategy containing $z_{i1}, z_{i2}, ..., z_{i\tilde{p}}$. The submodel is based on fewer predictors selected by the ALASSO strategy which is $z_{i1}, z_{i2}, ..., z_{i\tilde{p}_1}$.

To construct the shrinkage strategy, we divide the regression coefficients into two subsets $\mathcal{S}_1$ and $\mathcal{S}_2$, which are coefficients from the full model and submodel with $\tilde{k} = \tilde{p} + 2$ and $\tilde{k}_1 = \tilde{p}_1 + 2$ number of parameters, respectively. For the sparsity condition $\boldsymbol{\theta}_2 = \mathbf{0}_{\tilde{k}-\tilde{k}_1}$, we set $\boldsymbol{\theta}_2 = (\theta_1, \theta_2, ..., \theta_{\tilde{k}-\tilde{k}_1})^\top$ which is the coefficient of $\tilde{k} - \tilde{k}_1 = \tilde{p} - \tilde{p}_1$ predictors that exist in the full model but not in the submodel, i.e. $\mathcal{S}_1 \cap \mathcal{S}_2^c$. Under the sparsity assumption and

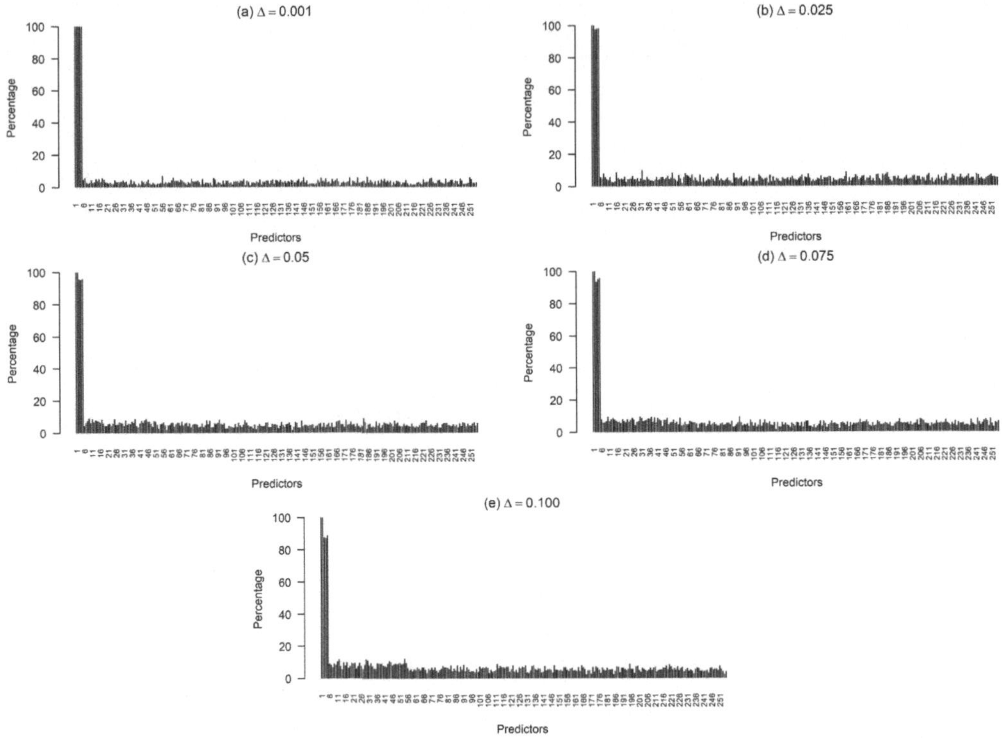

FIGURE 8.3: Percentage of Selection of each Predictor Variable for $(n, p_1, p_2, p_3) = (75, 5, 50, 200)$ using ALASSO strategy.

TABLE 8.3: RMSE of Estimators for a High-Dimensional Data.

κ	SM	PS1	PS2	PS3
0.001	3.3812	1.6912	1.6696	2.0982
0.025	0.0133	1.3544	1.3470	1.4797
0.050	0.0005	1.1461	1.1447	1.2174
0.075	0.0004	1.1360	1.1351	1.1794
0.100	0.0003	1.1159	1.1152	1.1350

$n \to \infty$, the distribution of W_n in (8.2) converges to a chi-square distribution with $\tilde{k} - \tilde{k}_1$ degrees of freedom, providing a theoretical justification.

The RMSE of the shrinkage estimators are reported in Table 8.3.

According to Table 8.3, all estimators perform best when $\kappa = 0.001$ but their efficiency decreases as κ increases. The submodel estimator has the highest RMSE at the smallest value of κ. This indicates that the submoel selected by ALASSO is the right one, whereas LASSO identifies an over-fitted model. On the other hand, ALASSO produces an under-fitted model when κ increases and the RMSE of the submodel estimator decreases.

The performance of the post-selection shrinkage estimators is consistent with the results presented in a low-dimensional setting. Based on this simulation study, the RMSE of shrinkage estimators can be placed in ascending order as RMSE(PS2) < RMSE(PS1) < RMSE(PS3).

8.5 Application to Rice Yield Data

We use the aforementioned strategies to an agricultural research application. We consider a sample of farmer households in the benefit areas of Kwae Noy Dam, Thailand in an effort to analyze the effects on rice cultivation in the planted year of 2014-2015. The was data obtained from socioeconomic monitoring and evaluation under His Majesty the King's initiation for the budget year 2015 Chaowagul et al. (2015).

We aim to predict the average rice yield (kg/0.16 ha) given 140 predictors from 105 sample households that planted rice twice a year. This is high-dimensional case since the number of predictors exceed the sample size.

We apply the penalized estimation strategy for dimensional reduction and model selection. The LASSO strategy picked five significant predictors, in our notation ($\tilde{p} = 5$. ALASSO choose only two predictors for the prediction purpose, $\tilde{p}_1 = 2$.

The model selection results are portrayed in Table 8.4. The selected predictors are given as follows:

- `yield` = average yield of rice harvested (kg/0.16 ha),
- `chem.p` = average price (Thai baht) of herbicides in powder or tablet form,
- `chem.q` = average amount of liquid pesticide used (L/0.16 ha),
- `chem.c` = average cost of liquid pesticides (Thai baht/0.16 ha),
- `labr.p` = average labor rate for fertilizing rice (Thai baht), and
- `machine` = number (units) of available rice spray seeding machines in 2014.

TABLE 8.4: Variable Selection Results for Rice Yield Data.

Model	Method	Number of parameters	Number of predictor variables	Predictor variables
Full	LASSO	$\tilde{k} = 7$	$\tilde{p} = 5$	`chem.p`, `chem.q`, `chem.c`, `labr.p`, `machine`
Reduced	ALASSO	$\tilde{k}_1 = 4$	$\tilde{p}_1 = 2$	`chem.q`, `chem.c`

The mono-molecular nonlinear model provided a decent fit to the data, see Piladaeng et al. (2022) and is demonstrated in Figure 8.4.

The performance of the shrinkage estimators is evaluated by its mean squared prediction error for each bootstrap replicate. In order to facilitate easy comparison we also calculated the relative mean squared prediction error of the estimators with respect to the estimator from the model selected by LASSO. By definition,

$$\text{RMSPE}\left(\widehat{\boldsymbol{\theta}}_1^{\text{FM}}, \widehat{\boldsymbol{\theta}}_1^{*}\right) = \frac{\left[\boldsymbol{y} - f(\boldsymbol{z}, \widehat{\boldsymbol{\theta}}^{\text{FM}})\right]^{\top}\left[\boldsymbol{y} - f(\boldsymbol{z}, \widehat{\boldsymbol{\theta}}^{\text{FM}})\right]}{\left[\boldsymbol{y} - f(\boldsymbol{z}, \widehat{\boldsymbol{\theta}}^{*})\right]^{\top}\left[\boldsymbol{y} - f(\boldsymbol{z}, \widehat{\boldsymbol{\theta}}^{*})\right]}.$$

We see that when RMSPE > 1, $\widehat{\boldsymbol{\theta}}_1^{*}$ outperforms $\widehat{\boldsymbol{\theta}}_1^{\text{FM}}$.

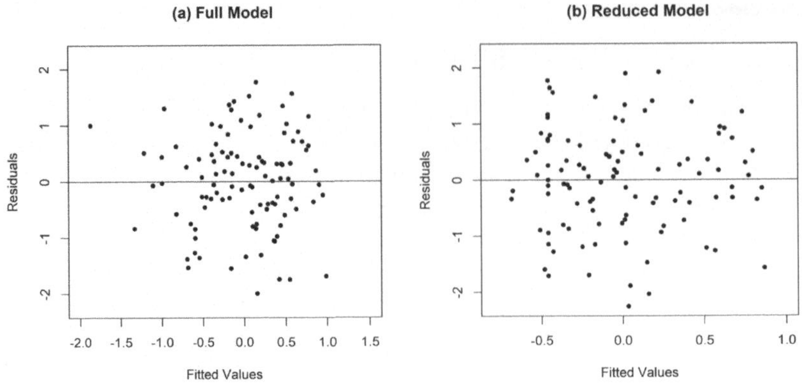

FIGURE 8.4: Plot of Residuals against Fitted Values.

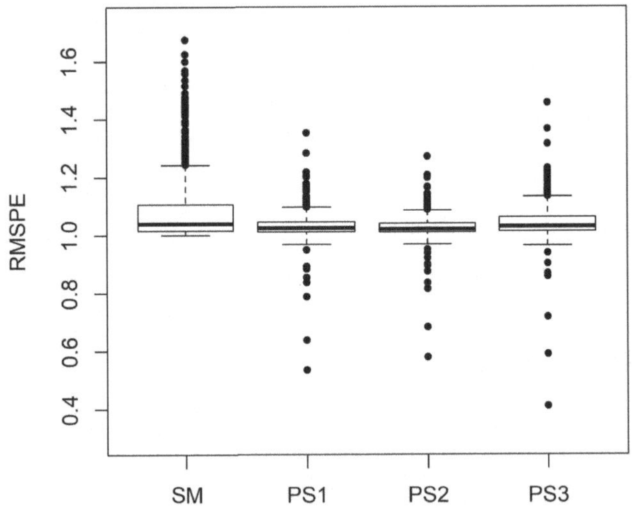

FIGURE 8.5: Boxplot of RMSPE of Estimators for Rice Yield Data.

TABLE 8.5: RMSPE of Estimators for Rice Yield Data.

	SM	PS1	PS2	PS3
RMSPE	1.0842	1.0318	1.0275	1.0414

To calculate the prediction error, we sample $m = 75$ with replacement for $N = 1,000$ iterations. Figure 8.5 shows a plot of the simulated MSPEs of each Monte Carlo replication for the submodel and shrinkage strategies and the RMSPEs of the listed estimators are given in Table 8.5. The results indicate that shrinkage estimators perform better than the LASSO based estimator. It seems like shrinkage estimators shrink in the direction of the submodel (based on ALSSO) indicating that ALASSO chooses the correct submodel.

8.6 R-Codes

```
> # We clearly state that the codes are written by Dr. Janjira
   Piladaeng.
> # nlsLasso Function
> nlsLasso =
   function(x,y,formulaOld,formulaNew,lambda,initialLasso,p)
+ {
+   I = 5000
+   T = 5000
+   n = nrow(x)
+   Alpha = numeric(I)
+   phi.beta = numeric(T)
+   initial.LASSO = initialLasso
+   expression.old = formulaOld
+   expression.new = formulaNew
+   # Setting initial for find beta hat of LASSO #
+   B.hat_LASSO = matrix(c(initial.LASSO,rep(0,(I*p))), p, I+1,
   byrow = F)
+   b.hat_LASSO = matrix(0, p, T+1, byrow = T)
+   y.hat = matrix(0, n, I+1, byrow = T)
+   L.beta = numeric(I+1)
+   GL.beta = matrix(0, p, I+1, byrow = T)
+   delta = matrix(0, p, I, byrow = T)
+   g = matrix(0, p, I, byrow = T)
+   u = matrix(0, p, I, byrow = T)
+   for(i in 1:I)
+   {
+     B.old = B.hat_LASSO[,i]
+     for(c in 1:p){
+       assign(paste("b", c, sep = ""), as.numeric(B.hat_LASSO[c,i]))
+     }
+     for(d in 1:(p-2))
+     {
+       assign(paste("x", d, sep = ""), x[,d])
+     }
+     # y.hat and least squares loss function #
+     y.hat[,i] = Y(x,B.old,n,p)
+     L.beta[i] = (1/(2*n))*t(y-y.hat[,i])%*%(y-y.hat[,i])
+     # Find gradient of least squares loss funciton #
+     # (differentiate with respect to beta) #
+     dL.beta = deriv(expression.old, name.1)
+     eval(dL.beta)
+     dL = numeric(p)
+     for(f in 1:p)
+     {
+       dL[f] = (1/(2*n))*sum(attr(.value, "gradient")[,f])
+     }
+     GL.beta[,i] = dL
+     #  Choosing alpha #
+     if(i == 1){Alpha[i] = 1}else{
+       Alpha[i] = abs((t(delta[,i-1])%*%g[,i-1])/
```

```
+        (t(delta[,i-1])%*%delta[,i-1]))}
+      alpha = Alpha[i]
+      b.hat_LASSO[,1] = B.hat_LASSO[,i]
+      for(t in 1:T)
+      {
+        u[,t] = as.numeric(b.hat_LASSO[,t])-(1/as.vector(alpha))*
+                  GL.beta[,i]
+        # check Soft.Cri[h] = (abs(u[h,t])-(lambda/alpha) > 0) for
   all beta#
+        for(h in 1:p)
+        {
+          if(assign(paste("Soft.C", h, sep = ""), u[h,t] >
   (lambda/alpha)))
+          {
+            b.hat_LASSO[h,t+1] = u[h,t]-(lambda/as.vector(alpha))
+          }else if(assign(paste("Soft.C", h, sep = ""), u[h,t] <
   (-lambda/alpha))){
+            b.hat_LASSO[h,t+1] = u[h,t]+(lambda/as.vector(alpha))
+          }else{
+            b.hat_LASSO[h,t+1] = 0
+          }
+        }
+        B.new = b.hat_LASSO[,t+1]
+        for(h in 1:p)
+        {
+          assign(paste("b", h, sep = "", ".new"),
   as.numeric(b.hat_LASSO[h,t+1]))
+        }
+        # y.hat and least squares loss function for new beta #
+        y.hat.new = Y(x,B.new,n,p)#B.new[1]*product.x.B.new
+        L.beta.new = (1/(2*n))*t(y-y.hat.new)%*%(y-y.hat.new)
+        # L1 least squares #
+        phi.beta.old =
   L.beta[i]+lambda*sum(abs(as.numeric(b.hat_LASSO[,t])))
+        phi.beta.new =
   L.beta.new+lambda*sum(abs(as.numeric(b.hat_LASSO[,t+1])))
+        criterion = numeric(T)
+        sum.diff.beta = sum((b.hat_LASSO[,t+1]-b.hat_LASSO[,t])^2)
+        phi.beta[t] = phi.beta.new
+        phi.b = c(phi.beta.old, phi.beta)
+        for(j in (max(t-M,0)):t)
+        {
+          criterion[j] = phi.b[j]-const*(alpha/2)*sum.diff.beta
+        }
+        # Acceptance criterion #
+        if(phi.beta.new <= max(criterion))
+        {
+          b.hat_LASSO[,t+1] = b.hat_LASSO[,t+1]
+          break
+        }else
+        {
+          b.hat_LASSO[,t+1] = b.hat_LASSO[,t]
+          alpha = eta*alpha
+        }
+      }
```

```
+       B.hat_LASSO[,i+1] = b.hat_LASSO[,t+1]
+       dL.beta.new = deriv(expression.new, name.2)
+       eval(dL.beta.new)
+       dL.new = numeric(p)
+       for(f in 1:p)
+       {
+         dL.new[f] = (1/(2*n))*sum(attr(.value, "gradient")[,f])
+       }
+       GL.beta[,i+1] = dL.new
+       delta[,i] =
    as.numeric(B.hat_LASSO[,i+1])-as.numeric(B.hat_LASSO[,i])
+       g[,i] = GL.beta[,i+1]-GL.beta[,i]
+       # check criterion sufficiently small #
+       # check Acc.C[h] = abs(as.numeric(B.hat_LASSO[h,i+1])
    -as.numeric(B.hat_LASSO[h,i]))/
+       # as.numeric(B.hat_LASSO[h,i+1])) < 1e-05 for all beta#
+       Numi = numeric(p)
+       Deno = numeric(p)
+       for(h in 1:p)
+       {
+         Numi[h] = (as.numeric(B.hat_LASSO[h,i+1])-
+          as.numeric(B.hat_LASSO[h,i]))^2
+         Deno[h] = (as.numeric(B.hat_LASSO[h,i+1]))^2
+       }
+       Acc.Cri = sqrt(sum(Numi))/sqrt(sum(Deno))
+       if(Acc.Cri < 1e-6)
+       {
+         B.hat_LASSO[,i+1] = B.hat_LASSO[,i]
+         break
+       }else
+       {
+         B.hat_LASSO[,i+1] = B.hat_LASSO[,i+1]
+       }
+     }
+     beta.hat_LASSO = as.numeric(B.hat_LASSO[, i+1])
+     Output = matrix(c(beta.hat_LASSO), 1, p, byrow = T)
+     colnames(Output, do.NULL = FALSE)
+     colnames(Output) = name.1
+     Output
+ }
> # nlsaLasso function
> nlsaLasso = function(x,y,formulaOld,formulaNew,lambda,
+                      initialLasso,p,B.weight)
+ {
+   I = 5000
+   T = 5000
+   n = nrow(x)
+   Alpha = numeric(I)
+   phi.beta = numeric(T)
+   initial.LASSO = initialLasso
+   expression.old = formulaOld
+   expression.new = formulaNew
+   # Setting initial for find beta hat of LASSO #
+   B.hat_LASSO = matrix(c(initial.LASSO,rep(0,(I*p))), p, I+1,
    byrow = F)
```

```
+    b.hat_LASSO = matrix(0, p, T+1, byrow = T)
+    y.hat = matrix(0, n, I+1, byrow = T)
+    L.beta = numeric(I+1)
+    GL.beta = matrix(0, p, I+1, byrow = T)
+    delta = matrix(0, p, I, byrow = T)
+    g = matrix(0, p, I, byrow = T)
+    u = matrix(0, p, I, byrow = T)
+    for(i in 1:I)
+    {
+      B.old = B.hat_LASSO[,i]
+      for(c in 1:p){
+        assign(paste("b", c, sep = ""), as.numeric(B.hat_LASSO[c,i]))
+      }
+      for(d in 1:(p-2))
+      {
+        assign(paste("x", d, sep = ""), x[,d])
+      }
+      # y.hat and least squares loss function #
+      y.hat[,i] = Y(x,B.old,n,p)#B.old[1]*product.x.B
+      L.beta[i] = (1/(2*n))*t(y-y.hat[,i])%*%(y-y.hat[,i])
+      # Find gradient of least squares loss funciton #
+      # (differentiate with respect to beta) #
+      dL.beta = deriv(expression.old, name.1)
+      eval(dL.beta)
+      dL = numeric(p)
+      for(f in 1:p)
+      {
+        dL[f] = (1/(2*n))*sum(attr(.value, "gradient")[,f])
+      }
+      GL.beta[,i] = dL
+      #  Choosing alpha #
+      if(i == 1){Alpha[i] = 1}else{
+        Alpha[i] = abs((t(delta[,i-1])%*%g[,i-1])/
+                     (t(delta[,i-1])%*%delta[,i-1]))}
+      alpha = Alpha[i]
+      b.hat_LASSO[,1] = B.hat_LASSO[,i]
+      for(t in 1:T)
+      {
+        u[,t] = as.numeric(b.hat_LASSO[,t])-(1/as.vector(alpha))*
+                GL.beta[,i]
+        # check Soft.Cri[h] = (abs(u[h,t])-(lambda/alpha) > 0) for
     all beta#
+        for(h in 1:p)
+        {
+          if(assign(paste("Soft.C", h, sep = ""), u[h,t] >
     (lambda/alpha)))
+          {
+            b.hat_LASSO[h,t+1] = u[h,t]-(lambda/as.vector(alpha))
+          }else if(assign(paste("Soft.C", h, sep = ""), u[h,t] <
     (-lambda/alpha))){
+            b.hat_LASSO[h,t+1] = u[h,t]+(lambda/as.vector(alpha))
+          }else{
+            b.hat_LASSO[h,t+1] = 0
+          }
+        }
```

```
+         B.new = b.hat_LASSO[,t+1]
+         for(h in 1:p)
+         {
+           assign(paste("b", h, sep = "", ".new"),
    as.numeric(b.hat_LASSO[h,t+1]))
+         }
+         # y.hat and least squares loss function for new beta #
+         y.hat.new = Y(x,B.new,n,p)
+         L.beta.new = (1/(2*n))*t(y-y.hat.new)%*%(y-y.hat.new)
+         # L1 least squares #
+         weight = 1/(abs(B.weight)^0.1)
+         phi.beta.old = L.beta[i]+lambda*
+                   sum(weight*abs(as.numeric(b.hat_LASSO[,t])))
+         phi.beta.new = L.beta.new+lambda*
+                  sum(weight*abs(as.numeric(b.hat_LASSO[,t+1])))
+         criterion = numeric(T)
+         sum.diff.beta = sum((b.hat_LASSO[,t+1]-b.hat_LASSO[,t])^2)
+         phi.beta[t] = phi.beta.new
+         phi.b = c(phi.beta.old, phi.beta)
+         for(j in (max(t-M,0)):t)
+         {
+           criterion[j] = phi.b[j]-const*(alpha/2)*sum.diff.beta
+         }
+         # Acceptance criterion #
+         if(phi.beta.new <= max(criterion))
+         {
+           b.hat_LASSO[,t+1] = b.hat_LASSO[,t+1]
+           break
+         }else
+         {
+           b.hat_LASSO[,t+1] = b.hat_LASSO[,t]
+           alpha = eta*alpha
+         }
+       }
+       B.hat_LASSO[,i+1] = b.hat_LASSO[,t+1]
+       dL.beta.new = deriv(expression.new, name.2)
+       eval(dL.beta.new)
+       dL.new = numeric(p)
+       for(f in 1:p)
+       {
+         dL.new[f] = (1/(2*n))*sum(attr(.value, "gradient")[,f])
+       }
+       GL.beta[,i+1] = dL.new
+       delta[,i] =
    as.numeric(B.hat_LASSO[,i+1])-as.numeric(B.hat_LASSO[,i])
+       g[,i] = GL.beta[,i+1]-GL.beta[,i]
+       # check criterion sufficiently small #
+       # check Acc.C[h] = abs(as.numeric(B.hat_LASSO[h,i+1])-
+            as.numeric(B.hat_LASSO[h,i]))/
+            as.numeric(B.hat_LASSO[h,i+1])) < 1e-05 for all beta#
+       Numi = numeric(p)
+       Deno = numeric(p)
+       for(h in 1:p)
+       {
+         Numi[h] = (as.numeric(B.hat_LASSO[h,i+1])-
```

```
+           as.numeric(B.hat_LASSO[h,i]))^2
+         Deno[h] = (as.numeric(B.hat_LASSO[h,i+1]))^2
+       }
+       Acc.Cri = sqrt(sum(Numi))/sqrt(sum(Deno))
+       if(Acc.Cri < 1e-6)
+       {
+         B.hat_LASSO[,i+1] = B.hat_LASSO[,i]
+         break
+       }else
+       {
+         B.hat_LASSO[,i+1] = B.hat_LASSO[,i+1]
+       }
+     }
+     beta.hat_LASSO = as.numeric(B.hat_LASSO[, i+1])
+     Output = matrix(c(beta.hat_LASSO), 1, p, byrow = T)
+     colnames(Output, do.NULL = FALSE)
+     colnames(Output) = name.1
+     Output
+ }
> Low Dimensional Case
> # Required Packages
> library(MASS)
> library(mvtnorm)
> library(minpack.lm)
> library(mosaic)
> library(glmnet)
> library(cvTools)
> library(modelr)
> library(msgps)
> set.seed(2022)
> p1 = 4
> p2 = 8
> p = p1+p2
> # Expression [y-f(x,beta)]^2 #
> A1 = NULL
> A2 = NULL
> for(u in 0:(p-2))
+ {
+   A1[u] = paste("-(b", u+2, "*x", u, ")", sep = "")
+   B1 = paste("(y-(b1*(1-(b2*exp(", sep = "")
+   C1 = paste(")))))^2", sep = "")
+   D1 = c(B1,A1,C1)
+   E1 = paste(D1, collapse = "")
+   #
+   A2[u] = paste("-(b", u+2, ".new", "*x", u, ")", sep = "")
+   B2 = paste("(y-(b1.new*(1-(b2.new*exp(", sep = "")
+   C2 = paste(")))))^2", sep = "")
+   D2 = c(B2,A2,C2)
+   E2 = paste(D2, collapse = "")
+ }
> expression.old = as.expression(parse(text = E1)[[1]])
> expression.new = as.expression(parse(text = E2)[[1]])
> # Expression f(x,beta) for nls function #
> A3 = NULL
> A4 = NULL
```

```
> for(u in 0:(p-2))
+ {
+   A3[u] = paste("-(b", u+2, "*x[,", u, "])", sep = "")
+   B3 = paste("y ~ b1*(1-(b2*exp(", sep = "")
+   C3 = paste(")))", sep = "")
+   D3 = c(B3,A3,C3)
+   E3 = paste(D3, collapse = "")
+ }
> for(u in 0:(p1-2))
+ {
+   A4[u] = paste("-(b", u+2, "*x[,", u, "])", sep = "")
+   B4 = paste("y ~ b1*(1-(b2*exp(", sep = "")
+   C4 = paste(")))", sep = "")
+   D4 = c(B4,A4,C4)
+   E4 = paste(D4, collapse = "")
+ }
> expression.nls.full = parse(text = E3)[[1]]
> expression.nls.sub = parse(text = E4)[[1]]
> # Naming a vector (b1, b2, ...) #
> name.1 = NULL
> name.2 = NULL
> for(m in 1:p)
+ {
+   name.1[m] = paste("b", m, sep = "")
+   name.2[m] = paste("b", m, sep = "", ".new")
+ }
> #
> eta = 100
> const = 1e-3
> M = 3
> Y = function(x,beta,n,p)
+ {
+   x.beta = matrix(0,n,p-2)
+   sum.x.beta = c(rep(0,n))
+   for(k in 1:(p-2))
+   {
+     x.beta[,k] = (beta[k+2]*x[,k])
+     sum.x.beta = sum.x.beta-x.beta[,k]
+   }
+   Y = beta[1]*(1-(beta[2]*exp(sum.x.beta)))
+   Y
+ }
>   n = 75
>   beta = c(2.5,1.5,1.5,0.5,0,rep(0,p-5))
>   beta1 = beta[c(rep(1:p1))]
>   e = rnorm(n, 0, 1)
>   sigma = diag(p-2)
>   mu = rep(0,p-2)
>   x = mvrnorm(n, mu, sigma)
>   y = Y(x,beta,n,p)+e
>   # Tranform variable x and y #
>   max.y = max(y)+0.1
>   x.new = x
>   y.new = log(max.y/(max.y-y))
>   # Find initial value #
```

```
>   model.0 = lm(log(max.y/(max.y-y)) ~ x)
>   #
>   b1 = max.y
>   b2 = 1/exp(coef(model.0)[1])
>   start.0_UE = c(b1,b2,rep(0,p-2))
>   #
>   for(l in 1:(p-2))
+   {
+     start.0_UE[l+2] = assign(paste("b", l+2, sep = ""),
  coef(model.0)[l+1])
+   }
>   names(start.0_UE) = name.1
>   start.0_RE = start.0_UE[c(rep(1:p1))]
>   # Full Model
>   model_UE = nls(as.formula(expression.nls.full),
+                  start = start.0_UE,
+                  control = nls.control(maxiter = 50000, warnOnly =
  TRUE, minFactor = 2e-10))
>   # Sub Model
>   model_RE = nls(as.formula(expression.nls.sub),
+                  start = start.0_RE,
+                  control = nls.control(maxiter = 50000, warnOnly =
  TRUE, minFactor = 2e-10))
>   #tuning parameters of Lasso and ALasso
>   lambda_LASSO = msgps(x.new, y.new)
>   lambda_aLASSO = msgps(x.new, y.new, penalty = "alasso", gamma =
  1, lambda = 0)
>   Lambda_LASSO  = lambda_LASSO$dfbic_result$tuning
>   Lambda_aLASSO = lambda_aLASSO$dfbic_result$tuning
>   # Ridge
>   RIDGE.cv = cv.glmnet(x.new, y.new, type.measure = "mse", alpha =
  0)
>   RIDGE.coef = as.numeric(coef(RIDGE.cv, s = RIDGE.cv$lambda.min))
>   start.weight = c(max.y,(1/exp(RIDGE.coef[1])),RIDGE.coef[-1])
>   #Lasso
>   opt.lambda_LASSO = cv.glmnet(x.new, y.new, family = 'gaussian',
  type.measure = "mse",
+                              alpha = 1)$lambda.min
>   fit.LASSO = glmnet(x.new, y.new, family = 'gaussian', lambda =
  opt.lambda_LASSO, alpha=1)
>   LASSO.coef = drop(predict(fit.LASSO, type="coef", s =
  fit.LASSO$lambda.min))
>   start_LASSO = c(max.y,(1/exp(LASSO.coef[1])),LASSO.coef[-1])
>   #ALASSO
>   opt.lambda_aLASSO = cv.glmnet(x.new, y.new, penalty.factor =
  1/(abs(RIDGE.coef[-1]))^1,
+                               family = 'gaussian', type.measure
  = "mse",
+                               alpha = 1)$lambda.min
>   fit.aLASSO = glmnet(x.new, y.new, penalty.factor =
  1/(abs(RIDGE.coef[-1]))^0.5,
+                        family = 'gaussian', lambda =
  opt.lambda_aLASSO, alpha=1)
>   aLASSO.coef = drop(predict(fit.aLASSO, type="coef", s =
  fit.aLASSO$lambda.min))
```

```
>   start_aLASSO = c(max.y,(1/exp(aLASSO.coef[1])),aLASSO.coef[-1])
>   #
>   model_LASSO = nlsLasso(x, y, expression.old, expression.new,
+                         Lambda_LASSO, start_LASSO, p)
>   model_aLASSO = nlsaLasso(x ,y ,expression.old, expression.new,
+                           Lambda_aLASSO, start_aLASSO, p,
   start.weight)
>   pi1 = 0.25
>   pi2 = 0.50
>   pi3 = 0.75
>   alpha1 = 0.01
>   alpha2 = 0.05
>   alpha3 = 0.10
>   #
>   beta.hat_UE = coef(model_UE)[c(rep(1:p1))]
>   beta.hat_RE = coef(model_RE)
>   #
>   beta2.hat = coef(model_UE)[-c(rep(1:p1))]
>   Sigma2.hat = deviance(model_UE)/(n-p)
>   var.beta.hat.UE = vcov(model_UE)
>   I = ginv(vcov(model_UE))
>   #
>   XX = I*Sigma2.hat
>   X11 = XX[seq(1,p1),seq(1,p1)]
>   X22 = XX[seq(p1+1,p),seq(p1+1,p)]
>   X12 = XX[seq(1,p1),seq(p1+1,p)]
>   X21 = t(X12)
>   #
>   C = X22-(X21%*%ginv(X11)%*%X12)
>   Ln = (1/Sigma2.hat)*t(beta2.hat)%*%C%*%beta2.hat
>   #
>   beta.hat_JS1 = beta.hat_RE+
+
   drop(1-((p2-2)/Ln))*(beta.hat_UE-beta.hat_RE)
>   beta.hat_JS2 = beta.hat_RE+
+
   drop(1-((p2-2)/(Ln+1)))*(beta.hat_UE-beta.hat_RE)
>   beta.hat_JS3 = beta.hat_RE+
+
   drop(1-((p2-2)*(atan(Ln))/Ln))*(beta.hat_UE-beta.hat_RE)
>
>   if((1-((p2-2)/Ln)) > 0){beta.hat_PJS1 = beta.hat_JS1
+   } else{beta.hat_PJS1 = beta.hat_RE}
>
>   if((1-((p2-2)/(Ln+1))) > 0){beta.hat_PJS2 = beta.hat_JS2
+   } else{beta.hat_PJS2 = beta.hat_RE}
>
>   if((1-((p2-2)*(atan(Ln))/Ln)) > 0){beta.hat_PJS3 = beta.hat_JS3
+   }else{beta.hat_PJS3 = beta.hat_RE}
>
>   beta.hat_LASSO = model_LASSO[c(rep(1:p1))]
>   beta.hat_aLASSO = model_aLASSO[c(rep(1:p1))]
>
>   # Calculate MSEs
>   MSE<-
```

```
+      c(FM = t(beta1-beta.hat_UE)%*%(beta1-beta.hat_UE),
+    SM = t(beta1-beta.hat_RE)%*%(beta1-beta.hat_RE),
+    PS1 = t(beta1-beta.hat_PJS1)%*%(beta1-beta.hat_PJS1),
+    PS2 = t(beta1-beta.hat_PJS2)%*%(beta1-beta.hat_PJS2),
+    PS3 = t(beta1-beta.hat_PJS3)%*%(beta1-beta.hat_PJS3),
+    LASSO = t(beta1-beta.hat_LASSO)%*%(beta1-beta.hat_LASSO),
+    ALASSO = t(beta1-beta.hat_aLASSO)%*%(beta1-beta.hat_aLASSO))
>    cbind(MSE=MSE,Best_Ranking= rank(MSE))
              MSE Best_Ranking
FM      0.09213770            5
SM      0.02474195            1
PS1     0.05336101            3
PS2     0.05582298            4
PS3     0.03867662            2
LASSO   0.98783689            6
ALASSO  1.48138415            7
```

8.7 Concluding Remarks

In this chapter, we presented a class of positive-part shrinkage estimators to improve the performance of classical estimators in a nonlinear model when the sparsity assumption may or may not hold. For the low-dimensional case, we provide a meaningful asymptotic bias and risk of the full model, submodel, and shrinkage estimators. We established that the shrinkage strategy is far superior than the classical estimators in several situations. The asymptotic results are justified with a simulation study of moderate sample size.

In our simulations, we also considered the high-dimensional case and include two penalty procedures, LASSO and ALASSO. We assessed the relative performance of penalty and shrinkage estimators using the mean squared criterion. The simulation study clearly indicated that shrinkage successfully combines two models and has an edge over the penalized estimation theory, which may do well under the stringent sparsity assumption. However, in high-dimensional cases we rely on the penalized procedure to obtain two models and use to construct shrinkage estimators. For dimensional reduction of the high-dimensional data, our simulation results confirmed that the LASSO and ALASSO methods may not select the optimal submodel with significant predictors in all situations. As expected, the LASSO strategy selected a model that contained many predictors with weak signals, resulting in over-fitting. On the other hand, the ALASSO strategy eliminated many significant predictor variables when the weak signals have moderate strength, resulting in underfitting. Generally, ALASSO selected a smaller number of predictor variables than LASSO thus ALASSO was determined as the submodel. One can form the shrinkage strategy in high-dimensional cases by using these two penalized procedures. There are other penalized methods that are readily available for model selection and parameter estimation, such as ENET, Ridge and SCAD, the most popular one.

In addition, we applied the suggested estimators to real data to confirm the benefits of shrinkage methods. The data analysis clearly shows the supremacy of the shrinkage estimators and validates the findings of the theoretical results.

In a nutshell, the performance of the shrinkage estimators are superior than the full model, submodel, and penalized estimators, in both low- and high-dimensional settings. We recommend the shrinkage strategy when the assumption of sparsity is in question.

The shrinkage strategy successfully combines two models that contain strong and weak-to-moderate signals and may reduce the prediction error drastically, a winning strategy!

Appendix

The following lemma and theorems facilitate computation of QADB and ADR.

Lemma 8.3 *Assuming appropriate regularity conditions of nonlinear least squares Ahmed and Nicol (2012), if* $\boldsymbol{u} \sim N(\boldsymbol{0}, \sigma^2 \boldsymbol{I}_n)$ *and* n *is large then, approximately,* $\widehat{\boldsymbol{\theta}}^{\rm FM} \xrightarrow{D} N_k\left(\boldsymbol{\theta}, \frac{\sigma^2}{n}\boldsymbol{G}^{-1}\right)$, *where* $\boldsymbol{G} = \lim_{n\to\infty} \frac{1}{n}\boldsymbol{P}^\top \boldsymbol{P}$ *and* $\xrightarrow{D}$ *indicates convergence in distribution.*

Lemma 8.4 *Under the assumed regularity conditions, and* $n \to \infty$ *is large, then from Theorem 8.6 in Muller and Stewart (2006), the marginal distribution of* $\widehat{\boldsymbol{\theta}}_1^{\rm FM}$ *is* $N_{k_1}\left(\boldsymbol{\theta}_1, \frac{\sigma^2}{n}\boldsymbol{G}_{11.2}^{-1}\right)$ *and that of* $\widehat{\boldsymbol{\theta}}_2^{\rm FM}$ *is* $N_{k_2}\left(\boldsymbol{\theta}_2, \frac{\sigma^2}{n}\boldsymbol{G}_{22.1}^{-1}\right)$. *Here,* $\boldsymbol{G} = \lim_{n\to\infty} \frac{1}{n}(\boldsymbol{P}^\top \boldsymbol{P})$, *and*

$$\boldsymbol{G}^{-1} = \begin{bmatrix} \boldsymbol{G}_{11} & \boldsymbol{G}_{12} \\ \boldsymbol{G}_{21} & \boldsymbol{G}_{22} \end{bmatrix}^{-1} = \begin{bmatrix} \boldsymbol{G}_{11.2}^{-1} & -\boldsymbol{G}_{11.2}^{-1}\boldsymbol{G}_{12}\boldsymbol{G}_{22}^{-1} \\ -\boldsymbol{G}_{22}^{-1}\boldsymbol{G}_{21}\boldsymbol{G}_{11.2}^{-1} & \boldsymbol{G}_{22.1}^{-1} \end{bmatrix},$$

where $\boldsymbol{G}_{11.2}^{-1} = \left(\boldsymbol{G}_{11} - \boldsymbol{G}_{12}\boldsymbol{G}_{22}^{-1}\boldsymbol{G}_{21}\right)^{-1}$ *and* $\boldsymbol{G}_{22.1}^{-1} = \left(\boldsymbol{G}_{22} - \boldsymbol{G}_{21}\boldsymbol{G}_{11}^{-1}\boldsymbol{G}_{12}\right)^{-1}$.

Lemma 8.5 *Under the sequence of local alternatives* $\{K_n\}$ *and assumed regularity conditions, as* $n \to \infty$, *we have*

$$\begin{bmatrix} \boldsymbol{\zeta}_n \\ \boldsymbol{\varrho}_n \end{bmatrix} \xrightarrow{D} \begin{bmatrix} \boldsymbol{\zeta} \\ \boldsymbol{\varrho} \end{bmatrix} \sim N\left(\begin{bmatrix} \boldsymbol{0} \\ \boldsymbol{\gamma\delta} \end{bmatrix}, \sigma^2 \begin{bmatrix} \boldsymbol{G}_{11.2}^{-1} & \boldsymbol{\Sigma} \\ \boldsymbol{\Sigma} & \boldsymbol{\Sigma} \end{bmatrix}\right),$$
$$\begin{bmatrix} \boldsymbol{\kappa}_n \\ \boldsymbol{\varrho}_n \end{bmatrix} \xrightarrow{D} \begin{bmatrix} \boldsymbol{\kappa} \\ \boldsymbol{\varrho} \end{bmatrix} \sim N\left(\begin{bmatrix} -\boldsymbol{\gamma\delta} \\ \boldsymbol{\gamma\delta} \end{bmatrix}, \sigma^2 \begin{bmatrix} \boldsymbol{G}_{11}^{-1} & \boldsymbol{0} \\ \boldsymbol{0} & \boldsymbol{\Sigma} \end{bmatrix}\right),$$

where $\boldsymbol{\gamma} = -\boldsymbol{G}_{11}^{-1}\boldsymbol{G}_{12}$, $\boldsymbol{\Sigma} = \boldsymbol{G}_{11}^{-1}\boldsymbol{G}_{12}\boldsymbol{G}_{22.1}^{-1}\boldsymbol{G}_{21}\boldsymbol{G}_{11}^{-1}$, *and* $\xrightarrow{D}$ *implies converge in distribution. Here,* $\boldsymbol{\zeta}_n = \sqrt{n}(\widehat{\boldsymbol{\theta}}_1^{\rm FM} - \boldsymbol{\theta}_1)$, $\boldsymbol{\kappa}_n = \sqrt{n}(\widehat{\boldsymbol{\theta}}_1^{\rm SM} - \boldsymbol{\theta}_1)$, *and* $\boldsymbol{\varrho}_n = \sqrt{n}(\widehat{\boldsymbol{\theta}}_1^{\rm FM} - \widehat{\boldsymbol{\theta}}_1^{\rm SM})$.

Lemma 8.6 *Under the assumed regularity conditions and local alternatives* $\{K_n\}$, *as* $n \to \infty$, W_n *converges to a non-central chi-squared distribution with* k_2 *degrees of freedom and the non-centrality parameter* $\Delta = \boldsymbol{\delta}^\top \boldsymbol{G}_{22.1}\boldsymbol{\delta}/\sigma^2$, $\boldsymbol{G}_{22.1} = \lim_{n\to\infty} \frac{1}{n}\boldsymbol{Q}_{22.1}$.

Corollary 8.1 *Under assumed regularity conditions and the sequence of local alternatives, as* $n \to \infty$,

$$\boldsymbol{\varrho}_n^* = \sqrt{n}\sigma^{-1}\boldsymbol{\Sigma}_n^{-\frac{1}{2}}(\widehat{\boldsymbol{\theta}}_1^{\rm FM} - \widehat{\boldsymbol{\theta}}_1^{\rm SM}) \xrightarrow{D} \boldsymbol{\varrho}^* \sim N(\sigma^{-1}\boldsymbol{\Sigma}^{-\frac{1}{2}}\boldsymbol{\gamma\delta}, \boldsymbol{I}_{k_2}),$$

where $\boldsymbol{\Sigma}_n = \boldsymbol{Q}_{11}^{-1}\boldsymbol{Q}_{12}\boldsymbol{Q}_{22.1}^{-1}\boldsymbol{Q}_{21}\boldsymbol{Q}_{11}^{-1}$ *and* $\boldsymbol{\Sigma}_n \xrightarrow{P} \boldsymbol{\Sigma}$.

Noting $\boldsymbol{\varrho}_n^*$ has covariance matrix $\boldsymbol{I}_{k_2}$, so that we may use Lemma 3.2 for computing QADB and ADR of the estimators. The relation between $\boldsymbol{\varrho}$ and $\boldsymbol{\varrho}^*$ is

$$\boldsymbol{\varrho} = \sigma\boldsymbol{\Sigma}^{\frac{1}{2}}\boldsymbol{\varrho}^* = (\sigma^2\boldsymbol{\Sigma})^{\frac{1}{2}}\boldsymbol{\varrho}^*. \tag{8.7}$$

Proof of Theorem 8.1 *By definition, the asymptotic distribution bias (ADB) of an estimator is $\widehat{\boldsymbol{\theta}}_1^*$ is defined as*

$$ADB(\widehat{\boldsymbol{\theta}}_1^*) = \lim_{n\to\infty} E\left[\sqrt{n}(\widehat{\boldsymbol{\theta}}_1^* - \boldsymbol{\theta}_1)\right].$$

Under the assumed regularity conditions, and sequence of local alternatives $\{K_n\}$, by using Lemma 8.5 and Lemma 3.2, the ADB of the estimators are obtained as follows:

$$\begin{aligned}
ADB(\widehat{\boldsymbol{\theta}}_1^{\mathrm{FM}}) &= \lim_{n\to\infty} E[\sqrt{n}(\widehat{\boldsymbol{\theta}}_1^{\mathrm{FM}} - \boldsymbol{\theta}_1)] = E(\boldsymbol{\zeta}) = \mathbf{0},\\
ADB(\widehat{\boldsymbol{\theta}}_1^{\mathrm{SM}}) &= \lim_{n\to\infty} E[\sqrt{n}(\widehat{\boldsymbol{\theta}}_1^{\mathrm{SM}} - \boldsymbol{\theta}_1)] = E(\boldsymbol{\kappa}) = -\boldsymbol{\gamma\delta} = \boldsymbol{G}_{11}^{-1}\boldsymbol{G}_{12}\boldsymbol{\delta},
\end{aligned}$$

We provide some important steps for deriving he ADB of the shrinkage estimator as follows:

$$\begin{aligned}
ADB(\widehat{\boldsymbol{\theta}}_1^{\mathrm{S}}) &= \lim_{n\to\infty} E[\sqrt{n}(\widehat{\boldsymbol{\theta}}_1^{\mathrm{S}} - \boldsymbol{\theta}_1)]\\
&= \lim_{n\to\infty} E\left[\sqrt{n}\left(\widehat{\boldsymbol{\theta}}_1^{\mathrm{SM}} + \left(1 - \frac{r\cdot g(W_n)}{W_n}\right)(\widehat{\boldsymbol{\theta}}_1^{\mathrm{FM}} - \widehat{\boldsymbol{\theta}}_1^{\mathrm{SM}}) - \boldsymbol{\theta}_1\right)\right]\\
&= \lim_{n\to\infty} E\left[\sqrt{n}(\widehat{\boldsymbol{\theta}}_1^{\mathrm{FM}} - \boldsymbol{\theta}_1) - \frac{r\cdot g(W_n)}{W_n}\sqrt{n}(\widehat{\boldsymbol{\theta}}_1^{\mathrm{FM}} - \widehat{\boldsymbol{\theta}}_1^{\mathrm{SM}})\right]\\
&= E\lim_{n\to\infty}(\boldsymbol{\zeta}_n) - rE\lim_{n\to\infty}(W_n^{-1}\boldsymbol{\varrho}_n) = E(\boldsymbol{\zeta}) - rE\left[\frac{g(\chi^2_{k_2}(\Delta))}{\chi^2_{k_2}(\Delta)}\boldsymbol{\varrho}\right]\\
&= E(\boldsymbol{\zeta}) - rE\left[\frac{g(\chi^2_{k_2}(\Delta))}{\chi^2_{k_2}(\Delta)}(\sigma^2\boldsymbol{\Sigma})^{\frac{1}{2}}\boldsymbol{\varrho}^*\right] && \because By\,(8.7)\\
&= -r(\sigma^2\boldsymbol{\Sigma})^{\frac{1}{2}}(\sigma^2\boldsymbol{\Sigma})^{-\frac{1}{2}}\boldsymbol{\gamma\delta}E\left[\frac{g(\chi^2_{k_2+2}(\Delta))}{\chi^2_{k_2+2}(\Delta)}\right] && \because By\, Lemma\ 3.2\\
&= r\boldsymbol{G}_{11}^{-1}\boldsymbol{G}_{12}\boldsymbol{\delta}E\left[\frac{g(W_1)}{W_1}\right].
\end{aligned}$$

Here $W_1 = \chi^2_{k_2+2}(\Delta)$. Then, the key steps for the ADB of positive shrinkage estimator is given as follows:

$$\begin{aligned}
ADB(\widehat{\boldsymbol{\theta}}_1^{\mathrm{PS}}) &= \lim_{n\to\infty} E[\sqrt{n}(\widehat{\boldsymbol{\theta}}_1^{\mathrm{PS}} - \boldsymbol{\theta}_1)]\\
&= \lim_{n\to\infty} E\left[\sqrt{n}\left(\widehat{\boldsymbol{\theta}}_1^{\mathrm{S}} - \left(1 - \frac{r\cdot g(W_n)}{W_n}\right)(\widehat{\boldsymbol{\theta}}_1^{\mathrm{FM}} - \widehat{\boldsymbol{\theta}}_1^{\mathrm{SM}})\right.\right.\\
&\quad \left.\left. \times\, I\left(\frac{W_n}{g(W_n)} \le r\right) - \boldsymbol{\theta}_1\right)\right]\\
&= \lim_{n\to\infty} E\left[\begin{array}{c}\sqrt{n}(\widehat{\boldsymbol{\theta}}_1^{\mathrm{S}} - \boldsymbol{\theta}_1)\\ -\left(1 - \frac{r\cdot g(W_n)}{W_n}\right)\sqrt{n}(\widehat{\boldsymbol{\theta}}_1^{\mathrm{FM}} - \widehat{\boldsymbol{\theta}}_1^{\mathrm{SM}})\, I\left(\frac{W_n}{g(W_n)} \le r\right)\end{array}\right]\\
&= ADB(\widehat{\boldsymbol{\theta}}_1^{\mathrm{S}}) - E\left[\begin{array}{c}\left(1 - \frac{r\cdot g(\chi^2_{k_2}(\Delta))}{\chi^2_{k_2}(\Delta)}\right)\\ (\sigma^2\boldsymbol{\Sigma})^{\frac{1}{2}}\boldsymbol{\varrho}^*\, I\left(\frac{\chi^2_{k_2}(\Delta)}{g(\chi^2_{k_2}(\Delta))} \le r\right)\end{array}\right] && \because By\,(8.7)
\end{aligned}$$

$$= ADB(\widehat{\boldsymbol{\theta}}_1^{\mathrm{S}}) - \boldsymbol{\gamma\delta} E\left[\begin{array}{c}\left(1-\frac{r\cdot g(\chi^2_{k_2+2}(\Delta))}{\chi^2_{k_2+2}(\Delta)}\right)\\ I\left(\frac{\chi^2_{k_2+2}(\Delta)}{g(\chi^2_{k_2+2}(\Delta))}\le r\right)\end{array}\right] \qquad \because By\, Lemma\ 3.2$$

$$= ADB(\widehat{\boldsymbol{\theta}}_1^{\mathrm{S}}) + \boldsymbol{G}_{11}^{-1}\boldsymbol{G}_{12}\boldsymbol{\delta} E\left[\left(1-\frac{r\cdot g(W_1)}{W_1}\right) I\left(\frac{W_1}{g(W_1)}\le r\right)\right],$$

where $W_1 = \chi^2_{k_2+2}(\Delta)$.

The proof of QADB of the estimators can be easily derived by following the above ADB and Equation (8.3).

Proof of Theorem 8.2 *We first derive the asymptotic covariance matrix of an estimator* $\boldsymbol{\theta}_1^*$. *By defination and help of Lemma 8.5 and Lemma 3.2, the asymptotic covariance matrix of full model and submodel estimators are given, respectively as follows:*

$$\begin{aligned}
\boldsymbol{\Gamma}^*(\widehat{\boldsymbol{\theta}}_1^{\mathrm{FM}}) &= E\lim_{n\to\infty}\left[\sqrt{n}(\widehat{\boldsymbol{\theta}}_1^{\mathrm{FM}}-\boldsymbol{\theta}_1)\sqrt{n}(\widehat{\boldsymbol{\theta}}_1^{\mathrm{FM}}-\boldsymbol{\theta}_1)^\top\right]\\
&= E(\boldsymbol{\zeta\zeta}^\top) = var(\boldsymbol{\zeta}) + \mathbb{E}(\boldsymbol{\zeta})E(\boldsymbol{\zeta}^\top) = \sigma^2\boldsymbol{G}_{11.2}^{-1},\\
\boldsymbol{\Gamma}^*(\widehat{\boldsymbol{\theta}}_1^{\mathrm{SM}}) &= E\lim_{n\to\infty}\left[\sqrt{n}(\widehat{\boldsymbol{\theta}}_1^{\mathrm{SM}}-\boldsymbol{\theta}_1)\sqrt{n}(\widehat{\boldsymbol{\theta}}_1^{\mathrm{SM}}-\boldsymbol{\theta}_1)^\top\right]\\
&= E(\boldsymbol{\kappa\kappa}^\top) = var(\boldsymbol{\kappa}) + E(\boldsymbol{\kappa})E(\boldsymbol{\kappa}^\top) = \sigma^2\boldsymbol{G}_{11}^{-1} + \boldsymbol{G}_{11}^{-1}\boldsymbol{G}_{12}\boldsymbol{\delta\delta}^\top\boldsymbol{G}_{21}\boldsymbol{G}_{11}^{-1},
\end{aligned}$$

we first consider $\boldsymbol{\Gamma}^*(\widehat{\boldsymbol{\theta}}_1^{\mathrm{S}})$ *as follows:*

$$\begin{aligned}
\boldsymbol{\Gamma}^*(\widehat{\boldsymbol{\theta}}_1^{\mathrm{S}}) &= E\lim_{n\to\infty}\left[\sqrt{n}(\widehat{\boldsymbol{\theta}}_1^{\mathrm{S}}-\boldsymbol{\theta}_1)\sqrt{n}(\widehat{\boldsymbol{\theta}}_1^{\mathrm{S}}-\boldsymbol{\theta}_1)^\top\right]\\
&= E\lim_{n\to\infty}\left[\left(\boldsymbol{\zeta}_n - r\boldsymbol{\varrho}_n\frac{g(W_n)}{W_n}\right)\left(\boldsymbol{\zeta}_n - r\boldsymbol{\varrho}_n\frac{g(W_n)}{W_n}\right)^\top\right]\\
&= \underbrace{E(\boldsymbol{\zeta\zeta}^\top)}_{\boldsymbol{\Gamma}^*(\widehat{\boldsymbol{\theta}}_1^{\mathrm{FM}})} - 2r\underbrace{\mathbb{E}\left[\boldsymbol{\zeta\varrho}^\top\frac{g(\chi^2_{k_2}(\Delta))}{\chi^2_{k_2}(\Delta)}\right]}_{A_3} + r^2\underbrace{\mathbb{E}\left[\boldsymbol{\varrho\varrho}^\top\left(\frac{g(\chi^2_{k_2}(\Delta))}{\chi^2_{k_2}(\Delta)}\right)^2\right]}_{A_4}.
\end{aligned}$$

Applying Lemma 3.2 and (8.7), we get

$$\begin{aligned}
A_4 &= E\left[(\sigma^2\boldsymbol{\Sigma})^{\frac12}\boldsymbol{\varrho}^*[(\sigma^2\boldsymbol{\Sigma})^{\frac12}\boldsymbol{\varrho}^*]^\top\left(\frac{g(\chi^2_{k_2}(\Delta))}{\chi^2_{k_2}(\Delta)}\right)^2\right]\\
&= (\sigma^2\boldsymbol{\Sigma})^{\frac12}\left\{\begin{array}{c}\boldsymbol{I}_{k_1}E\left[\left(\frac{g(\chi^2_{k_2+2}(\Delta))}{\chi^2_{k_2+2}(\Delta)}\right)^2\right]\\ +(\sigma^2\boldsymbol{\Sigma})^{-\frac12}(\boldsymbol{\gamma\delta})((\sigma^2\boldsymbol{\Sigma})^{-\frac12}\boldsymbol{\gamma\delta})^\top E\left[\left(\frac{g(\chi^2_{k_2+4}(\Delta))}{\chi^2_{k_2+4}(\Delta)}\right)^2\right]\end{array}\right\}[(\sigma^2\boldsymbol{\Sigma})^{\frac12}]^\top\\
&= \sigma^2\boldsymbol{\Sigma}E\left[\left(\frac{g(W_1)}{W_1}\right)^2\right] + (\boldsymbol{\gamma\delta})(\boldsymbol{\gamma\delta})^\top E\left[\left(\frac{g(W_2)}{W_2}\right)^2\right],
\end{aligned}$$

where $W_1 = \chi^2_{k_2+2}(\Delta)$ *and* $W_2 = \chi^2_{k_2+4}(\Delta)$. *By using the definition of conditional expectation of the Lemma 3.2, and (8.7), therefore,* A_3 *becomes,*

$$\begin{aligned}
A_3 &= E\left[\mathbb{E}\left(\boldsymbol{\zeta}\boldsymbol{\varrho}^\top \frac{g(\chi^2_{k_2}(\Delta))}{\chi^2_{k_2}(\Delta)} \mid \boldsymbol{\varrho}\right)\right] = E\left[E(\boldsymbol{\zeta}|\boldsymbol{\varrho})\boldsymbol{\varrho}^\top \frac{g(\chi^2_{k_2}(\Delta))}{\chi^2_{k_2}(\Delta)}\right] \\
&= E\left[(\boldsymbol{\varrho} - \boldsymbol{\gamma\delta})\boldsymbol{\varrho}^\top \frac{g(\chi^2_{k_2}(\Delta))}{\chi^2_{k_2}(\Delta)}\right] \\
&= E\left[\boldsymbol{\varrho}\boldsymbol{\varrho}^\top \frac{g(\chi^2_{k_2}(\Delta))}{\chi^2_{k_2}(\Delta)}\right] - E\left[\boldsymbol{\gamma\delta}\boldsymbol{\varrho}^\top \frac{g(\chi^2_{k_2}(\Delta))}{\chi^2_{k_2}(\Delta)}\right] \\
&= \sigma^2 \boldsymbol{\Sigma} E\left[\frac{g(\chi^2_{k_2+2}(\Delta))}{\chi^2_{k_2+2}(\Delta)}\right] + (\boldsymbol{\gamma\delta})(\boldsymbol{\gamma\delta})^\top E\left[\frac{g(\chi^2_{k_2+4}(\Delta))}{\chi^2_{k_2+4}(\Delta)}\right] \\
&- (\boldsymbol{\gamma\delta})(\boldsymbol{\gamma\delta})^\top E\left[\frac{g(\chi^2_{k_2+2}(\Delta))}{\chi^2_{k_2+2}(\Delta)}\right].
\end{aligned}$$

Finally, the asymptotic covariance matrix of $\widehat{\boldsymbol{\theta}}_1^{\mathrm{S}}$ *is given as follows:*

$$\begin{aligned}
\boldsymbol{\Gamma}(\widehat{\boldsymbol{\theta}}_1^{\mathrm{S}}) &= \sigma^2 \boldsymbol{G}_{11.2}^{-1} - 2r\left\{\sigma^2\boldsymbol{\Sigma}E\left[\frac{g(W_1)}{W_1}\right] - (\boldsymbol{\gamma\delta})(\boldsymbol{\gamma\delta})^\top \right. \\
&\times \left.\left(E\left[\frac{g(W_1)}{W_1}\right] - E\left[\frac{g(W_2)}{W_2}\right]\right)\right\} \\
&+ r^2\left\{\sigma^2\boldsymbol{\Sigma}E\left[\left(\frac{g(W_1)}{W_1}\right)^2\right] + (\boldsymbol{\gamma\delta})(\boldsymbol{\gamma\delta})^\top E\left[\left(\frac{g(W_2)}{W_2}\right)^2\right]\right\} \\
&= \sigma^2 \boldsymbol{G}_{11.2}^{-1} - r\sigma^2 \boldsymbol{G}_{11}^{-1}\boldsymbol{G}_{12}\boldsymbol{G}_{22.1}^{-1}\boldsymbol{G}_{21}\boldsymbol{G}_{11}^{-1} \\
&\times \left(2E\left[\frac{g(W_1)}{W_1}\right] - rE\left[\left(\frac{g(W_1)}{W_1}\right)^2\right]\right) \\
&+ r\boldsymbol{G}_{11}^{-1}\boldsymbol{G}_{12}\boldsymbol{\delta}\boldsymbol{\delta}^\top \boldsymbol{G}_{21}\boldsymbol{G}_{11}^{-1} \\
&\times \left(2E\left[\frac{g(W_1)}{W_1}\right] - 2E\left[\frac{g(W_2)}{W_2}\right] + rE\left[\left(\frac{g(W_2)}{W_2}\right)^2\right]\right).
\end{aligned}$$

Finally, we derive the asymptotic covariance matrix of the general form of $\widehat{\boldsymbol{\theta}}_1^{\mathrm{PS}}$ *as follows:*

$$\begin{aligned}
\boldsymbol{\Gamma}^*(\widehat{\boldsymbol{\theta}}_1^{\mathrm{PS}}) &= E\lim_{n\to\infty}\left[\sqrt{n}(\widehat{\boldsymbol{\theta}}_1^{\mathrm{PS}} - \boldsymbol{\beta}_1)\sqrt{n}(\widehat{\boldsymbol{\theta}}_1^{\mathrm{PS}} - \boldsymbol{\beta}_1)^\top\right] \\
&= E\lim_{n\to\infty}\left[\begin{array}{c}\left\{\left(\boldsymbol{\zeta}_n - r\boldsymbol{\varrho}_n\frac{g(W_n)}{W_n}\right) - (1 - \frac{r\cdot g(W_n)}{W_n})\boldsymbol{\varrho}_n I\left(\frac{W_n}{g(W_n)} \le r\right)\right\} \\ \left\{\left(\boldsymbol{\zeta}_n - r\boldsymbol{\varrho}_n\frac{g(W_n)}{W_n}\right) - (1 - \frac{r\cdot g(W_n)}{W_n})\boldsymbol{\varrho}_n I\left(\frac{W_n}{g(W_n)} \le r\right)\right\}^\top\end{array}\right] \\
&= \underbrace{E\left[\left(\boldsymbol{\zeta} - r\boldsymbol{\varrho}\frac{g(\chi^2_{k_2}(\Delta))}{\chi^2_{k_2}(\Delta)}\right)\left(\boldsymbol{\zeta} - r\boldsymbol{\varrho}\frac{g(\chi^2_{k_2}(\Delta))}{\chi^2_{k_2}(\Delta)}\right)^\top\right]}_{\boldsymbol{\Gamma}^*(\widehat{\boldsymbol{\theta}}_1^{\mathrm{S}})} \\
&- 2E\left[\left(\boldsymbol{\zeta} - r\boldsymbol{\varrho}\frac{g(\chi^2_{k_2}(\Delta))}{\chi^2_{k_2}(\Delta)}\right)\left(1 - \frac{r\cdot g(\chi^2_{k_2}(\Delta))}{\chi^2_{k_2}(\Delta)}\right)\boldsymbol{\varrho}^\top\right.
\end{aligned}$$

$$
\times I\left(\frac{\chi^2_{k_2}(\Delta)}{g(\chi^2_{k_2}(\Delta))} \le r\right)\Bigg]
+ E\left[\left(1 - \frac{r\cdot g(\chi^2_{k_2}(\Delta))}{\chi^2_{k_2}(\Delta)}\right)^2 \boldsymbol{\varrho}\boldsymbol{\varrho}^\top I\left(\frac{\chi^2_{k_2}(\Delta)}{g(\chi^2_{k_2}(\Delta))} \le r\right)\right]
$$

$$
\begin{aligned}
\boldsymbol{\Gamma}^*(\widehat{\boldsymbol{\theta}}_1^{\mathrm{PS}}) = \boldsymbol{\Gamma}^*(\widehat{\boldsymbol{\theta}}_1^{\mathrm{S}}) - 2 &\underbrace{E\left[\left(1 - \frac{r\cdot g(\chi^2_{k_2}(\Delta))}{\chi^2_{k_2}(\Delta)}\right) \boldsymbol{\zeta}\boldsymbol{\varrho}^\top I\left(\frac{\chi^2_{k_2}(\Delta)}{g(\chi^2_{k_2}(\Delta))} \le r\right)\right]}_{A_5} \\
&\underbrace{+2\,E\left[\left(\frac{r\cdot g(\chi^2_{k_2}(\Delta))}{\chi^2_{k_2}(\Delta)}\right)\left(1 - \frac{r\cdot g(\chi^2_{k_2}(\Delta))}{\chi^2_{k_2}(\Delta)}\right) \boldsymbol{\varrho}\boldsymbol{\varrho}^\top I\left(\frac{\chi^2_{k_2}(\Delta)}{g(\chi^2_{k_2}(\Delta))} \le r\right)\right]}_{A_6} \\
&\underbrace{+\,E\left[\left(1 - \frac{r\cdot g(\chi^2_{k_2}(\Delta))}{\chi^2_{k_2}(\Delta)}\right)^2 \boldsymbol{\varrho}\boldsymbol{\varrho}^\top I\left(\frac{\chi^2_{k_2}(\Delta)}{g(\chi^2_{k_2}(\Delta))} \le r\right)\right]}_{A_7}.
\end{aligned} \tag{8.8}
$$

Using conditional expectation of the Lemma 3.2 and (8.7), *we obtain*

$$
\begin{aligned}
A_5 &= E\left[\left(1 - \frac{r\cdot g(\chi^2_{k_2}(\Delta))}{\chi^2_{k_2}(\Delta)}\right) E\left\{\boldsymbol{\zeta}\boldsymbol{\varrho}^\top I\left(\frac{\chi^2_{k_2}(\Delta)}{g(\chi^2_{k_2}(\Delta))} \le r\right) \Big| \boldsymbol{\varrho}\right\}\right] \\
&= E\left[\left(1 - \frac{r\cdot g(\chi^2_{k_2}(\Delta))}{\chi^2_{k_2}(\Delta)}\right) \boldsymbol{\varrho}\boldsymbol{\varrho}^\top I\left(\frac{\chi^2_{k_2}(\Delta)}{g(\chi^2_{k_2}(\Delta))} \le r\right)\right] \\
&- E\left[\left(1 - \frac{r\cdot g(\chi^2_{k_2}(\Delta))}{\chi^2_{k_2}(\Delta)}\right) \boldsymbol{\gamma}\boldsymbol{\delta}\boldsymbol{\varrho}^\top I\left(\frac{\chi^2_{k_2}(\Delta)}{g(\chi^2_{k_2}(\Delta))} \le r\right)\right] \\
&= \sigma^2\boldsymbol{\Sigma} E\left[\left(1 - \frac{r\cdot g(\chi^2_{k_2+2}(\Delta))}{\chi^2_{k_2+2}(\Delta)}\right) I\left(\frac{\chi^2_{k_2+2}(\Delta)}{g(\chi^2_{k_2+2}(\Delta))} \le r\right)\right] \\
&+ (\boldsymbol{\gamma}\boldsymbol{\delta})(\boldsymbol{\gamma}\boldsymbol{\delta})^\top E\left[\left(1 - \frac{r\cdot g(\chi^2_{k_2+4}(\Delta))}{\chi^2_{k_2+4}(\Delta)}\right) I\left(\frac{\chi^2_{k_2+4}(\Delta)}{g(\chi^2_{k_2+4}(\Delta))} \le r\right)\right] \\
&- (\boldsymbol{\gamma}\boldsymbol{\delta})(\boldsymbol{\gamma}\boldsymbol{\delta})^\top E\left[\left(1 - \frac{r\cdot g(\chi^2_{k_2+2}(\Delta))}{\chi^2_{k_2+2}(\Delta)}\right) I\left(\frac{\chi^2_{k_2+2}(\Delta)}{g(\chi^2_{k_2+2}(\Delta))} \le r\right)\right],
\end{aligned} \tag{8.9}
$$

$$
\begin{aligned}
A_6 &= \sigma^2\boldsymbol{\Sigma} E\left[\left(\frac{r\cdot g(\chi^2_{k_2+2}(\Delta))}{\chi^2_{k_2+2}(\Delta)}\right)\left(1 - \frac{r\cdot g(\chi^2_{k_2+2}(\Delta))}{\chi^2_{k_2+2}(\Delta)}\right)\right. \\
&\times I\left.\left(\frac{\chi^2_{k_2+2}(\Delta)}{g(\chi^2_{k_2+2}(\Delta))} \le r\right)\right] \\
&+ (\boldsymbol{\gamma}\boldsymbol{\delta})(\boldsymbol{\gamma}\boldsymbol{\delta})^\top E\left[\begin{array}{c}\left(\frac{r\cdot g(\chi^2_{k_2+4}(\Delta))}{\chi^2_{k_2+4}(\Delta)}\right) \\ \left(1 - \frac{r\cdot g(\chi^2_{k_2+4}(\Delta))}{\chi^2_{k_2+4}(\Delta)}\right) I\left(\frac{\chi^2_{k_2+4}(\Delta)}{g(\chi^2_{k_2+4}(\Delta))} \le r\right)\end{array}\right],
\end{aligned} \tag{8.10}
$$

$$A_7 = \sigma^2 \boldsymbol{\Sigma} E\left[\left(1 - \frac{r \cdot g(\chi^2_{k_2+2}(\Delta))}{\chi^2_{k_2+2}(\Delta)}\right)^2 I\left(\frac{\chi^2_{k_2+2}(\Delta)}{g(\chi^2_{k_2+2}(\Delta))} \le r\right)\right]$$
$$+ (\gamma\boldsymbol{\delta})(\gamma\boldsymbol{\delta})^\top E\left[\left(1 - \frac{r \cdot g(\chi^2_{k_2+4}(\Delta))}{\chi^2_{k_2+4}(\Delta)}\right)^2 I\left(\frac{\chi^2_{k_2+4}(\Delta)}{g(\chi^2_{k_2+4}(\Delta))} \le r\right)\right]. \quad (8.11)$$

Substituting (8.9), (8.10), *and* (8.11) *into* (8.8) *and rearranging the terms, we get*

$$\begin{aligned}
\boldsymbol{\Gamma}^*(\widehat{\boldsymbol{\theta}}_1^{\text{PS}}) = \boldsymbol{\Gamma}^*(\widehat{\boldsymbol{\theta}}_1^{\text{S}}) &- \sigma^2 \boldsymbol{G}_{11}^{-1}\boldsymbol{G}_{12}\boldsymbol{G}_{22.1}^{-1}\boldsymbol{G}_{21}\boldsymbol{G}_{11}^{-1} \\
&\times E\left[\left(1 - \frac{r \cdot g(W_1)}{W_1}\right)^2 I\left(\frac{W_1}{g(W_1)} \le r\right)\right] \\
&- \boldsymbol{G}_{11}^{-1}\boldsymbol{G}_{12}\boldsymbol{\delta}\boldsymbol{\delta}^\top\boldsymbol{G}_{21}\boldsymbol{G}_{11}^{-1} E\left[\left(1 - \frac{r \cdot g(W_2)}{W_2}\right)^2 I\left(\frac{W_2}{g(W_2)} \le r\right)\right] \\
&+ 2\boldsymbol{G}_{11}^{-1}\boldsymbol{G}_{12}\boldsymbol{\delta}\boldsymbol{\delta}^\top\boldsymbol{G}_{21}\boldsymbol{G}_{11}^{-1} E\left[\left(1 - \frac{r \cdot g(W_1)}{W_1}\right) I\left(\frac{W_1}{g(W_1)} \le r\right)\right].
\end{aligned}$$

The expression ADR of the estimators can be readily obtained from (8.4) using the asymptotic covariance matrix.

9

Shrinkage Strategies in Sparse Robust Regression Models

9.1 Introduction

In this chapter, we consider shrinkage estimation strategies in a sparse multiple regression containing some outlying observations. The classical least squares estimation strategy is sensitive to outliers and cannot be used as a base estimator; we refer to Montgomery et al. (2011) for some insights. Let us consider a data set $\{(\boldsymbol{x}_i^\top, y_i)\}$, $i = 1, 2, \ldots, n$, where $\boldsymbol{x}_i \in \mathbb{R}^p$ are predicting variables and $y_i \in \mathbb{R}$ is the response variable. We have the following multiple regression model:

$$y_i = \boldsymbol{x}_i^\top \boldsymbol{\beta} + \epsilon_i, \quad i = 1, 2, \ldots, n, \tag{9.1}$$

where $\boldsymbol{\beta} = (\beta_1, \beta_2, \ldots, \beta_p)^\top$ is an unknown regression parameter vector, and ϵ_i is normally distributed with mean 0, variance σ^2, and independent of $\boldsymbol{x}_i$. In the usual situation, we estimate $\boldsymbol{\beta}$ by the least squares estimator which minimizes the sum of squared residuals, $\sum_{i=1}^n (y_i - \boldsymbol{x}_i^\top \boldsymbol{\beta})^2$. The least squares estimator has some optimal theoretical properties and is computationally desirable, see Chapter 3 for detailed information. However, when the distribution of the errors is non-normal and the data has outliers, the LSE is sensitive to such assumptions. In this chapter, we consider a more realistic problem when the response variable contains outliers and assume that the model is sparse. Therefore, we consider a robust estimator, the least absolute deviation (LAD) estimator, for estimating the regression parameters. This estimator minimizes the sum of the absolute values of the residuals,

$$\sum_{i=1}^n |y_i - \boldsymbol{x}_i^\top \boldsymbol{\beta}|. \tag{9.2}$$

It has been documented in the literature that the LAD estimator is relatively insensitive to outliers. However, the analytical minimization problem for (9.2) is not feasible. There are some methods available in reviewed literature that solve the problem via an iterative algorithm. One popular method is a simplex, based on the Barrodale-Roberts algorithm that can be found in R software (e.g. function rq in *quantreg* package).

Let us turn our attention to sparsity in the model. Model (9.1) can be written in matrix notation as

$$\boldsymbol{y} = \boldsymbol{X}\boldsymbol{\beta} + \boldsymbol{\epsilon}, \tag{9.3}$$

where $\boldsymbol{y} = (y_1, y_2, \ldots, y_n)^\top$ is the vector of responses, $\boldsymbol{X} = (\boldsymbol{x}_1^\top, \boldsymbol{x}_2^\top, \ldots, \boldsymbol{x}_n^\top)$ is an $n \times p$ fixed design matrix, and $\boldsymbol{\epsilon} = (\epsilon_1, \epsilon_2, \ldots, \epsilon_n)^\top$ is the vector of unobservable random errors that has a cumulative distribution function $F(\boldsymbol{\epsilon})$ with median zero. Let us partition the regression parameters as $\boldsymbol{\beta} = (\boldsymbol{\beta}_1^\top, \boldsymbol{\beta}_2^\top)^\top$, where $\boldsymbol{\beta}_1 = (\beta_1, \beta_2, \ldots, \beta_{p_1})^\top$ and $\boldsymbol{\beta}_2 = (\beta_{p_1+1}, \beta_{p_1+2}, \ldots, \beta_{p_1+p_2})^\top$, $p = p_1 + p_2$, where p_1 and p_2 are the dimensions of the active and inactive predictors, respectively. We are primarily interested in the estimation of $\boldsymbol{\beta}_1$ associated with the active predictors when $\boldsymbol{\beta}_2$ may be close to zero. The equation

DOI: 10.1201/9781003170259-9

(9.3) can be rewritten as:

$$\boldsymbol{y} = \boldsymbol{X}_1\boldsymbol{\beta}_1 + \boldsymbol{X}_2\boldsymbol{\beta}_2 + \boldsymbol{\epsilon}, \tag{9.4}$$

where $\boldsymbol{X}_1$ and $\boldsymbol{X}_2$ are assumed to have dimensions $n \times p_1$ and $n \times p_2$, respectively. Again, the aim is to estimate $\boldsymbol{\beta}_1$ when $\boldsymbol{\beta}_2 = \mathbf{0}$.

The problem of selecting a model under sparsity in low-dimensional cases is studied extensively in the reviewed literature, which includes AIC, BIC, and the Mallows-Cp statistic. However, most selection criteria are based on the maximum likelihood/least square estimation and are robust-lacking estimation strategies. Under a heavy-tailed error distribution, the performance of these estimators may not be desirable and may lead to incorrect conclusions. Hurvish and Tsai (1990) proposed some useful model selection methods based on LAD estimates. However, these methods have some limitations. Noting that, if the number of the predictors is relatively large then finding the best submodel by considering all possible candidate models will be extremely difficult and inefficient. The number of possible candidate models increases exponentially as the number of predictors increases. For such high-dimensional cases, Wang et al. (2007) developed a robust shrinkage and selection method that can perform the same tasks as the least absolute shrinkage and selection operator. This procedure performs robustly in the presence of outliers and/or heavy-tailed errors like the LAD estimator. Yüzbaşi et al. (2018) proposed a combined correlation-based estimation with Lasso penalty for quantile regression in high-dimensional sparse models.

Yüzbaşı et al. (2019) suggested shrinkage strategies by using the LAD as a benchmark estimator in the presence of outliers when the model is sparse. Also, we refer to some non-parametric shrinkage regression estimation, Ahmed (1997b, 1998); Ahmed and Md. E. Saleh (1999); Norouzirad et al. (2017); Yüzbaşı et al. (2018); Arashi et al. (2018); Ahmed et al. (2006).

In Section 9.2, the shrinkage LAD estimators are defined, and their asymptotic properties are given in Section 9.2.1. Section 9.3 contains a Monte Carlo simulation study to numerically appraise the relative performance of the listed estimators. The real data example is considered in Section 9.5. A high-dimensional case is introduced in section 9.6, followed by numerical work in Subsections 9.6.1- 9.6.2. The R codes can be found in Section 9.7. We conclude our results in Section 9.8. Proof of the theorems is provided in the appendix.

9.2 LAD Shrinkage Strategies

The main objective is to estimate $\boldsymbol{\beta}_1$ when $\boldsymbol{\beta}_2 = \mathbf{0}$ more precisely when $\boldsymbol{\beta}_2$ maybe a null vector. We consider the decomposed model in (9.4). The full model LAD estimator of $\boldsymbol{\beta}$ denoted by $\widehat{\boldsymbol{\beta}}^{\mathrm{FM}}$ is the minimum of the objective function,

$$\left\|\boldsymbol{y} - \boldsymbol{X}\boldsymbol{\beta}\right\|_1, \tag{9.5}$$

where $\|\boldsymbol{v}\|_1 = \sum_{i=1}^{p}|v_i|$ is the L_1 norm, for $\boldsymbol{v} = (v_1, v_2, \ldots, v_p)^\top$. Thus, $\widehat{\boldsymbol{\beta}}_1^{\mathrm{FM}}$ is a full model LAD estimator of $\boldsymbol{\beta}_1$. However, under the sparsity assumption of $\boldsymbol{\beta}_2 = \mathbf{0}$, the model in Eq. (9.4) reduces to

$$\boldsymbol{y} = \boldsymbol{X}_1\boldsymbol{\beta}_1 + \boldsymbol{\epsilon}. \tag{9.6}$$

Thus, the submodel LAD estimator of $\boldsymbol{\beta}_1$, $\widehat{\boldsymbol{\beta}}_1^{\mathrm{SM}}$ is the minimizer of the following objective function

$$\left\|\boldsymbol{y} - \boldsymbol{X}_1\boldsymbol{\beta}_1\right\|_1. \tag{9.7}$$

The following assumptions are imposed for ensuring the $\sqrt{n}$-consistency and other asymptotic properties of the LAD estimator, see Bassett and Koenker (1978).

(A) $F(\epsilon)$ is continuous and has continuous positive density $f(\epsilon)$ at the median,

(B) For a positive definite (p.d.) matrix $\boldsymbol{C}$, $\lim_{n\to\infty} \frac{1}{n}\boldsymbol{X}^\top \boldsymbol{X} = \boldsymbol{C}$, and $\frac{1}{n}\max_{1\le i\le n} \boldsymbol{x}_i^\top \boldsymbol{x}_i \to 0$, where $\boldsymbol{C} = \begin{bmatrix} \boldsymbol{C}_{11} & \boldsymbol{C}_{12} \\ \boldsymbol{C}_{21} & \boldsymbol{C}_{22} \end{bmatrix}$.

Generally speaking, when $\boldsymbol{\beta}_2 = \boldsymbol{0}$, the submodel LAD estimator will have a smaller asymptotic dispersion than the full model LAD estimator. However, for $\boldsymbol{\beta}_2 \neq \boldsymbol{0}$ the $\widehat{\boldsymbol{\beta}}_1^{\mathrm{SM}}$ may be biased and inconsistent in many cases.

For this reason, we combine full model and submodel estimators, using the shrinkage strategy to improve the performance of the submodel estimator. To develop the shrinkage strategy we use a normalized distance:

$$\mathcal{L}_n = \frac{\left(\widehat{\boldsymbol{\beta}}_2^{\mathrm{LAD}}\right)^\top \boldsymbol{C}_{22.1}\left(\widehat{\boldsymbol{\beta}}_2^{\mathrm{LAD}}\right)}{\tau^2}, \quad \tau^2 = [2f(0)]^{-2}, \tag{9.8}$$

where $\boldsymbol{C}_{rr.s} = \boldsymbol{C}_{rr} - \boldsymbol{C}_{rs}\boldsymbol{C}_{ss}^{-1}\boldsymbol{C}_{sr}$, for $s \neq r = 1, 2$.

Theorem 9.1 *Assuming (A) and (B) hold, the normalized distance $\mathcal{L}_n$, is given by The statistics $\mathcal{L}_n$ has a χ^2-distribution with p_2 degrees of freedom (d.f.), when $\boldsymbol{\beta}_2 = \boldsymbol{0}$.*

Proof is found in the appendix.

Obviously, the $\mathcal{L}_n$ depends on the error density through τ^2. This density is unknown, so we use a non-parametric estimator in a general form:

$$\widehat{f}(e) = \frac{1}{h}K\left(\frac{e - e_i}{h}\right),$$

where $h = h_n$ is the bandwidth that approaches 0 as $n \to \infty$ and the kernel function $K(\cdot)$ satisfies $\int K(u)du = 1$.

Following Ahmed (2014), the shrinkage LAD estimator of $\boldsymbol{\beta}_1$ denoted by $\widehat{\boldsymbol{\beta}}_1^{\mathrm{S}}$ is defined as

$$\begin{aligned} \widehat{\boldsymbol{\beta}}_1^{\mathrm{S}} &= \widehat{\boldsymbol{\beta}}_1^{\mathrm{FM}} - (p_2 - 2)\mathcal{L}_n^{-1}\left(\widehat{\boldsymbol{\beta}}_1^{\mathrm{FM}} - \widehat{\boldsymbol{\beta}}_1^{\mathrm{SM}}\right), \\ &= \widehat{\boldsymbol{\beta}}_1^{\mathrm{SM}} + (1 - (p_2 - 2)\mathcal{L}_n^{-1})\left(\widehat{\boldsymbol{\beta}}_1^{\mathrm{FM}} - \widehat{\boldsymbol{\beta}}_1^{\mathrm{SM}}\right), \quad p_2 \geq 3. \end{aligned} \tag{9.9}$$

To avoid over-shrinkage, the positive-rule shrinkage LAD estimator is given by

$$\widehat{\boldsymbol{\beta}}_1^{\mathrm{PS}} = \widehat{\boldsymbol{\beta}}_1^{\mathrm{SM}} + (1 - (p_2 - 2)\mathcal{L}_n^{-1})^+\left(\widehat{\boldsymbol{\beta}}_1^{\mathrm{FM}} - \widehat{\boldsymbol{\beta}}_1^{\mathrm{SM}}\right), \quad p_2 \geq 3,$$

where $s^+ = \max\{0, s\}$. This estimator can alternatively written as

$$\begin{aligned} \widehat{\boldsymbol{\beta}}_1^{\mathrm{PS}} &= \widehat{\boldsymbol{\beta}}_1^{\mathrm{SM}} + (1 - (p_2 - 2)\mathcal{L}_n^{-1})\left(\widehat{\boldsymbol{\beta}}_1^{\mathrm{FM}} - \widehat{\boldsymbol{\beta}}_1^{\mathrm{SM}}\right) I(\mathcal{L}_n > (p_2 - 2)), \\ &= \widehat{\boldsymbol{\beta}}_1^{\mathrm{S}} - (1 - (p_2 - 2)\mathcal{L}_n^{-1})\left(\widehat{\boldsymbol{\beta}}_1^{\mathrm{FM}} - \widehat{\boldsymbol{\beta}}_1^{\mathrm{SM}}\right) I(\mathcal{L}_n \leq (p_2 - 2)). \end{aligned} \tag{9.10}$$

9.2.1 Asymptotic Properties

Here we present the expressions for the asymptotic distributional bias (ADB), quadratic asymptotic distributional bias (QADB), and asymptotic distributional risk (ADR) of the suggested estimators. The following theorem enables us to achieve desirable results.

Theorem 9.2 *Assume (A) and (B) are held, then*

$$\begin{aligned}\sqrt{n}\left(\widehat{\boldsymbol{\beta}}_1^{\mathrm{FM}}-\boldsymbol{\beta}_1\right) &\sim \mathcal{N}_{p_1}\left(\mathbf{0},\tau^2\boldsymbol{C}_{11.2}^{-1}\right),\\ \sqrt{n}\left(\widehat{\boldsymbol{\beta}}_1^{\mathrm{SM}}-\boldsymbol{\beta}_1\right) &\sim \mathcal{N}_{p_1}\left(\mathbf{0},\tau^2\boldsymbol{C}_{11}^{-1}\right).\end{aligned}$$

The proof is given in the Appendix.

Consider the following sequence to obtain asymptotic results when the sparsity condition may not hold.

$$\mathcal{K}_{(n)}:\boldsymbol{\beta}_2=\boldsymbol{\beta}_{2(n)}=n^{-1/2}\boldsymbol{\xi},\quad \boldsymbol{\xi}=(\xi_{p_1+1},\ldots,\xi_{p_1+p_2})^\top\in\mathbb{R}^{p_2}. \tag{9.11}$$

Proposition 9.3 *Under* $\left\{\mathcal{K}_{(n)}\right\}$,

$$\widehat{\boldsymbol{\beta}}_1^{\mathrm{SM}}=\widehat{\boldsymbol{\beta}}_1^{\mathrm{FM}}+\boldsymbol{C}_{11}^{-1}C_{12}\widehat{\boldsymbol{\beta}}_2^{\mathrm{LAD}}+o(n^{-1/2}). \tag{9.12}$$

The results follow by using similar arguments as Lawless and Singhal (1978).

9.2.2 Bias of the Estimators

The ADB of an estimator $\widehat{\boldsymbol{\beta}}_1^*$ is defined as

$$\mathrm{ADB}(\widehat{\boldsymbol{\beta}}_1^*)=\lim_{n\to\infty}E\left[\sqrt{n}(\widehat{\boldsymbol{\beta}}_1^*-\boldsymbol{\beta}_1)\right]. \tag{9.13}$$

Theorem 9.4 *Assume (A) and (B) hold. Under* $\left\{\mathcal{K}_{(n)}\right\}$ *for* $p_2\geq 3$, *the* ADB*s of the estimators are given by*

$$\begin{aligned}\mathrm{ADB}(\widehat{\boldsymbol{\beta}}_1^{\mathrm{FM}}) &= \mathbf{0},\\ \mathrm{ADB}(\widehat{\boldsymbol{\beta}}_1^{\mathrm{SM}}) &= -\boldsymbol{C}_{11}^{-1}C_{12}\boldsymbol{\xi}=-\boldsymbol{\delta},\\ \mathrm{ADB}(\widehat{\boldsymbol{\beta}}_1^{\mathrm{S}}) &= -(p_2-2)\boldsymbol{\delta}E\left[\chi_{p_2+2}^{-2}(\Delta)\right],\\ \mathrm{ADB}(\widehat{\boldsymbol{\beta}}_1^{\mathrm{PS}}) &= ADB(\widehat{\boldsymbol{\beta}}_1^{\mathrm{S}})-\boldsymbol{\delta}E\left[\left(1-(p_2-2)\chi_{p_2+2}^{-2}(\Delta)\right)\right.\\ &\times\ \left.I\left(\chi_{p_2+2}^{2}(\Delta)\leq(p_2-2)\right)\right],\end{aligned}$$

where $E\left[\chi_\nu^{-2}(\Delta)\right]$ *is the expected value of the inverse of a non-central* χ_ν^2 *distribution with* ν *d.f. and non-centrality parameter* Δ *and* $H_\nu(\cdot,\Delta)$ *is the cumulative distribution function (c.d.f.) of a non-central* χ_ν^2 *distribution.*

The bias expressions for all estimators are not in scalar form. Hence, we use the QADB.

$$\mathrm{QADB}(\widehat{\boldsymbol{\beta}}_1^*)=\tau^{-2}[\mathrm{ADB}(\widehat{\boldsymbol{\beta}}_1^*)]^\top C_{11.2}[\mathrm{ADB}(\widehat{\boldsymbol{\beta}}_1^*)]. \tag{9.14}$$

Thus ,the QADBs of the estimators are given by

$$\begin{aligned}\mathrm{QADB}(\widehat{\boldsymbol{\beta}}_1^{\mathrm{FM}}) &= 0,\\ \mathrm{QADB}(\widehat{\boldsymbol{\beta}}_1^{\mathrm{SM}}) &= \boldsymbol{\delta}^\top\boldsymbol{C}_{11.2}\boldsymbol{\delta},\\ \mathrm{QADB}(\widehat{\boldsymbol{\beta}}_1^{\mathrm{S}}) &= (p_2-2)^2\boldsymbol{\delta}^\top\boldsymbol{C}_{11.2}\boldsymbol{\delta}\left\{E\left[\chi_{p_2+2}^{-2}(\Delta)\right]\right\}^2,\\ \mathrm{QADB}(\widehat{\boldsymbol{\beta}}_1^{\mathrm{PS}}) &= \boldsymbol{\delta}^\top\boldsymbol{C}_{11.2}\boldsymbol{\delta}\left\{(p_2-2)E\left[\chi_{p_2+2}^{-2}(\Delta)\right]\right.\\ &-\ \left.E\left[\left(1-(p_2-2)\chi_{p_2+2}^{-2}(\Delta^2)\right)I\left(\chi_{p_2+2}^{2}(\Delta)<(p_2-2)\right)\right]\right\}^2.\end{aligned}$$

For $\boldsymbol{C}_{12}=\mathbf{0}$, the above formula concludes $\boldsymbol{C}_{21}\boldsymbol{C}_{11}^{-1}\boldsymbol{C}_{11}^{-1}\boldsymbol{C}_{12}=\mathbf{0}$ and $\boldsymbol{C}_{11.2}=\boldsymbol{C}_{11}$. Consequently, all the QADBs reduce to the common value zero for all $\boldsymbol{\xi}$. Thus, all these expressions become QADB-equivalent. We then consider $\boldsymbol{C}_{12}\neq\mathbf{0}$ and the asymptotic bias properties of these estimators are the same as reported in earlier chapters.

9.2.3 Risk of Estimators

Consider the quadratic error loss of the form $n\left(\widehat{\boldsymbol{\beta}}_1^* - \boldsymbol{\beta}_1\right)^\top \boldsymbol{W}\left(\widehat{\boldsymbol{\beta}}_1^* - \boldsymbol{\beta}_1\right)$. For a positive definite matrix $\boldsymbol{W}$, the ADR of $\widehat{\boldsymbol{\beta}}_1^*$ is defined by

$$\mathrm{ADR}(\widehat{\boldsymbol{\beta}}_1^*; \boldsymbol{\beta}_1) = \lim_{n\to\infty} E\left[n\left(\widehat{\boldsymbol{\beta}}_1^* - \boldsymbol{\beta}_1\right)^\top \boldsymbol{W}\left(\widehat{\boldsymbol{\beta}}_1^* - \boldsymbol{\beta}_1\right)\right]. \tag{9.15}$$

Theorem 9.5 *Assume (A) and (B) are held. Under* $\{\mathcal{K}_{(n)}\}$, *the* ADR*s of the estimators are given by*

$$\begin{aligned}
\mathrm{ADR}(\widehat{\boldsymbol{\beta}}_1^{\mathrm{FM}}; \boldsymbol{W}) &= \tau^2 tr(\boldsymbol{W}\boldsymbol{C}_{11.2}^{-1}),\\
\mathrm{ADR}(\widehat{\boldsymbol{\beta}}_1^{\mathrm{SM}}; \boldsymbol{W}) &= \tau^2 tr(\boldsymbol{W}\boldsymbol{C}_{11}^{-1}) + \boldsymbol{\delta}^\top \boldsymbol{W}\boldsymbol{\delta},\\
\mathrm{ADR}(\widehat{\boldsymbol{\beta}}_1^{\mathrm{S}}; \boldsymbol{W}) &= \tau^2 tr(\boldsymbol{W}\boldsymbol{C}_{11.2}^{-1}) + (p_2^2 - 4)\boldsymbol{\delta}^\top \boldsymbol{W}\boldsymbol{\delta} E\left[\chi_{p_2+4}^{-4}(\Delta)\right]\\
&\quad -\tau^2 (p_2 - 2) tr(\boldsymbol{C}_{21}\boldsymbol{C}_{11}^{-1}\boldsymbol{W}\boldsymbol{C}_{11}^{-1}\boldsymbol{C}_{12})\\
&\quad \times\left\{2E\left[\chi_{p_2+2}^{-2}(\Delta)\right] - (p_2-2)E\left[\chi_{p_2+2}^{-4}(\Delta)\right]\right\},\\
\mathrm{ADR}(\widehat{\boldsymbol{\beta}}_1^{\mathrm{PS}}; \boldsymbol{W}) &= ADR(\widehat{\boldsymbol{\beta}}_1^{\mathrm{S}}; \boldsymbol{W}) - \tau^2 tr(\boldsymbol{C}_{21}\boldsymbol{C}_{11}^{-1}\boldsymbol{W}\boldsymbol{C}_{11}^{-1}\boldsymbol{C}_{12}\boldsymbol{C}_{22.1}^{-1})\\
&\quad \times H_{p_2+2}(p_2 - 2; \Delta)\\
&\quad +\tau^2 (p_2 - 2) tr(\boldsymbol{C}_{21}\boldsymbol{C}_{11}^{-1}\boldsymbol{W}\boldsymbol{C}_{11}^{-1}\boldsymbol{C}_{12})\\
&\quad \times\left\{2E\left[\chi_{p_2+2}^{-2}(\Delta) I(\chi_{p_2+2}^{2}(\Delta) \le p_2 - 2)\right]\right.\\
&\quad\quad \left.-(p_2-2)E\left[\chi_{p_2+2}^{-4}(\Delta) I(\chi_{p_2+2}^{2}(\Delta) \le p_2 - 2)\right]\right\}\\
&\quad -\boldsymbol{\delta}^\top \boldsymbol{W}\boldsymbol{\delta}\boldsymbol{I}g\{2H_{p_2+2}(p_2 - 2; \Delta) - H_{p_2+4}(p_2 - 2; \Delta)\boldsymbol{I}g\}\\
&\quad -(p_2-2)\boldsymbol{\delta}^\top \boldsymbol{W}\boldsymbol{\delta}\left\{2E\left[\chi_{p_2+2}^{-2}(\Delta) I(\chi_{p_2+2}^{-2}(\Delta) \le p_2 - 2)\right]\right.\\
&\quad -2E\left[\chi_{p_2+4}^{-2}(\Delta) I(\chi_{p_2+4}^{-2}(\Delta) \le p_2 - 2)\right]\\
&\quad \left.+(p_2-2)E\left[\chi_{p_2+2}^{-4}(\Delta) I(\chi_{p_2+2}^{-2}(\Delta) \le p_2 - 2)\right]\right\}.
\end{aligned}$$

For $\boldsymbol{C}_{12} = \boldsymbol{0}$ according to the above formula, $\boldsymbol{C}_{21}\boldsymbol{C}_{11}^{-1}\boldsymbol{W}\boldsymbol{C}_{11}^{-1}\boldsymbol{C}_{12} = \boldsymbol{0}$ and $\boldsymbol{C}_{11.2} = \boldsymbol{C}_{11}$ are resulted; therefore all the ADR reduce to common value $\tau^2 \mathrm{tr}(\boldsymbol{W}\boldsymbol{C}_{11}^{-1})$ for all $\boldsymbol{\xi}$. Hence, all these estimators become ADR-equivalent so $\boldsymbol{C}_{12} \neq \boldsymbol{0}$ can be assumed. According to the asymptotic results, the order of dominance of the listed estimators under the sparsity assumption may be ordered as follows:

$$\mathrm{ADR}(\widehat{\boldsymbol{\beta}}_1^{\mathrm{SM}}; \boldsymbol{W}) \le \mathrm{ADR}(\widehat{\boldsymbol{\beta}}_1^{\mathrm{PS}}; \boldsymbol{W}) \le \mathrm{ADR}(\widehat{\boldsymbol{\beta}}_1^{\mathrm{S}}; \boldsymbol{W}).$$

Conversely, if the strict sparsity assumption fails to hold, then

$$\mathrm{ADR}(\widehat{\boldsymbol{\beta}}_1^{\mathrm{PS}}; \boldsymbol{W}) \le \mathrm{ADR}(\widehat{\boldsymbol{\beta}}_1^{\mathrm{S}}; \boldsymbol{W}) \le \mathrm{ADR}(\widehat{\boldsymbol{\beta}}_1^{\mathrm{FM}}; \boldsymbol{W}).$$

When the sparsity condition may not hold, the behavior of the estimators is the same as reported in earlier chapters for other models.

9.3 Simulation Experiments

In this section, we present the details of the Monte Carlo simulation study. We simulate the response from the following model:

$$y_i = x_{1i}\beta_1 + x_{2i}\beta_2 + ... + x_{pi}\beta_p + \varepsilon_i\ ,\ i = 1, 2, ..., n, \tag{9.16}$$

where x_i and ε_i are i.i.d. $N(0,1)$. In this study, ε_i is considered to be a normal distribution, a heavy tailed t_5-distribution, and a χ^2_5 distribution.

We let $\boldsymbol{\beta} = \left(\boldsymbol{\beta}_1^\top, \boldsymbol{\beta}_2^\top, \boldsymbol{\beta}_3^\top\right)^\top$ with dimensions p_1, p_2 and p_3, respectively. In order to investigate the behavior of the estimators when $\boldsymbol{\beta}_3 = \mathbf{0}$ is violated, we define $\Delta = \|\boldsymbol{\beta} - \boldsymbol{\beta}_0\|$, where $\boldsymbol{\beta}_0 = \left(\mathbf{1}_{p_1}^\top, \mathbf{0.1}_{p_2}^\top, \mathbf{0}_{p_3}^\top\right)^\top$. Clearly, $\boldsymbol{\beta}_1$ represents strong signals when $\boldsymbol{\beta}_1$ is a vector of 1 values, $\boldsymbol{\beta}_2$ stands for the weak signals of 0.1 values, and $\boldsymbol{\beta}_3$ means no signals as $\boldsymbol{\beta}_3 = \mathbf{0}$.

In this simulation setting, 1000 data sets consisting of $n = 100, 500$, with $p_1 = 4, 8$, $p_2 = 0, 3, 6, 9$ and $p_3 = 4, 8, 12, 16$. The performance of an estimator is evaluated by using the mean absolute prediction error (MAPE):

$$\text{MAPE}\left(\widehat{\boldsymbol{\beta}}_1^*\right) = \frac{1}{p}\sum |\boldsymbol{\beta}_1 - \widehat{\boldsymbol{\beta}}_1^*|, \tag{9.17}$$

where $\widehat{\boldsymbol{\beta}}_1^*$ is one of the listed estimators.

The relative MAPE (RMAPE) of the $\widehat{\boldsymbol{\beta}}_1^*$ to the $\widehat{\boldsymbol{\beta}}_1^{\text{FM}}$ is also calculated and is given by

$$\text{RMAPE}(\widehat{\boldsymbol{\beta}}_1^{\text{FM}}; \widehat{\boldsymbol{\beta}}_1^*) = \frac{\text{MAPE}(\widehat{\boldsymbol{\beta}}_1^{\text{FM}})}{\text{MAPE}(\widehat{\boldsymbol{\beta}}_1^*)}. \tag{9.18}$$

If the value of RMAPE is greater than 1, it is indicative of the degree of superiority of the selected estimator over the full model estimator.

For some useful values of Δ, the results are reported in Figures 9.1–9.4.

Tables 9.1–9.12 present the RMAPE for $n = 100, 500$, and selected values of p_1, p_2, p_3. According to Tables 9.1–9.12 and Figures 9.1–9.4, we summarize the finding as follows:

- When $\Delta = 0$, the submodel estimator performs better than the shrinkage estimator for all the configurations considered in the simulation study, as expected.
- However, when $\Delta > 0$ meaning the sparsity assumption is not correct, the RMAPE of the submodel estimator decays sharply and converges toward 0. On the other hand, the RMAPE of the positive shrinkage estimator approaches. This indicates that in the event of an imprecise sparsity assumption, that is even if $\boldsymbol{\beta}_2 \neq 0$, the shrinkage estimator performance is preferable.
- More importantly, as the number of weak signals (p_2) and no signals (p_3) increase, the relative performance of the shrinkage strategy is notable. For example, when $\Delta = 0, p_2 = 0$ the RMAPE of shrinkage estimator is 1.458 when $p_3 = 4$, it increases to 3.460 when $p_3 = 16$, as shown in Table 9.1. A similar pattern is observed when (p_3) increases.

Tables 9.5–9.8 give the simulated RMAPE when the data is generated from a χ^2_5 distribution. The performance of the positive shrinkage estimator is observed to be as good as the normal case. We can safely conclude that the shrinkage estimator retains its dominant property and behaves as a robust estimator when samples are collected from a skewed distribution since the LAD estimator is used as the base estimator. The shrinkage estimator behaves robustly and efficiently when the parent population is heavy-tailed, as evident from the results reported in Tables 9.9–9.12.

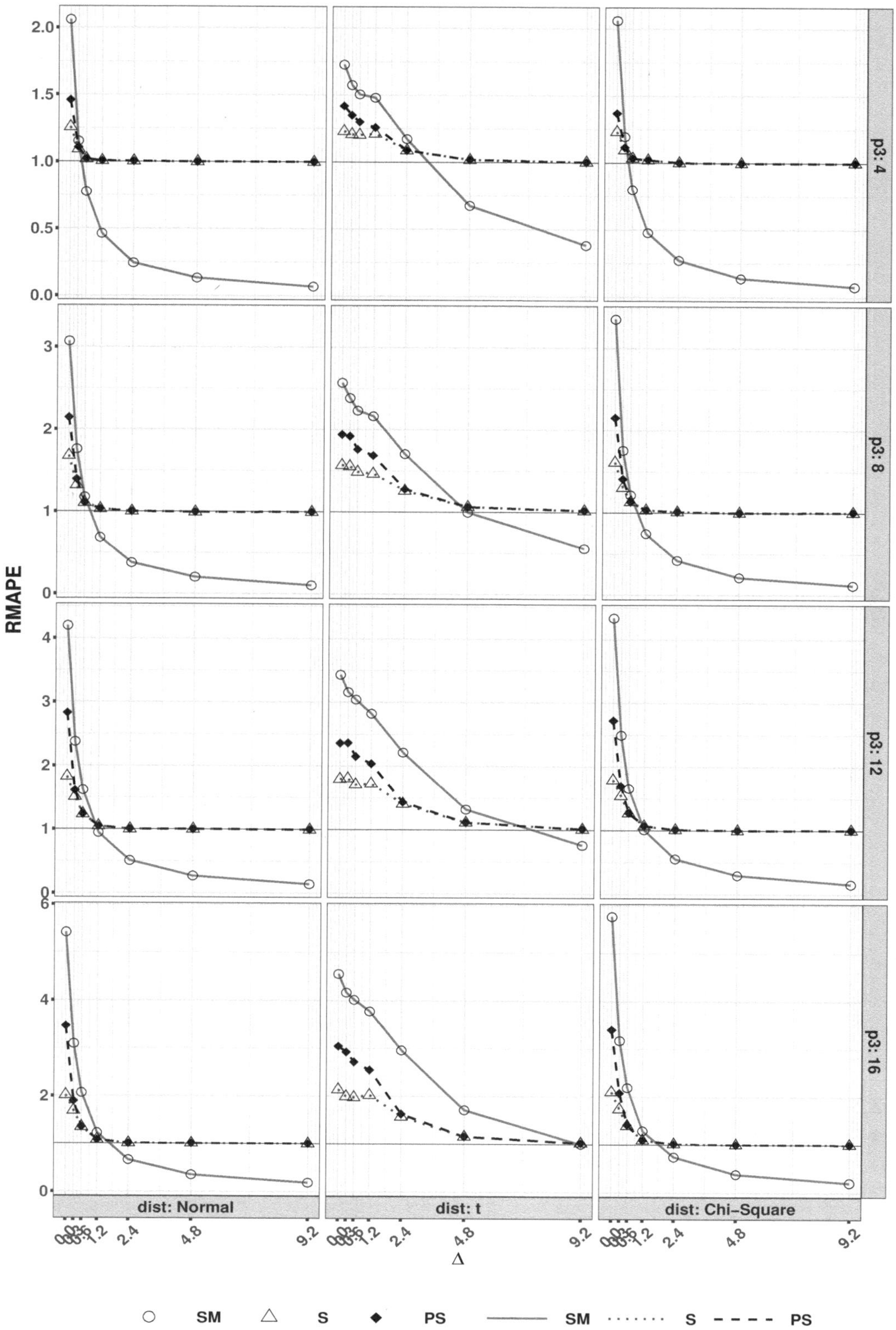

FIGURE 9.1: RMAPE of the Estimators for $n = 100$, $p_1 = 4$ and $p_2 = 0$.

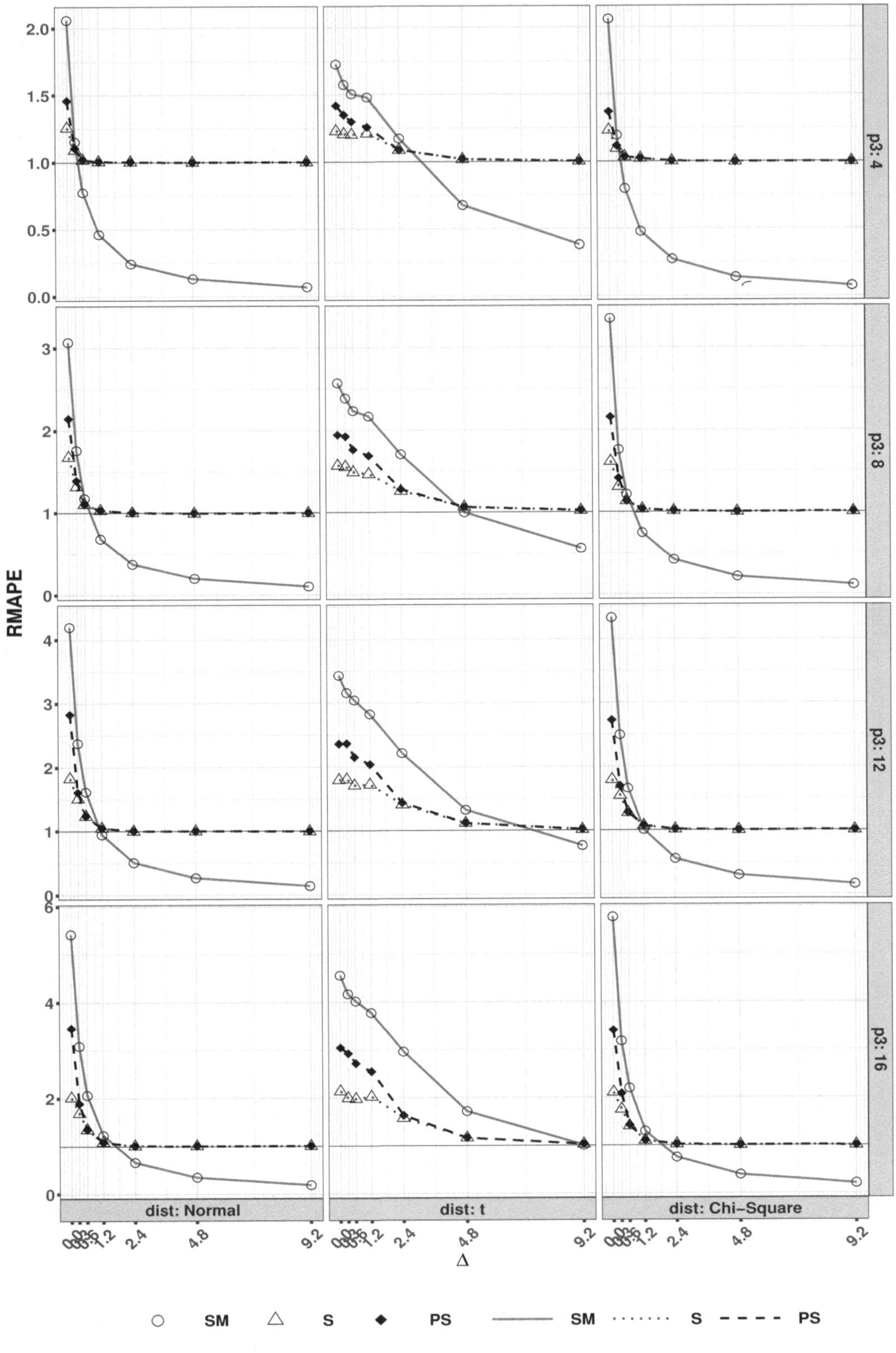

FIGURE 9.2: RMAPE of the Estimators for $n = 100$, $p_1 = 4$ and $p_2 = 6$.

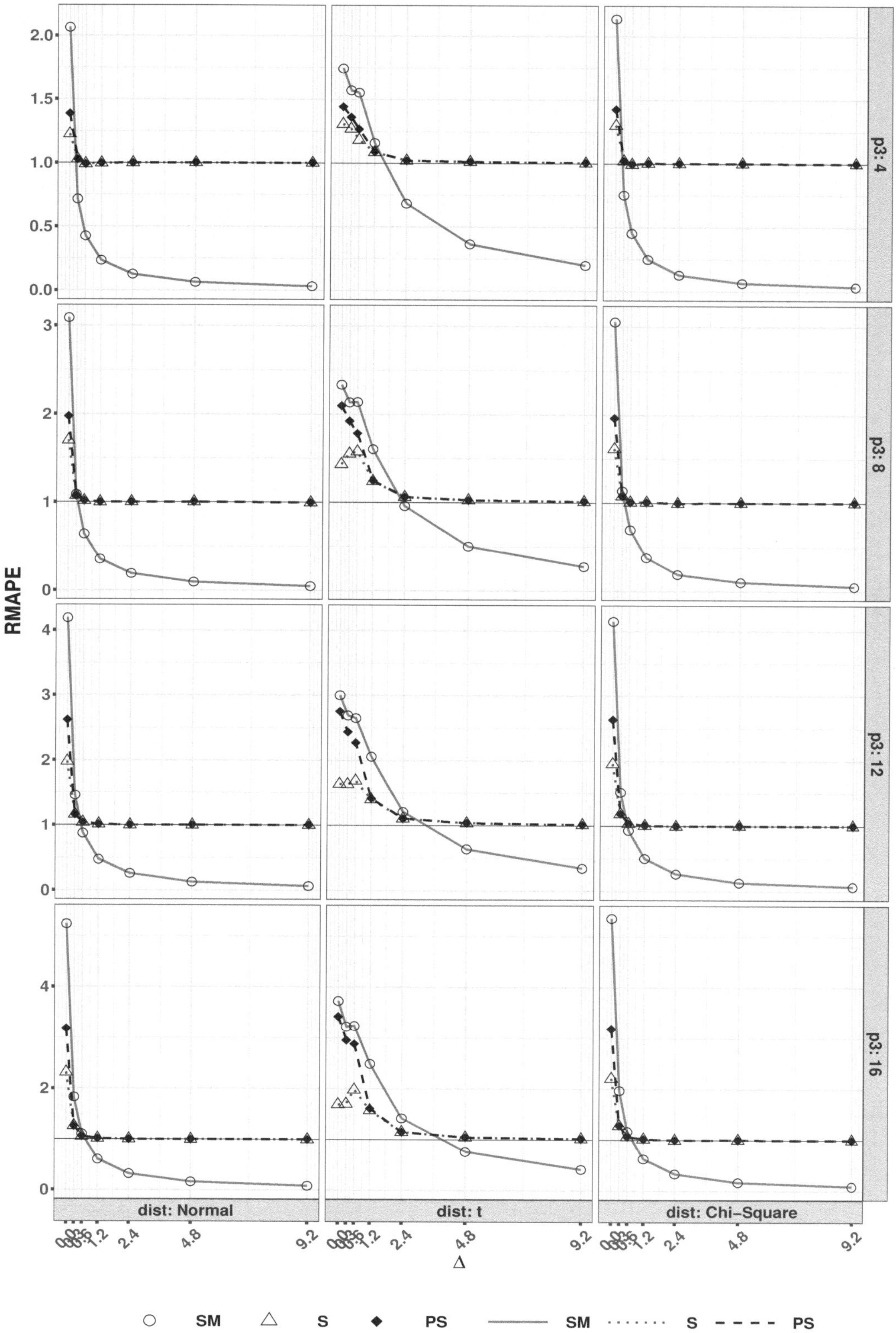

FIGURE 9.3: RMAPE of the Estimators for $n = 500$, $p_1 = 4$ and $p_2 = 0$.

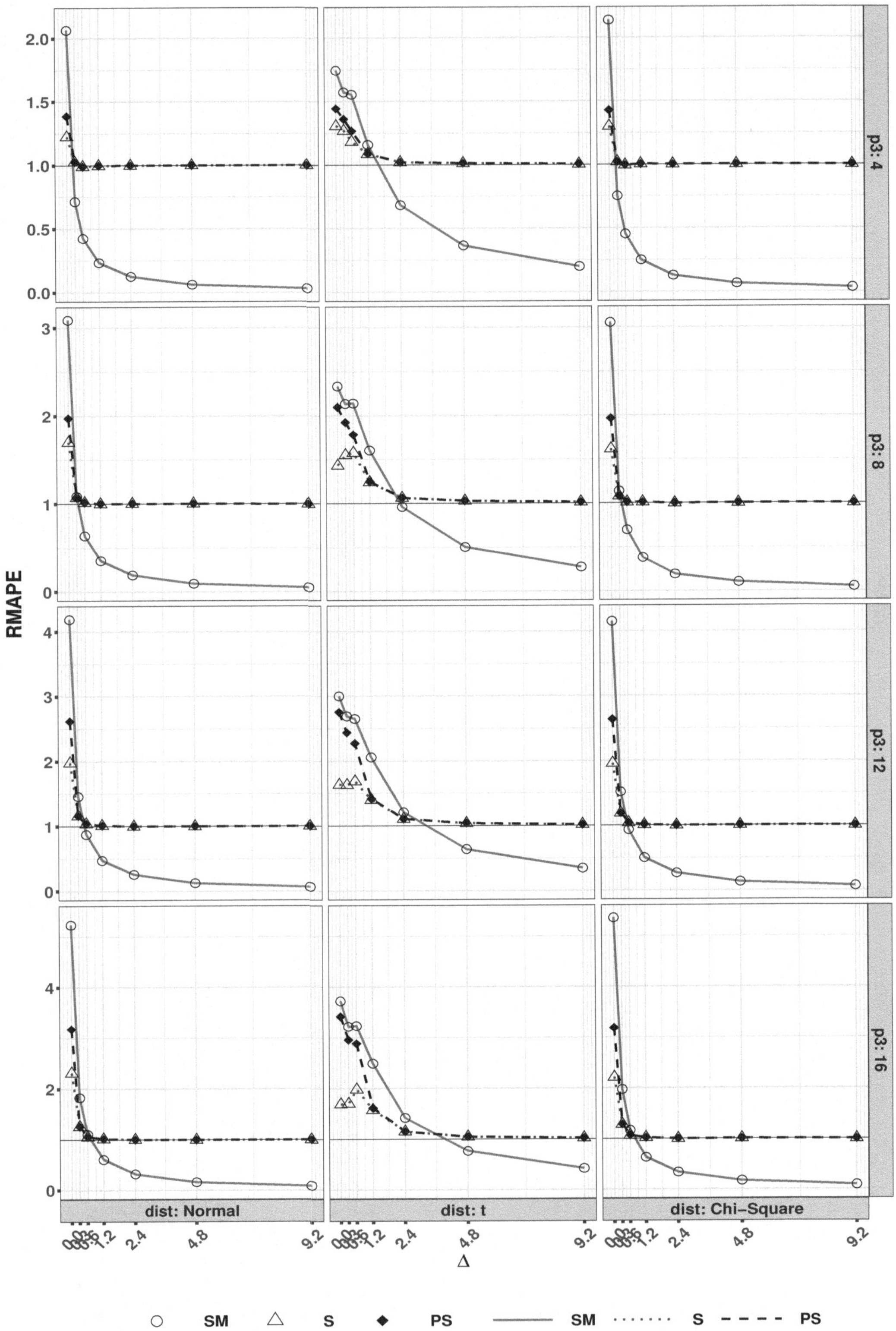

FIGURE 9.4: RMAPE of the Estimators for $n = 500$, $p_1 = 4$ and $p_2 = 6$.

TABLE 9.1: Normal Distribution: RMAPE of the Estimators for $n = 100$ and $p_1 = 4$.

		$p_2 = 0$		$p_2 = 3$		$p_2 = 6$		$p_2 = 9$	
p_3	Δ	SM	PS	SM	PS	SM	PS	SM	PS
4	0.0	2.060	1.458	1.591	1.296	1.445	1.237	1.367	1.196
	0.3	1.153	1.107	1.118	1.103	1.076	1.058	1.065	1.056
	0.6	0.775	1.019	0.775	1.033	0.806	1.027	0.845	1.024
	1.2	0.462	1.006	0.454	1.006	0.524	1.001	0.571	1.010
	2.4	0.244	1.003	0.239	0.999	0.287	1.004	0.295	1.004
	4.8	0.132	0.999	0.121	0.999	0.150	0.998	0.156	1.003
	9.2	0.069	0.999	0.063	0.998	0.079	0.997	0.083	0.999
8	0.0	3.070	2.145	2.224	1.694	1.887	1.514	1.711	1.481
	0.3	1.757	1.391	1.523	1.284	1.393	1.229	1.369	1.227
	0.6	1.176	1.105	1.071	1.097	1.068	1.083	1.073	1.087
	1.2	0.684	1.031	0.633	1.017	0.692	1.023	0.704	1.012
	2.4	0.375	1.002	0.328	1.007	0.382	1.008	0.382	1.009
	4.8	0.201	0.994	0.168	0.998	0.198	0.997	0.198	1.005
	9.2	0.104	0.997	0.088	1.002	0.105	0.998	0.107	1.000
12	0.0	4.197	2.826	2.921	2.080	2.392	1.831	2.112	1.811
	0.3	2.376	1.608	1.985	1.548	1.745	1.448	1.670	1.416
	0.6	1.615	1.236	1.418	1.202	1.332	1.173	1.315	1.172
	1.2	0.946	1.046	0.830	1.056	0.882	1.043	0.866	1.050
	2.4	0.509	1.002	0.431	1.012	0.477	0.997	0.471	1.015
	4.8	0.270	1.000	0.220	0.997	0.249	0.998	0.246	1.006
	9.2	0.143	0.997	0.116	1.002	0.129	0.997	0.132	1.002
16	0.0	5.416	3.460	3.659	2.621	2.952	2.337	2.530	2.146
	0.3	3.088	1.894	2.439	1.808	2.146	1.708	1.939	1.627
	0.6	2.065	1.355	1.759	1.302	1.645	1.307	1.573	1.288
	1.2	1.229	1.086	1.044	1.086	1.073	1.075	1.030	1.078
	2.4	0.662	1.010	0.539	1.006	0.590	1.018	0.557	1.022
	4.8	0.351	1.008	0.276	1.003	0.307	1.004	0.293	1.001
	9.2	0.187	1.006	0.144	1.001	0.161	1.003	0.159	1.003

TABLE 9.2: Normal Distribution: RMAPE of the Estimators for $n = 100$ and $p_1 = 8$.

		$p_2 = 0$		$p_2 = 3$		$p_2 = 6$		$p_2 = 9$	
p_3	Δ	SM	PS	SM	PS	SM	PS	SM	PS
4	0.0	1.486	1.267	1.399	1.211	1.307	1.153	1.256	1.137
	0.3	1.111	1.075	1.069	1.046	1.049	1.051	1.072	1.065
	0.6	0.838	1.025	0.850	1.015	0.883	1.022	0.869	1.021
	1.2	0.547	1.009	0.571	1.002	0.610	1.003	0.587	1.005
	2.4	0.305	1.002	0.316	1.003	0.370	0.998	0.329	0.997
	4.8	0.169	0.997	0.171	1.001	0.206	1.000	0.178	1.002
	9.2	0.089	0.999	0.090	1.002	0.108	1.001	0.095	1.001
8	0.0	2.030	1.637	1.836	1.490	1.661	1.436	1.554	1.399
	0.3	1.505	1.264	1.388	1.237	1.306	1.190	1.302	1.212
	0.6	1.153	1.115	1.130	1.094	1.104	1.083	1.066	1.080
	1.2	0.756	1.018	0.749	1.030	0.776	1.025	0.723	1.030
	2.4	0.414	1.002	0.416	1.003	0.462	1.003	0.405	1.004
	4.8	0.227	1.004	0.223	0.997	0.261	1.000	0.221	1.003
	9.2	0.122	1.000	0.119	0.999	0.133	0.998	0.119	1.003
12	0.0	2.624	2.056	2.295	1.859	2.046	1.773	1.861	1.654
	0.3	1.948	1.473	1.705	1.425	1.607	1.387	1.515	1.376
	0.6	1.474	1.189	1.405	1.174	1.368	1.182	1.276	1.179
	1.2	0.988	1.048	0.941	1.056	0.944	1.043	0.857	1.062
	2.4	0.538	1.013	0.520	1.004	0.572	1.009	0.479	1.012
	4.8	0.295	1.007	0.280	1.004	0.321	1.003	0.264	1.007
	9.2	0.159	0.999	0.147	1.000	0.166	1.002	0.142	1.000
16	0.0	3.248	2.508	2.768	2.259	2.450	2.134	2.220	2.012
	0.3	2.387	1.747	2.051	1.646	1.909	1.637	1.823	1.605
	0.6	1.845	1.307	1.713	1.302	1.620	1.295	1.503	1.272
	1.2	1.202	1.068	1.133	1.069	1.133	1.093	1.032	1.083
	2.4	0.673	1.018	0.633	1.017	0.676	1.018	0.571	1.016
	4.8	0.363	0.997	0.338	1.005	0.387	1.006	0.312	1.009
	9.2	0.194	1.001	0.178	0.999	0.194	1.004	0.166	1.000

TABLE 9.3: Normal Distribution: RMAPE of the Estimators for $n = 500$ and $p_1 = 4$.

		$p_2 = 0$		$p_2 = 3$		$p_2 = 6$		$p_2 = 9$	
p_3	Δ	SM	PS	SM	PS	SM	PS	SM	PS
4	0.0	2.064	1.386	1.582	1.244	1.376	1.184	1.325	1.144
	0.3	0.714	1.027	0.800	1.007	0.814	1.004	0.842	1.017
	0.6	0.425	0.991	0.513	1.008	0.562	1.001	0.595	1.005
	1.2	0.234	0.998	0.301	0.999	0.342	0.998	0.359	1.001
	2.4	0.126	1.000	0.166	1.001	0.191	0.996	0.206	1.000
	4.8	0.062	1.001	0.082	0.998	0.100	1.000	0.107	0.998
	9.2	0.032	1.001	0.043	1.000	0.053	1.000	0.056	0.999
8	0.0	3.089	1.975	2.160	1.645	1.823	1.477	1.671	1.406
	0.3	1.086	1.072	1.088	1.059	1.059	1.059	1.028	1.054
	0.6	0.638	1.015	0.721	1.016	0.735	1.017	0.743	1.025
	1.2	0.354	1.002	0.415	1.000	0.447	1.001	0.453	1.004
	2.4	0.190	1.003	0.229	0.999	0.250	1.006	0.253	0.996
	4.8	0.093	1.003	0.114	1.001	0.128	1.001	0.132	1.001
	9.2	0.049	0.997	0.059	0.999	0.068	0.999	0.071	1.002
12	0.0	4.188	2.619	2.768	2.088	2.229	1.812	1.991	1.622
	0.3	1.459	1.159	1.394	1.142	1.283	1.128	1.228	1.108
	0.6	0.874	1.038	0.922	1.025	0.901	1.034	0.893	1.031
	1.2	0.473	1.010	0.535	1.009	0.550	1.009	0.536	1.005
	2.4	0.257	1.000	0.290	1.003	0.303	1.003	0.304	1.001
	4.8	0.126	1.000	0.145	0.997	0.159	0.997	0.160	0.998
	9.2	0.066	1.001	0.075	0.999	0.084	1.000	0.084	0.998
16	0.0	5.232	3.176	3.329	2.441	2.660	2.024	2.298	1.894
	0.3	1.829	1.259	1.664	1.217	1.532	1.185	1.443	1.179
	0.6	1.105	1.054	1.132	1.064	1.077	1.043	1.038	1.056
	1.2	0.604	1.015	0.648	1.004	0.651	1.015	0.630	1.008
	2.4	0.319	1.002	0.349	1.004	0.361	1.007	0.354	1.001
	4.8	0.159	0.997	0.178	1.005	0.189	1.002	0.183	1.002
	9.2	0.083	1.000	0.090	1.000	0.099	1.001	0.097	0.999

TABLE 9.4: Normal Distribution: RMAPE of the Estimators for $n = 500$ and $p_1 = 8$.

		$p_2 = 0$		$p_2 = 3$		$p_2 = 6$		$p_2 = 9$	
p_3	Δ	SM	PS	SM	PS	SM	PS	SM	PS
4	0.0	1.495	1.205	1.355	1.179	1.320	1.146	1.259	1.135
	0.3	0.802	1.011	0.841	1.009	0.829	1.012	0.840	1.008
	0.6	0.523	1.004	0.582	1.009	0.591	1.007	0.619	1.004
	1.2	0.313	1.004	0.358	0.999	0.358	1.002	0.395	1.007
	2.4	0.176	1.001	0.200	1.000	0.202	0.998	0.215	1.001
	4.8	0.090	1.002	0.100	1.000	0.104	1.002	0.111	1.000
	9.2	0.048	0.999	0.053	1.000	0.055	1.001	0.060	1.000
8	0.0	2.023	1.590	1.748	1.466	1.612	1.372	1.498	1.315
	0.3	1.078	1.061	1.077	1.051	1.005	1.042	1.003	1.038
	0.6	0.716	1.016	0.745	1.015	0.725	1.011	0.744	1.010
	1.2	0.418	1.003	0.462	1.005	0.441	1.003	0.466	0.997
	2.4	0.237	1.000	0.253	1.002	0.246	1.003	0.258	1.001
	4.8	0.121	0.999	0.128	1.003	0.130	0.999	0.134	1.000
	9.2	0.065	1.001	0.067	1.000	0.068	1.000	0.071	1.000
12	0.0	2.543	1.933	2.106	1.730	1.919	1.570	1.739	1.530
	0.3	1.351	1.133	1.284	1.117	1.200	1.093	1.178	1.095
	0.6	0.908	1.028	0.916	1.035	0.867	1.031	0.865	1.029
	1.2	0.533	1.011	0.559	1.013	0.523	1.006	0.547	1.009
	2.4	0.294	1.002	0.305	1.001	0.293	1.000	0.300	0.998
	4.8	0.154	1.001	0.157	1.003	0.153	1.003	0.154	0.999
	9.2	0.081	0.999	0.081	1.000	0.080	1.003	0.082	1.000
16	0.0	3.048	2.234	2.522	1.970	2.201	1.823	1.975	1.713
	0.3	1.614	1.208	1.525	1.193	1.406	1.171	1.350	1.164
	0.6	1.102	1.050	1.073	1.044	0.999	1.044	0.961	1.038
	1.2	0.637	1.011	0.658	1.015	0.608	1.009	0.621	1.007
	2.4	0.355	1.000	0.360	1.000	0.337	1.002	0.343	1.001
	4.8	0.186	1.005	0.181	0.999	0.175	1.002	0.174	1.000
	9.2	0.097	1.002	0.095	1.000	0.093	1.000	0.095	0.999

TABLE 9.5: χ_5^2 Distribution: RMAPE of the Estimators for $n = 100$ and $p_1 = 4$.

		$p_2 = 0$		$p_2 = 3$		$p_2 = 6$		$p_2 = 9$	
p_3	Δ	SM	PS	SM	PS	SM	PS	SM	PS
	0.0	1.728	1.418	1.536	1.311	1.386	1.221	1.354	1.226
	0.3	1.574	1.349	1.422	1.266	1.321	1.200	1.286	1.206
	0.6	1.505	1.299	1.382	1.234	1.265	1.176	1.269	1.201
	1.2	1.480	1.259	1.352	1.235	1.221	1.141	1.289	1.211
4	2.4	1.173	1.088	1.117	1.082	1.094	1.091	1.159	1.118
	4.8	0.675	1.019	0.692	1.024	0.770	1.027	0.763	1.029
	9.2	0.381	1.003	0.388	1.006	0.453	1.004	0.420	1.012
	0.0	2.571	1.940	2.068	1.692	1.840	1.560	1.774	1.478
	0.3	2.386	1.919	1.921	1.627	1.755	1.525	1.716	1.473
	0.6	2.234	1.760	1.882	1.595	1.668	1.500	1.710	1.456
	1.2	2.166	1.686	1.809	1.544	1.666	1.445	1.654	1.455
8	2.4	1.704	1.279	1.527	1.240	1.478	1.251	1.499	1.288
	4.8	0.996	1.063	0.947	1.078	1.009	1.082	0.986	1.071
	9.2	0.560	1.021	0.530	1.012	0.579	1.010	0.543	1.023
	0.0	3.430	2.360	2.728	2.105	2.445	1.719	2.197	1.535
	0.3	3.158	2.364	2.584	2.067	2.322	1.767	2.174	1.548
	0.6	3.042	2.153	2.522	2.031	2.162	1.677	2.106	1.571
	1.2	2.823	2.039	2.385	1.846	2.159	1.641	2.057	1.586
12	2.4	2.214	1.434	2.005	1.410	2.003	1.404	1.915	1.422
	4.8	1.321	1.115	1.267	1.116	1.312	1.133	1.285	1.124
	9.2	0.763	1.015	0.694	1.018	0.769	1.017	0.685	1.033
	0.0	4.553	3.043	3.626	2.166	3.029	1.818	2.666	1.669
	0.3	4.168	2.922	3.442	2.219	2.886	1.718	2.604	1.646
	0.6	4.016	2.718	3.281	2.115	2.739	1.726	2.501	1.600
	1.2	3.774	2.550	3.055	2.125	2.745	1.713	2.599	1.677
16	2.4	2.960	1.633	2.601	1.567	2.524	1.465	2.340	1.518
	4.8	1.718	1.164	1.634	1.172	1.663	1.172	1.568	1.193
	9.2	1.004	1.029	0.887	1.037	0.969	1.040	0.849	1.048

TABLE 9.6: χ_5^2 Distribution: RMAPE of the Estimators for $n = 100$ and $p_1 = 8$.

		$p_2 = 0$		$p_2 = 3$		$p_2 = 6$		$p_2 = 9$	
p_3	Δ	SM	PS	SM	PS	SM	PS	SM	PS
4	0.0	1.379	1.256	1.318	1.231	1.309	1.232	1.270	1.189
	0.3	1.344	1.228	1.293	1.220	1.224	1.186	1.248	1.198
	0.6	1.318	1.201	1.239	1.173	1.218	1.182	1.232	1.193
	1.2	1.339	1.203	1.224	1.164	1.234	1.171	1.196	1.157
	2.4	1.235	1.079	1.117	1.087	1.220	1.097	1.178	1.085
	4.8	0.822	1.021	0.807	1.023	0.929	1.028	0.886	1.021
	9.2	0.495	1.010	0.485	1.013	0.566	1.004	0.518	1.008
8	0.0	1.853	1.608	1.700	1.514	1.665	1.456	1.611	1.444
	0.3	1.776	1.572	1.637	1.476	1.596	1.471	1.556	1.395
	0.6	1.712	1.531	1.614	1.475	1.557	1.426	1.515	1.378
	1.2	1.692	1.446	1.530	1.405	1.559	1.375	1.494	1.352
	2.4	1.636	1.268	1.430	1.238	1.552	1.269	1.447	1.232
	4.8	1.106	1.065	1.016	1.083	1.170	1.076	1.123	1.066
	9.2	0.645	1.020	0.617	1.026	0.727	1.024	0.668	1.025
12	0.0	2.345	1.948	2.166	1.802	2.070	1.645	1.936	1.596
	0.3	2.240	1.886	2.051	1.730	2.017	1.638	1.856	1.524
	0.6	2.171	1.839	2.043	1.706	1.909	1.580	1.797	1.583
	1.2	2.182	1.730	1.962	1.632	1.915	1.559	1.850	1.511
	2.4	2.064	1.433	1.807	1.357	1.899	1.361	1.771	1.331
	4.8	1.384	1.112	1.301	1.112	1.466	1.129	1.335	1.118
	9.2	0.836	1.035	0.780	1.037	0.907	1.035	0.792	1.041
16	0.0	2.978	2.225	2.709	1.911	2.503	1.815	2.377	1.791
	0.3	2.890	2.168	2.547	1.948	2.431	1.796	2.266	1.735
	0.6	2.761	2.107	2.565	1.861	2.303	1.743	2.252	1.748
	1.2	2.678	1.984	2.376	1.741	2.311	1.725	2.223	1.700
	2.4	2.582	1.542	2.249	1.479	2.327	1.517	2.183	1.459
	4.8	1.733	1.154	1.626	1.184	1.801	1.179	1.618	1.178
	9.2	1.045	1.048	0.965	1.049	1.083	1.057	0.975	1.063

TABLE 9.7: χ_5^2 Distribution: RMAPE of the Estimators for $n = 500$ and $p_1 = 4$.

		$p_2 = 0$		$p_2 = 3$		$p_2 = 6$		$p_2 = 9$	
p_3	Δ	SM	PS	SM	PS	SM	PS	SM	PS
4	0.0	1.744	1.442	1.398	1.304	1.317	1.273	1.210	1.201
	0.3	1.571	1.360	1.325	1.246	1.223	1.204	1.171	1.169
	0.6	1.553	1.267	1.336	1.223	1.191	1.177	1.156	1.159
	1.2	1.159	1.089	1.229	1.087	1.227	1.098	1.186	1.101
	2.4	0.682	1.021	0.783	1.028	0.842	1.026	0.825	1.025
	4.8	0.365	1.013	0.440	1.008	0.502	1.009	0.479	1.007
	9.2	0.199	1.005	0.243	1.002	0.274	1.002	0.280	1.001
8	0.0	2.335	2.097	1.800	1.677	1.599	1.527	1.478	1.420
	0.3	2.135	1.924	1.662	1.592	1.514	1.461	1.387	1.357
	0.6	2.140	1.782	1.694	1.555	1.453	1.405	1.359	1.349
	1.2	1.603	1.248	1.571	1.266	1.490	1.256	1.351	1.261
	2.4	0.962	1.064	1.002	1.063	1.054	1.071	0.971	1.062
	4.8	0.504	1.029	0.564	1.019	0.607	1.017	0.578	1.022
	9.2	0.275	1.014	0.311	1.006	0.330	1.007	0.336	1.000
12	0.0	3.000	2.753	2.141	2.067	1.887	1.798	1.735	1.684
	0.3	2.690	2.437	2.055	1.934	1.754	1.707	1.612	1.571
	0.6	2.650	2.272	2.034	1.915	1.716	1.640	1.605	1.557
	1.2	2.059	1.415	1.920	1.456	1.730	1.406	1.579	1.422
	2.4	1.207	1.104	1.191	1.102	1.213	1.107	1.123	1.111
	4.8	0.639	1.039	0.685	1.039	0.713	1.033	0.667	1.032
	9.2	0.346	1.015	0.375	1.004	0.377	1.011	0.384	1.010
16	0.0	3.724	3.419	2.563	2.403	2.208	2.086	2.003	1.936
	0.3	3.221	2.954	2.444	2.272	2.039	1.975	1.853	1.795
	0.6	3.234	2.881	2.375	2.235	1.993	1.900	1.836	1.790
	1.2	2.490	1.613	2.215	1.609	1.976	1.600	1.781	1.561
	2.4	1.419	1.150	1.389	1.143	1.375	1.136	1.293	1.150
	4.8	0.763	1.048	0.790	1.045	0.821	1.045	0.758	1.040
	9.2	0.415	1.018	0.437	1.013	0.439	1.018	0.432	1.009

TABLE 9.8: χ_5^2 Distribution: RMAPE of the Estimators for $n = 500$ and $p_1 = 8$.

		$p_2 = 0$		$p_2 = 3$		$p_2 = 6$		$p_2 = 9$	
p_3	Δ	SM	PS	SM	PS	SM	PS	SM	PS
4	0.0	1.456	1.307	1.371	1.283	1.263	1.243	1.208	1.191
	0.3	1.370	1.280	1.242	1.226	1.196	1.193	1.159	1.187
	0.6	1.361	1.221	1.264	1.176	1.187	1.178	1.149	1.147
	1.2	1.290	1.092	1.335	1.097	1.258	1.099	1.239	1.084
	2.4	0.862	1.029	0.916	1.024	0.949	1.028	1.003	1.026
	4.8	0.483	1.007	0.552	1.009	0.584	1.005	0.595	1.009
	9.2	0.273	1.002	0.304	0.999	0.321	1.001	0.320	1.004
8	0.0	1.878	1.673	1.660	1.540	1.522	1.470	1.379	1.366
	0.3	1.752	1.635	1.547	1.487	1.401	1.384	1.321	1.308
	0.6	1.772	1.526	1.581	1.481	1.429	1.392	1.337	1.316
	1.2	1.695	1.248	1.623	1.270	1.535	1.273	1.466	1.264
	2.4	1.101	1.064	1.150	1.073	1.147	1.070	1.141	1.081
	4.8	0.662	1.022	0.688	1.022	0.688	1.018	0.696	1.020
	9.2	0.351	1.010	0.383	1.010	0.386	1.007	0.381	1.012
12	0.0	2.323	2.073	1.953	1.827	1.716	1.657	1.566	1.538
	0.3	2.149	1.949	1.849	1.757	1.617	1.583	1.507	1.481
	0.6	2.167	1.835	1.852	1.699	1.656	1.595	1.517	1.461
	1.2	2.065	1.431	1.913	1.435	1.737	1.425	1.661	1.411
	2.4	1.402	1.101	1.385	1.125	1.314	1.120	1.295	1.120
	4.8	0.788	1.028	0.817	1.029	0.791	1.036	0.782	1.037
	9.2	0.436	1.008	0.454	1.009	0.449	1.013	0.428	1.018
16	0.0	2.682	2.509	2.233	2.116	1.912	1.850	1.715	1.681
	0.3	2.542	2.287	2.178	2.047	1.833	1.774	1.678	1.652
	0.6	2.568	2.221	2.144	1.995	1.804	1.738	1.686	1.659
	1.2	2.459	1.651	2.242	1.653	2.027	1.603	1.836	1.546
	2.4	1.627	1.150	1.606	1.163	1.504	1.160	1.447	1.166
	4.8	0.943	1.047	0.943	1.048	0.897	1.046	0.868	1.052
	9.2	0.515	1.013	0.525	1.015	0.508	1.016	0.478	1.020

TABLE 9.9: t_5 Distribution: RMAPE of the Estimators for $n = 100$ and $p_1 = 4$.

		$p_2 = 0$		$p_2 = 3$		$p_2 = 6$		$p_2 = 9$	
p_3	Δ	SM	PS	SM	PS	SM	PS	SM	PS
4	0.0	2.061	1.369	1.626	1.260	1.445	1.223	1.326	1.168
	0.3	1.195	1.117	1.105	1.088	1.113	1.071	1.093	1.076
	0.6	0.799	1.034	0.803	1.029	0.824	1.021	0.864	1.023
	1.2	0.479	1.023	0.479	1.006	0.547	1.005	0.585	1.001
	2.4	0.273	1.002	0.260	0.998	0.314	0.997	0.332	1.001
	4.8	0.138	0.997	0.130	1.001	0.162	1.004	0.173	1.005
	9.2	0.076	0.999	0.068	1.001	0.086	1.000	0.091	0.998
8	0.0	3.352	2.156	2.291	1.730	1.875	1.519	1.731	1.512
	0.3	1.757	1.410	1.545	1.316	1.439	1.261	1.357	1.240
	0.6	1.216	1.133	1.145	1.104	1.099	1.103	1.093	1.084
	1.2	0.746	1.034	0.689	1.023	0.730	1.020	0.747	1.021
	2.4	0.420	1.012	0.368	1.003	0.424	1.005	0.420	1.005
	4.8	0.212	0.999	0.181	1.006	0.221	0.994	0.219	0.996
	9.2	0.116	1.005	0.097	0.999	0.114	1.004	0.121	1.000
12	0.0	4.333	2.724	2.992	2.147	2.330	1.865	2.118	1.788
	0.3	2.495	1.698	2.056	1.591	1.906	1.518	1.740	1.459
	0.6	1.659	1.275	1.459	1.224	1.402	1.209	1.367	1.203
	1.2	1.009	1.072	0.874	1.066	0.937	1.069	0.907	1.054
	2.4	0.556	1.014	0.481	1.022	0.535	1.010	0.521	1.019
	4.8	0.297	1.001	0.239	1.008	0.271	1.000	0.269	1.000
	9.2	0.159	1.005	0.125	1.001	0.146	1.003	0.146	1.002
16	0.0	5.763	3.401	3.668	2.684	3.005	2.380	2.521	2.134
	0.3	3.176	2.081	2.498	1.904	2.275	1.827	2.053	1.749
	0.6	2.195	1.426	1.880	1.409	1.717	1.373	1.660	1.353
	1.2	1.296	1.103	1.111	1.105	1.176	1.107	1.099	1.107
	2.4	0.745	1.021	0.602	1.017	0.668	1.032	0.615	1.021
	4.8	0.385	1.005	0.298	0.996	0.342	1.006	0.324	1.008
	9.2	0.207	1.003	0.161	1.004	0.175	0.999	0.178	0.997

TABLE 9.10: t_5 Distribution: RMAPE of the Estimators for $n = 100$ and $p_1 = 8$.

		$p_2 = 0$		$p_2 = 3$		$p_2 = 6$		$p_2 = 9$	
p_3	Δ	SM	PS	SM	PS	SM	PS	SM	PS
4	0.0	1.548	1.250	1.383	1.198	1.308	1.166	1.288	1.164
	0.3	1.104	1.074	1.080	1.059	1.068	1.063	1.063	1.081
	0.6	0.839	1.024	0.884	1.030	0.895	1.029	0.883	1.019
	1.2	0.590	1.008	0.597	1.007	0.640	1.006	0.614	1.004
	2.4	0.342	1.006	0.350	1.000	0.398	1.003	0.354	1.003
	4.8	0.178	0.999	0.181	1.002	0.223	1.002	0.190	0.999
	9.2	0.099	1.001	0.098	0.999	0.119	0.999	0.107	0.999
8	0.0	2.115	1.624	1.810	1.508	1.618	1.414	1.564	1.399
	0.3	1.530	1.293	1.424	1.251	1.407	1.245	1.348	1.230
	0.6	1.165	1.124	1.131	1.096	1.121	1.086	1.087	1.102
	1.2	0.778	1.032	0.760	1.031	0.805	1.030	0.749	1.034
	2.4	0.458	1.011	0.458	1.012	0.509	1.007	0.439	1.006
	4.8	0.250	1.007	0.236	1.004	0.279	1.008	0.237	1.004
	9.2	0.135	1.002	0.129	1.003	0.151	1.001	0.131	1.002
12	0.0	2.699	2.068	2.231	1.811	2.027	1.783	1.878	1.657
	0.3	1.942	1.538	1.795	1.503	1.710	1.490	1.585	1.414
	0.6	1.540	1.224	1.432	1.216	1.361	1.187	1.313	1.195
	1.2	1.030	1.053	0.977	1.061	1.017	1.061	0.905	1.062
	2.4	0.605	1.010	0.568	1.022	0.629	1.023	0.526	1.016
	4.8	0.321	1.010	0.299	1.002	0.346	1.004	0.282	1.005
	9.2	0.176	0.999	0.165	1.000	0.180	0.999	0.156	1.002
16	0.0	3.338	2.583	2.849	2.330	2.524	2.091	2.281	2.009
	0.3	2.435	1.801	2.120	1.784	2.075	1.730	1.902	1.660
	0.6	1.891	1.365	1.798	1.360	1.656	1.373	1.554	1.339
	1.2	1.284	1.096	1.220	1.108	1.189	1.104	1.081	1.097
	2.4	0.739	1.023	0.692	1.025	0.749	1.018	0.641	1.022
	4.8	0.400	1.006	0.374	1.012	0.422	1.007	0.340	1.006
	9.2	0.216	1.001	0.192	1.000	0.221	1.001	0.188	1.004

TABLE 9.11: t_5 Distribution: RMAPE of the Estimators for $n = 400$ and $p_1 = 4$.

		$p_2 = 0$		$p_2 = 3$		$p_2 = 6$		$p_2 = 9$	
p_3	Δ	SM	PS	SM	PS	SM	PS	SM	PS
4	0.0	2.136	1.426	1.529	1.242	1.399	1.198	1.316	1.147
	0.3	0.753	1.021	0.812	1.016	0.854	1.021	0.853	1.011
	0.6	0.453	0.996	0.533	1.004	0.592	1.010	0.609	1.006
	1.2	0.249	1.004	0.317	1.003	0.358	1.002	0.373	1.004
	2.4	0.127	1.002	0.174	1.003	0.194	1.002	0.210	1.001
	4.8	0.064	1.002	0.087	1.001	0.106	1.000	0.111	1.002
	9.2	0.033	1.000	0.045	1.001	0.054	0.999	0.059	0.999
8	0.0	3.054	1.966	2.206	1.671	1.853	1.526	1.666	1.393
	0.3	1.138	1.078	1.147	1.071	1.092	1.077	1.069	1.069
	0.6	0.696	1.010	0.747	1.014	0.774	1.017	0.777	1.013
	1.2	0.380	1.010	0.431	1.014	0.458	1.002	0.460	1.003
	2.4	0.190	0.999	0.234	0.995	0.256	1.001	0.263	1.007
	4.8	0.101	1.002	0.119	0.999	0.134	0.997	0.138	1.000
	9.2	0.051	1.002	0.059	1.002	0.071	0.998	0.074	1.002
12	0.0	4.148	2.635	2.796	2.139	2.202	1.804	1.934	1.643
	0.3	1.519	1.187	1.425	1.143	1.344	1.142	1.284	1.123
	0.6	0.933	1.036	0.946	1.024	0.914	1.030	0.918	1.034
	1.2	0.502	1.009	0.552	1.010	0.566	1.002	0.558	1.010
	2.4	0.265	0.997	0.304	1.008	0.312	0.999	0.310	1.006
	4.8	0.131	1.001	0.154	0.999	0.163	1.000	0.166	0.999
	9.2	0.069	0.998	0.078	0.999	0.087	1.000	0.088	1.000
16	0.0	5.355	3.184	3.475	2.523	2.700	2.181	2.305	1.965
	0.3	1.971	1.283	1.765	1.256	1.610	1.214	1.520	1.199
	0.6	1.166	1.066	1.147	1.054	1.108	1.047	1.051	1.055
	1.2	0.629	1.020	0.675	1.010	0.670	1.014	0.657	1.017
	2.4	0.333	1.000	0.369	1.008	0.369	0.996	0.368	1.002
	4.8	0.166	1.003	0.186	1.003	0.196	1.001	0.195	1.003
	9.2	0.086	0.997	0.095	1.002	0.103	0.999	0.103	0.999

TABLE 9.12: t_5 Distribution: RMAPE of the Estimators for $n = 500$ and $p_1 = 8$.

		$p_2 = 0$		$p_2 = 3$		$p_2 = 6$		$p_2 = 9$	
p_3	Δ	SM	PS	SM	PS	SM	PS	SM	PS
	0.0	1.498	1.222	1.388	1.179	1.310	1.146	1.225	1.119
	0.3	0.808	1.014	0.872	1.022	0.858	1.011	0.878	1.008
	0.6	0.553	1.003	0.608	1.006	0.621	1.002	0.647	1.008
	1.2	0.331	0.995	0.365	1.001	0.374	1.002	0.395	1.001
4	2.4	0.175	1.000	0.204	0.999	0.207	1.001	0.223	1.001
	4.8	0.096	0.999	0.104	0.999	0.110	0.999	0.117	1.001
	9.2	0.049	1.000	0.053	1.000	0.057	1.000	0.062	0.999
	0.0	2.031	1.630	1.768	1.494	1.559	1.365	1.446	1.303
	0.3	1.103	1.059	1.089	1.060	1.047	1.057	1.049	1.058
	0.6	0.747	1.021	0.773	1.001	0.749	1.016	0.782	1.014
	1.2	0.438	1.006	0.465	1.006	0.458	1.003	0.473	1.005
8	2.4	0.243	1.005	0.266	1.001	0.254	1.002	0.264	1.001
	4.8	0.125	1.003	0.134	0.999	0.131	0.999	0.140	0.998
	9.2	0.067	0.999	0.070	0.999	0.071	0.999	0.074	1.000
	0.0	2.547	1.972	2.166	1.765	1.929	1.634	1.770	1.549
	0.3	1.407	1.137	1.323	1.136	1.265	1.125	1.235	1.116
	0.6	0.942	1.043	0.931	1.033	0.886	1.027	0.896	1.028
	1.2	0.549	1.008	0.573	1.009	0.541	1.006	0.567	1.011
12	2.4	0.306	1.005	0.322	1.001	0.300	1.004	0.314	1.001
	4.8	0.159	1.001	0.164	1.001	0.158	1.000	0.164	0.998
	9.2	0.084	1.000	0.085	1.000	0.084	1.000	0.088	1.000
	0.0	3.093	2.361	2.574	2.025	2.167	1.800	1.990	1.682
	0.3	1.665	1.248	1.561	1.219	1.451	1.199	1.404	1.171
	0.6	1.150	1.065	1.118	1.061	1.024	1.043	1.035	1.053
	1.2	0.670	1.002	0.697	1.012	0.629	1.009	0.642	1.015
16	2.4	0.364	1.003	0.377	1.005	0.354	1.006	0.355	1.003
	4.8	0.191	1.005	0.194	0.998	0.186	1.003	0.189	1.004
	9.2	0.100	1.002	0.100	1.002	0.099	1.002	0.101	1.001

9.4 Penalized Estimation

In this section, we provide a comparative study of shrinkage estimators with some penalized methods in an effort to provide a complete statistical package to practitioners for modeling, parameter estimation, and prediction for robust regression.

Let us consider the LASSO strategy proposed by Tibshirani (1996) that minimizes the penalized likelihood function:

$$\sum_{i=1}^{n}(y_i - \boldsymbol{x}_i^\top \boldsymbol{\beta})^2 + n\lambda \sum_{j=1}^{p} |\beta_j|, \tag{9.19}$$

where $\lambda > 0$ is the tuning parameter. When the errors in (9.1) are distributed in a heavy-tailed manner, the performance of the LASSO becomes unsatisfactory since it is not defined for the heavy-tailed error distribution and/or outliers. The least absolute deviation (LAD) regression is a useful strategy for robust regression, Wang et al. (2007) suggested combining LAD and LASSO together to produce a LAD-LASSO technique. As expected, LAD-LASSO also performs both variable selection and parameter estimation simultaneously. Under some conditions, the LAD-LASSO strategy shares the same asymptotic efficiency as classical LAD. However, we are interested in investigating the relative properties of shrinkage estimators using the penalized methods under the sparsity assumption in the presence of weak regression coefficients.

The LAD-LASSO strategy is defined by replacing the quadratic function with the L_1 function from LASSO, we denote it by $\widehat{\beta}^{\text{LAD-LASSO}}$ that minimizes

$$\sum_{i=1}^{p} \left|y_i - \boldsymbol{x}_i^\top \boldsymbol{\beta}\right| + n \sum_{j=1}^{p} \lambda_j |\beta_j|. \tag{9.20}$$

Using the equation (9.16), we conduct a simulation experiment from different distributions of ε_i: (i) standard normal distribution, (ii) χ_5^2 distribution, (iii) t_5 distribution, (iv) standard Cauchy distribution, (v) Laplace distribution, and (vi) skewed-normal distribution. Finally, we consider the bivariate mixture (BM) distribution with $\gamma = 0.1, 0.25$:

$$F_\epsilon(t, \gamma) = \gamma \Phi(t) + (1-\gamma)\left[\frac{1}{\pi}\arctan(t) + \frac{1}{2}\right],$$

where $\Phi(t)$ and the expression in the square brackets denote the standard normal and the standard Cauchy distribution, respectively. The proportion γ is often useful to verify the effect of outliers, and small values of γ lead to a contaminated normal distribution.

In the simulation experiment, we consider the regression coefficients set to $\boldsymbol{\beta} = \left(\boldsymbol{\beta}_1^\top, \boldsymbol{\beta}_2^\top, \boldsymbol{\beta}_3^\top\right)^\top = \left(\mathbf{1}_{p_1}^\top, \mathbf{0.1}_{p_2}^\top, \mathbf{0}_{p_3}^\top\right)^\top$ with dimensions p_1, p_2 and p_3, respectively. We simulate 500 data sets consisting of $n = 100, 500$, with $p_1 = 4$, $p_2 = 0, 3, 6, 9$ and $p_3 = 4, 8, 12, 16$.

We extend the LAD-LASSO formulation to ridge regression, ALASSO, and ENET penalized procedures. We calculate the RMAPE of these estimators to the full model LAD estimator.

The results of the simulation are given in Tables 9.13–9.20 and some results are graphically displayed in Figures 9.5–9.8. Both graphical and tabulated results show the supremacy of the shrinkage estimators over penalized methods in all reported scenarios. We can safely conclude that our suggested shrinkage strategy is the preferred choice over LAD-LASSO and other penalty methods in normal and non-normal cases.

TABLE 9.13: The RMAPE of the Estimators for $n = 100$ and $p_1 = 4$.

		$p_2 = 0$			$p_2 = 3$		
	Method	$p_3 = 4$	$p_3 = 8$	$p_3 = 16$	$p_3 = 4$	$p_3 = 8$	$p_3 = 16$
Normal	SM	2.073	3.125	5.398	1.610	2.206	3.623
	S	1.256	1.707	2.168	1.196	1.377	1.992
	PS	1.408	2.169	3.797	1.308	1.664	2.569
	Ridge	0.656	0.777	0.920	0.744	0.823	0.931
	ENET	1.333	1.420	1.508	1.266	1.337	1.473
	LASSO	1.206	1.250	1.308	1.150	1.219	1.325
	ALASSO	1.991	2.648	3.803	1.582	1.993	2.936
χ^2_5	SM	1.702	2.565	4.515	1.520	2.075	3.530
	S	1.221	1.581	2.164	1.184	1.499	1.880
	PS	1.390	2.018	3.249	1.315	1.752	2.226
	Ridge	1.491	1.784	2.492	1.745	1.976	2.784
	ENET	1.779	2.350	3.351	2.053	2.393	3.671
	LASSO	1.863	2.470	3.693	2.242	2.589	3.999
	ALASSO	1.770	2.469	3.892	2.100	2.622	4.305
t_5	SM	2.109	3.394	5.757	1.595	2.327	3.806
	S	1.237	1.632	2.111	1.206	1.425	1.930
	PS	1.392	2.128	3.679	1.276	1.756	2.690
	Ridge	0.669	0.779	0.940	0.752	0.884	0.978
	ENET	1.264	1.349	1.486	1.252	1.359	1.442
	LASSO	1.139	1.225	1.372	1.154	1.212	1.321
	ALASSO	1.893	2.516	3.746	1.591	2.174	3.000
BM (0.1)	SM	2.008	3.245	5.448	1.645	2.137	3.799
	S	1.250	1.648	2.301	1.174	1.397	2.022
	PS	1.421	2.085	3.692	1.270	1.769	2.741
	Ridge	0.633	0.759	0.894	0.746	0.818	0.936
	ENET	1.267	1.371	1.450	1.245	1.316	1.462
	LASSO	1.108	1.247	1.298	1.144	1.196	1.321
	ALASSO	1.834	2.821	3.916	1.527	2.111	2.996
BM (0.25)	SM	2.153	3.365	5.969	1.611	2.263	3.967
	S	1.221	1.575	2.160	1.164	1.452	1.841
	PS	1.445	2.151	3.389	1.281	1.700	2.779
	Ridge	0.634	0.755	0.909	0.729	0.843	0.943
	ENET	1.286	1.348	1.461	1.268	1.343	1.483

	LASSO	1.174	1.227	1.322	1.150	1.239	1.356
	ALASSO	1.890	2.762	3.856	1.566	2.138	3.010
Cauchy	SM	2.303	3.616	7.399	1.715	2.575	4.599
	S	1.193	1.563	2.000	1.156	1.440	1.962
	PS	1.396	2.091	4.035	1.305	1.911	3.114
	Ridge	0.762	0.929	1.212	0.892	1.054	1.233
	ENET	1.266	1.373	1.572	1.275	1.416	1.570
	LASSO	1.156	1.227	1.399	1.160	1.269	1.434
	ALASSO	1.713	2.281	3.016	1.643	2.162	2.700
Laplace	SM	2.169	3.339	6.383	1.642	2.350	3.930
	S	1.201	1.644	2.111	1.172	1.454	1.912
	PS	1.440	2.131	3.712	1.329	1.859	2.890
	Ridge	0.535	0.634	0.822	0.636	0.734	0.836
	ENET	1.255	1.361	1.504	1.271	1.337	1.474
	LASSO	1.171	1.223	1.374	1.167	1.225	1.351
	ALASSO	2.026	2.876	5.144	1.456	1.988	3.174
Lognormal	SM	2.009	3.315	5.819	1.700	2.425	3.934
	S	1.334	1.677	2.181	1.251	1.542	2.079
	PS	1.474	2.062	3.827	1.376	1.877	3.030
	Ridge	0.616	0.777	0.932	0.754	0.856	0.966
	ENET	2.286	3.219	4.424	2.192	2.691	3.521
	LASSO	1.846	2.668	3.681	1.810	2.225	2.938
	ALASSO	2.786	4.444	7.722	2.298	3.223	5.157
Skewed	SM	2.105	3.127	5.359	1.641	2.224	3.628
	S	1.242	1.572	2.029	1.155	1.386	1.857
	PS	1.392	1.984	3.314	1.311	1.712	2.783
	Ridge	0.629	0.748	0.890	0.735	0.811	0.923
	ENET	1.238	1.330	1.435	1.271	1.315	1.428
	LASSO	1.143	1.196	1.307	1.124	1.188	1.304
	ALASSO	1.895	2.595	3.628	1.578	2.048	2.874

TABLE 9.14: The RMAPE of the Estimators for $n = 100$ and $p_1 = 4$.

		$p_2 = 6$			$p_2 = 9.$		
	Method	$p_3 = 4$	$p_3 = 8$	$p_3 = 16$	$p_3 = 4$	$p_3 = 8$	$p_3 = 16$
	SM	1.404	1.815	2.860	1.361	1.699	2.547

Normal	S	1.139	1.300	1.786	1.106	1.289	1.642
	PS	1.219	1.469	2.304	1.176	1.484	2.198
	Ridge	0.833	0.890	1.021	0.893	0.948	1.049
	ENET	1.290	1.326	1.501	1.273	1.361	1.489
	LASSO	1.138	1.199	1.336	1.156	1.212	1.338
	ALASSO	1.450	1.801	2.601	1.400	1.701	2.393
χ^2_5	SM	1.390	1.840	3.008	1.357	1.778	2.742
	S	1.147	1.406	1.694	1.167	1.335	1.627
	PS	1.233	1.580	1.816	1.242	1.502	1.632
	Ridge	1.999	2.339	3.205	2.232	2.709	3.577
	ENET	2.295	2.753	4.019	2.467	3.114	4.273
	LASSO	2.435	2.851	4.386	2.569	3.239	4.406
	ALASSO	2.402	2.920	4.542	2.610	3.327	4.545
t_5	SM	1.450	1.831	2.974	1.338	1.740	2.523
	S	1.134	1.354	1.685	1.092	1.289	1.686
	PS	1.235	1.501	2.355	1.176	1.480	2.144
	Ridge	0.837	0.935	1.033	0.917	0.992	1.070
	ENET	1.275	1.333	1.451	1.281	1.384	1.490
	LASSO	1.157	1.194	1.274	1.167	1.252	1.351
	ALASSO	1.517	1.891	2.546	1.481	1.845	2.497
BM (0.1)	SM	1.444	1.922	3.060	1.359	1.716	2.560
	S	1.154	1.321	1.836	1.124	1.311	1.698
	PS	1.222	1.519	2.407	1.174	1.481	2.202
	Ridge	0.796	0.919	0.990	0.873	0.958	1.016
	ENET	1.239	1.349	1.449	1.262	1.355	1.455
	LASSO	1.133	1.200	1.344	1.143	1.223	1.322
	ALASSO	1.404	1.848	2.633	1.409	1.749	2.349
BM (0.25)	SM	1.490	1.938	3.109	1.346	1.759	2.682
	S	1.160	1.377	1.721	1.126	1.296	1.700
	PS	1.237	1.496	2.477	1.206	1.507	2.163
	Ridge	0.826	0.920	0.980	0.863	0.979	1.011
	ENET	1.264	1.351	1.442	1.261	1.400	1.452
	LASSO	1.169	1.239	1.332	1.153	1.230	1.327
	ALASSO	1.457	1.927	2.594	1.380	1.799	2.445
	SM	1.490	2.104	3.632	1.442	1.904	3.101
	S	1.093	1.435	1.935	1.131	1.348	1.763

	PS	1.212	1.627	2.664	1.182	1.523	2.613
Cauchy	Ridge	1.027	1.146	1.404	1.117	1.229	1.445
	ENET	1.344	1.432	1.645	1.387	1.467	1.671
	LASSO	1.225	1.316	1.482	1.264	1.333	1.489
	ALASSO	1.708	2.120	2.740	1.767	2.014	2.666
Laplace	SM	1.460	1.984	3.199	1.391	1.811	2.705
	S	1.119	1.393	1.750	1.118	1.302	1.594
	PS	1.206	1.584	2.580	1.188	1.508	2.371
	Ridge	0.711	0.799	0.914	0.775	0.856	0.920
	ENET	1.207	1.336	1.512	1.254	1.324	1.463
	LASSO	1.129	1.231	1.375	1.124	1.190	1.373
	ALASSO	1.225	1.691	2.569	1.219	1.542	2.237
Lognormal	SM	1.484	1.933	3.088	1.346	1.710	2.654
	S	1.193	1.391	1.906	1.169	1.307	1.846
	PS	1.246	1.667	2.421	1.268	1.576	2.262
	Ridge	0.901	0.966	1.102	0.937	1.017	1.165
	ENET	2.165	2.568	3.438	2.057	2.420	3.206
	LASSO	1.749	2.067	2.841	1.693	2.031	2.700
	ALASSO	2.103	2.722	4.307	1.914	2.436	3.764
Skewed	SM	1.380	1.797	2.841	1.348	1.693	2.510
	S	1.130	1.324	1.774	1.072	1.263	1.756
	PS	1.214	1.478	2.320	1.148	1.435	2.088
	Ridge	0.808	0.879	1.006	0.850	0.924	1.017
	ENET	1.259	1.316	1.456	1.244	1.315	1.468
	LASSO	1.150	1.215	1.338	1.156	1.182	1.303
	ALASSO	1.479	1.812	2.607	1.418	1.693	2.349

TABLE 9.15: The RMAPE of the Estimators for $n = 500$ and $p_1 = 4$.

		$p_2 = 0$			$p_2 = 3$		
	Method	$p_3 = 4$	$p_3 = 8$	$p_3 = 16$	$p_3 = 4$	$p_3 = 8$	$p_3 = 16$
Normal	SM	2.084	3.170	5.296	1.584	2.145	3.283
	S	1.245	1.696	2.353	1.175	1.487	1.924
	PS	1.405	1.972	3.199	1.246	1.654	2.471
	Ridge	0.373	0.479	0.637	0.460	0.543	0.672
	ENET	1.283	1.467	1.680	1.230	1.365	1.527
	LASSO	1.255	1.427	1.551	1.180	1.308	1.432

	ALASSO	1.976	3.003	5.077	1.060	1.439	2.223
	SM	1.756	2.290	3.783	1.409	1.813	2.631
	S	1.314	1.409	1.750	1.181	1.328	1.610
	PS	1.427	2.066	3.566	1.336	1.686	2.433
	Ridge	0.934	1.122	1.398	1.094	1.236	1.456
χ^2_5	ENET	1.662	2.078	2.833	1.715	2.035	2.666
	LASSO	2.329	2.963	4.096	2.336	2.737	3.698
	ALASSO	1.827	2.339	3.609	1.876	2.444	3.461
	SM	2.204	3.042	5.464	1.515	2.189	3.472
	S	1.321	1.597	2.250	1.155	1.418	1.966
	PS	1.478	2.024	3.247	1.229	1.659	2.551
	Ridge	0.399	0.504	0.669	0.463	0.568	0.690
t_5	ENET	1.330	1.449	1.698	1.158	1.351	1.531
	LASSO	1.312	1.414	1.586	1.140	1.279	1.449
	ALASSO	2.056	2.965	5.127	1.086	1.483	2.366
	SM	2.116	3.184	5.326	1.606	2.179	3.357
	S	1.250	1.714	2.433	1.162	1.440	1.979
	PS	1.377	2.177	3.168	1.256	1.718	2.438
	Ridge	0.381	0.482	0.635	0.458	0.536	0.673
BM (0.1)	ENET	1.273	1.468	1.665	1.196	1.334	1.488
	LASSO	1.235	1.426	1.562	1.150	1.284	1.409
	ALASSO	1.926	3.138	4.872	1.028	1.413	2.167
	SM	1.982	3.069	5.256	1.592	2.187	3.484
	S	1.234	1.709	2.365	1.218	1.470	1.917
	PS	1.382	2.194	3.090	1.264	1.698	2.533
	Ridge	0.375	0.476	0.653	0.455	0.532	0.687
BM (0.25)	ENET	1.279	1.453	1.678	1.196	1.325	1.559
	LASSO	1.271	1.392	1.545	1.168	1.291	1.466
	ALASSO	1.934	3.020	5.051	1.012	1.397	2.222
	SM	2.107	3.275	5.633	1.602	2.255	3.531
	S	1.310	1.706	2.407	1.176	1.531	2.042
	PS	1.438	1.990	3.241	1.248	1.655	2.564
	Ridge	0.447	0.569	0.724	0.531	0.630	0.754
Cauchy	ENET	1.278	1.447	1.544	1.248	1.324	1.451
	LASSO	1.245	1.360	1.475	1.188	1.270	1.397
	ALASSO	2.013	3.103	5.166	1.257	1.707	2.665

Laplace	SM	2.131	3.332	5.792	1.611	2.304	3.620
	S	1.293	1.726	2.366	1.159	1.566	1.979
	PS	1.391	2.017	3.103	1.273	1.693	2.572
	Ridge	0.295	0.390	0.524	0.367	0.453	0.574
	ENET	1.265	1.456	1.619	1.135	1.329	1.509
	LASSO	1.258	1.433	1.599	1.153	1.326	1.454
	ALASSO	2.049	3.198	5.414	0.762	1.086	1.713
Lognormal	SM	1.909	2.877	4.749	1.599	2.167	3.414
	S	1.387	1.852	2.436	1.336	1.626	1.980
	PS	1.507	2.091	3.528	1.428	1.821	2.784
	Ridge	0.304	0.405	0.558	0.379	0.466	0.627
	ENET	1.600	2.377	3.769	1.382	1.830	2.801
	LASSO	1.326	1.977	3.184	1.166	1.561	2.402
	ALASSO	2.038	3.078	5.113	1.106	1.494	2.351
Skewed	SM	1.974	3.027	5.099	1.644	2.198	3.454
	S	1.239	1.641	2.252	1.169	1.458	1.897
	PS	1.412	2.005	3.317	1.259	1.598	2.515
	Ridge	0.373	0.483	0.640	0.466	0.554	0.687
	ENET	1.307	1.482	1.674	1.197	1.356	1.508
	LASSO	1.287	1.400	1.581	1.152	1.287	1.421
	ALASSO	2.023	3.112	5.265	1.070	1.427	2.246

TABLE 9.16: The RMAPE of the Estimators for $n = 500$ and $p_1 = 4$.

		$p_2 = 6$			$p_2 = 9$		
	Method	$p_3 = 4$	$p_3 = 8$	$p_3 = 16$	$p_3 = 4$	$p_3 = 8$	$p_3 = 16$
Normal	SM	1.371	1.816	2.688	1.323	1.662	2.283
	S	1.114	1.356	1.684	1.102	1.259	1.584
	PS	1.188	1.484	2.050	1.171	1.415	1.898
	Ridge	0.512	0.601	0.716	0.568	0.636	0.727
	ENET	1.130	1.277	1.445	1.137	1.232	1.399
	LASSO	1.094	1.238	1.374	1.112	1.179	1.321
	ALASSO	0.820	1.091	1.577	0.730	0.908	1.250
	SM	1.305	1.580	2.187	1.225	1.473	2.032
	S	1.111	1.202	1.438	1.085	1.144	1.349
	PS	1.281	1.514	2.080	1.213	1.408	1.976
	Ridge	1.194	1.308	1.505	1.264	1.375	1.580

χ^2_5	ENET	1.747	2.026	2.522	1.747	2.020	2.565
	LASSO	2.182	2.569	3.305	2.161	2.615	3.277
	ALASSO	1.869	2.353	3.128	1.927	2.338	3.126
	SM	1.389	1.867	2.727	1.325	1.656	2.292
	S	1.150	1.366	1.855	1.111	1.303	1.685
	PS	1.205	1.510	2.127	1.135	1.394	1.971
	Ridge	0.524	0.623	0.730	0.593	0.646	0.727
t_5	ENET	1.136	1.315	1.450	1.126	1.218	1.353
	LASSO	1.114	1.252	1.361	1.093	1.181	1.308
	ALASSO	0.861	1.152	1.667	0.774	0.951	1.292
	SM	1.414	1.795	2.638	1.311	1.684	2.283
	S	1.147	1.307	1.707	1.116	1.309	1.656
	PS	1.207	1.502	2.169	1.158	1.443	1.914
	Ridge	0.523	0.585	0.710	0.569	0.636	0.721
BM (0.1)	ENET	1.160	1.268	1.414	1.122	1.228	1.370
	LASSO	1.127	1.213	1.340	1.074	1.187	1.324
	ALASSO	0.810	1.046	1.514	0.714	0.885	1.239
	SM	1.401	1.810	2.695	1.308	1.654	2.309
	S	1.145	1.337	1.768	1.106	1.314	1.609
	PS	1.193	1.492	2.211	1.169	1.421	1.884
	Ridge	0.522	0.593	0.720	0.574	0.622	0.728
BM (0.25)	ENET	1.149	1.265	1.467	1.111	1.216	1.410
	LASSO	1.094	1.225	1.380	1.079	1.172	1.325
	ALASSO	0.795	1.033	1.544	0.700	0.858	1.238
	SM	1.443	1.840	2.693	1.290	1.652	2.344
	S	1.148	1.424	1.762	1.088	1.302	1.659
	PS	1.208	1.553	2.129	1.149	1.402	1.923
	Ridge	0.627	0.687	0.790	0.640	0.721	0.817
Cauchy	ENET	1.201	1.241	1.381	1.104	1.225	1.357
	LASSO	1.148	1.223	1.316	1.079	1.189	1.299
	ALASSO	1.051	1.346	1.951	0.916	1.172	1.636
	SM	1.464	1.915	2.918	1.353	1.669	2.423
	S	1.135	1.375	1.775	1.106	1.286	1.639
	PS	1.177	1.477	2.240	1.172	1.397	2.008
	Ridge	0.427	0.495	0.615	0.470	0.527	0.622
Laplace	ENET	1.092	1.224	1.432	1.074	1.166	1.378

	LASSO	1.096	1.208	1.394	1.067	1.149	1.308
	ALASSO	0.574	0.756	1.133	0.498	0.617	0.894
Lognormal	SM	1.404	1.783	2.608	1.307	1.609	2.279
	S	1.286	1.490	1.818	1.236	1.354	1.706
	PS	1.346	1.712	2.381	1.298	1.545	2.106
	Ridge	0.447	0.525	0.669	0.500	0.574	0.694
	ENET	1.231	1.564	2.202	1.227	1.491	2.053
	LASSO	1.069	1.341	1.912	1.076	1.296	1.805
	ALASSO	0.855	1.097	1.613	0.740	0.910	1.295
Skewed	SM	1.389	1.820	2.651	1.331	1.649	2.324
	S	1.101	1.366	1.733	1.122	1.269	1.644
	PS	1.174	1.510	2.128	1.152	1.393	1.935
	Ridge	0.526	0.612	0.718	0.573	0.631	0.722
	ENET	1.156	1.298	1.425	1.123	1.210	1.376
	LASSO	1.126	1.246	1.361	1.102	1.173	1.303
	ALASSO	0.826	1.084	1.569	0.721	0.896	1.253

TABLE 9.17: The RMAPE of the Estimators for $n = 100$ and $p_1 = 4$ – SM with Strong Signals.

		$p_2 = 0$			$p_2 = 3$		
	Method	$p_3 = 4$	$p_3 = 8$	$p_3 = 16$	$p_3 = 4$	$p_3 = 8$	$p_3 = 16$
Normal	SM	2.073	3.125	5.398	1.619	2.212	3.633
	S	1.256	1.707	2.168	1.372	1.477	2.114
	PS	1.408	2.169	3.797	1.453	1.797	2.639
	Ridge	0.656	0.777	0.920	0.744	0.823	0.931
	ENET	1.333	1.420	1.508	1.266	1.337	1.473
	LASSO	1.206	1.250	1.308	1.150	1.219	1.325
	ALASSO	1.991	2.648	3.803	1.582	1.993	2.936
χ^2_5	SM	1.702	2.565	4.515	2.065	2.908	4.995
	S	1.221	1.581	2.164	1.473	1.868	2.076
	PS	1.390	2.018	3.249	1.743	2.262	2.687
	Ridge	1.491	1.784	2.492	1.745	1.976	2.784
	ENET	1.779	2.350	3.351	2.053	2.393	3.671
	LASSO	1.863	2.470	3.693	2.242	2.589	3.999
	ALASSO	1.770	2.469	3.892	2.100	2.622	4.305

	SM	2.109	3.394	5.757	1.704	2.479	4.032
	S	1.237	1.632	2.111	1.291	1.608	2.066
	PS	1.392	2.128	3.679	1.437	1.867	2.871
	Ridge	0.669	0.779	0.940	0.752	0.884	0.978
t_5	ENET	1.264	1.349	1.486	1.252	1.359	1.442
	LASSO	1.139	1.225	1.372	1.154	1.212	1.321
	ALASSO	1.893	2.516	3.746	1.591	2.174	3.000
	SM	2.008	3.245	5.448	1.557	2.239	3.612
	S	1.250	1.648	2.301	1.283	1.482	2.085
	PS	1.421	2.085	3.692	1.424	1.803	2.830
	Ridge	0.633	0.759	0.894	0.746	0.818	0.936
BM (0.1)	ENET	1.267	1.371	1.450	1.245	1.316	1.462
	LASSO	1.108	1.247	1.298	1.144	1.196	1.321
	ALASSO	1.834	2.821	3.916	1.527	2.111	2.996
	SM	2.153	3.365	5.969	1.547	2.328	3.749
	S	1.221	1.575	2.160	1.301	1.611	2.010
	PS	1.445	2.151	3.389	1.413	1.840	2.830
BM (0.25)	Ridge	0.634	0.755	0.909	0.729	0.843	0.943
	ENET	1.286	1.348	1.461	1.268	1.343	1.483
	LASSO	1.174	1.227	1.322	1.150	1.239	1.356
	ALASSO	1.890	2.762	3.856	1.566	2.138	3.010
	SM	2.303	3.616	7.399	2.025	3.006	5.408
	S	1.193	1.563	2.000	1.350	1.765	2.208
	PS	1.396	2.091	4.035	1.566	2.226	3.827
	Ridge	0.762	0.929	1.212	0.892	1.054	1.233
Cauchy	ENET	1.266	1.373	1.572	1.275	1.416	1.570
	LASSO	1.156	1.227	1.399	1.160	1.269	1.434
	ALASSO	1.713	2.281	3.016	1.643	2.162	2.700
	SM	2.169	3.339	6.383	1.359	1.945	3.285
	S	1.201	1.644	2.111	1.241	1.484	1.905
	PS	1.440	2.131	3.712	1.281	1.711	2.492
	Ridge	0.535	0.634	0.822	0.636	0.734	0.836
Laplace	ENET	1.255	1.361	1.504	1.271	1.337	1.474
	LASSO	1.171	1.223	1.374	1.167	1.225	1.351
	ALASSO	2.026	2.876	5.144	1.456	1.988	3.174
	SM	2.009	3.315	5.819	2.111	3.012	4.901

	S	1.334	1.677	2.181	1.448	1.783	2.335
	PS	1.474	2.062	3.827	1.560	2.116	3.641
	Ridge	0.616	0.777	0.932	0.754	0.856	0.966
Lognormal	ENET	2.286	3.219	4.424	2.192	2.691	3.521
	LASSO	1.846	2.668	3.681	1.810	2.225	2.938
	ALASSO	2.786	4.444	7.722	2.298	3.223	5.157
	SM	2.105	3.127	5.359	1.659	2.246	3.615
	S	1.242	1.572	2.029	1.314	1.538	1.886
	PS	1.392	1.984	3.314	1.422	1.826	2.733
	Ridge	0.629	0.748	0.890	0.735	0.811	0.923
Skewed	ENET	1.238	1.330	1.435	1.271	1.315	1.428
	LASSO	1.143	1.196	1.307	1.124	1.188	1.304
	ALASSO	1.895	2.595	3.628	1.578	2.048	2.874

TABLE 9.18: The RMAPE of the Estimators for $n = 100$ and $p_1 = 4$ – SM with Strong Signals.

		$p_2 = 6$			$p_2 = 9$		
	Method	$p_3 = 4$	$p_3 = 8$	$p_3 = 16$	$p_3 = 4$	$p_3 = 8$	$p_3 = 16$
	SM	1.440	1.868	2.934	1.385	1.732	2.598
	S	1.365	1.503	1.933	1.333	1.493	1.902
	PS	1.420	1.652	2.447	1.398	1.629	2.268
	Ridge	0.833	0.890	1.021	0.893	0.948	1.049
Normal	ENET	1.290	1.326	1.501	1.273	1.361	1.489
	LASSO	1.138	1.199	1.336	1.156	1.212	1.338
	ALASSO	1.450	1.801	2.601	1.400	1.701	2.393
	SM	2.498	3.460	5.373	2.794	3.701	5.729
	S	1.643	2.210	2.156	1.877	2.079	2.035
	PS	2.047	2.590	2.216	2.229	2.586	2.041
	Ridge	1.999	2.339	3.205	2.232	2.709	3.577
χ^2_5	ENET	2.295	2.753	4.019	2.467	3.114	4.273
	LASSO	2.435	2.851	4.386	2.569	3.239	4.406
	ALASSO	2.402	2.920	4.542	2.610	3.327	4.545
	SM	1.558	2.020	3.150	1.472	1.910	2.793
	S	1.367	1.598	1.995	1.458	1.597	1.922
t_5	PS	1.482	1.756	2.583	1.522	1.786	2.477
	Ridge	0.837	0.935	1.033	0.917	0.992	1.070

	ENET	1.275	1.333	1.451	1.281	1.384	1.490
	LASSO	1.157	1.194	1.274	1.167	1.252	1.351
	ALASSO	1.517	1.891	2.546	1.481	1.845	2.497
BM (0.1)	SM	1.395	1.904	2.952	1.373	1.807	2.556
	S	1.346	1.524	2.033	1.348	1.569	1.896
	PS	1.394	1.688	2.487	1.413	1.698	2.298
	Ridge	0.796	0.919	0.990	0.873	0.958	1.016
	ENET	1.239	1.349	1.449	1.262	1.355	1.455
	LASSO	1.133	1.200	1.344	1.143	1.223	1.322
	ALASSO	1.404	1.848	2.633	1.409	1.749	2.349
BM (0.25)	SM	1.452	2.003	3.023	1.367	1.846	2.667
	S	1.350	1.572	1.998	1.310	1.580	1.905
	PS	1.415	1.682	2.521	1.376	1.702	2.309
	Ridge	0.826	0.920	0.980	0.863	0.979	1.011
	ENET	1.264	1.351	1.442	1.261	1.400	1.452
	LASSO	1.169	1.239	1.332	1.153	1.230	1.327
	ALASSO	1.457	1.927	2.594	1.380	1.799	2.445
Cauchy	SM	2.020	2.844	4.883	2.064	2.764	4.469
	S	1.530	1.842	2.548	1.579	1.863	2.463
	PS	1.699	2.301	3.789	1.912	2.355	3.976
	Ridge	1.027	1.146	1.404	1.117	1.229	1.445
	ENET	1.344	1.432	1.645	1.387	1.467	1.671
	LASSO	1.225	1.316	1.482	1.264	1.333	1.489
	ALASSO	1.708	2.120	2.740	1.767	2.014	2.666
Laplace	SM	1.138	1.567	2.518	1.131	1.467	2.190
	S	1.226	1.391	1.828	1.201	1.395	1.766
	PS	1.246	1.489	2.201	1.217	1.451	2.019
	Ridge	0.711	0.799	0.914	0.775	0.856	0.920
	ENET	1.207	1.336	1.512	1.254	1.324	1.463
	LASSO	1.129	1.231	1.375	1.124	1.190	1.373
	ALASSO	1.225	1.691	2.569	1.219	1.542	2.237
Lognormal	SM	2.054	2.668	4.222	1.844	2.350	3.647
	S	1.601	1.807	2.400	1.610	1.887	2.323
	PS	1.696	2.121	3.017	1.713	2.127	2.928
	Ridge	0.901	0.966	1.102	0.937	1.017	1.165
	ENET	2.165	2.568	3.438	2.057	2.420	3.206

	LASSO	1.749	2.067	2.841	1.693	2.031	2.700
	ALASSO	2.103	2.722	4.307	1.914	2.436	3.764
	SM	1.455	1.893	2.977	1.389	1.743	2.589
	S	1.363	1.531	1.954	1.381	1.550	1.943
	PS	1.420	1.671	2.411	1.433	1.648	2.296
	Ridge	0.808	0.879	1.006	0.850	0.924	1.017
Skewed	ENET	1.259	1.316	1.456	1.244	1.315	1.468
	LASSO	1.150	1.215	1.338	1.156	1.182	1.303
	ALASSO	1.479	1.812	2.607	1.418	1.693	2.349

TABLE 9.19: The RMAPE of the Estimators for n = 500 and p_1 = 4 – SM with Strong Signals.

		$p_2 = 0$			$p_2 = 3$		
	Method	$p_3 = 4$	$p_3 = 8$	$p_3 = 16$	$p_3 = 4$	$p_3 = 8$	$p_3 = 16$
	SM	2.084	3.170	5.296	1.045	1.417	2.183
	S	1.245	1.696	2.353	1.130	1.298	1.596
	PS	1.405	1.972	3.199	1.131	1.312	1.632
	Ridge	0.373	0.479	0.637	0.460	0.543	0.672
Normal	ENET	1.283	1.467	1.680	1.230	1.365	1.527
	LASSO	1.255	1.427	1.551	1.180	1.308	1.432
	ALASSO	1.976	3.003	5.077	1.060	1.439	2.223
	SM	1.756	2.290	3.783	1.892	2.404	3.458
	S	1.314	1.409	1.750	1.410	1.658	1.921
	PS	1.427	2.066	3.566	1.722	2.228	3.204
	Ridge	0.934	1.122	1.398	1.094	1.236	1.456
χ^2_5	ENET	1.662	2.078	2.833	1.715	2.035	2.666
	LASSO	2.329	2.963	4.096	2.336	2.737	3.698
	ALASSO	1.827	2.339	3.609	1.876	2.444	3.461
	SM	2.204	3.042	5.464	1.060	1.469	2.323
	S	1.321	1.597	2.250	1.126	1.302	1.670
	PS	1.478	2.024	3.247	1.126	1.317	1.721
	Ridge	0.399	0.504	0.669	0.463	0.568	0.690
t_5	ENET	1.330	1.449	1.698	1.158	1.351	1.531
	LASSO	1.312	1.414	1.586	1.140	1.279	1.449
	ALASSO	2.056	2.965	5.127	1.086	1.483	2.366
	SM	2.116	3.184	5.326	0.999	1.387	2.151
	S	1.250	1.714	2.433	1.102	1.254	1.576

	PS	1.377	2.177	3.168	1.108	1.299	1.620
	Ridge	0.381	0.482	0.635	0.458	0.536	0.673
BM (0.1)	ENET	1.273	1.468	1.665	1.196	1.334	1.488
	LASSO	1.235	1.426	1.562	1.150	1.284	1.409
	ALASSO	1.926	3.138	4.872	1.028	1.413	2.167
	SM	1.982	3.069	5.256	0.984	1.358	2.169
	S	1.234	1.709	2.365	1.118	1.255	1.582
	PS	1.382	2.194	3.090	1.121	1.275	1.663
	Ridge	0.375	0.476	0.653	0.455	0.532	0.687
BM (0.25)	ENET	1.279	1.453	1.678	1.196	1.325	1.559
	LASSO	1.271	1.392	1.545	1.168	1.291	1.466
	ALASSO	1.934	3.020	5.051	1.012	1.397	2.222
	SM	2.107	3.275	5.633	1.246	1.735	2.696
	S	1.310	1.706	2.407	1.197	1.401	1.830
	PS	1.438	1.990	3.241	1.216	1.451	1.965
	Ridge	0.447	0.569	0.724	0.531	0.630	0.754
Cauchy	ENET	1.278	1.447	1.544	1.248	1.324	1.451
	LASSO	1.245	1.360	1.475	1.188	1.270	1.397
	ALASSO	2.013	3.103	5.166	1.257	1.707	2.665
	SM	2.131	3.332	5.792	0.749	1.068	1.680
	S	1.293	1.726	2.366	1.056	1.129	1.340
	PS	1.391	2.017	3.103	1.056	1.130	1.345
	Ridge	0.295	0.390	0.524	0.367	0.453	0.574
Laplace	ENET	1.265	1.456	1.619	1.135	1.329	1.509
	LASSO	1.258	1.433	1.599	1.153	1.326	1.454
	ALASSO	2.049	3.198	5.414	0.762	1.086	1.713
	SM	1.909	2.877	4.749	1.204	1.628	2.571
	S	1.387	1.852	2.436	1.275	1.449	1.842
	PS	1.507	2.091	3.528	1.275	1.450	1.873
	Ridge	0.304	0.405	0.558	0.379	0.466	0.627
Lognormal	ENET	1.600	2.377	3.769	1.382	1.830	2.801
	LASSO	1.326	1.977	3.184	1.166	1.561	2.402
	ALASSO	2.038	3.078	5.113	1.106	1.494	2.351
	SM	1.974	3.027	5.099	1.049	1.399	2.206
	S	1.239	1.641	2.252	1.145	1.286	1.606
	PS	1.412	2.005	3.317	1.145	1.298	1.676

Skewed	Ridge	0.373	0.483	0.640	0.466	0.554	0.687
	ENET	1.307	1.482	1.674	1.197	1.356	1.508
	LASSO	1.287	1.400	1.581	1.152	1.287	1.421
	ALASSO	2.023	3.112	5.265	1.070	1.427	2.246

TABLE 9.20: The RMAPE of the Estimators for $n = 500$ and $p_1 = 4$ – SM with Strong Signals.

		$p_2 = 6$			$p_2 = 9$		
	Method	$p_3 = 4$	$p_3 = 8$	$p_3 = 16$	$p_3 = 4$	$p_3 = 8$	$p_3 = 16$
Normal	SM	0.798	1.061	1.539	0.720	0.900	1.237
	S	1.088	1.200	1.395	1.140	1.178	1.282
	PS	1.088	1.200	1.411	1.140	1.178	1.283
	Ridge	0.512	0.601	0.716	0.568	0.636	0.727
	ENET	1.130	1.277	1.445	1.137	1.232	1.399
	LASSO	1.094	1.238	1.374	1.112	1.179	1.321
	ALASSO	0.820	1.091	1.577	0.730	0.908	1.250
χ^2_5	SM	1.937	2.279	3.300	1.950	2.391	3.179
	S	1.508	1.625	1.856	1.577	1.679	1.901
	PS	1.839	2.212	3.112	1.925	2.290	3.069
	Ridge	1.194	1.308	1.505	1.264	1.375	1.580
	ENET	1.747	2.026	2.522	1.747	2.020	2.565
	LASSO	2.182	2.569	3.305	2.161	2.615	3.277
	ALASSO	1.869	2.353	3.128	1.927	2.338	3.126
t_5	SM	0.854	1.118	1.643	0.764	0.940	1.278
	S	1.117	1.215	1.450	1.115	1.174	1.304
	PS	1.117	1.216	1.471	1.115	1.174	1.305
	Ridge	0.524	0.623	0.730	0.593	0.646	0.727
	ENET	1.136	1.315	1.450	1.126	1.218	1.353
	LASSO	1.114	1.252	1.361	1.093	1.181	1.308
	ALASSO	0.861	1.152	1.667	0.774	0.951	1.292
BM (0.1)	SM	0.801	1.026	1.502	0.703	0.874	1.223
	S	1.125	1.173	1.372	1.099	1.138	1.280
	PS	1.125	1.173	1.390	1.099	1.138	1.281
	Ridge	0.523	0.585	0.710	0.569	0.636	0.721
	ENET	1.160	1.268	1.414	1.122	1.228	1.370
	LASSO	1.127	1.213	1.340	1.074	1.187	1.324
	ALASSO	0.810	1.046	1.514	0.714	0.885	1.239

	SM	0.780	1.024	1.517	0.686	0.845	1.206
	S	1.087	1.174	1.395	1.092	1.137	1.287
	PS	1.087	1.174	1.402	1.092	1.137	1.287
	Ridge	0.522	0.593	0.720	0.574	0.622	0.728
BM (0.25)	ENET	1.149	1.265	1.467	1.111	1.216	1.410
	LASSO	1.094	1.225	1.380	1.079	1.172	1.325
	ALASSO	0.795	1.033	1.544	0.700	0.858	1.238
	SM	1.020	1.307	1.911	0.883	1.128	1.599
	S	1.188	1.307	1.569	1.135	1.250	1.492
	PS	1.188	1.323	1.653	1.135	1.252	1.522
	Ridge	0.627	0.687	0.790	0.640	0.721	0.817
Cauchy	ENET	1.201	1.241	1.381	1.104	1.225	1.357
	LASSO	1.148	1.223	1.316	1.079	1.189	1.299
	ALASSO	1.051	1.346	1.951	0.916	1.172	1.636
	SM	0.565	0.743	1.121	0.492	0.608	0.880
	S	1.054	1.076	1.199	1.060	1.056	1.148
	PS	1.054	1.076	1.199	1.060	1.056	1.148
	Ridge	0.427	0.495	0.615	0.470	0.527	0.622
Laplace	ENET	1.092	1.224	1.432	1.074	1.166	1.378
	LASSO	1.096	1.208	1.394	1.067	1.149	1.308
	ALASSO	0.574	0.756	1.133	0.498	0.617	0.894
	SM	0.904	1.162	1.712	0.772	0.950	1.356
	S	1.267	1.335	1.544	1.236	1.246	1.371
	PS	1.267	1.335	1.584	1.236	1.246	1.372
	Ridge	0.447	0.525	0.669	0.500	0.574	0.694
Lognormal	ENET	1.231	1.564	2.202	1.227	1.491	2.053
	LASSO	1.069	1.341	1.912	1.076	1.296	1.805
	ALASSO	0.855	1.097	1.613	0.740	0.910	1.295
	SM	0.815	1.066	1.553	0.709	0.879	1.234
	S	1.099	1.199	1.414	1.100	1.129	1.277
	PS	1.099	1.199	1.417	1.100	1.129	1.279
	Ridge	0.526	0.612	0.718	0.573	0.631	0.722
Skewed	ENET	1.156	1.298	1.425	1.123	1.210	1.376
	LASSO	1.126	1.246	1.361	1.102	1.173	1.303
	ALASSO	0.826	1.084	1.569	0.721	0.896	1.253

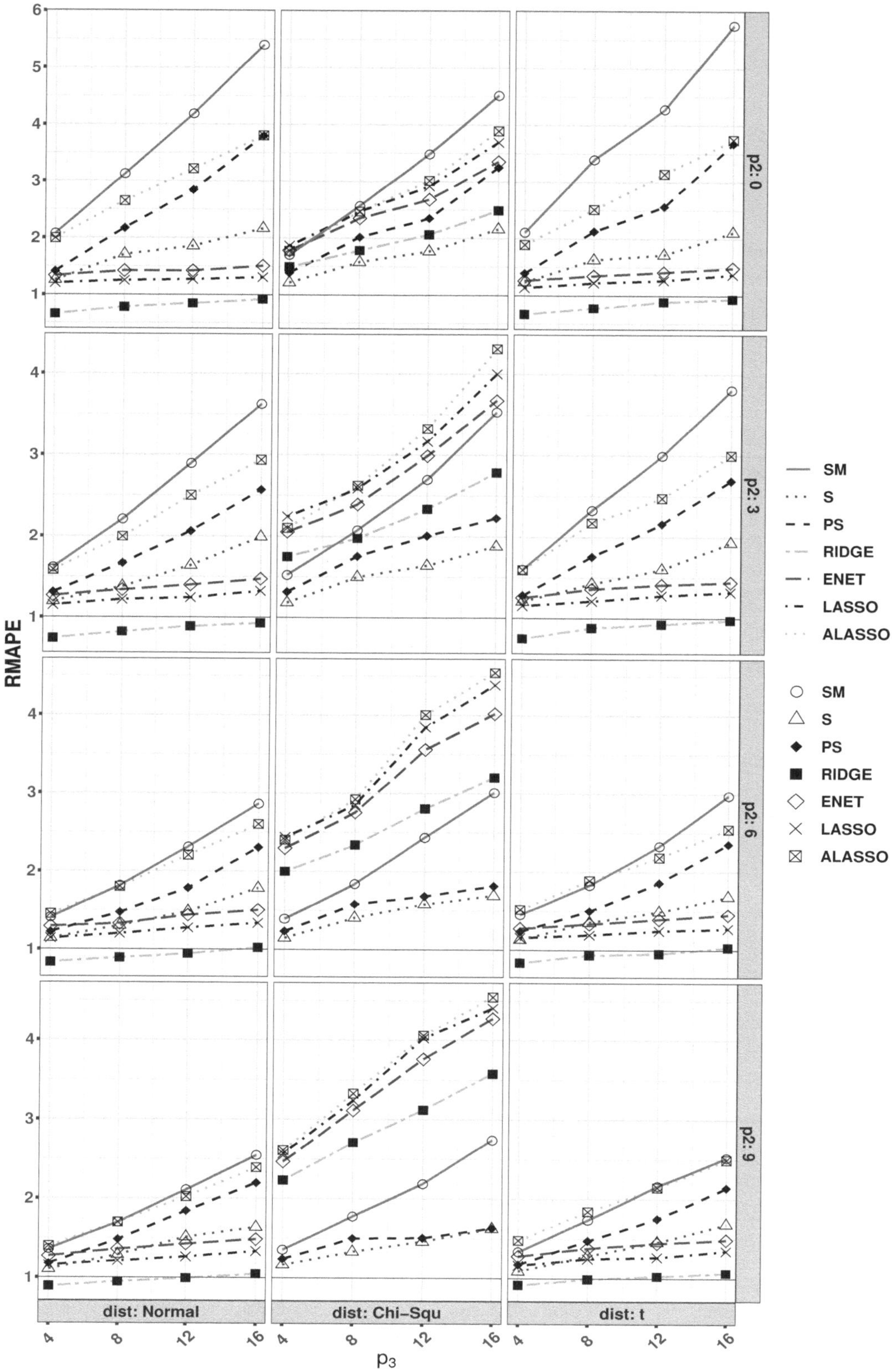

FIGURE 9.5: The RMAPE of Estimators for $n = 100$ and $p_1 = 4$.

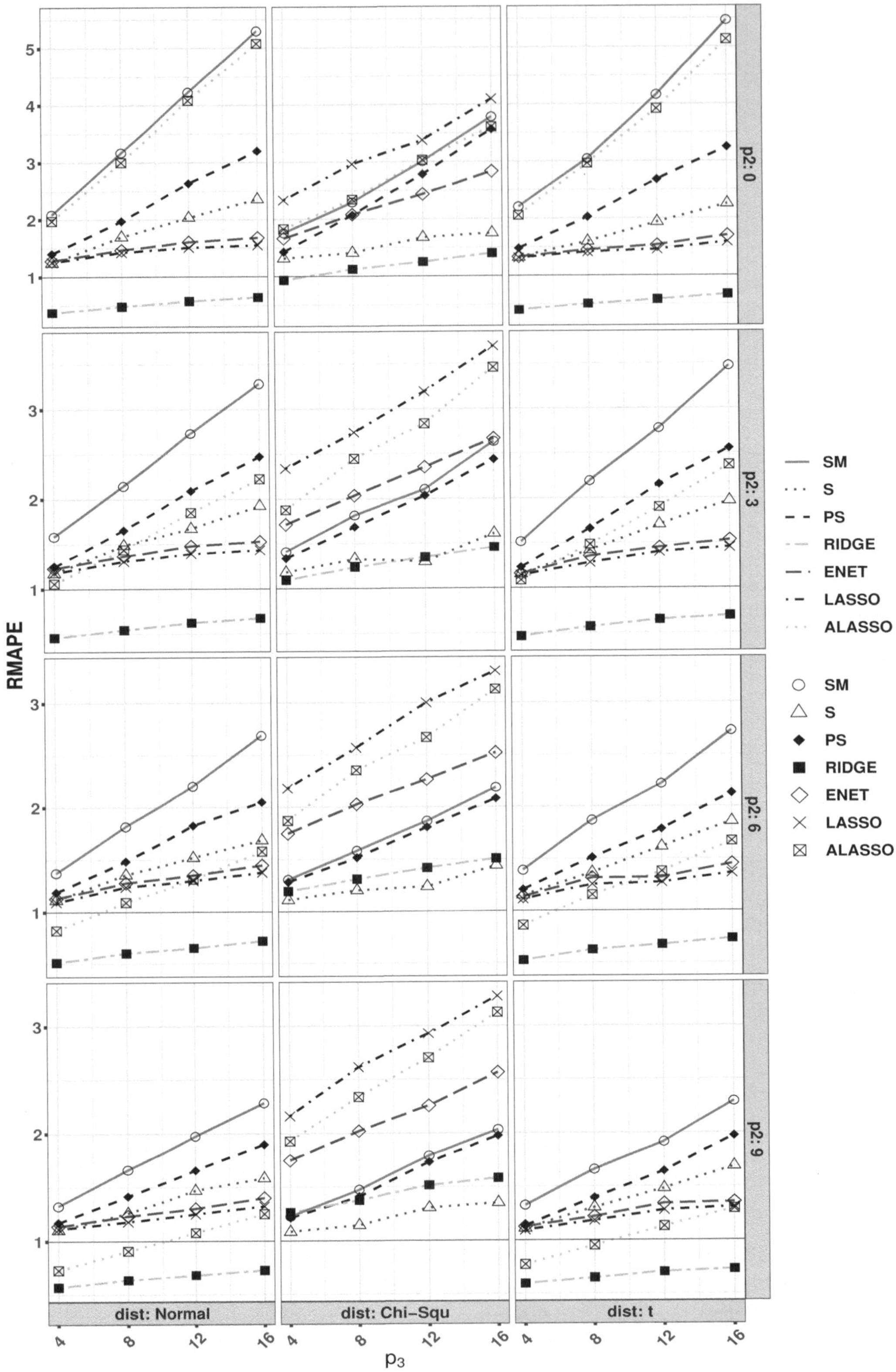

FIGURE 9.6: RMAPE of the PLS versus Shrinkage for $n = 500$ and $p_1 = 4$.

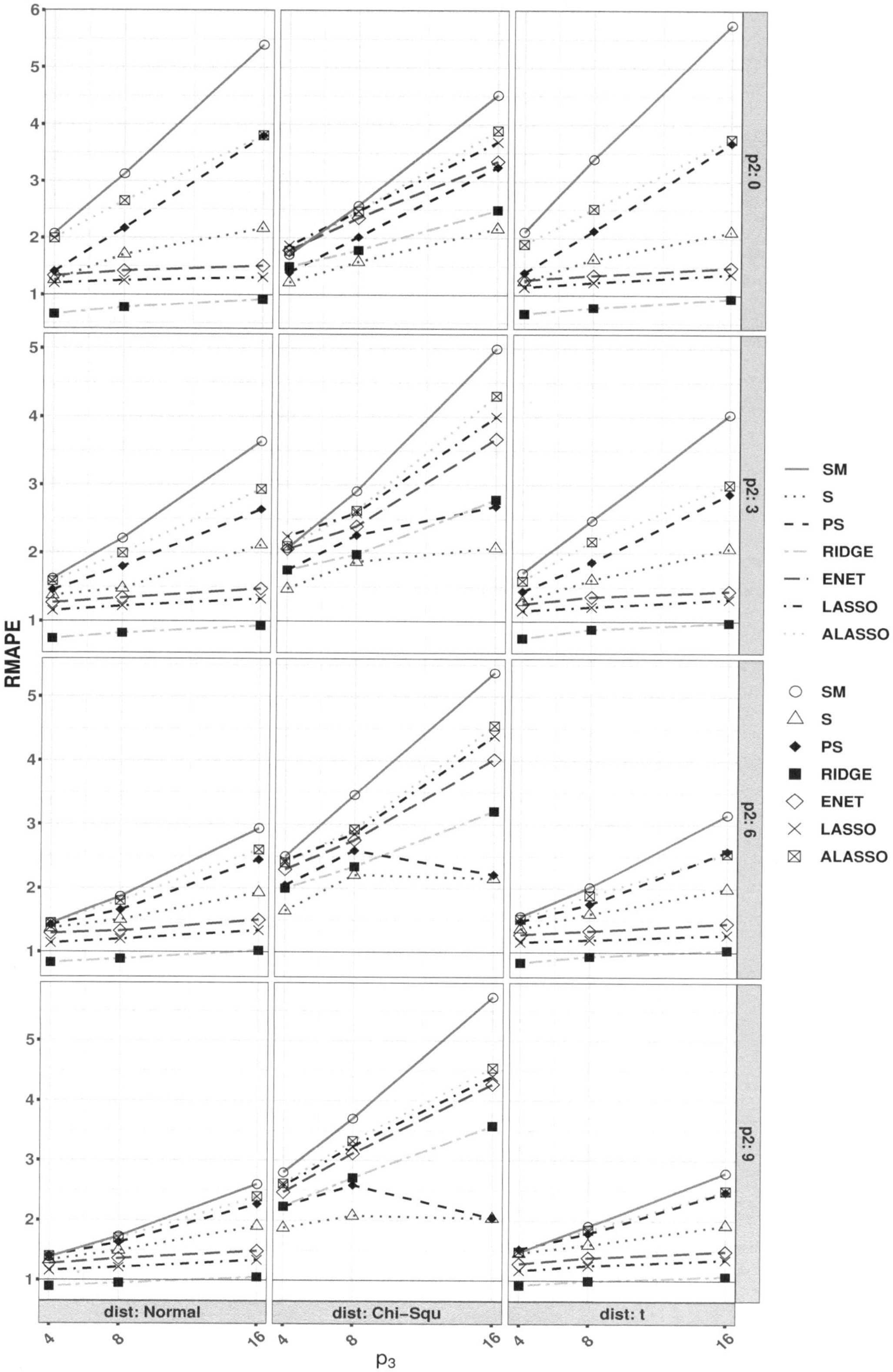

FIGURE 9.7: The RMAPE of Estimators for $n = 100$ and $p_1 = 4$ – SM with Strong Signals.

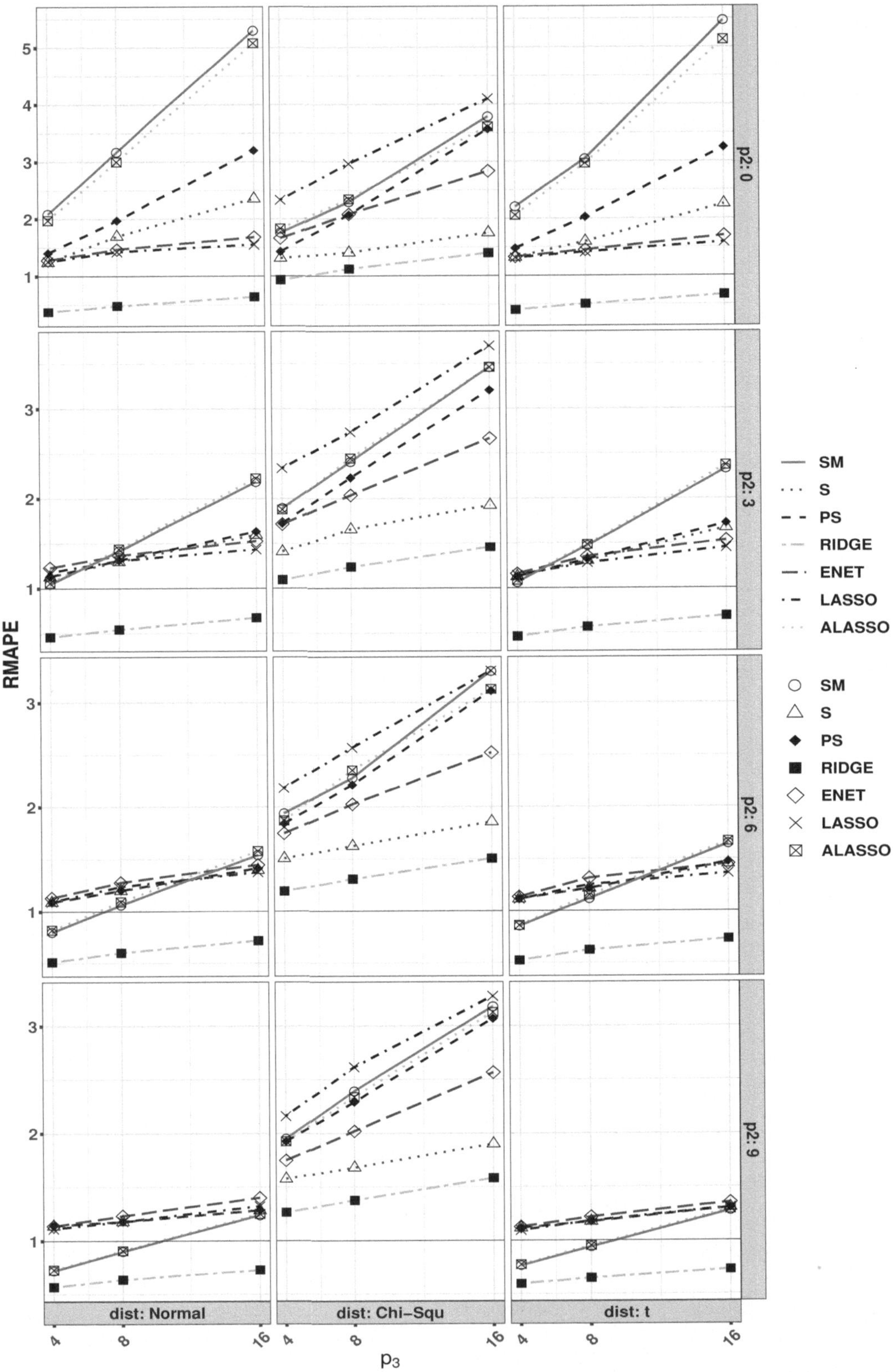

FIGURE 9.8: The RMAPE of Estimators for $n = 500$ and $p_1 = 4$ – SM with Strong Signals.

9.5 Real Data Applications

We apply the LAD strategies to real data sets to illustrate the application of these methods. We calculate the trimmed mean squared prediction error (TMSPE) and the relative TMSPE (RTMSPE) with respect to the full model estimator in appraising the performance of the procedures. A proportion of the largest squared differences between the observed and fitted values are trimmed. Here we apply 15% trimming, and the *tmspe* function *cvTools* package in R software is used in this analysis. To calculate the prediction error of estimators, we randomly split the data into a training set that has 75% of the observations and a testing set that has the remaining 25% of the observations. The response and the predictors are centered and scaled based on the training data set before fitting the model.

9.5.1 US Crime Data

Criminologists are often interested in the effect of punishment regimes on crime rates. We use criminal data available from *MASS* package in R software. The description of the data is given in Table 9.21. The researchers are interested in predicting the rate of crimes in a particular category per head of population using 15 different predictors. Here, $(n, p) = (47, 15)$.

TABLE 9.21: The Description of the US Crime Data.

Variable	Description
M	percentage of males aged 14-24
So	indicator variable for a Southern state
Ed	mean years of schooling
Po1	police expenditure in 1960
Po2	police expenditure in 1959
LF	labour force participation rate
MF	number of males per 1000 females
Pop	state population
NW	number of non-whites per 1000 people
U1	unemployment rate of urban males 14-24
U2	unemployment rate of urban males 35-39
GDP	gross domestic product per head
Ineq	income inequality
Prob	probability of imprisonment
Time	average time served in state prisons
rateofcr (Response)	rate of crimes in a particular category per head of population

For brevity, the variables have been re-scaled. The predictive model is thus:

$$\begin{aligned}\text{rateofcr}_i &= \beta_0 + \beta_1 \text{M}_i + \beta_2 \text{So}_i + \beta_3 \text{Ed}_i + \beta_4 \text{Po1}_i + \beta_5 \text{Po2}_i + \beta_6 \text{LF}_i \\ &+ \beta_7 \text{MF}_i + \beta_8 \text{Pop}_i + \beta_9 \text{NW}_i + \beta_{10} \text{U1}_i + \beta_{11} \text{U2}_i + \beta_{12} \text{GDP}_i \\ &+ \beta_{13} \text{Ineq}_i + \beta_{14} \text{Prob}_i + \beta_{15} \text{Time}_i + \epsilon_i.\end{aligned}$$

When all linear regression assumptions are met, the ordinary least squares method yields the most accurate estimates. Inaccurate assumptions can result in unsatisfactory least squares regression findings. Using residual diagnostics, one might determine where some incorrect assumptions originated. Typically, we deploy the following four diagnostic graphs:

- Residual vs. Fitted: Used to verify linear relationship hypotheses. A horizontal line devoid of different patterns indicates a linear relationship, which is desirable.
- Normal Q-Q: Used to determine whether residuals follow a normal distribution. It is desirable for residuals points to align with the straight dashed line.
- Scale-Location: Used to examine the homogeneity of residual variance (homoscedasticity). A horizontal line with evenly distributed points is an excellent indicator of homoscedasticity.
- Residuals vs Leverage: Used to detect influential cases, or extreme values that may influence the regression results based on whether they are included or excluded from the study. This plot may assist us in locating influential observations, if any exist. On this graph, outlier values are typically found in the upper right or lower right corners. These are the locations where data points can have influence over a regression line.

Observing Figure 9.9, it is evident that all assumptions have been violated. The failure of the most basic assumptions in least squares regression necessitates the adoption of a robust regression analysis as an alternative.

Since prior information is not provided, we must apply the approaches in two stages. The initial step is to choose the best submodel that can be done with standard variable selection. The BIC variable selection strategy identifies six predictors (M, Ed, Po1, U2, Ineq, Prob) for further study, and the submodel is built as follows:

$$\begin{aligned}\text{rateofcr}_i &= \beta_0 + \beta_1 \text{M}_i + \beta_3 \text{Ed}_i + \beta_4 \text{Po1}_i \beta_{11} \text{U2}_i + \beta_{13} \text{Ineq}_i \\ &+ \beta_{14} \text{Prob}_i + \epsilon_i.\end{aligned}$$

The RTMSPE of the listed estimators are reported in Table 9.22. Clearly, the positive shrinkage strategy is the best at reducing prediction error. As aforementioned, if the selected submodel is the correct one, we suggest using the submodel strategy over shrinkage for prediction purposes. In reality, no one knows if the selected submodel is truly correct.

TABLE 9.22: The RTMSPE of the Estimators for US Crime Data.

	RTMSPE
SM	2.062
S	1.159
PS	**1.905**
Ridge	0.820
ENET	1.036
LASSO	1.087
ALASSO	1.175

9.5.2 Barro Data

In this example, we use the data given by Barro and Lee (1994) which consists of $n = 161$ observations of averaged national growth rates over a five-year period from 1960 to 1985.

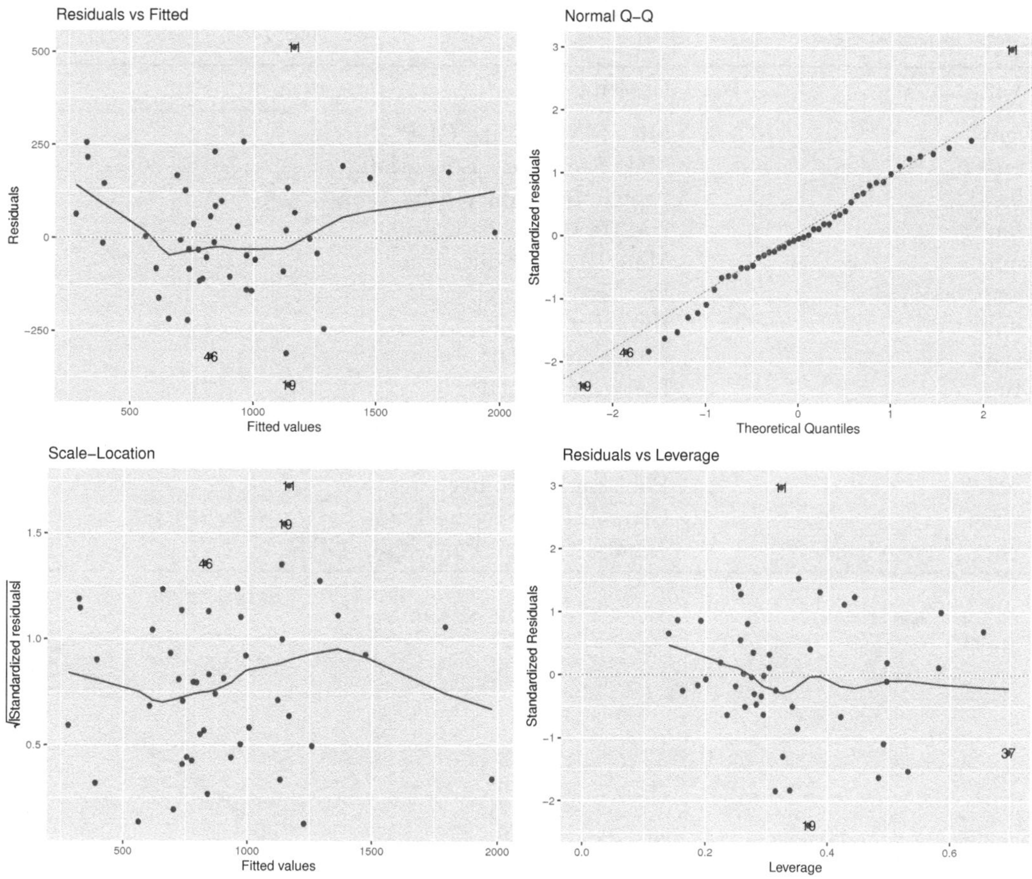

FIGURE 9.9: Residual Diagnosis of Us Crime Data.

This data set is freely available using the *quentreg* package in R. The description of data is given in Table 9.23. Suppose the investigators are interested in predicting an annual change in per capita GDP using 13 predictors. Hence, the model is:

$$\begin{aligned}
\text{ynet}_i &= \beta_0 + \beta_1 \text{lgdp}_i + \beta_2 \text{mse}_i + \beta_3 \text{fse}_i + \beta_4 \text{fhe}_i + \beta_5 \text{mhe}_i + \beta_6 \text{lexp}_i \\
&+ \beta_7 \text{lintr}_i + \beta_8 \text{gedy}_i + \beta_9 \text{Iy}_i + \beta_{10} \text{gcony}_i + \beta_{11} \text{lblakp}_i \\
&+ \beta_{12} \text{pol}_i + \beta_{13} \text{ttrad}_i + \epsilon_i.
\end{aligned}$$

Observing Figure 9.10, the normality assumption has been broken and the data contains outliers. The failure of the two hypotheses suggests that robust regression analysis should be utilized instead. The BIC variable selection approach identifies four predictors (fse, fhe, mhe, and gedy) that can be eliminated from the entire or initial model in order to identify a candidate submodel. Therefore, the submodel using the remaining predictors is expressed as follows:

$$\begin{aligned}
\text{ynet}_i &= \beta_0 + \beta_1 \text{lgdp}_i + \beta_2 \text{mse}_i + \beta_6 \text{lexp}_i + \beta_7 \text{lintr}_i + \beta_9 \text{Iy}_i + \beta_{10} \text{gcony}_i \\
&+ \beta_{11} \text{lblakp}_i + \beta_{12} \text{pol}_i + \beta_{13} \text{ttrad}_i + \epsilon_i.
\end{aligned}$$

The RTMSPE of the listed estimators are reported in Table 9.24. In this data example, the performance of positive shrinkage and ridge strategies is highly competitive. It can be

TABLE 9.23: The Description of Barro Data.

Variable	Description
lgdp	Initial Per Capita GDP
mse	Male Secondary Education
fse	Female Secondary Education
fhe	Female Higher Education
mhe	Male Higher Education
lexp	Life Expectancy
lintr	Human Capital
gedy	Education/GDP
Iy	Investment/GDP
gcony	Public Consumption/GDP
lblakp	Black Market Premium
pol	Political Instability
ttrad	Growth Rate Terms Trade
ynet (Dependent)	Annual Change Per Capita GDP

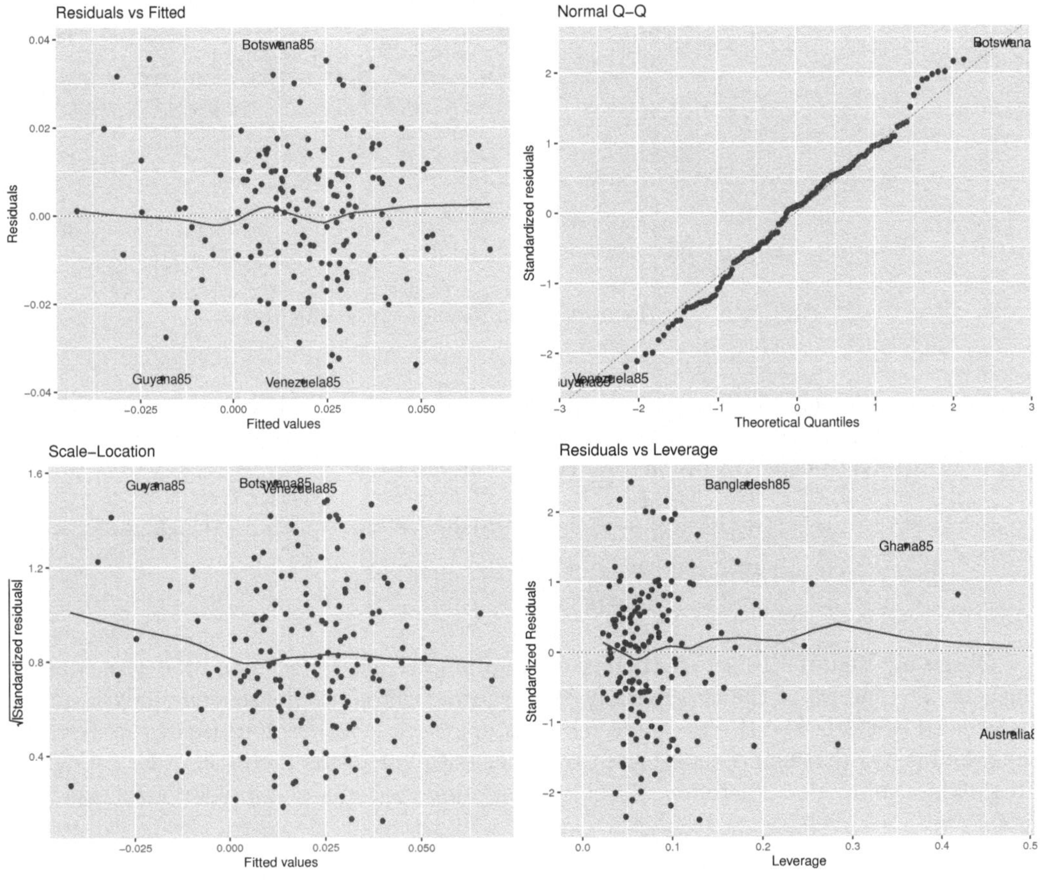

FIGURE 9.10: Residual Diagnosis of Barro Data.

TABLE 9.24: The RTMSPE of the Estimators for Barro Data.

	RTMSPE
SM	1.137
S	1.030
PS	**1.086**
Ridge	**1.088**
ENET	0.994
LASSO	0.979
ALASSO	0.978

TABLE 9.25: The Description of Murder Rate Data.

Variable	Description
murders	murders per 1,000,000
pctmetro	the percent of the population living in metropolitan areas
pctwhite	the percent of the population that is white
pcths	percent of population with a high school education or above
poverty	percent of population living under poverty line
single	percent of population that are single parents
crime (Dependent)	violent crimes per 100,000 people

assumed that the predictors are subject to multicollinearity, thus the ridge estimator performs well. The performance of the penalized methods is not satisfactory for this particular data set.

9.5.3 Murder Rate Data

The data used for this investigation comes from UCLAs Institute for Digital Research and Education. The description of the data is given in Table 9.25. Here, $(n, p) = (51, 7)$. The purpose of the study is to predict violent crimes per 100,000 people based on seven predictors.

A full or initial regression model is given below:

$$\begin{aligned} \text{crime}_i &= \beta_0 + \beta_1 \text{murders}_i + \beta_2 \text{pctmetro}_i + \beta_3 \text{pctwhite}_i \\ &+ \beta_4 \text{pcths}_i + \beta_5 \text{poverty}_i + \beta_6 \text{single}_i \epsilon_i. \end{aligned}$$

Observing Figure 9.11, again, the linear regression model assumption has been broken. Due to the invalidity of the assumptions, robust regression analysis should be utilized instead. In order to identify a candidate submodel, we apply the BIC variable selection method, which ignores three predictors, namely pctwhite, pcths, and poverty, and selects the remaining three predictors to form the submodel given below.

$$\text{crime}_i = \beta_0 + \beta_1 \text{murder}_i + \beta_2 \text{pctmetro}_i + \beta_6 \text{single}_i + \epsilon_i.$$

The RTMSPEs of the listed estimators are reported in Table 9.26. Interestingly, in this data example, we find LASSO and ALASSO are performing better than shrinkage estimators, which is possible. In some cases, if a selected submodel is far from being the right one, then the distance measure will heavily shrink toward the full model, resulting in a larger prediction error. While LASSO and ALASSO seem to select the right submodel.

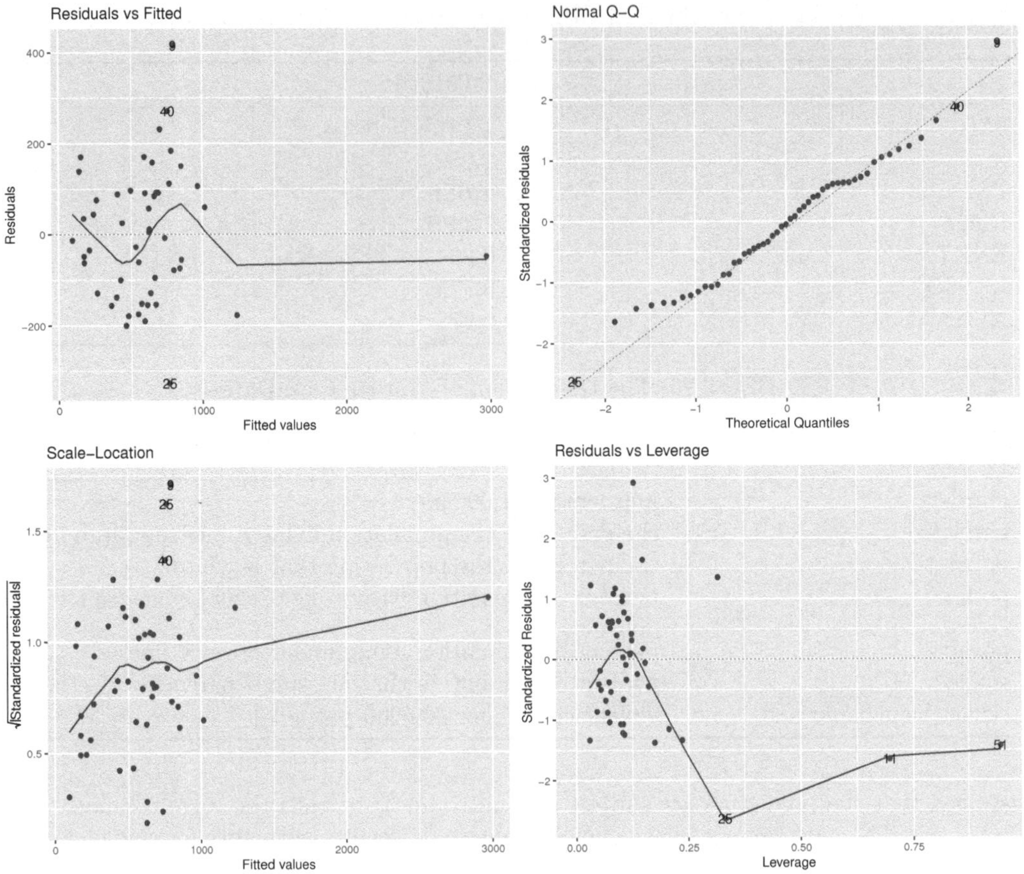

FIGURE 9.11: Residual Diagnosis of Murder Rate Data.

TABLE 9.26: The RTMSPE of the Estimators for Murder Rate Data.

	RTMSPE
SM	1.263
S	1.048
PS	1.054
Ridge	0.423
ENET	0.961
LASSO	1.234
ALASSO	1.221

9.6 High-Dimensional Data

Here we provide some numerical comparisons of the listed estimators when the number of predictors (p) exceeds the sample size (n). In this situation, it would be prohibited to obtain full model parameter estimation based on classical estimation methods. We will take recourse, using penalized methods to select two models, treating a model with a larger number of predictors as a full model and the model with a smaller number of predictors as a submodel. Using these two models, one can construct shrinkage estimators.

Different penalized methods may produce different models, resulting in more than one shrinkage estimators based on different combinations. A data analyst familiar with the data may decide to select two specific penalized methods to build a shrinkage estimator. Essentially, we are integrating two submodels to obtain a shrinkage strategy through a distance measure.

9.6.1 Simulation Experiments

In our simulation experiment, we use ENET to produce a model with a large number of predictors and use ALASSO to yield a submodel with a relatively smaller number of predictors. We combine the ENET model estimators with ALASSO model estimators to build a shrinkage estimator using the same distance measure as used in the low-dimensional case. We generate data from normal and t_5 distributions. We partition the regression parameter as $\boldsymbol{\beta} = \left(\boldsymbol{\beta}_1^\top, \boldsymbol{\beta}_2^\top, \boldsymbol{\beta}_3^\top\right)^\top$, with dimensions p_1, p_2 and p_3, respectively. Here $p = p_1 + p_2, +p_3$, where p_1 represents the strong signals, p_2 denotes the weak signals, and p_3 represents the number of zeros in the model. $\boldsymbol{\beta}_1$ is associated with strong signals and $\boldsymbol{\beta}_1$ is a vector of 1 values in our simulation experiment. The regression coefficient $\boldsymbol{\beta}_2$ corresponds with the weak signals, with signal strength $\boldsymbol{\kappa} = 0, 0.05, 0.01, 0.02$ and we set $\boldsymbol{\beta}_3 = \mathbf{0}$. We conduct an extensive simulation study to examine the behavior of the various estimators in some configurations of (n, p_1, p_2, p_3).

In Tables 9.27 and 9.31, we report the RMAPE of the estimators relative to the ENET model. The results are consistent with case of low-dimensions as the shrinkage estimator is superior to penalized based estimators. However, the submodel estimators are expected to be superior when the selected submodel is assumed to be the correct one.

9.6.2 Real Data Application

In this data set, laboratory rats (Rattus norvegicus) were studied to learn about gene expression and regulation in the mammalian eye. Inbred rat strains were crossed, and tissue was extracted from the eyes of $n = 120$ animals from the F2 generation. Microarrays were used to measure levels of RNA expression in the isolated eye tissues of each subject. $p = 31,041$ different probes were detected at a sufficient level to be considered expressed in the mammalian eye. The data was downloaded from the Gene Expression Omnibus, accession number GSE5680.

For the purposes of this analysis, we treat one gene, Trim32, as the outcome. Trim32 is known to be linked to a genetic disorder called Bardet-Biedl Syndrome (BBS) as the mutation (P130S) in Trim32 gives rise to BBS. Table 9.32 reports the relative TMSPE based on ENET and LASSO with the full model, respectively. Clearly, the shrinkage strategy using ENET and ALASSO has better performance.

9.7 R-Codes

```
> library('MASS')    # It is for 'mvrnorm' function
> library('lmtest')  # It is for 'lrtest' function
> library('caret')   # It is for 'split' function
```

TABLE 9.27: RMAPE of the Estimators for $p_1 = 4$ and $p_3 = 1000$.

	n	p_2	κ	SM	PS	LASSO	Ridge
Normal	75	50	0.00	1.796	1.346	1.049	0.221
			0.05	1.508	1.234	1.106	0.268
			0.10	1.174	1.108	1.053	0.396
			0.20	0.947	1.017	1.005	0.704
	150	100	0.00	6.115	1.938	1.146	0.183
			0.05	2.305	1.448	1.089	0.270
			0.10	1.375	1.174	1.052	0.433
			0.20	1.010	1.037	1.018	0.758
	750	500	0.00	67.577	2.663	0.973	0.223
			0.05	1.372	1.301	1.018	0.490
			0.10	0.889	1.092	0.978	0.803
			0.20	0.745	1.008	0.922	1.039
t_5	75	50	0.00	1.529	1.300	1.131	0.311
			0.05	1.255	1.170	1.095	0.376
			0.10	1.196	1.139	1.105	0.463
			0.20	0.930	1.016	1.015	0.801
	150	100	0.00	4.778	1.936	1.198	0.300
			0.05	2.579	1.561	1.168	0.370
			0.10	1.410	1.200	1.093	0.544
			0.20	1.002	1.043	1.030	0.815
	750	500	0.00	84.488	2.906	1.058	0.291
			0.05	1.709	1.404	1.053	0.545
			0.10	0.969	1.112	0.957	0.809
			0.20	0.747	1.008	0.894	1.014

```
> library('quantreg')# It is for 'rq' function
> library('hqreg') # # It is for 'hqreg_raw' function
> set.seed(2023)
> ####
> # Optimum beta for all penalty terms
> opt.beta<-function(coef.matrix, newx, newy){
+   # calculate the mse
+   mse.validate<-NULL
+   for(i in 1:ncol(coef.matrix)){
+     mse.validate<-c(mse.validate,
+                     MAPE(newy,newx%*%coef.matrix[,i]))
+   }
+   # find the optimal coefficients set
```

TABLE 9.28: RMAPE of the Estimators for $n = 100$ and $p_1 = 4$.

			Normal Distribution				t_5 Distribution			
p_2	p_3	κ	SM	PS	LASSO	Ridge	SM	PS	LASSO	Ridge
30	200	0.2	1.350	1.127	1.048	0.611	1.431	1.162	1.076	0.742
		0.4	0.917	1.016	0.963	0.768	0.959	1.029	0.993	0.869
		0.6	0.897	1.007	0.963	0.800	0.913	1.009	0.970	0.870
		0.8	0.955	1.010	0.989	0.769	0.914	1.007	1.006	0.853
	500	0.2	1.174	1.083	1.041	0.524	1.192	1.107	1.090	0.647
		0.4	0.918	1.008	0.995	0.844	0.921	1.008	0.999	0.911
		0.6	0.812	0.993	0.904	0.920	0.861	0.997	0.941	0.958
		0.8	0.807	0.992	0.899	0.916	0.806	0.991	0.902	0.946
	1000	0.2	1.126	1.064	1.019	0.521	1.114	1.073	1.061	0.612
		0.4	0.943	1.006	1.017	0.851	0.929	1.006	1.010	0.923
		0.6	0.927	1.000	1.000	0.935	0.914	1.000	0.994	0.987
		0.8	0.893	0.997	0.966	0.957	0.906	0.999	0.985	0.990
50	200	0.2	1.070	1.073	1.024	0.685	1.107	1.088	1.039	0.819
		0.4	0.802	0.997	0.908	0.893	0.820	1.000	0.927	0.988
		0.6	0.756	0.988	0.879	0.978	0.780	0.993	0.895	1.014
		0.8	0.764	0.992	0.885	1.006	0.769	0.991	0.894	1.022
	500	0.2	1.022	1.046	1.019	0.684	1.013	1.052	1.039	0.808
		0.4	0.832	0.993	0.929	1.011	0.825	0.992	0.924	1.038
		0.6	0.805	0.990	0.905	1.080	0.825	0.992	0.919	1.088
		0.8	0.789	0.989	0.889	1.072	0.819	0.993	0.911	1.085
	1000	0.2	1.017	1.036	1.003	0.692	0.987	1.033	1.028	0.782
		0.4	0.855	0.992	0.954	1.022	0.858	0.995	0.951	1.041
		0.6	0.807	0.988	0.913	1.054	0.816	0.989	0.922	1.065
		0.8	0.797	0.989	0.907	1.059	0.810	0.991	0.921	1.061

```
+    coef.opt<-coef.matrix[,which.min(mse.validate)]
+    return(list(coef_beta = coef.opt,
+           MAPE=mse.validate[which.min(mse.validate)]))
+ }
> # Mean Absolute Prediction Error
> MAPE = function (y_pred, y_true)
+ {
+    MAPE <- mean(abs((y_true - y_pred)))
+    return(MAPE)
+ }
```

TABLE 9.29: RMAPE of the Estimators for $n = 100$ and $p_1 = 8$.

			Normal Distribution				t_5 Distribution			
p_2	p_3	κ	SM	PS	LASSO	Ridge	SM	PS	LASSO	Ridge
30	200	0.2	1.417	1.129	1.054	0.459	1.453	1.154	1.072	0.558
		0.4	0.966	1.021	0.990	0.640	0.979	1.026	1.017	0.801
		0.6	0.915	1.008	0.977	0.742	0.874	1.004	0.951	0.867
		0.8	0.921	1.006	0.995	0.794	0.914	1.005	0.995	0.872
	500	0.2	1.324	1.096	1.061	0.414	1.311	1.120	1.098	0.526
		0.4	1.001	1.018	1.053	0.758	0.965	1.017	1.012	0.797
		0.6	0.836	0.995	0.929	0.892	0.845	0.997	0.934	0.928
		0.8	0.773	0.989	0.884	0.926	0.779	0.991	0.883	0.939
	1000	0.2	1.286	1.085	1.046	0.435	1.210	1.085	1.081	0.513
		0.4	0.934	1.004	1.002	0.804	0.944	1.007	1.020	0.847
		0.6	0.884	0.996	0.972	0.948	0.886	0.997	0.979	0.988
		0.8	0.837	0.994	0.940	0.989	0.866	0.995	0.965	1.016
50	200	0.2	1.185	1.093	1.024	0.531	1.162	1.094	1.044	0.633
		0.4	0.783	0.997	0.905	0.863	0.801	1.001	0.918	0.904
		0.6	0.744	0.993	0.882	0.959	0.773	0.994	0.901	0.992
		0.8	0.767	0.994	0.917	1.030	0.776	0.994	0.922	1.046
	500	0.2	1.186	1.080	1.038	0.543	1.160	1.090	1.086	0.652
		0.4	0.905	1.007	1.002	0.947	0.904	1.008	1.001	0.963
		0.6	0.843	0.997	0.945	1.079	0.842	0.998	0.952	1.087
		0.8	0.831	0.995	0.924	1.109	0.844	0.996	0.938	1.104
	1000	0.2	1.153	1.067	1.046	0.558	1.121	1.068	1.054	0.607
		0.4	0.881	1.000	0.984	0.967	0.867	0.999	0.971	0.992
		0.6	0.791	0.988	0.913	1.083	0.802	0.990	0.917	1.071
		0.8	0.771	0.988	0.880	1.084	0.780	0.988	0.889	1.080

```
> #
> n  <- 200  # The number of sample
> p1 <- 5   # The number of strong signals
> p2 <- 10   # The number of weak signals
> p3 <- 20  # The number of zeros , no signal
> p <- p1+p2+p3
> beta1_true <- rep(1  ,p1)
> beta2_true <- rep(0.1,p2)
> beta3_true <- rep(0  ,p3)
> # The ture value of covariates
```

TABLE 9.30: RMAPE of the Estimators for $n = 200$ and $p_1 = 4$.

			Normal Distribution				t_5 Distribution			
p_2	p_3	κ	SM	PS	LASSO	Ridge	SM	PS	LASSO	Ridge
30	200	0.2	1.776	1.150	0.911	0.641	1.903	1.185	0.777	0.692
		0.4	2.305	1.078	0.928	0.581	1.918	1.082	0.861	0.693
		0.6	3.592	1.068	0.957	0.358	3.264	1.071	0.972	0.491
		0.8	4.515	1.060	0.889	0.218	4.335	1.066	0.988	0.340
	500	0.2	1.576	1.123	1.051	0.430	1.899	1.166	1.134	0.589
		0.4	1.658	1.055	1.022	0.343	1.569	1.063	1.129	0.512
		0.6	2.897	1.058	1.012	0.212	2.492	1.060	1.122	0.352
		0.8	4.418	1.059	1.032	0.133	3.842	1.061	1.110	0.226
	1000	0.2	1.412	1.093	1.046	0.360	1.508	1.118	1.101	0.494
		0.4	1.251	1.032	1.015	0.376	1.228	1.035	1.077	0.508
		0.6	2.334	1.046	0.998	0.271	1.550	1.035	1.057	0.405
		0.8	4.651	1.057	0.991	0.180	3.183	1.053	1.058	0.285
50	200	0.2	1.289	1.147	0.891	0.685	1.396	1.172	0.779	0.734
		0.4	1.247	1.070	0.956	0.610	1.124	1.061	0.899	0.752
		0.6	2.648	1.086	0.991	0.393	1.906	1.076	0.987	0.543
		0.8	3.419	1.082	0.921	0.209	3.147	1.086	1.039	0.346
	500	0.2	1.184	1.100	1.034	0.518	1.402	1.136	1.067	0.655
		0.4	0.975	1.026	0.976	0.525	1.044	1.035	1.036	0.667
		0.6	1.126	1.028	0.840	0.366	1.049	1.024	0.963	0.530
		0.8	1.919	1.051	0.752	0.271	1.225	1.032	0.839	0.383
	1000	0.2	1.208	1.078	1.044	0.518	1.263	1.097	1.071	0.629
		0.4	0.991	1.015	1.011	0.693	0.979	1.015	1.018	0.757
		0.6	0.940	1.005	0.981	0.704	0.929	1.005	0.986	0.764
		0.8	0.914	1.002	0.967	0.707	0.923	1.004	0.981	0.739

```
> beta_true <- c(beta1_true,beta2_true,beta3_true)
> # The design matrix
> X <- matrix(0, n, p)
> for (i in 1:p){
+   X[,i]= rnorm(n)
+ }
> ## assigning colnames of X to "X1","X2",....
> v<-NULL
> for(i in 1:p){
+   v[i] <- paste("X",i,sep="")
+   assign(v[i],X[,i])
+ }
```

TABLE 9.31: RMAPE of the Estimators for $n = 200$ and $p_1 = 8$.

			Normal Distribution				t_5 Distribution			
p_2	p_3	κ	SM	PS	LASSO	Ridge	SM	PS	LASSO	Ridge
30	200	0.2	1.804	1.165	0.945	0.455	2.126	1.215	0.944	0.597
		0.4	1.934	1.077	0.980	0.456	1.814	1.084	0.963	0.595
		0.6	3.789	1.072	0.943	0.300	3.347	1.076	1.039	0.467
		0.8	4.301	1.061	0.874	0.187	4.227	1.067	0.978	0.278
	500	0.2	1.598	1.130	1.039	0.287	1.961	1.178	1.105	0.395
		0.4	1.402	1.050	1.004	0.300	1.491	1.062	1.087	0.430
		0.6	3.050	1.062	1.005	0.219	2.560	1.062	1.125	0.356
		0.8	4.989	1.061	0.982	0.153	3.654	1.061	1.070	0.253
	1000	0.2	1.453	1.104	1.042	0.246	1.634	1.142	1.094	0.336
		0.4	1.194	1.035	0.998	0.322	1.296	1.044	1.068	0.433
		0.6	1.976	1.043	0.973	0.286	1.370	1.032	1.013	0.415
		0.8	4.919	1.058	0.961	0.208	2.553	1.048	1.005	0.320
50	200	0.2	1.425	1.169	0.960	0.550	1.625	1.214	0.977	0.650
		0.4	1.271	1.071	0.988	0.512	1.227	1.069	0.973	0.668
		0.6	2.462	1.081	0.944	0.323	1.919	1.076	0.987	0.487
		0.8	3.194	1.079	0.866	0.190	2.755	1.080	0.980	0.302
	500	0.2	1.248	1.105	1.034	0.371	1.503	1.157	1.103	0.473
		0.4	1.006	1.025	1.008	0.504	1.057	1.034	1.045	0.614
		0.6	1.028	1.017	0.942	0.491	0.984	1.015	0.978	0.599
		0.8	1.179	1.023	0.870	0.430	1.061	1.017	0.974	0.567
	1000	0.2	1.310	1.093	1.055	0.362	1.361	1.116	1.073	0.428
		0.4	1.024	1.018	1.029	0.628	0.995	1.018	1.025	0.675
		0.6	0.969	1.007	1.018	0.764	0.966	1.008	1.011	0.793
		0.8	0.954	1.004	1.022	0.838	0.938	1.004	1.018	0.853

```
> # The errors
> epsilon <- rt(n, df=5) # t-distribution
> #rchisq(n, df=5, ncp = 0) #Chi-square distribution
> sigma <- 1
> y <- X%*%beta_true+epsilon
> # Split data into train and test set
> all.folds <- split (sample (1 : n), rep (1 : 2, length = n))
> train_ind <- all.folds$'1'
> test_ind  <- all.folds$'2'
> y_train <- y[train_ind]
```

TABLE 9.32: The RTMSPE of the Estimators for Trim 32 Data.

	TMSPE	RTMSPE(ENET)	RTMSPE(LASSO)
ENET	0.00511	1.00000	0.97413
LASSO	0.00498	1.02656	1.00000
ALASSO	0.00478	1.06817	1.04054
Ridge	0.00470	1.08776	1.05962
PS(ENET)	0.00462	**1.10622**	**1.07760**
PS(LASSO)	0.00494	1.03406	1.00731

```
> X_train <- X[train_ind,]
> ### Centering train data of y and X
> X_train_mean  <- apply(X_train,2,mean)
> X_train_scale <- scale(X_train,X_train_mean,F)
> y_train_mean  <- mean(y_train)
> y_train_scale  <- y_train - y_train_mean
> train_scale_df <- data.frame(y_train_scale,X_train_scale)
> #test data
> y_test <- y[test_ind]
> X_test <- X[test_ind,]
> ### Centering test data of y and X based on train means
> y_test_scale <- y_test - y_train_mean
> X_test_scale <- scale(X_test,X_train_mean,F)
> # Formula of the Full model
> xcount.FM <- c(0,paste("X", 1:(p1+p2+p3), sep=""))
> Formula_FM<- as.formula(paste("y_train~",
+                               paste(xcount.FM, collapse= "+")))
> # Formula of the Sub model
> xcount.SM <- c(0,paste("X", 1:(p1), sep=""))
> Formula_SM<- as.formula(paste("y_train~",
+                               paste(xcount.SM, collapse= "+")))
> # Full Model
> Full_model   <- rq(Formula_FM, tau=.5,
+                    data = train_scale_df)
> beta_FM <- array(c(Full_model$coefficients), dim = c(p,1))
> # Sub Model
> Sub_model    <- rq(Formula_SM, tau=.5,
+                    data = train_scale_df)
> beta_SM <- c(Sub_model$ coefficients,rep(0,p2+p3))
> # Estimation of the sub model estimator
> getcov <- summary.rq(Full_model, se = "iid", covariance = TRUE)
> covm <- (getcov$cov)
> H <- diag(1,p,p)[-c(1:(p-p2-p3)),]
> h <- array(0, dim = c(p2+p3,1))
> Hbeta <- H %*% beta_FM
> #test stat
> tn <- drop(t(Hbeta) %*% solve(H %*% covm %*% t(H)) %*% (Hbeta))
> #Shrinkage Estimation
> beta.S <-beta_FM-(beta_FM-beta_SM)*(p2+p3-2)/tn
> beta.PS<-beta_SM+ max(0,(1-(p2+p3-2)/tn))*(beta_FM-beta_SM)
> ##PENALIZED Methods
> # LAD-ENET
> alphas=seq(0.1,0.9,0.2)
```

```
> fits.enet <- list()
> for (ind in 1:length(alphas)){
+   fits.enet[[ind]] <- hqreg_raw(X_train_scale, y_train_scale,
+                                 alpha = alphas[ind],
+                                 tau = .5,
+                                 intercept = FALSE)
+ }
> cvs.enet <- sapply(fits.enet, function(x) {
+   enet.train = coef(x,s='lambda.min')
+   Optumum_vals= opt.beta(enet.train, X_test_scale,
+                          y_test_scale)
+   return(list(MAPE=Optumum_vals$MAPE,
+               coef=Optumum_vals$coef_beta))
+ })
> ind_opt  <- which.min(do.call(rbind,cvs.enet[1,]))
> beta.enet <- do.call(rbind,cvs.enet[2,])[ind_opt,]
> # LAD-Lasso
> cv.lasso.train <- hqreg_raw(X_train_scale, y_train_scale,
+                             method = c("quantile"),
+                             tau = .5, alpha = 1,
+                             intercept = FALSE)
> lasso.train <- as.matrix(coef(cv.lasso.train))
> beta.lasso <- opt.beta(lasso.train, X_test_scale,
+                        y_test_scale)$coef_beta
> #LAD-ALASSO
> weight <- beta.lasso
> weight <- ifelse(weight == 0, 0.00001, weight)
> alasso.fit <- hqreg_raw(X_train_scale, y_train_scale,
+                         alpha = 1,
+                         penalty.factor=1/abs(weight),
+                         tau = .5,intercept = FALSE)
> alasso.train <- as.matrix(coef(alasso.fit))
> beta.alasso <- opt.beta(alasso.train, X_test_scale,
+                         y_test_scale)$coef_beta
> # LAD-RIDGE
> cv.ridge.train <- hqreg_raw(X_train_scale, y_train_scale,
+                             method = c("quantile"),
+                             tau = .5, alpha = 0,
+                             intercept = FALSE)
> ridge.train <- as.matrix(coef(cv.ridge.train))
> beta.ridge <- opt.beta(ridge.train, X_test_scale,
+                        y_test_scale)$coef_beta
> # Calculte MAPE of estimators for betas
> MAPE_beta<-
+   c(FM=MAPE(beta_true,beta_FM),
+     SM=MAPE(beta_true,beta_SM),
+     PS=MAPE(beta_true,beta.PS),
+     ENET=MAPE(beta_true,beta.enet),
+     LASSO=MAPE(beta_true,beta.lasso),
+     ALASSO=MAPE(beta_true,beta.alasso),
+     RIDGE=MAPE(beta_true,beta.ridge))
> cbind(MAPE=MAPE_beta,Best_Ranking= rank(MAPE_beta))
             MAPE Best_Ranking
FM     0.16224412            7
SM     0.04469823            1
```

```
PS      0.04794439             2
ENET    0.09531864             5
LASSO   0.08562754             4
ALASSO  0.05891371             3
RIDGE   0.14390473             6
> # Calculte MAPE of estimators based on test data
> MAPE_y<-
+   c(FM=MAPE(y_test_scale,X_test_scale%*%beta_FM),
+     SM=MAPE(y_test_scale,X_test_scale%*%beta_SM),
+     PS=MAPE(y_test_scale,X_test_scale%*%beta.PS),
+     ENET=MAPE(y_test_scale,X_test_scale%*%beta.enet),
+     LASSO=MAPE(y_test_scale,X_test_scale%*%beta.lasso),
+     ALASSO=MAPE(y_test_scale,X_test_scale%*%beta.alasso),
+     RIDGE=MAPE(y_test_scale,X_test_scale%*%beta.ridge))
> cbind(MAPE=MAPE_y,Best_Ranking= rank(MAPE_beta))
            MAPE Best_Ranking
FM      1.1885974            7
SM      0.8697956            1
PS      0.8633215            2
ENET    1.0083398            5
LASSO   1.0272345            4
ALASSO  0.9242035            3
RIDGE   1.2329353            6
> # High Dimensional Case
> set.seed(2023)
> n  <- 200  # The number of sample
> p1 <- 5   # The number of strong signals
> p2 <- 30   # The number of weak signals
> p3 <- 1000  # The number of zeros, no signal
> p <- p1+p2+p3
> beta1_true <- rep(1  ,p1)
> beta2_true <- rep(0.1,p2)
> beta3_true <- rep(0  ,p3)
> # The ture value of covariates
> beta_true <- c(beta1_true,beta2_true,beta3_true)
> # The design matrix
> X <- matrix(0, n, p)
> for (i in 1:p){
+   X[,i]= rnorm(n)
+ }
> ## assigning colnames of X to "X1","X2",....
> v<-NULL
> for(i in 1:p){
+   v[i] <- paste("X",i,sep="")
+   assign(v[i],X[,i])
+ }
> # The errors
> epsilon <- rt(n, df=5) # t-distribution
> #rchisq(n, df=5, ncp = 0) #Chi-square distribution
> sigma <- 1
> y <- X%*%beta_true+epsilon
> # Split data into train and test set
> all.folds <- split (sample (1 : n), rep (1 : 2, length = n))
> train_ind <- all.folds$`1`
> test_ind  <- all.folds$`2`
```

```
> y_train <- y[train_ind]
> X_train <- X[train_ind,]
> ### Centering train data of y and X
> X_train_mean  <- apply(X_train,2,mean)
> X_train_scale <- scale(X_train,X_train_mean,F)
> y_train_mean  <- mean(y_train)
> y_train_scale  <- y_train - y_train_mean
> train_scale_df <- data.frame(y_train_scale,X_train_scale)
> #test data
> y_test <- y[test_ind]
> X_test <- X[test_ind,]
> ### Centering test data of y and X based on train means
> y_test_scale <- y_test - y_train_mean
> X_test_scale <- scale(X_test,X_train_mean,F)
> # Formula of the Full model
> xcount.FM <- c(0,paste("X", 1:(p1+p2), sep=""))
> Formula_FM<- as.formula(paste("y_train~",
+                              paste(xcount.FM, collapse= "+")))
> # Formula of the Sub model
> xcount.SM <- c(0,paste("X", 1:p1, sep=""))
> Formula_SM<- as.formula(paste("y_train~",
+                              paste(xcount.SM, collapse= "+")))
> # Calculation of test stat
> Full_model <- rq(Formula_FM, data = train_scale_df, tau=.5)
> Sub_model  <- rq(Formula_SM, data = train_scale_df, tau=.5)
> tn <- -2*(logLik(Sub_model)-logLik(Full_model))[1]
> ##PENALIZED Methods
> # LAD-ENET
> alphas=seq(0.1,0.9,0.2)
> fits.enet <- list()
> for (ind in 1:length(alphas)){
+   fits.enet[[ind]] <- hqreg_raw(X_train_scale, y_train_scale,
+                                alpha = alphas[ind],
+                                tau = .5,
+                                intercept = FALSE)
+ }
> cvs.enet <- sapply(fits.enet, function(x) {
+   enet.train = coef(x,s='lambda.min')
+   Optumum_vals= opt.beta(enet.train, X_test_scale,
+                          y_test_scale)
+   return(list(MAPE=Optumum_vals$MAPE,
+               coef=Optumum_vals$coef_beta))
+ })
> ind_opt  <- which.min(do.call(rbind,cvs.enet[1,]))
> beta.enet <- do.call(rbind,cvs.enet[2,])[ind_opt,]
> # LAD-Lasso
> cv.lasso.train <- hqreg_raw(X_train_scale, y_train_scale,
+                            method = c("quantile"),
+                            tau = .5, alpha = 1,
+                            intercept = FALSE)
> lasso.train <- as.matrix(coef(cv.lasso.train))
> beta.lasso <- opt.beta(lasso.train, X_test_scale,
+                        y_test_scale)$coef_beta
> #LAD-ALASSO
> weight <- beta.lasso
```

```
> weight <- ifelse(weight == 0, 0.00001, weight)
> alasso.fit <- hqreg_raw(X_train_scale, y_train_scale,
+                         alpha = 1,
+                         penalty.factor=1/abs(weight),
+                         tau = .5,intercept = FALSE)
> alasso.train <- as.matrix(coef(alasso.fit))
> beta.alasso <- opt.beta(alasso.train, X_test_scale,
+                         y_test_scale)$coef_beta
> # LAD-RIDGE
> cv.ridge.train <- hqreg_raw(X_train_scale, y_train_scale,
+                             method = c("quantile"),
+                             tau = .5, alpha = 0,
+                             intercept = FALSE)
> ridge.train <- as.matrix(coef(cv.ridge.train))
> beta.ridge <- opt.beta(ridge.train, X_test_scale,
+                        y_test_scale)$coef_beta
> #Shrinkage Estimation
> beta.S <-beta.enet-(beta.enet-beta.alasso)*(p2-2)/tn
> beta.PS<-beta.alasso+ max(0,(1-(p2-2)/tn))*(beta.enet-beta.alasso)
> # Calculte MAPE of estimators for betas
> MAPE_beta<-
+   c(FM=MAPE(beta_true,beta.enet),
+     SM=MAPE(beta_true,beta.alasso),
+     PS=MAPE(beta_true,beta.PS),
+     LASSO=MAPE(beta_true,beta.lasso),
+     RIDGE=MAPE(beta_true,beta.ridge))
> cbind(MAPE=MAPE_beta,Best_Ranking= rank(MAPE_beta))
             MAPE Best_Ranking
FM    0.010672154            3
SM    0.006935548            1
PS    0.009656741            2
LASSO 0.010759548            4
RIDGE 0.023564144            5
> # Calculte MAPE of estimators based on test data
> MAPE_y<-
+   c(FM=MAPE(y_test_scale,X_test_scale%*%beta.enet),
+     SM=MAPE(y_test_scale,X_test_scale%*%beta.alasso),
+     PS=MAPE(y_test_scale,X_test_scale%*%beta.PS),
+     LASSO=MAPE(y_test_scale,X_test_scale%*%beta.lasso),
+     RIDGE=MAPE(y_test_scale,X_test_scale%*%beta.ridge))
> cbind(MAPE=MAPE_y,Best_Ranking= rank(MAPE_beta))
          MAPE Best_Ranking
FM    1.521627            3
SM    1.426808            1
PS    1.481357            2
LASSO 1.564957            4
RIDGE 2.047501            5
```

9.8 Conclusion Remarks

In this chapter, we extended the shrinkage strategy to a robust regression when the model was sparse. We assess the relative properties of shrinkage and penalized estimators both in high- and low-dimensional cases. In low-dimensional cases, we use the LAD procedure to obtain a base model, and then using BIC, we obtain a submodel. We combined the base model and submodel to build shrinkage estimators. As expected, shrinkage estimators perform better than the base model regardless of the correctness of the submodel, based on both theoretical and numerical results. Real data examples support the findings.

For high-dimensional cases, the penalized procedure is based on Wang et al. (2007), namely LAD-LASSO, which is extended to other penalized methods. We obtain two submodels based on the LAD-penalized procedure and combine them to obtain the shrinkage estimators. We conducted a simulation experiment to assess the relative performance of the estimators. Based on the simulation study and real data analysis, we found that the prediction error of our shrinkage estimators is less than that of the LAD and LAD-penalized estimators in important cases.

Appendix

Lemma 9.6 ***Bassett and Koenker (1978)*** *Whenever the assumptions (A) and (B) hold, $\sqrt{n}(\widehat{\boldsymbol{\beta}}^{\mathrm{FM}}-\boldsymbol{\beta})$ convergences in distribution to a p-dimensional normal distribution with mean $\mathbf{0}$ and covariance matrix $\tau^2\boldsymbol{C}^{-1}$, where τ^2 is the asymptotic variance of the sample median of symmetric F distribution with median 0, i.e. $\tau^2=[2f(0)]^{-2}$. On the other hand,*

$$\sqrt{n}\left(\widehat{\boldsymbol{\beta}}^{\mathrm{FM}}-\boldsymbol{\beta}\right)\xrightarrow{d}\mathcal{N}_p\left(\mathbf{0},\tau^2\boldsymbol{C}^{-1}\right). \tag{9.21}$$

Corollary 9.1 *Assume (A) and (B) are held, thus*

$$\widehat{\boldsymbol{\beta}}_1^{\mathrm{FM}} \sim \mathcal{N}_{p_1}\left(\boldsymbol{\beta}_1,\tau^2\boldsymbol{C}_{11.2}^{-1}\right), \tag{9.22}$$

$$\widehat{\boldsymbol{\beta}}_2^{\mathrm{LAD}} \sim \mathcal{N}_{q}\left(\boldsymbol{\beta}_2,\tau^2\boldsymbol{C}_{22.1}^{-1}\right). \tag{9.23}$$

Proof *Suppose that the matrix $\boldsymbol{X}$ is partitioned as $\boldsymbol{X}_1$ and $\boldsymbol{X}_2$ with sizes $n\times p_1$ and $n\times p_2$, respectively. Accordingly, $\boldsymbol{\mu}$ and $\boldsymbol{\Sigma}$ partitioned as follows:*

$$\boldsymbol{\mu}=\begin{bmatrix}\boldsymbol{\mu}_1\\ \boldsymbol{\mu}_2\end{bmatrix} \quad \textit{with dimension} \begin{bmatrix}p_1\times 1\\ p_2\times 1\end{bmatrix},$$

and

$$\boldsymbol{\Sigma}=\begin{bmatrix}\boldsymbol{\Sigma}_{11} & \boldsymbol{\Sigma}_{12}\\ \boldsymbol{\Sigma}_{21} & \boldsymbol{\Sigma}_{22}\end{bmatrix} \quad \textit{with dimension} \quad \begin{bmatrix}p_1\times p_1 & p_1\times p_2\\ p_2\times p_1 & p_2\times p_2\end{bmatrix}.$$

Now, based on assumption (A) consider the covariance matrix as $\tau^2\boldsymbol{C}^{-1}$. Using Schure complement, the inverted blockwise is as below

$$\boldsymbol{\Sigma}^{-1}=\tau^2\begin{bmatrix}\boldsymbol{C}_{11.2}^{-1} & -\boldsymbol{C}_{11.2}^{-1}\boldsymbol{C}_{12}\boldsymbol{C}_{22}^{-1}\\ -\boldsymbol{C}_{22.1}^{-1}\boldsymbol{C}_{21}\boldsymbol{C}_{11}^{-1} & \boldsymbol{C}_{22.1}^{-1}\end{bmatrix}. \tag{9.24}$$

Base on Lemma 9.6, the result is concluded.

Theorem 9.1 *Based on Eq.* (9.23), *we can construct Hotelling's T-squared statistic as*

$$\frac{\left(\widehat{\boldsymbol{\beta}}_2^{\mathrm{LAD}} - \boldsymbol{\beta}_2\right)^{\top} \boldsymbol{C}_{22.1} \left(\widehat{\boldsymbol{\beta}}_2^{\mathrm{LAD}} - \boldsymbol{\beta}_2\right)}{\tau^2}, \tag{9.25}$$

which follows of a χ^2*-distribution with* p_2 *d.f. and non-centrality parameter* Δ. *Under the null hypothesis, the test statistic,* $\mathcal{L}_n$, *is a central* χ^2 *distribution with* p_2 *d.f.*

Theorem 9.2 *The first equation is derived from* (9.22). *To obtain the asymptotic distribution of the restricted LAD estimator and complete the proof, we must use the conditional normal* $\left(\widehat{\boldsymbol{\beta}}_1^{\mathrm{FM}} | \widehat{\boldsymbol{\beta}}_2^{\mathrm{LAD}} = \boldsymbol{\beta}_2 = \mathbf{0}\right)$.

The distribution $X_1 | X_2 = x_2$ *is*

$$\mathcal{N}_{p_1 - p_2}\left(\boldsymbol{\mu}_1 + \boldsymbol{C}_{12}\boldsymbol{C}_{22}^{-1}(x_2 - \boldsymbol{\mu}_2), \boldsymbol{C}_{11} - \boldsymbol{C}_{12}\boldsymbol{C}_{22}^{-1}\boldsymbol{C}_{21}\right). \tag{9.26}$$

Now, with substituting the Eqs. (9.22) *to* (9.24) *into* (9.26) *the result is obtained.*

The mean vector is as

$$\boldsymbol{\beta}_1 + \left(-\boldsymbol{C}_{11.2}^{-1}\boldsymbol{C}_{12}\boldsymbol{C}_{22}^{-1}\right)\left(\boldsymbol{C}_{22.1}^{-1}\right)^{-1}(\boldsymbol{\beta}_2 - \boldsymbol{\beta}_2) = \boldsymbol{\beta}_1. \tag{9.27}$$

Also, the covariance matrix has the form

$$\begin{aligned} \mathrm{Cov}\left(\widehat{\boldsymbol{\beta}}_1^{\mathrm{FM}} | \widehat{\boldsymbol{\beta}}_2^{\mathrm{LAD}} = \boldsymbol{\beta}_2 = \mathbf{0}\right) &= \boldsymbol{C}_{11.2}^{-1} - \left(-\boldsymbol{C}_{11.2}^{-1}\boldsymbol{C}_{12}\boldsymbol{C}_{22}^{-1}\right) \\ &\times \boldsymbol{C}_{22.1}\left(-\boldsymbol{C}_{22.1}^{-1}\boldsymbol{C}_{21}\boldsymbol{C}_{11}^{-1}\right) \\ &= \boldsymbol{C}_{11.2}^{-1} - \boldsymbol{C}_{11.2}^{-1}\boldsymbol{C}_{12}\boldsymbol{C}_{22}^{-1}\boldsymbol{C}_{21}\boldsymbol{C}_{11}^{-1}. \end{aligned} \tag{9.28}$$

Using this fact that $\boldsymbol{C}_{12}\boldsymbol{C}_{22}^{-1}\boldsymbol{C}_{21} = \boldsymbol{C}_{11} - \boldsymbol{C}_{11.2}$, *the Eq.* (9.29) *equals to*

$$\begin{aligned} \boldsymbol{C}_{11.2}^{-1} - \boldsymbol{C}_{11.2}^{-1}\left(\boldsymbol{C}_{11} - \boldsymbol{C}_{11.2}\right)\boldsymbol{C}_{11}^{-1} &= \boldsymbol{C}_{11.2}^{-1} - \boldsymbol{C}_{11.2}^{-1} + \boldsymbol{C}_{11}^{-1}. \\ &= \boldsymbol{C}_{11}^{-1}. \end{aligned} \tag{9.29}$$

The proof is complete.

10

Liu-type Shrinkage Estimations in Linear Sparse Models

10.1 Introduction

A fundamental assumption of the multiple linear regression model is that the predicting variables are independent of each other. This is not always the case, especially when too many predictors are entered into the model at the building stage and some may not be independent to each other. This phenomenon is known as multicollinearity in the related literature. Multicollinearity may exist due to improper experimental design during data collection on various predictors in addition to a lack of understanding regarding the relationships between predictors. It is well known that the least square estimation theory is built on the assumption that predictors are independent of one another and very sensitive when this assumption is violated. To deal with inherited multicollinearity issues, some estimation procedures, such as ridge estimation (Hoerl and Kennard, 1970), partial least squares estimation, Liu estimation (Liu, 1993) and Liu-type estimation (Liu, 2003) have been proposed.

In this chapter, we consider a multiple linear regression model (10.1) subject to multicollinearity. Keeping in mind that multicollinearity is a characteristic of the data matrix and does not concern the underlying regression model. We assume that the underlying model is sparse, a common assumption in recent studies. Under sparsity, the vector of regression coefficients $\boldsymbol{\beta}$ can be partitioned as $(\boldsymbol{\beta}_1, \boldsymbol{\beta}_2)$, where $\boldsymbol{\beta}_1$ is the coefficient vector related to active predictors and $\boldsymbol{\beta}_2$ is the vector of nuisance parameters related to insignificant predictors. Under the sparsity condition $\boldsymbol{\beta}_2$ is set to a zero vector. The sparsity condition maybe very stringent at times and in some cases may not be judiciously justified. For this reason, we are interested in the estimation of $\boldsymbol{\beta}_1$ when $\boldsymbol{\beta}_2$ may or may not close to a vector zero. In a sense, we are dealing with a classical model misspecification problem. The model estimation based on all predictors may be subject to high variability and may not be easily interpretable. Conversely, a submodel strategy based on a few predictors may results in an under-fitted model with large bias. Thus, we consider a shrinkage strategy to control the magnitude of the bias. Ahmed (2001) gave a detailed definition of shrinkage estimation and discussed large sample estimation techniques in a regression model. For more recent work on the subject; we refer to Ahmed et al. (2012) and Ahmed and Fallahpour (2012); Ahmed (2014).

Let us consider a multiple regression model

$$y_i = \boldsymbol{x}_i^\top \boldsymbol{\beta} + \varepsilon_i, \quad i = 1, 2, ..., n, \tag{10.1}$$

where $\boldsymbol{y} = (y_1, y_2, \ldots, y_n)$ is the vector of responses, $\boldsymbol{x}_i = (x_{i1}, x_{i2}, ..., x_{ip})^\top$ are observation vectors, $\boldsymbol{\beta} = (\beta_1, \beta_2, ..., \beta_p)^\top$ is the vector of unknown regression coefficients, ε_i's are unobservable random errors and the superscript $(^\top)$ denotes the transpose of a vector or a matrix. Further, $\boldsymbol{\varepsilon} = (\varepsilon_1, \varepsilon_2, ..., \varepsilon_n)^\top$ has a cumulative distribution function $\mathrm{F}(\cdot)$; $\mathrm{E}(\boldsymbol{\varepsilon}) = \mathbf{0}$

DOI: 10.1201/9781003170259-10

and $\mathrm{Var}(\boldsymbol{\varepsilon}) = \sigma^2 \mathbf{I}_n$, where σ^2 is finite and $\mathbf{I}_n$ is an identity matrix of dimension $n \times n$. In this chapter, we consider that the data matrix $\boldsymbol{X} = [\boldsymbol{x}_1, \boldsymbol{x}_2, \ldots, \boldsymbol{x}_n]^\top$ has rank of p $(p \leq n)$. The ordinary least squares estimator is not suitable as a base estimator to form a shrinkage strategy. We take recourse with a Liu-type estimator as the base estimator for estimation of the regression parameters of the full and submodel. Finally, we combine the estimators of each model to construct the shrinkage strategy.

The rest of the chapter is organized as follows. The full and submodel estimators based on Liu regression alongside the shrinkage estimators are given in Section 10.2. The large-sample properties of the estimators are given in Section 10.3. The results of a Monte Carlo simulation experiment are presented in Section 10.4. A numerical comparison with some penalized estimators are also given in Section 10.4. For illustrative purposes, a real data example is available in Section 10.5. The R codes are available in Section 10.6. We conclude the chapter in Section 10.7.

10.2 Estimation Strategies

For the above multiple regression model, the ordinary least squares estimator (LSE) minimizes the objective function

$$S(\boldsymbol{\beta}) = (\boldsymbol{y} - \boldsymbol{X}\boldsymbol{\beta})^\top (\boldsymbol{y} - \boldsymbol{X}\boldsymbol{\beta}) \tag{10.2}$$

and it is obtained as $\widehat{\boldsymbol{\beta}}^{\mathrm{LSE}} = \left(\boldsymbol{X}^\top \boldsymbol{X}\right)^{-1} \boldsymbol{X}^\top \boldsymbol{y}$.

The ridge estimator proposed by Hoerl and Kennard (1970) is generally obtained as a solution to the following objective function

$$S\left(\boldsymbol{\beta}, \lambda^R\right) = (\boldsymbol{y} - \boldsymbol{X}\boldsymbol{\beta})^\top (\boldsymbol{y} - \boldsymbol{X}\boldsymbol{\beta}) + \lambda^R \boldsymbol{\beta}^\top \boldsymbol{\beta},$$

where $\lambda^R > 0$ is known as the regularization constant or ridge parameter. Alternatively,

$$\underset{\boldsymbol{\beta}}{\arg\min} \left\{ \sum_{i=1}^{n} \left(y_i - \boldsymbol{x}_i^\top \boldsymbol{\beta}\right)^2 + \lambda^R \sum_{j=1}^{p} \beta_j^2 \right\}$$

yielding a ridge full model estimator $\widehat{\boldsymbol{\beta}}^{\mathrm{RFM}}$

$$\widehat{\boldsymbol{\beta}}^{\mathrm{RFM}} = \left(\boldsymbol{X}^\top \boldsymbol{X} + \lambda^R \mathbf{I}_p\right)^{-1} \boldsymbol{X}^\top \boldsymbol{y}.$$

Clearly, if $\lambda^R = 0$ then $\widehat{\boldsymbol{\beta}}^{\mathrm{RFM}}$ is the least square estimator and if $\lambda^R = \infty$, then $\widehat{\boldsymbol{\beta}}^{\mathrm{RFM}} = \mathbf{0}$. Generally, we are interested in a moderate value of the ridge parameter λ. The, Liu estimator suggested by Liu (1993) is obtained by augmenting $d\widehat{\boldsymbol{\beta}}^{\mathrm{LSE}} = \beta + \epsilon'$ to (10.1)

$$\widehat{\boldsymbol{\beta}}^{\mathrm{FM}} = \left(\boldsymbol{X}^\top \boldsymbol{X} + \mathbf{I}_p\right)^{-1} \left(\boldsymbol{X}^\top \boldsymbol{X} + d\mathbf{I}_p\right) \widehat{\boldsymbol{\beta}}^{\mathrm{LSE}}$$

where $0 < d < 1$. It can also be obtained as a solution to the following objective function

$$S(\boldsymbol{\beta}, d) = (\boldsymbol{y} - \boldsymbol{X}\boldsymbol{\beta})^\top (\boldsymbol{y} - \boldsymbol{X}\boldsymbol{\beta}) + \left(\boldsymbol{\beta} - d\widehat{\boldsymbol{\beta}}^{\mathrm{LSE}}\right)^\top \left(\boldsymbol{\beta} - d\widehat{\boldsymbol{\beta}}^{\mathrm{LSE}}\right).$$

The advantage of $\widehat{\boldsymbol{\beta}}^{\mathrm{FM}}$ over $\widehat{\boldsymbol{\beta}}^{\mathrm{RFM}}$ is that $\widehat{\boldsymbol{\beta}}^{\mathrm{FM}}$ is a linear function of d (Liu (1993)). As a special case, for $d = 1$ the estimator reduces to the LSE.

Many researchers have considered the Liu estimator so far. Among them, Liu (2003) suggested a two-parameter Liu estimator to deal with multicollinearity, Akdeniz and Erol (2003) gave mean squared error comparisons of some estimators including the Liu estimator, Arashi et al. (2021) combined Liu estimator by Lasso penalty, and Johnson et al. (2014) considered a stochastic restricted Liu estimator in the presence of stochastic prior information, to name a few.

The main theme of this chapter to build a shrinkage estimation strategy when the model may be sparse. In the following subsections, we formulate the problem and define the shrinkage estimators.

10.2.1 Estimation Under a Sparsity Assumption

Let $\boldsymbol{X} = (\boldsymbol{X_1}, \boldsymbol{X_2})$, where $\boldsymbol{X_1}$ is an $n \times p_1$ sub-matrix containing active predictors and $\boldsymbol{X_2}$ is an $n \times p_2$ sub-matrix that may or may not be useful in the analysis of the main regressors. Similarly, let $\boldsymbol{\beta} = \left(\boldsymbol{\beta}_1^\top, \boldsymbol{\beta}_2^\top\right)^\top$ be the vector of parameters, where $\boldsymbol{\beta}_1$ and $\boldsymbol{\beta}_2$ have dimensions p_1 and p_2, respectively, with $p_1 + p_2 = p$. Thus, our parameter of interest is $\boldsymbol{\beta_1}$ when $bm\beta_2$ may or may not equal to $\mathbf{0}$.

For brevity sake, we first define a submodel ridge regression under the sparsity condition:

$$\boldsymbol{y} = \boldsymbol{X\beta} + \boldsymbol{\varepsilon}, \quad \text{subject to } \boldsymbol{\beta}^\top \boldsymbol{\beta} \ \leq \phi \text{ and } \boldsymbol{\beta_2} = \mathbf{0},$$

Alternatively,

$$\boldsymbol{y} = \boldsymbol{X}_1 \boldsymbol{\beta}_1 + \boldsymbol{\varepsilon}, \text{ subject to } \boldsymbol{\beta}_1^\top \boldsymbol{\beta}_1 \leq \phi. \tag{10.3}$$

Let $\widehat{\boldsymbol{\beta}}_1^{\text{RFM}}$ be the full model ridge estimator of $\boldsymbol{\beta}_1$ given by

$$\widehat{\boldsymbol{\beta}}_1^{\text{RFM}} = \left(\boldsymbol{X}_1^\top \mathbf{M}_2^R \boldsymbol{X}_1 + \lambda^R \mathbf{I}_{p_1}\right)^{-1} \boldsymbol{X}_1^\top \mathbf{M}_2^R \boldsymbol{y},$$

where $\mathbf{M}_2^R = \mathbf{I}_n - \boldsymbol{X}_2 \left(\boldsymbol{X}_2^\top \boldsymbol{X}_2 + \lambda^R \mathbf{I}_{p_2}\right)^{-1} \boldsymbol{X}_2^\top$, and λ_1^R is the usual ridge parameter.

The submodel estimator $\widehat{\boldsymbol{\beta}}_1^{\text{RSM}}$ of $\boldsymbol{\beta}_1$ for model (10.3) has the form

$$\widehat{\boldsymbol{\beta}}_1^{\text{RSM}} = \left(\boldsymbol{X}_1^\top \boldsymbol{X}_1 + \lambda_1^R \mathbf{I}_{p_1}\right)^{-1} \boldsymbol{X}_1^\top \boldsymbol{y}.$$

Similarly, we define the full model Liu estimator $\widehat{\boldsymbol{\beta}}_1^{\text{FM}}$ as follows:

$$\widehat{\boldsymbol{\beta}}_1^{\text{FM}} = \left(\boldsymbol{X}_1^\top \mathbf{M}_2^{\text{L}} \boldsymbol{X}_1 + \mathbf{I}_{p_1}\right)^{-1} \left(\boldsymbol{X}_1^\top \mathbf{M}_2^{\text{L}} \boldsymbol{X}_1 + d\mathbf{I}_{p_1}\right) \widehat{\boldsymbol{\beta}}_1^{\text{LSE}},$$

where

$$\mathbf{M}_2^{\text{L}} = \mathbf{I}_n - \boldsymbol{X}_2 \left(\boldsymbol{X}_2' \boldsymbol{X}_2 + \mathbf{I}_{p_2}\right)^{-1} \left(\boldsymbol{X}_2^\top \boldsymbol{X}_2 + d\mathbf{I}_{p_2}\right) \boldsymbol{X}_2^\top,$$

$0 < d_1 < 1$ and $\widehat{\boldsymbol{\beta}}_1^{\text{LSE}} = \left(\boldsymbol{X}_1^\top \boldsymbol{X}_1\right)^{-1} \boldsymbol{X}_1^\top \boldsymbol{y}$.

The submodel Liu estimator $\widehat{\boldsymbol{\beta}}_1^{\text{SM}}$ is defined as

$$\widehat{\boldsymbol{\beta}}_1^{\text{SM}} = \left(\boldsymbol{X}_1^\top \boldsymbol{X}_1 + \mathbf{I}_{p_1}\right)^{-1} \left(\boldsymbol{X}_1^\top \boldsymbol{X}_1 + d_1 \mathbf{I}_{p_1}\right) \widehat{\boldsymbol{\beta}}_1^{\text{LSE}}$$

By design, $\widehat{\boldsymbol{\beta}}_1^{\text{SM}}$ performs better than $\widehat{\boldsymbol{\beta}}_1^{\text{FM}}$ when $\boldsymbol{\beta}_2$ is close to the null vector. However, when $\boldsymbol{\beta}_2$ falls away from the null vector $\widehat{\boldsymbol{\beta}}_1^{\text{SM}}$ can be biased and inefficient.

10.2.2 Shrinkage Liu Estimation

The shrinkage Liu estimator $\widehat{\boldsymbol{\beta}}_1^{\text{S}}$ of $\boldsymbol{\beta}_2$ is defined as

$$\widehat{\boldsymbol{\beta}}_1^{\text{S}} = \widehat{\boldsymbol{\beta}}_1^{\text{SM}} + \left(\widehat{\boldsymbol{\beta}}_1^{\text{FM}} - \widehat{\boldsymbol{\beta}}_1^{\text{SM}}\right) \left(1 - (p_2 - 2)\mathscr{L}_n^{-1}\right).$$

where

$$\mathscr{L}_n = \frac{n}{\widehat{\sigma}^2}\left(\widehat{\boldsymbol{\beta}}_2^{\text{LSE}}\right)^\top \boldsymbol{X}_2^\top \mathbf{M}_1 \boldsymbol{X}_2 \left(\widehat{\boldsymbol{\beta}}_2^{\text{LSE}}\right),$$

where $\widehat{\sigma}^2 = \frac{1}{n-p}(\boldsymbol{y} - \boldsymbol{X}\widehat{\boldsymbol{\beta}}^{\text{FM}})^\top(\boldsymbol{y} - \boldsymbol{X}\widehat{\boldsymbol{\beta}}^{\text{FM}})$ is a consistent estimator of σ^2, $\mathbf{M}_1 = \mathbf{I}_n - \boldsymbol{X}_1\left(\boldsymbol{X}_1^\top \boldsymbol{X}_1\right)^{-1}\boldsymbol{X}_1^\top$ and $\widehat{\boldsymbol{\beta}}_2^{\text{LSE}} = \left(\boldsymbol{X}_2^\top \mathbf{M}_1 \boldsymbol{X}_2\right)^{-1}\boldsymbol{X}_2^\top \mathbf{M}_1 \boldsymbol{y}$. The estimator $\widehat{\boldsymbol{\beta}}_1^{\text{S}}$ is the general form of the Stein-rule estimator which shrinks the benchmark estimator toward the submodel estimator $\widehat{\boldsymbol{\beta}}_1^{\text{SM}}$.

To overcome the over-shrinkage problem in the shrinkage estimator, we define the positive part of the shrinkage Liu regression estimator $\widehat{\boldsymbol{\beta}}_1^{\text{PS}}$ of $\boldsymbol{\beta}_1$ as

$$\widehat{\boldsymbol{\beta}}_1^{\text{PS}} = \widehat{\boldsymbol{\beta}}_1^{\text{SM}} + \left(\widehat{\boldsymbol{\beta}}_1^{\text{FM}} - \widehat{\boldsymbol{\beta}}_1^{\text{SM}}\right)\left(1 - (p_2 - 2)\mathscr{L}_n^{-1}\right)^+,$$

where $z^+ = max(0, z)$.

Now, we present some asymptotic properties of the estimators in the following section.

10.3 Asymptotic Analysis

To obtain meaningful asymptotic results, we consider a sequence of local alternatives $\{K_n\}$ as:

$$K_n : \boldsymbol{\beta}_2 = \boldsymbol{\beta}_{2(n)} = \frac{\boldsymbol{\omega}}{\sqrt{n}},$$

where $\boldsymbol{\omega} = (\omega_1, \omega_2, ..., \omega_{p_2})^\top$ is a fixed vector. The asymptotic bias of an estimator $\widehat{\boldsymbol{\beta}}_1^*$ is defined as

$$\text{ADB}\left(\widehat{\boldsymbol{\beta}}_1^*\right) = E \lim_{n\to\infty}\left\{\sqrt{n}\left(\widehat{\boldsymbol{\beta}}_1^* - \boldsymbol{\beta}_1\right)\right\},$$

and the asymptotic covariance of an estimator $\widehat{\boldsymbol{\beta}}_1^*$ is given by

$$\boldsymbol{\Gamma}\left(\widehat{\boldsymbol{\beta}}_1^*\right) = E \lim_{n\to\infty}\left\{n\left(\widehat{\boldsymbol{\beta}}_1^* - \boldsymbol{\beta}_1\right)\left(\widehat{\boldsymbol{\beta}}_1^* - \boldsymbol{\beta}_1\right)^\top\right\},$$

thus the asymptotic risk of an estimator $\widehat{\boldsymbol{\beta}}_1^*$ is

$$\text{ADR}\left(\widehat{\boldsymbol{\beta}}_1^*\right) = \text{tr}\left(\boldsymbol{W}\boldsymbol{\Gamma}\left(\widehat{\boldsymbol{\beta}}_1^*\right)\right),$$

where $\boldsymbol{W}$ is a positive-definite matrix of weights with dimensions of $p \times p$, and $\widehat{\boldsymbol{\beta}}_1^*$ is a suitable estimator.

To assess the asymptotic properties of the estimators we consider the following regularity conditions that are required.

1. $\frac{1}{n}\max\limits_{1\leq i\leq n} \boldsymbol{x}_i^\top(\boldsymbol{X}^\top\boldsymbol{X})^{-1}\boldsymbol{x}_i \to 0$ as $n \to \infty$, where $\boldsymbol{x}_i^\top$ is the ith row of $\boldsymbol{X}$

2. $\lim\limits_{n\to\infty} n^{-1}(\boldsymbol{X}^\top\boldsymbol{X}) = \lim\limits_{n\to\infty}\mathbf{C}_n = \mathbf{C} = \boldsymbol{X}^\top\boldsymbol{X}$, where $\mathbf{C}$ is finite and positive definite.

3. $\lim\limits_{n\to\infty}\mathbf{F}_n(d) = \mathbf{F}_d$, for finite $\mathbf{F}_d$, where $\mathbf{F}_n(d) = (\mathbf{C}_n + \mathbf{I}_p)^{-1}(\mathbf{C}_n + d\mathbf{I}_p)$ and $\mathbf{F}_d = (\mathbf{C} + \mathbf{I}_p)^{-1}(\mathbf{C} + d\mathbf{I}_p)$.

Then

$$\sqrt{n}\left(\widehat{\boldsymbol{\beta}}^{\text{LSE}}-\boldsymbol{\beta}\right)\sim\mathcal{N}_p\left(\mathbf{0},\,\sigma^2\mathbf{C}^{-1}\right) \tag{10.4}$$

Consequently,

Theorem 10.1 *If* $0<d<1$ *and* $\mathbf{C}$ *is non-singular, then*

$$\sqrt{n}\left(\widehat{\boldsymbol{\beta}}^{\text{FM}}-\boldsymbol{\beta}\right)\sim\mathcal{N}_p\left(-(1-d)\left(\mathbf{C}+\mathbf{I}_p\right)^{-1}\boldsymbol{\beta},\,\sigma^2\mathbf{S}\right)$$

where $\mathbf{S}=\mathbf{F}_d\mathbf{C}^{-1}\mathbf{F}_d^{\top}$.

Proof of Theorem 10.1 *Since* $\widehat{\boldsymbol{\beta}}^{\text{FM}}$ *is a linear function of* $\widehat{\boldsymbol{\beta}}^{\text{LSE}}$ *it is asymptotically normally distributed. The asymptotic bias of* $\widehat{\boldsymbol{\beta}}^{FM}$ *is computed as*

$$\begin{aligned}
& E\left(\sqrt{n}\left(\widehat{\boldsymbol{\beta}}^{\text{FM}}-\boldsymbol{\beta}\right)\right)\\
= & E\lim_{n\to\infty}\left\{\sqrt{n}\left(\left(\mathbf{C}+\mathbf{I}_p\right)^{-1}\left(\mathbf{C}+d\mathbf{I}_p\right)\widehat{\boldsymbol{\beta}}^{\text{LSE}}-\boldsymbol{\beta}\right)\right\}\\
= & E\lim_{n\to\infty}\left\{\sqrt{n}\left[\left(\left(\mathbf{C}+\mathbf{I}_p\right)^{-1}\mathbf{C}+d\left(\mathbf{C}+\mathbf{I}_p\right)^{-1}\right)\widehat{\boldsymbol{\beta}}^{\text{LSE}}-\boldsymbol{\beta}\right]\right\}\\
= & E\lim_{n\to\infty}\left\{\sqrt{n}\left[\left(\left(\mathbf{I}_p+\mathbf{C}^{-1}\right)^{-1}+d\left(\mathbf{C}+\mathbf{I}_p\right)^{-1}\right)\widehat{\boldsymbol{\beta}}^{\text{LSE}}-\boldsymbol{\beta}\right]\right\}\\
= & E\lim_{n\to\infty}\left\{\sqrt{n}\left[\left(\mathbf{I}_p-\left(\mathbf{C}+\mathbf{I}_p\right)^{-1}+d\left(\mathbf{C}+\mathbf{I}_p\right)^{-1}\right)\widehat{\boldsymbol{\beta}}^{\text{LSE}}-\boldsymbol{\beta}\right]\right\}\\
= & E\lim_{n\to\infty}\left\{\sqrt{n}\left[\left(\widehat{\boldsymbol{\beta}}^{LSE}-\boldsymbol{\beta}\right)-(1-d)\left(\mathbf{C}+\mathbf{I}_p\right)\widehat{\boldsymbol{\beta}}^{LSE}\right]\right\}\\
= & -(1-d)\left(\mathbf{C}+\mathbf{I}_p\right)^{-1}\boldsymbol{\beta}.
\end{aligned}$$

Further,

$$\begin{aligned}
Cov\left(\widehat{\boldsymbol{\beta}}^{\text{FM}}\right) &= Cov\left(\mathbf{F}_d\widehat{\boldsymbol{\beta}}^{LSE}\right)\\
&= \mathbf{F}_d Cov\left(\widehat{\boldsymbol{\beta}}^{LSE}\right)\mathbf{F}_d^{\top}\\
&= \sigma^2\mathbf{F}_d\mathbf{C}^{-1}\mathbf{F}_d^{\top}.
\end{aligned}$$

Now, we use the Lemma 3.2 for the proof.

Proposition 10.2 *Let* $\boldsymbol{\vartheta}_1=\sqrt{n}\left(\widehat{\boldsymbol{\beta}}_1^{\text{FM}}-\boldsymbol{\beta}_1\right)$, $\boldsymbol{\vartheta}_2=\sqrt{n}\left(\widehat{\boldsymbol{\beta}}_1^{\text{SM}}-\boldsymbol{\beta}_1\right)$ *and* $\boldsymbol{\vartheta}_3=\sqrt{n}\left(\widehat{\boldsymbol{\beta}}_1^{\text{FM}}-\widehat{\boldsymbol{\beta}}_1^{\text{SM}}\right)$.

Under the regularity conditions (i)-(iii) and $\{K_n\}$ *as* $n\to\infty$

$$\begin{pmatrix}\boldsymbol{\vartheta}_1\\ \boldsymbol{\vartheta}_3\end{pmatrix}\sim\mathcal{N}\left[\begin{pmatrix}-\boldsymbol{\mu}_{11.2}\\ \boldsymbol{\delta}\end{pmatrix},\begin{pmatrix}\sigma^2\mathbf{S}_{11.2}^{-1} & \boldsymbol{\Phi}\\ \boldsymbol{\Phi} & \boldsymbol{\Phi}\end{pmatrix}\right],$$

$$\begin{pmatrix}\boldsymbol{\vartheta}_3\\ \boldsymbol{\vartheta}_2\end{pmatrix}\sim\mathcal{N}\left[\begin{pmatrix}\boldsymbol{\delta}\\ -\boldsymbol{\gamma}\end{pmatrix},\begin{pmatrix}\boldsymbol{\Phi} & \mathbf{0}\\ \mathbf{0} & \sigma^2\mathbf{S}_{11}^{-1}\end{pmatrix}\right],$$

where $\mathbf{C}=\begin{pmatrix}\mathbf{C}_{11} & \mathbf{C}_{12}\\ \mathbf{C}_{21} & \mathbf{C}_{22}\end{pmatrix}$, $\mathbf{S}=\begin{pmatrix}\mathbf{S}_{11} & \mathbf{S}_{12}\\ \mathbf{S}_{21} & \mathbf{S}_{22}\end{pmatrix}$, $\boldsymbol{\gamma}=\boldsymbol{\mu}_{11.2}+\boldsymbol{\delta}$ *and* $\boldsymbol{\delta}=\left(\mathbf{C}_{11}+\mathbf{I}_{p_1}\right)^{-1}\left(\mathbf{C}_{11}+d\mathbf{I}_{p_1}\right)\mathbf{C}_{12}\boldsymbol{\omega}$, $\boldsymbol{\Phi}=\sigma^2\mathbf{F}_d^{11}\mathbf{C}_{12}\mathbf{S}_{22.1}^{-1}\mathbf{C}_{21}\mathbf{F}_d^{11}$, *where* $\mathbf{F}_d^{11}=\left(\mathbf{C}_{11}+\mathbf{I}_{p_1}\right)^{-1}\left(\mathbf{C}_{11}+d\mathbf{I}_{p_1}\right)$ *and* $\boldsymbol{\mu}=-(1-d)\left(\mathbf{C}+\mathbf{I}\right)^{-1}\boldsymbol{\beta}=\begin{pmatrix}\boldsymbol{\mu}_1\\ \boldsymbol{\mu}_2\end{pmatrix}$ *and* $\boldsymbol{\mu}_{11.2}=\boldsymbol{\mu}_1-\mathbf{C}_{12}\mathbf{C}_{22}^{-1}\left(\left(\boldsymbol{\beta}_2-\boldsymbol{\omega}\right)-\boldsymbol{\mu}_2\right)$ *such that* $\boldsymbol{\mu}_{11.2}$, *is the conditional mean of* $\boldsymbol{\beta}_1$, *given* $\boldsymbol{\beta}_2=\mathbf{0}_{p_2}$, *and* $\sigma^2\mathbf{S}_{11.2}^{-1}$ *is the covariance matrix.*

Proof of Proposition 10.2 *Recognizing* $\widehat{\boldsymbol{\beta}}^{FM}$ *is a linear combination of* $\widehat{\boldsymbol{\beta}}_1^{\mathrm{SM}}$ *and* $\widehat{\boldsymbol{\beta}}_2^{\mathrm{FM}}$*, let,* $\widetilde{\boldsymbol{y}} = \boldsymbol{y} - \boldsymbol{X}_2\widehat{\boldsymbol{\beta}}_2^{\mathrm{FM}}$*, then we have*

$$\begin{aligned}
\widehat{\boldsymbol{\beta}}_1^{\mathrm{FM}} &= \left(\boldsymbol{X}_1^\top \boldsymbol{X}_1 + \mathbf{I}_{p_1}\right)^{-1}\left(\boldsymbol{X}_1^\top \boldsymbol{X}_1 + d\mathbf{I}_{p_1}\right)\boldsymbol{X}_1^\top \widetilde{\boldsymbol{y}} \\
&= \left(\boldsymbol{X}_1^\top \boldsymbol{X}_1 + \mathbf{I}_{p_1}\right)^{-1}\left(\boldsymbol{X}_1^\top \boldsymbol{X}_1 + d\mathbf{I}_{p_1}\right)\boldsymbol{X}_1^\top \boldsymbol{y} \\
&\quad - \left(\boldsymbol{X}_1^\top \boldsymbol{X}_1 + \mathbf{I}_{p_1}\right)^{-1}\left(\boldsymbol{X}_1^\top \boldsymbol{X}_1 + d\mathbf{I}_{p_1}\right)\boldsymbol{X}_1^\top \boldsymbol{X}_2\widehat{\boldsymbol{\beta}}_2^{\mathrm{FM}} \\
&= \widehat{\boldsymbol{\beta}}_1^{\mathrm{SM}} - \left(\boldsymbol{X}_1^\top \boldsymbol{X}_1 + \mathbf{I}_{p_1}\right)^{-1}\left(\boldsymbol{X}_1^\top \boldsymbol{X}_1 + d\mathbf{I}_{p_1}\right)\boldsymbol{X}_1^\top \boldsymbol{X}_2\widehat{\boldsymbol{\beta}}_2^{\mathrm{FM}}. \qquad (10.5)
\end{aligned}$$

Now, under the local alternatives $\{K_n\}$ *using the Equation* (10.5) *we can obtain* $\boldsymbol{\Phi}$ *as follows:*

$$\begin{aligned}
\boldsymbol{\Phi} &= Cov\left(\widehat{\boldsymbol{\beta}}_1^{\mathrm{FM}} - \widehat{\boldsymbol{\beta}}_1^{\mathrm{SM}}\right) \\
&= E\left[\left(\widehat{\boldsymbol{\beta}}_1^{\mathrm{FM}} - \widehat{\boldsymbol{\beta}}_1^{\mathrm{SM}}\right)\left(\widehat{\boldsymbol{\beta}}_1^{\mathrm{FM}} - \widehat{\boldsymbol{\beta}}_1^{\mathrm{SM}}\right)^\top\right] \\
&= E\left[\left(\left(\mathbf{C}_{11} + \mathbf{I}_{p1}\right)^{-1}\left(\mathbf{C}_{11} + d\mathbf{I}_{p1}\right)\mathbf{C}_{12}\widehat{\boldsymbol{\beta}}_2^{\mathrm{FM}}\right)\right. \\
&\times \left.\left(\left(\mathbf{C}_{11} + \mathbf{I}_{p1}\right)^{-1}\left(\mathbf{C}_{11} + d\mathbf{I}_{p1}\right)\mathbf{C}_{12}\widehat{\boldsymbol{\beta}}_2^{\mathrm{FM}}\right)^\top\right] \\
&= \left(\mathbf{C}_{11} + \mathbf{I}_{p1}\right)^{-1}\left(\mathbf{C}_{11} + d\mathbf{I}_{p1}\right)\mathbf{C}_{12}E\left[\widehat{\boldsymbol{\beta}}_2^{\mathrm{FM}}\left(\widehat{\boldsymbol{\beta}}_2^{\mathrm{FM}}\right)^\top\right] \\
&\times \mathbf{C}_{21}\left(\mathbf{C}_{11} + d\mathbf{I}_{p1}\right)\left(\mathbf{C}_{11} + \mathbf{I}_{p1}\right)^{-1} \\
&= \sigma^2\mathbf{F}_d^{11}\mathbf{C}_{12}\mathbf{S}_{22.1}^{-1}\mathbf{C}_{21}\mathbf{F}_d^{11}
\end{aligned}$$

where $\mathbf{F}_d^{11} = \left(\mathbf{C}_{11} + \mathbf{I}_{p1}\right)^{-1}\left(\mathbf{C}_{11} + d\mathbf{I}_{p1}\right)$*. Also,*

$$\begin{aligned}
&E\left\{\lim_{n\to\infty}\sqrt{n}\left(\widehat{\boldsymbol{\beta}}_1^{\mathrm{SM}} - \boldsymbol{\beta}_1\right)\right\} \\
&= E\left\{\lim_{n\to\infty}\sqrt{n}\left(\widehat{\boldsymbol{\beta}}_1^{\mathrm{FM}} - \left(\mathbf{C}_{11} + \mathbf{I}_{p_1}\right)^{-1}\left(\mathbf{C}_{11} + d\mathbf{I}_{p_1}\right)\mathbf{C}_{12}\widehat{\boldsymbol{\beta}}_2^{\mathrm{FM}} - \boldsymbol{\beta}_1\right)\right\} \\
&= E\left\{\lim_{n\to\infty}\sqrt{n}\left(\widehat{\boldsymbol{\beta}}_1^{\mathrm{FM}} - \boldsymbol{\beta}_1\right)\right\} \\
&- E\left\{\lim_{n\to\infty}\sqrt{n}\left(\left(\mathbf{C}_{11} + \mathbf{I}_{p_1}\right)^{-1}\left(\mathbf{C}_{11} + d\mathbf{I}_{p_1}\right)\mathbf{C}_{12}\widehat{\boldsymbol{\beta}}_2^{\mathrm{FM}}\right)\right\} \\
&= -\boldsymbol{\mu}_{11.2} - \mathbf{F}_{11}(d)\mathbf{C}_{12}\boldsymbol{\omega} = -\left(\boldsymbol{\mu}_{11.2} + \boldsymbol{\delta}\right) = -\boldsymbol{\gamma}.
\end{aligned}$$

From Johnson et al. (2014), page 160, Result 4.6, we have:

$$\sqrt{n}\left(\widehat{\boldsymbol{\beta}}_1^{\mathrm{FM}} - \boldsymbol{\beta}_1\right) \sim \mathcal{N}_{p_1}\left(-\boldsymbol{\mu}_{11.2}, \sigma^2\mathbf{S}_{11.2}^{-1}\right),$$

Since $\boldsymbol{\vartheta}_2$ *and* $\boldsymbol{\vartheta}_3$ *are linear functions of* $\widehat{\boldsymbol{\beta}}^{\mathrm{LSE}}$ *they are also asymptotically normally distributed.*

$$\begin{aligned}
\sqrt{n}\left(\widehat{\boldsymbol{\beta}}_1^{\mathrm{SM}} - \boldsymbol{\beta}\right) &\sim \mathcal{N}_{p_1}\left(-\boldsymbol{\gamma}, \sigma^2\mathbf{S}_{11}^{-1}\right) \\
\sqrt{n}\left(\widehat{\boldsymbol{\beta}}_1^{\mathrm{FM}} - \widehat{\boldsymbol{\beta}}_1^{\mathrm{SM}}\right) &\sim \mathcal{N}_{p_1}\left(\boldsymbol{\delta}, \boldsymbol{\Phi}\right).
\end{aligned}$$

Now, we present the bias expressions of the estimators in the following theorem.

Theorem 10.3

$$\begin{aligned}
\mathrm{ADB}\left(\widehat{\boldsymbol{\beta}}_1^{\mathrm{FM}}\right) &= -\boldsymbol{\mu}_{11.2}\\
\mathrm{ADB}\left(\widehat{\boldsymbol{\beta}}_1^{\mathrm{SM}}\right) &= -\boldsymbol{\gamma}\\
\mathrm{ADB}\left(\widehat{\boldsymbol{\beta}}_1^{\mathrm{S}}\right) &= -\boldsymbol{\mu}_{11.2}-(p_2-2)\boldsymbol{\delta} E\left(\chi_{p_2+2}^{-2}(\Delta)\right)\\
\mathrm{ADB}\left(\widehat{\boldsymbol{\beta}}_1^{\mathrm{PS}}\right) &= -\boldsymbol{\mu}_{11.2}-\boldsymbol{\delta} H_{p_2+2}\left(\chi_{p_2,\alpha}^{2};\Delta\right),\\
&\quad -(p_2-2)\boldsymbol{\delta} E\left\{\chi_{p_2+2}^{-2}(\Delta)\, I\left(\chi_{p_2+2}^{2}(\Delta)>p_2-2\right)\right\}
\end{aligned}$$

where $\Delta=\left(\boldsymbol{\omega}^{\top}\mathbf{C}_{22.1}^{-1}\boldsymbol{\omega}\right)\sigma^{-2}$, $\mathbf{C}_{22.1}=\mathbf{C}_{22}-\mathbf{C}_{21}\mathbf{C}_{11}^{-1}C_{12}$, *and* $H_v(x,\Delta)$ *is the cumulative distribution function of the non-central chi-squared distribution with non-centrality parameter* Δ, *degrees of freedom* v, *and*

$$E\left(\chi_v^{-2j}(\Delta)\right)=\int_0^{\infty}x^{-2j}dH_v(x,\Delta).$$

Proof of Theorem 10.3 $\mathrm{ADB}\left(\widehat{\boldsymbol{\beta}}_1^{\mathrm{FM}}\right)$ *and* $\mathrm{ADB}\left(\widehat{\boldsymbol{\beta}}_1^{\mathrm{SM}}\right)$ *are provided by Proposition 10.2. By using Lemma 3.2, it can be written as follows:*

$$\begin{aligned}
\mathrm{ADB}\left(\widehat{\boldsymbol{\beta}}_1^{\mathrm{S}}\right) &= E\left\{\lim_{n\to\infty}\sqrt{n}\left(\widehat{\boldsymbol{\beta}}_1^{\mathrm{S}}-\boldsymbol{\beta}_1\right)\right\}\\
&= E\left\{\lim_{n\to\infty}\sqrt{n}\left(\widehat{\boldsymbol{\beta}}_1^{\mathrm{FM}}-\left(\widehat{\boldsymbol{\beta}}_1^{\mathrm{FM}}-\widehat{\boldsymbol{\beta}}_1^{\mathrm{SM}}\right)(p_2-2)\mathscr{L}_n^{-1}-\boldsymbol{\beta}_1\right)\right\}\\
&= E\left\{\lim_{n\to\infty}\sqrt{n}\left(\widehat{\boldsymbol{\beta}}_1^{\mathrm{FM}}-\boldsymbol{\beta}_1\right)\right\}\\
&- E\left\{\lim_{n\to\infty}\sqrt{n}\left(\left(\widehat{\boldsymbol{\beta}}_1^{\mathrm{FM}}-\widehat{\boldsymbol{\beta}}_1^{\mathrm{SM}}\right)(p_2-2)\mathscr{L}_n^{-1}\right)\right\}\\
&= -\boldsymbol{\mu}_{11.2}-(p_2-2)\boldsymbol{\delta} E\left(\chi_{p_2+2}^{-2}(\Delta)\right).
\end{aligned}$$

$$\begin{aligned}
\mathrm{ADB}\left(\widehat{\boldsymbol{\beta}}_1^{\mathrm{PS}}\right) &= E\left\{\lim_{n\to\infty}\sqrt{n}\left(\widehat{\boldsymbol{\beta}}_1^{\mathrm{PS}}-\boldsymbol{\beta}_1\right)\right\}\\
&= E\left\{\lim_{n\to\infty}\sqrt{n}(\widehat{\boldsymbol{\beta}}_1^{\mathrm{SM}}+\left(\widehat{\boldsymbol{\beta}}_1^{\mathrm{FM}}-\widehat{\boldsymbol{\beta}}_1^{\mathrm{SM}}\right)\left(1-(p_2-2)\mathscr{L}_n^{-1}\right)\right.\\
&\times\ I\left(\mathscr{L}_n>p_2-2\right)-\boldsymbol{\beta}_1)\}\\
&= E\left\{\lim_{n\to\infty}\sqrt{n}\left[\widehat{\boldsymbol{\beta}}_1^{\mathrm{SM}}+\left(\widehat{\boldsymbol{\beta}}_1^{\mathrm{FM}}-\widehat{\boldsymbol{\beta}}_1^{\mathrm{SM}}\right)\left(1-I\left(\mathscr{L}_n\le p_2-2\right)\right)\right.\right.\\
&\quad\left.\left.-\left(\widehat{\boldsymbol{\beta}}_1^{\mathrm{FM}}-\widehat{\boldsymbol{\beta}}_1^{\mathrm{SM}}\right)(p_2-2)\mathscr{L}_n^{-1}I\left(\mathscr{L}_n>p_2-2\right)-\boldsymbol{\beta}_1\right]\right\}\\
&= E\left\{\lim_{n\to\infty}\sqrt{n}\left(\widehat{\boldsymbol{\beta}}_1^{\mathrm{FM}}-\boldsymbol{\beta}_1\right)\right\}\\
&- E\left\{\lim_{n\to\infty}\sqrt{n}\left(\widehat{\boldsymbol{\beta}}_1^{\mathrm{FM}}-\widehat{\boldsymbol{\beta}}_1^{\mathrm{RSM}}\right)I\left(\mathscr{L}_n\le p_2-2\right)\right\}\\
&\quad -E\left\{\lim_{n\to\infty}\sqrt{n}\left(\widehat{\boldsymbol{\beta}}_1^{\mathrm{FM}}-\widehat{\boldsymbol{\beta}}_1^{\mathrm{SM}}\right)(p_2-2)\mathscr{L}_n^{-1}I\left(\mathscr{L}_n>p_2-2\right)\right\}\\
&= -\boldsymbol{\mu}_{11.2}-\boldsymbol{\delta} H_{p_2+2}\left(p_2-2;\Delta\right)-\\
&\quad \boldsymbol{\delta}(p_2-2)E\left\{\chi_{p_2+2}^{-2}(\Delta)\,I\left(\chi_{p_{2+2}}^{2}(\Delta)>p_2-2\right)\right\}.
\end{aligned}$$

By definition the quadratic asymptotic distributional bias of an estimator $\widehat{\boldsymbol{\beta}}_1^{*}$ is

$$\mathrm{QADB}\left(\widehat{\boldsymbol{\beta}}_1^{*}\right)=\left(\mathrm{ADB}\left(\widehat{\boldsymbol{\beta}}_1^{*}\right)\right)^{\top}\mathbf{S}_{11.2}\left(\mathrm{ADB}\left(\widehat{\boldsymbol{\beta}}_1^{*}\right)\right),$$

where $\mathbf{S}_{11.2} = \mathbf{S}_{11} - \mathbf{S}_{12}\mathbf{S}_{22}^{-1}\mathbf{S}_{21}$.

Thus,

$$\begin{aligned}
\text{QADB}\left(\widehat{\boldsymbol{\beta}}_1^{\text{FM}}\right) &= \boldsymbol{\mu}_{11.2}^\top \mathbf{S}_{11.2}\boldsymbol{\mu}_{11.2}, \\
\text{QADB}\left(\widehat{\boldsymbol{\beta}}_1^{\text{SM}}\right) &= \boldsymbol{\gamma}^\top \mathbf{S}_{11.2}\boldsymbol{\gamma}, \\
\text{QADB}\left(\widehat{\boldsymbol{\beta}}_1^{\text{S}}\right) &= \boldsymbol{\mu}_{11.2}^\top \mathbf{S}_{11.2}\boldsymbol{\mu}_{11.2} + (p_2-2)\boldsymbol{\mu}_{11.2}^\top \mathbf{S}_{11.2}\boldsymbol{\delta} E\left(\chi_{p_2+2}^{-2}(\Delta)\right) \\
&\quad + (p_2-2)\boldsymbol{\delta}^\top \mathbf{S}_{11.2}\boldsymbol{\mu}_{11.2} E\left(\chi_{p_2+2}^{-2}(\Delta)\right) \\
&\quad + (p_2-2)^2\boldsymbol{\delta}^\top \mathbf{S}_{11.2}\boldsymbol{\delta}\left(E\left(\chi_{p_2+2}^{-2}(\Delta)\right)\right)^2, \\
\text{QADB}\left(\widehat{\boldsymbol{\beta}}_1^{\text{PS}}\right) &= \boldsymbol{\mu}_{11.2}^\top \mathbf{S}_{11.2}\boldsymbol{\mu}_{11.2} + \left(\boldsymbol{\delta}^\top \mathbf{S}_{11.2}\boldsymbol{\mu}_{11.2} + \boldsymbol{\mu}_{11.2}^\top \mathbf{S}_{11.2}\boldsymbol{\delta}\right) \\
&\quad \cdot \left[H_{p_2+2}(p_2-2;\Delta)\right. \\
&\quad \left. + (p_2-2)E\left\{\chi_{p_2+2}^{-2}(\Delta) I\left(\chi_{p_2+2}^{-2}(\Delta) > p_2-2\right)\right\}\right] \\
&\quad + \boldsymbol{\delta}^\top \mathbf{S}_{11.2}\boldsymbol{\delta}\left[H_{p_2+2}(p_2-2;\Delta)\right. \\
&\quad \left. + (p_2-2)E\left\{\chi_{p_2+2}^{-2}(\Delta) I\left(\chi_{p_2+2}^{-2}(\Delta) > p_2-2\right)\right\}\right]^2.
\end{aligned}$$

The asymptotic distributional bias QADB of $\widehat{\boldsymbol{\beta}}_1^{\text{FM}}$ is constant in terms of the sparsity parameter. However, QADB of $\widehat{\boldsymbol{\beta}}_1^{\text{SM}}$ is an unbounded function of $\boldsymbol{\gamma}$. The magnitude of the bias of $\widehat{\boldsymbol{\beta}}_1^{\text{SM}}$ depends on the values $\boldsymbol{\gamma}$. On the other hand, QADB of $\widehat{\boldsymbol{\beta}}_1^{\text{S}}$ and $\widehat{\boldsymbol{\beta}}_1^{\text{PS}}$ is a function of the sparsity parameter Δ starting from $\boldsymbol{\mu}_{11.2}^\top \mathbf{S}_{11.2}\boldsymbol{\mu}_{11.2}$ increasing to some point then decreases and finally converges to QADB of $\widehat{\boldsymbol{\beta}}_1^{\text{FM}}$.

In order to compute the risk functions, we first compute the asymptotic covariance of the estimators.

The asymptotic covariance of the estimator $\widehat{\boldsymbol{\beta}}_1^{\text{FM}}$ s:

$$\boldsymbol{\Gamma}\left(\widehat{\boldsymbol{\beta}}_1^{\text{FM}}\right) = \sigma^2\mathbf{S}_{11.2}^{-1} + \boldsymbol{\mu}_{11.2}\boldsymbol{\mu}_{11.2}^\top. \tag{10.6}$$

Similarly, the asymptotic covariance of the estimator $\widehat{\boldsymbol{\beta}}_1^{\text{SM}}$ is obtained as

$$\boldsymbol{\Gamma}\left(\widehat{\boldsymbol{\beta}}_1^{\text{SM}}\right) = \sigma^2\mathbf{S}_{11.2}^{-1} + \boldsymbol{\gamma}_{11.2}\boldsymbol{\gamma}_{11.2}^\top. \tag{10.7}$$

The asymptotic covariance matrix of $\widehat{\boldsymbol{\beta}}_1^{\text{S}}$ is written as

$$\begin{aligned}
\boldsymbol{\Gamma}\left(\widehat{\boldsymbol{\beta}}_1^{\text{S}}\right) &= E\left\{\lim_{n\to\infty}\sqrt{n}\left(\widehat{\boldsymbol{\beta}}_1^{\text{S}} - \boldsymbol{\beta}_1\right)\sqrt{n}\left(\widehat{\boldsymbol{\beta}}_1^{\text{S}} - \boldsymbol{\beta}_1\right)^\top\right\} \\
&= E\left\{\lim_{n\to\infty} n\left[\left(\widehat{\boldsymbol{\beta}}_1^{\text{FM}} - \boldsymbol{\beta}_1\right) - \left(\widehat{\boldsymbol{\beta}}_1^{\text{FM}} - \widehat{\boldsymbol{\beta}}_1^{\text{SM}}\right)(p_2-2)\mathscr{L}_n^{-1}\right]\right. \\
&\quad \left.\left[\left(\widehat{\boldsymbol{\beta}}_1^{\text{FM}} - \boldsymbol{\beta}_1\right) - \left(\widehat{\boldsymbol{\beta}}_1^{\text{FM}} - \widehat{\boldsymbol{\beta}}_1^{\text{SM}}\right)(p_2-2)\mathscr{L}_n^{-1}\right]^\top\right\} \\
&= E\left\{\boldsymbol{\vartheta}_1\boldsymbol{\vartheta}_1^\top - 2(p_2-2)\boldsymbol{\vartheta}_3\boldsymbol{\vartheta}_1^\top\mathscr{L}_n^{-1} + (p_2-2)^2\boldsymbol{\vartheta}_3\boldsymbol{\vartheta}_3^\top\mathscr{L}_n^{-2}\right\}.
\end{aligned}$$

By using Lemma (3.2), we get:

$$E\left\{\boldsymbol{\vartheta}_3\boldsymbol{\vartheta}_3^\top\mathscr{L}_n^{-2}\right\} = \boldsymbol{\Phi} E\left(\chi_{p_2+2}^{-4}(\Delta)\right) + \boldsymbol{\delta}\boldsymbol{\delta}^\top E\left(\chi_{p_2+4}^{-4}(\Delta)\right),$$

and

$$\begin{aligned}
& E\left\{\boldsymbol{\vartheta}_3\boldsymbol{\vartheta}_1^\top \mathscr{L}_n^{-1}\right\} \\
= \; & E\left\{E\left(\boldsymbol{\vartheta}_3\boldsymbol{\vartheta}_1^\top \mathscr{L}_n^{-1}|\boldsymbol{\vartheta}_3\right)\right\} = E\left\{\boldsymbol{\vartheta}_3 E\left(\boldsymbol{\vartheta}_1^\top \mathscr{L}_n^{-1}|\boldsymbol{\vartheta}_3\right)\right\} \\
= \; & E\left\{\boldsymbol{\vartheta}_3\left[-\boldsymbol{\mu}_{11.2} + (\boldsymbol{\vartheta}_3 - \boldsymbol{\delta})\right]^\top \mathscr{L}_n^{-1}\right\} \\
= \; & -E\left\{\boldsymbol{\vartheta}_3\boldsymbol{\mu}_{11.2}^\top \mathscr{L}_n^{-1}\right\} + E\left\{\boldsymbol{\vartheta}_3\left(\boldsymbol{\vartheta}_3 - \boldsymbol{\delta}\right)^\top \mathscr{L}_n^{-1}\right\} \\
= \; & -\boldsymbol{\mu}_{11.2}^\top E\left\{\boldsymbol{\vartheta}_3 \mathscr{L}_n^{-1}\right\} + E\left\{\boldsymbol{\vartheta}_3\boldsymbol{\vartheta}_3^\top \mathscr{L}_n^{-1}\right\} - E\left\{\boldsymbol{\vartheta}_3\boldsymbol{\delta}^\top \mathscr{L}_n^{-1}\right\}.
\end{aligned}$$

Finally, $E\left\{\boldsymbol{\vartheta}_3\boldsymbol{\delta}^\top \mathscr{L}_n^{-1}\right\} = \boldsymbol{\delta}\boldsymbol{\delta}^\top E\left(\chi_{p_2+2}^{-2}(\Delta)\right)$ and $E\left\{\boldsymbol{\vartheta}_3 \mathscr{L}_n^{-1}\right\} = \boldsymbol{\delta} E\left(\chi_{p_2+2}^{-2}(\Delta)\right)$. After some algebraic manipulations, we obtain

$$\begin{aligned}
& \boldsymbol{\Gamma}\left(\widehat{\boldsymbol{\beta}}_1^{\mathrm{S}}\right) \\
= \; & \sigma^2 \mathbf{S}_{11.2}^{-1} + \boldsymbol{\mu}_{11.2}\boldsymbol{\mu}_{11.2}^\top + 2\left(p_2 - 2\right)\boldsymbol{\mu}_{11.2}^\top \boldsymbol{\delta} E\left(\chi_{p_2+2}^{-2}(\Delta)\right) \\
& - \left(p_2 - 2\right)\boldsymbol{\Phi}\left\{2E\left(\chi_{p_2+2}^{-2}(\Delta)\right) - \left(p_2 - 2\right)E\left(\chi_{p_2+2}^{-4}(\Delta)\right)\right\} \\
& + \left(p_2 - 2\right)\boldsymbol{\delta}\boldsymbol{\delta}^\top \\
& \times\left\{-2E\left(\chi_{p_2+4}^{-2}(\Delta)\right) + 2E\left(\chi_{p_2+2}^{-2}(\Delta)\right)\right. \\
& \left. + \left(p_2 - 2\right)E\left(\chi_{p_2+4}^{-4}(\Delta)\right)\right\}.
\end{aligned} \tag{10.8}$$

The asymptotic covariance of $\boldsymbol{\Gamma}\left(\widehat{\boldsymbol{\beta}}_1^{\mathrm{PS}}\right)$ derivation steps are given as follows:

$$\begin{aligned}
& \boldsymbol{\Gamma}\left(\widehat{\boldsymbol{\beta}}_1^{\mathrm{PS}}\right) \\
= \; & E\left\{\lim_{n\to\infty} n\left(\widehat{\boldsymbol{\beta}}_1^{\mathrm{PS}} - \boldsymbol{\beta}_1\right)\left(\widehat{\boldsymbol{\beta}}_1^{\mathrm{PS}} - \boldsymbol{\beta}_1\right)^\top\right\} \\
= \; & E\left\{\lim_{n\to\infty} n\left(\widehat{\boldsymbol{\beta}}_1^{\mathrm{S}} - \boldsymbol{\beta}_1\right)\left(\widehat{\boldsymbol{\beta}}_1^{\mathrm{S}} - \boldsymbol{\beta}_1\right)^\top\right\} \\
& -2E\left\{\lim_{n\to\infty} n\left(\widehat{\boldsymbol{\beta}}_1^{\mathrm{FM}} - \widehat{\boldsymbol{\beta}}_1^{\mathrm{SM}}\right)\left(\widehat{\boldsymbol{\beta}}_1^{\mathrm{S}} - \widehat{\boldsymbol{\beta}}_1\right)^\top\right. \\
& \times\left.\left\{1 - \left(p_2 - 2\right)\mathscr{L}_n^{-1}\right\} I\left(\mathscr{L}_n \le p_2 - 2\right)\right\} \\
& +E\left\{\lim_{n\to\infty} n\left(\widehat{\boldsymbol{\beta}}_1^{\mathrm{FM}} - \widehat{\boldsymbol{\beta}}_1^{\mathrm{SM}}\right)\left(\widehat{\boldsymbol{\beta}}_1^{\mathrm{FM}} - \widehat{\boldsymbol{\beta}}_1^{\mathrm{SM}}\right)^\top\right. \\
& \times\left.\left\{1 - \left(p_2 - 2\right)\mathscr{L}_n^{-1}\right\}^2 I\left(\mathscr{L}_n \le p_2 - 2\right)\right\} \\
= \; & \boldsymbol{\Gamma}\left(\widehat{\boldsymbol{\beta}}_1^{\mathrm{S}}\right) - 2E\left\{\boldsymbol{\vartheta}_3\boldsymbol{\vartheta}_1^\top\left[1 - \left(p_2 - 2\right)\mathscr{L}_n^{-1}\right] I\left(\mathscr{L}_n \le p_2 - 2\right)\right\} \\
& +2E\left\{\boldsymbol{\vartheta}_3\boldsymbol{\vartheta}_3^\top\left(p_2 - 2\right)\mathscr{L}_n^{-1}\left(1 - \left(p_2 - 2\right)\mathscr{L}_n^{-1}\right) I\left(\mathscr{L}_n \le p_2 - 2\right)\right\} \\
& +E\left\{\boldsymbol{\vartheta}_3\boldsymbol{\vartheta}_3^\top\left(1 - \left(p_2 - 2\right)\mathscr{L}_n^{-1}\right)^2 I\left(\mathscr{L}_n \le p_2 - 2\right)\right\}.
\end{aligned}$$

After some simplification,

$$\begin{aligned}
\boldsymbol{\Gamma}\left(\widehat{\boldsymbol{\beta}}_1^{\mathrm{PS}}\right) \; = \; & \boldsymbol{\Gamma}\left(\widehat{\boldsymbol{\beta}}_1^{\mathrm{S}}\right) - 2E\left\{\boldsymbol{\vartheta}_3\boldsymbol{\vartheta}_1^\top\left[1 - \left(p_2 - 2\right)\mathscr{L}_n^{-1}\right] I\left(\mathscr{L}_n \le p_2 - 2\right)\right\} \\
& -E\left\{\boldsymbol{\vartheta}_3\boldsymbol{\vartheta}_3^\top\left(p_2 - 2\right)^2 \mathscr{L}_n^{-2} I\left(\mathscr{L}_n \le p_2 - 2\right)\right\} \\
& +E\left\{\boldsymbol{\vartheta}_3\boldsymbol{\vartheta}_3^\top I\left(\mathscr{L}_n \le p_2 - 2\right)\right\}.
\end{aligned}$$

Noting that,

$$E\left\{\boldsymbol{\vartheta}_3\boldsymbol{\vartheta}_3^\top I\left(\mathscr{L}_n \le p_2-2\right)\right\} = \boldsymbol{\Phi} H_{p_2+2}\left(p_2-2;\Delta\right) + \boldsymbol{\delta}\boldsymbol{\delta}^\top H_{p_2+4}\left(p_2-2;\Delta\right).$$

By using Lemma 3.2 and using the formula of the conditional mean of a bivariate normal distribution, the first expectation becomes

$$\begin{aligned}
& E\left\{\boldsymbol{\vartheta}_3\boldsymbol{\vartheta}_1^\top\left[1-(p_1-2)\mathscr{L}_n^{-1}\right] I\left(\mathscr{L}_n \le p_2-2\right)\right\} \\
= & -\boldsymbol{\delta}\boldsymbol{\mu}_{11.2}^\top E\left(\left\{1-(p_2-2)\chi_{p_2+2}^{-2}(\Delta)\right\} I\left(\chi_{p_2+2}^{2}(\Delta) \le p_2-2\right)\right) \\
& +\boldsymbol{\Phi} E\left(\left\{1-(p_2-2)\chi_{p_2+2}^{-2}(\Delta)\right\} I\left(\chi_{p_2+2}^{2}(\Delta) \le p_2-2\right)\right) \\
& +\boldsymbol{\delta}\boldsymbol{\delta}^\top E\left(\left\{1-(p_2-2)\chi_{p_2+4}^{-2}(\Delta)\right\} I\left(\chi_{p_2+4}^{2}(\Delta) \le p_2-2\right)\right) \\
& -\boldsymbol{\delta}\boldsymbol{\delta}^\top E\left(\left\{1-(p_2-2)\chi_{p_2+2}^{-2}(\Delta)\right\} I\left(\chi_{p_2+2}^{2}(\Delta) \le p_2-2\right)\right).
\end{aligned}$$

Thus, the asymptotic covariance of $\boldsymbol{\Gamma}\left(\widehat{\boldsymbol{\beta}}_1^{\mathrm{PS}}\right)$ is given as:

$$\begin{aligned}
\boldsymbol{\Gamma}\left(\widehat{\boldsymbol{\beta}}_1^{\mathrm{PS}}\right) = & \boldsymbol{\Gamma}\left(\widehat{\boldsymbol{\beta}}_1^{\mathrm{RS}}\right) + 2\boldsymbol{\delta}\boldsymbol{\mu}_{11.2}^\top E\left(\left\{1-(p_2-2)\chi_{p_2+2}^{-2}(\Delta)\right\}\right. \\
& \left.\times I\left(\chi_{p_2+2}^{2}(\Delta) \le p_2-2\right)\right) \\
& -2\boldsymbol{\Phi} E\left(\left\{1-(p_2-2)\chi_{p_2+2}^{-2}(\Delta)\right\} I\left(\chi_{p_2+2}^{-2}(\Delta) \le p_2-2\right)\right) \\
& -2\boldsymbol{\delta}\boldsymbol{\delta}^\top E\left(\left\{1-(p_2-2)\chi_{p_2+4}^{-2}(\Delta)\right\} I\left(\chi_{p_2+4}^{2}(\Delta) \le p_2-2\right)\right) \\
& +2\boldsymbol{\delta}\boldsymbol{\delta}^\top E\left(\left\{1-(p_2-2)\chi_{p_2+2}^{-2}(\Delta)\right\} I\left(\chi_{p_2+2}^{2}(\Delta) \le p_2-2\right)\right) \\
& -(p_2-2)^2\boldsymbol{\Phi} E\left(\chi_{p_2+2,\alpha}^{-4}(\Delta) I\left(\chi_{p_2+2,\alpha}^{2}(\Delta) \le p_2-2\right)\right) \\
& -(p_2-2)^2\boldsymbol{\delta}\boldsymbol{\delta}^\top E\left(\chi_{p_2+4}^{-4}(\Delta) I\left(\chi_{p_2+2}^{2}(\Delta) \le p_2-2\right)\right) \\
& +\boldsymbol{\Phi} H_{p_2+2}\left(p_2-2;\Delta\right) + \boldsymbol{\delta}\boldsymbol{\delta}^\top H_{p_2+4}\left(p_2-2;\Delta\right).
\end{aligned} \tag{10.9}$$

By definition,

$$\mathrm{ADR}\left(\widehat{\boldsymbol{\beta}}_1^{*}\right) = \mathrm{tr}\left(\mathbf{W}\boldsymbol{\Gamma}\left(\widehat{\boldsymbol{\beta}}_1^{*}\right)\right) \tag{10.10}$$

where $\mathbf{W}$ is a positive definite matrix.

Theorem 10.4 *The asymptotic risks of the estimators are:*

$$\begin{aligned}
\mathrm{ADR}\left(\widehat{\boldsymbol{\beta}}_1^{\mathrm{FM}}\right) = & \sigma^2 tr\left(\boldsymbol{W}\boldsymbol{S}_{11.2}^{-1}\right) + \boldsymbol{\mu}_{11.2}^\top\boldsymbol{W}\boldsymbol{\mu}_{11.2} \\
\mathrm{ADR}\left(\widehat{\boldsymbol{\beta}}_1^{\mathrm{SM}}\right) = & \sigma^2 \mathrm{tr}\left(\boldsymbol{W}\boldsymbol{S}_{11}^{-1}\right) + \gamma^\top\boldsymbol{W}\gamma \\
\mathrm{ADR}\left(\widehat{\boldsymbol{\beta}}_1^{\mathrm{S}}\right) = & \mathrm{ADR}\left(\widehat{\boldsymbol{\beta}}_1^{\mathrm{FM}}\right) + 2(p_2-2)\boldsymbol{\mu}_{11.2}^\top\boldsymbol{W}\boldsymbol{\delta} E\left(\chi_{p_2+2}^{-2}(\Delta)\right) \\
& -(p_2-2)\mathrm{tr}\left(\boldsymbol{W}\boldsymbol{\Phi}\right)\left[E\left(\chi_{p_2+2}^{-2}(\Delta)\right)\right. \\
& \left.-(p_2-2)E\left(\chi_{p_2+2}^{-4}(\Delta)\right)\right] \\
& +(p_2-2)\boldsymbol{\delta}^\top\boldsymbol{W}\boldsymbol{\delta} \\
& \times\left[2E\left(\chi_{p_2+2}^{-2}(\Delta)\right) - 2E\left(\chi_{p_2+4}^{-2}(\Delta)\right)\right. \\
& \left.+(p_2-2)E\left(\chi_{p_2+4}^{-4}(\Delta)\right)\right],
\end{aligned}$$

$$
\begin{aligned}
& \mathrm{ADR}\left(\widehat{\boldsymbol{\beta}}_1^{\mathrm{PS}}\right) \\
= \; & \mathrm{ADR}\left(\widehat{\boldsymbol{\beta}}_1^{\mathrm{S}}\right)+2\boldsymbol{\mu}_{11.2}^{\top}\boldsymbol{W}\boldsymbol{\delta} \\
& \times E\left(\left\{1-(p_2-2)\chi_{p_2+2}^{-2}(\Delta)\right\} I\left(\chi_{p_2+2}^{2}(\Delta)\leq p_2-2\right)\right) \\
& +\operatorname{tr}(\boldsymbol{W}\boldsymbol{\Phi}) E\left(\left\{1-(p_2-2)\chi_{p_2+2}^{-2}(\Delta)\right\} I\left(\chi_{p_2+2}^{-2}(\Delta)\leq p_2-2\right)\right) \\
& -2\boldsymbol{\delta}^{\top}\boldsymbol{W}\boldsymbol{\delta} E\left(\left\{1-(p_2-2)\chi_{p_2+4}^{-2}(\Delta)\right\} I\left(\chi_{p_2+4}^{2}(\Delta)\leq p_2-2\right)\right) \\
& +2\boldsymbol{\delta}^{\top}\boldsymbol{W}\boldsymbol{\delta} E\left(\left\{1-(p_2-2)\chi_{p_2+2}^{-2}(\Delta)\right\} I\left(\chi_{p_2+2}^{2}(\Delta)\leq p_2-2\right)\right) \\
& -(p_2-2)^2\operatorname{tr}(\boldsymbol{W}\boldsymbol{\Phi}) E\left(\chi_{p_2+2,\alpha}^{-4}(\Delta) I\left(\chi_{p_2+2,\alpha}^{2}(\Delta)\leq p_2-2\right)\right) \\
& -(p_2-2)^2\boldsymbol{\delta}^{\top}\boldsymbol{W}\boldsymbol{\delta} E\left(\chi_{p_2+4}^{-4}(\Delta) I\left(\chi_{p_2+2}^{2}(\Delta)\leq p_2-2\right)\right) \\
& +\operatorname{tr}(\boldsymbol{W}\boldsymbol{\Phi}) H_{p_2+2}(p_2-2;\Delta)+\boldsymbol{\delta}^{\top}\boldsymbol{W}\boldsymbol{\delta} H_{p_2+4}(p_2-2;\Delta).
\end{aligned}
$$

Based on the above respective risk expressions, the shrinkage estimation retains its supremacy over submodel and full model estimators. Regardless of baseline estimator, the shrinkage strategy will retain its properties if constructed properly. The expression reveals that the sparsity assumption is true when the submodel estimator has an edge over the listed estimators. In contrast, the submodel estimator does not perform well when the sparsity assumption does not hold and it is an unbounded function of the sparsity parameter. Finally, the performance of the shrinkage estimators are superior with respect to the submodel estimator in most of the parameter space induced by the sparsity assumption and outclass the full model estimator in the entire parameter space. This indicates the power and beauty of the shrinkage strategy as it retains it dominating characteristics regardless of model and estimator type.

10.4 Simulation Experiments

In this section, we consider a Monte Carlo simulation to assess the estimator performance. The response is generated from the following multiple regression model:

$$y_i = x_{1i}\beta_1 + x_{2i}\beta_2 + \ldots + x_{pi}\beta_p + \varepsilon_i, \; i = 1, \ldots, n, \tag{10.11}$$

where ε_i are i.i.d. $\mathcal{N}(0,1)$. We use the following equation to inject varying levels of collinearity amongst the covariates.

$$x_{ij} = (1-\gamma^2)^{1/2} z_{ij} + \gamma z_{ip},$$

where z_{ij} are random numbers following a standard normal distribution such that $i = 1, 2, \ldots n$, $j = 1, 2, \ldots p$, where n is the sample size and p is the number of regressors Gibbons (1981). The degree of correlation γ is considered by $0.3, 0.6$ and 0.9. To quantify multicollinearity we also consider a condition number (CN) value, defined as the ratio of the largest eigenvalue to the smallest eigenvalue of matrix $\boldsymbol{X}^{\top}\boldsymbol{X}$. Belsley (2014) suggested that multicollinearity exists in the data if the CN value is greater than 30.

We consider the regression coefficients are set $\boldsymbol{\beta} = \left(\boldsymbol{\beta}_1^{\top}, \boldsymbol{\beta}_2^{\top}, \boldsymbol{\beta}_3^{\top}\right)^{\top}$ with dimensions p_1, p_2 and p_3, respectively. $\boldsymbol{\beta}_1$ represents strong signals when $\boldsymbol{\beta}_1$ is a vector of 1 values, $\boldsymbol{\beta}_2$ is a vector of 0.1 values, and $\boldsymbol{\beta}_3$ means no signals when $\boldsymbol{\beta}_3 = \mathbf{0}$. In this simulation setting, we simulated 1000 data sets consisting of $n = 100$, with $p_1 = 4$, $p_2 = 0, 3, 6, 9$

and $p_3 = 4, 8, 12, 16$. In order to investigate the behavior of the estimators, we define $\Delta = \|\boldsymbol{\beta} - \boldsymbol{\beta}_0\| \geq 0$, where $\boldsymbol{\beta}_0 = \left(\boldsymbol{\beta}_1^\top, \boldsymbol{\beta}_2^\top, \mathbf{0}_3^\top\right)^\top$ and $\|\cdot\|$ is the Euclidean norm. The biasing parameter d is obtained by minimizing the mean squared error function of $\widehat{\boldsymbol{\beta}}^{\mathrm{FM}}$ (see Liu (1993)) with respect to each individual parameter,

$$d_j = \frac{\lambda_j \left(\alpha_j^2 - \sigma^2\right)}{\sigma^2 + \lambda_j \alpha_j^2},$$

where λ_js ($\lambda_1 \geq \lambda_1 \geq \cdots \geq \lambda_p \geq 0$) are the eigenvalues of $\boldsymbol{X}^\top \boldsymbol{X}$ and $q_1, q_2, \ldots, q_p$ the corresponding eigenvectors. $\boldsymbol{\alpha} = \boldsymbol{Q}^\top \widehat{\boldsymbol{\beta}}^{\mathrm{LSE}}$ and $\boldsymbol{Q} = (q_1, q_2 \cdot \ldots, q_p)$. σ^2 is an unbiased estimator of the model variance. Following Månsson et al. (2012) we calculate of d by

$$d = \max(0, \max(d_j)), \tag{10.12}$$

and all computations were conducted using the statistical programming language Team (2021). The performance of the respective estimator was evaluated by using the mean squared error criterion. We report the relative mean squared error (RMSE) of an estimator $\boldsymbol{\beta}_1^*$ to the $\widehat{\boldsymbol{\beta}}_1^{\mathrm{FM}}$

By definition,

$$\mathrm{RMSE}\left(\widehat{\boldsymbol{\beta}}_1^{\mathrm{FM}} : \boldsymbol{\beta}_1^*\right) = \frac{\mathrm{MSE}\left(\widehat{\boldsymbol{\beta}}_1^{\mathrm{FM}}\right)}{\mathrm{MSE}\left(\boldsymbol{\beta}_1^*\right)},$$

where $\boldsymbol{\beta}_1^*$ is one of the listed estimators. Obviously, if the RMSE of an estimator is greater than one, it indicates that $\boldsymbol{\beta}_1^*$ is superior to the full model estimator.

The simulation results are shown in Tables 10.1–10.3. RMSEs against Δ are also plotted for easier comparison in Figures 10.1–10.3.

In summary, when the sparsity parameter $\Delta = 0$ is true, the submodel is superior to all remaining estimators, indicated by the highest RMSE values in the tables. However, the submodel does not perform well when the value of Δ increases. The RMSE decreases and converges to zero, an annoying feature of the submodel as the assumption of sparsity is vital to the submodel estimator. The full model estimator is not impacted by such departure. Both shrinkage estimators dominate the full model estimator for $\Delta \in [0, \infty)$. The shrinkage estimators outperform the submodel estimator for most values in the interval. As expected, the positive shrinkage estimator is uniformly better than shrinkage estimator. Evidently, the shrinkage estimators tend to one for large values of Δ.

The numerical analysis based on the simulation study essentially report the same conclusions from the asymptotic properties of the estimators. Thus, the numerical study supports the theoretical aspect of the estimators.

10.4.1 Comparisons with Penalty Estimators

In Tables 10.4–10.7 we compared the relative performance of shrinkage estimators with five estimators based on penalized procedures, namely ridge, ENET, LASSO, ALASSO, and SCAD at some selected values of given simulated parameters. We split p into three components, p_1 represents the strong signal, p_2 is referred to as weak signals, and p_3 is termed as no signals. The regression coefficients are set $\boldsymbol{\beta} = \left(\boldsymbol{\beta}_1^\top, \boldsymbol{\beta}_2^\top, \boldsymbol{\beta}_3^\top\right)^\top = \left(\mathbf{1}_{p_1}^\top, \mathbf{0.1}_{p_2}^\top, \mathbf{0}_{p_3}^\top\right)^\top$ with dimensions p_1, p_2 and p_3, respectively. We simulate 1000 data sets consisting of $n = 100, 200$, with $p_1 = 5, 10$, $p_2 = 0, 5, 10$, $p_3 = 5, 10, 15$ and $\gamma = 0.3, 0.6, 0.9$.

The results conclude that the shrinkage estimators outshine the penalized estimators in the presence of weak signals with few exceptions. Amongst penalized methods, the relative performance of ALASSO and SCAD are far superior and perhaps a better alternative to a

TABLE 10.1: The RMSE of the Estimators for $n = 100$, $p_1 = 4$, and $\gamma = 0.3$.

		$p_2 = 0$		$p_2 = 3$		$p_2 = 6$		$p_2 = 9$	
p_3	Δ	SM	PS	SM	PS	SM	PS	SM	PS
4	0.0	2.177	1.520	1.733	1.358	1.524	1.275	1.370	1.199
	0.3	0.748	1.108	0.829	1.078	0.922	1.080	0.893	1.051
	0.6	0.243	1.023	0.321	1.018	0.421	1.022	0.439	1.009
	1.2	0.066	1.007	0.092	1.002	0.131	1.005	0.148	0.999
	2.4	0.017	1.001	0.024	0.999	0.036	1.001	0.040	0.999
	4.8	0.004	1.000	0.006	1.000	0.009	1.000	0.010	1.000
	9.2	0.001	1.000	0.002	1.000	0.002	1.000	0.003	1.000
8	0.0	3.443	2.520	2.431	1.974	2.012	1.731	1.854	1.634
	0.3	1.139	1.411	1.176	1.353	1.231	1.312	1.206	1.286
	0.6	0.392	1.128	0.451	1.096	0.553	1.104	0.566	1.093
	1.2	0.106	1.033	0.128	1.018	0.176	1.024	0.188	1.020
	2.4	0.027	1.005	0.033	1.003	0.047	1.004	0.051	1.003
	4.8	0.007	1.001	0.008	0.999	0.012	1.001	0.013	0.999
	9.2	0.002	1.000	0.002	0.999	0.003	1.000	0.004	1.000
12	0.0	4.757	3.429	3.512	2.776	2.533	2.202	2.266	1.996
	0.3	1.691	1.827	1.662	1.753	1.546	1.572	1.467	1.500
	0.6	0.558	1.246	0.632	1.245	0.698	1.212	0.715	1.188
	1.2	0.154	1.065	0.183	1.059	0.221	1.054	0.241	1.047
	2.4	0.040	1.016	0.048	1.008	0.059	1.011	0.066	1.011
	4.8	0.010	1.003	0.012	1.000	0.015	1.002	0.017	1.001
	9.2	0.003	1.001	0.003	0.999	0.004	1.000	0.005	1.000
16	0.0	6.368	4.543	4.008	3.259	3.252	2.809	2.796	2.456
	0.3	2.267	2.278	1.977	2.025	1.948	1.938	1.846	1.798
	0.6	0.769	1.404	0.759	1.354	0.833	1.335	0.910	1.325
	1.2	0.215	1.097	0.224	1.101	0.259	1.097	0.304	1.094
	2.4	0.056	1.022	0.059	1.019	0.068	1.018	0.083	1.021
	4.8	0.014	1.006	0.015	1.003	0.017	1.003	0.021	1.003
	9.2	0.004	1.001	0.004	1.000	0.005	1.001	0.006	1.002

TABLE 10.2: The RMSE of the Estimators for $n = 100$, $p_1 = 4$, and $\gamma = 0.6$.

		$p_2 = 0$		$p_2 = 3$		$p_2 = 6$		$p_2 = 9$	
p_3	Δ	SM	PS	SM	PS	SM	PS	SM	PS
4	0.0	2.380	1.585	1.742	1.369	1.401	1.223	1.400	1.224
	0.3	1.078	1.191	1.082	1.145	1.033	1.086	1.091	1.101
	0.6	0.389	1.047	0.489	1.029	0.575	1.020	0.646	1.033
	1.2	0.113	1.012	0.154	1.004	0.207	1.003	0.245	1.007
	2.4	0.029	1.000	0.041	0.998	0.059	1.000	0.070	1.001
	4.8	0.007	1.000	0.010	0.999	0.015	1.000	0.018	1.000
	9.2	0.002	1.000	0.003	1.000	0.004	1.000	0.005	1.000
8	0.0	3.402	2.472	2.352	1.934	1.895	1.653	1.822	1.610
	0.3	1.638	1.631	1.553	1.490	1.372	1.361	1.402	1.356
	0.6	0.612	1.193	0.719	1.173	0.738	1.123	0.828	1.146
	1.2	0.174	1.047	0.228	1.044	0.266	1.027	0.319	1.037
	2.4	0.046	1.011	0.062	1.007	0.075	1.002	0.093	1.010
	4.8	0.012	1.001	0.016	1.001	0.019	0.998	0.024	1.001
	9.2	0.003	0.999	0.004	1.000	0.005	1.000	0.007	1.000
12	0.0	4.972	3.523	3.421	2.719	2.592	2.213	2.244	1.986
	0.3	2.399	2.176	2.150	1.968	1.886	1.772	1.696	1.630
	0.6	0.915	1.413	0.995	1.401	0.990	1.316	0.978	1.273
	1.2	0.264	1.113	0.311	1.104	0.353	1.093	0.374	1.086
	2.4	0.069	1.026	0.084	1.020	0.099	1.021	0.106	1.022
	4.8	0.017	1.005	0.022	1.004	0.025	1.004	0.027	1.005
	9.2	0.005	1.001	0.006	1.001	0.007	1.000	0.008	1.001
16	0.0	7.047	4.827	4.315	3.450	3.130	2.693	2.843	2.493
	0.3	3.184	2.834	2.768	2.500	2.156	2.030	2.135	2.010
	0.6	1.161	1.646	1.252	1.622	1.156	1.453	1.234	1.484
	1.2	0.331	1.171	0.391	1.177	0.419	1.137	0.459	1.134
	2.4	0.087	1.044	0.104	1.036	0.118	1.026	0.132	1.019
	4.8	0.022	1.008	0.026	1.004	0.031	1.005	0.034	1.002
	9.2	0.006	1.001	0.007	1.001	0.008	1.000	0.010	0.999

TABLE 10.3: The RMSE of the Estimators for $n = 100$, $p_1 = 4$, and $\gamma = 0.9$.

		$p_2 = 0$		$p_2 = 3$		$p_2 = 6$		$p_2 = 9$	
p_3	Δ	SM	PS	SM	PS	SM	PS	SM	PS
4	0.0	2.139	1.516	1.647	1.344	1.498	1.261	1.388	1.210
	0.3	1.768	1.375	1.461	1.254	1.393	1.215	1.279	1.167
	0.6	1.019	1.166	1.051	1.127	1.082	1.111	1.044	1.086
	1.2	0.378	1.044	0.469	1.023	0.563	1.020	0.610	1.008
	2.4	0.105	1.003	0.149	0.998	0.199	1.002	0.237	0.991
	4.8	0.027	1.001	0.040	0.999	0.056	0.999	0.070	0.993
	9.2	0.008	0.999	0.011	0.999	0.016	1.000	0.020	0.998
8	0.0	4.116	2.696	2.477	1.998	1.917	1.669	1.796	1.594
	0.3	3.388	2.397	2.252	1.862	1.832	1.591	1.676	1.519
	0.6	1.837	1.764	1.539	1.496	1.420	1.381	1.424	1.365
	1.2	0.609	1.215	0.686	1.169	0.740	1.130	0.820	1.126
	2.4	0.167	1.036	0.218	1.026	0.251	1.032	0.316	1.033
	4.8	0.043	0.999	0.059	1.002	0.069	1.005	0.093	1.003
	9.2	0.012	0.996	0.016	1.001	0.019	1.000	0.027	1.001
12	0.0	5.301	3.632	3.238	2.628	2.532	2.177	2.291	2.022
	0.3	4.192	3.075	2.778	2.354	2.362	2.052	2.196	1.954
	0.6	2.333	2.148	2.000	1.892	1.865	1.741	1.742	1.665
	1.2	0.839	1.392	0.903	1.342	0.983	1.309	0.996	1.282
	2.4	0.241	1.099	0.294	1.099	0.345	1.086	0.365	1.057
	4.8	0.063	1.020	0.079	1.025	0.098	1.021	0.104	1.002
	9.2	0.018	1.006	0.022	1.005	0.028	1.004	0.030	0.997
16	0.0	6.962	4.757	4.231	3.368	3.111	2.687	2.740	2.416
	0.3	5.450	4.052	3.697	3.096	2.901	2.534	2.565	2.289
	0.6	3.159	2.764	2.717	2.457	2.256	2.085	2.050	1.933
	1.2	1.103	1.626	1.221	1.593	1.206	1.506	1.211	1.445
	2.4	0.311	1.154	0.383	1.147	0.416	1.144	0.467	1.125
	4.8	0.082	1.035	0.103	1.026	0.117	1.035	0.140	1.023
	9.2	0.023	1.009	0.029	1.008	0.033	1.010	0.041	1.002

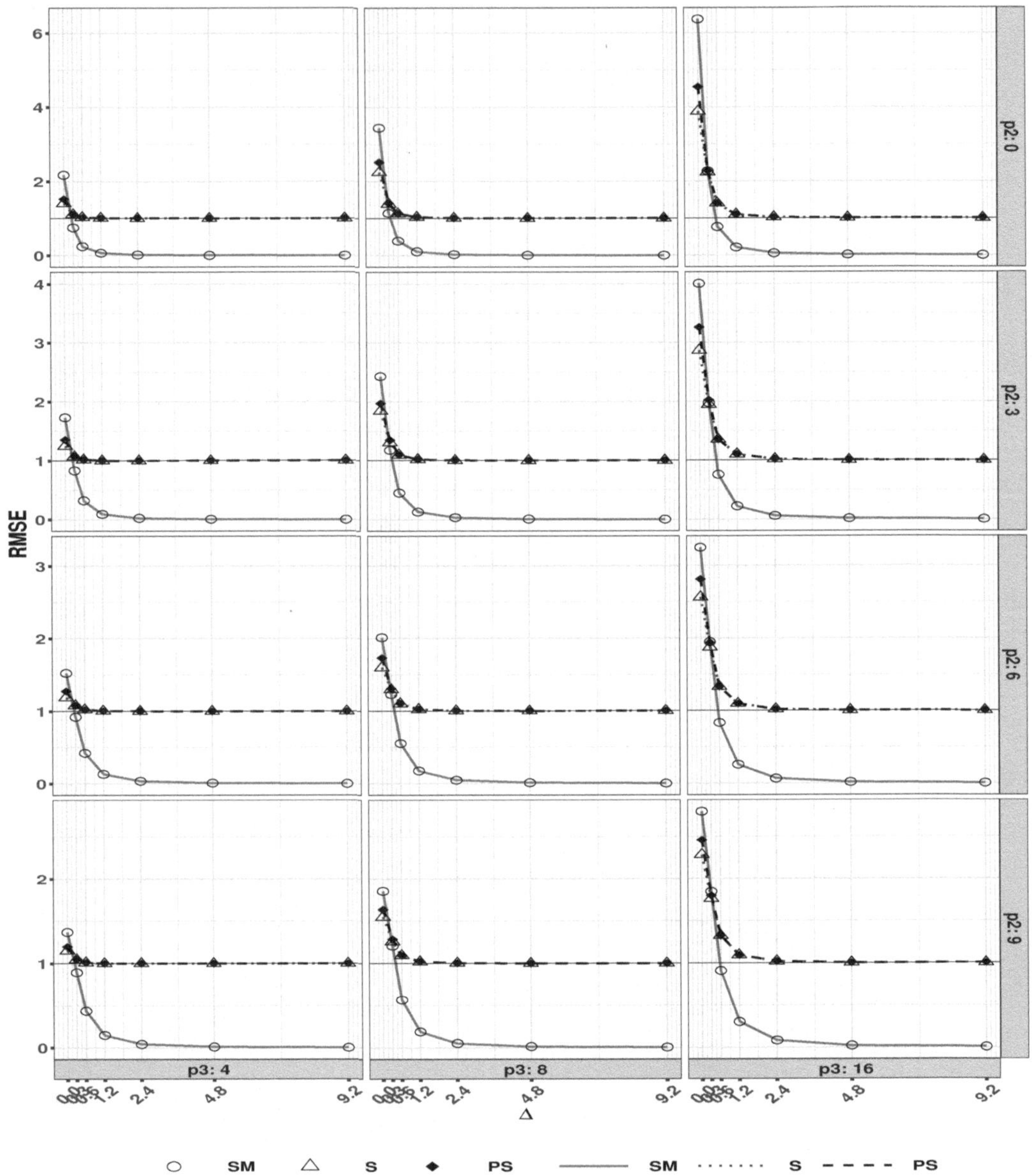

FIGURE 10.1: RMSEs of the Estimators as a Function of Δ when $n = 100$, $p_1 = 4$ and $\gamma = 0.3$.

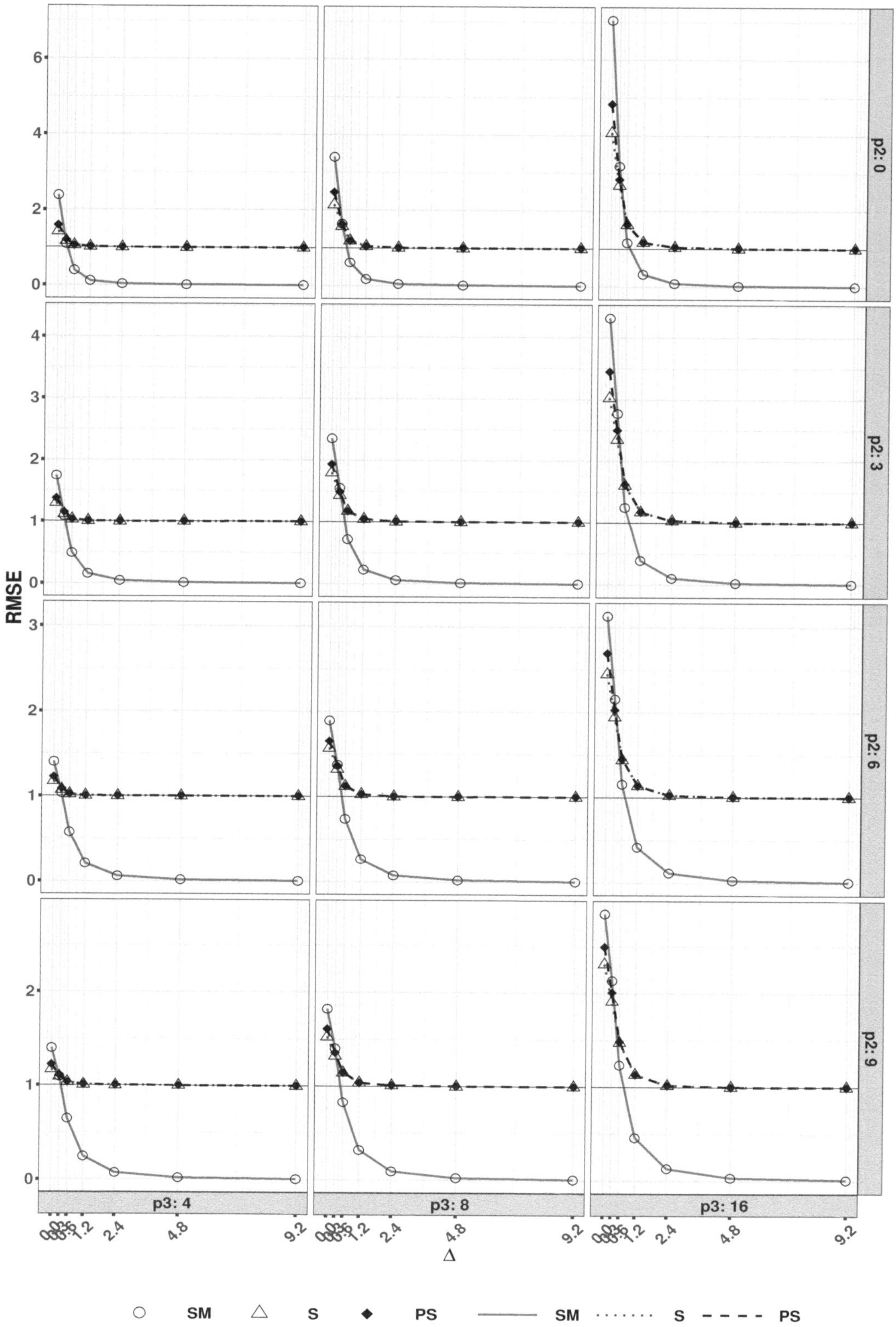

FIGURE 10.2: RMSEs of the Estimators as a Function of Δ when $n = 100$, $p_1 = 4$ and $\gamma = 0.6$.

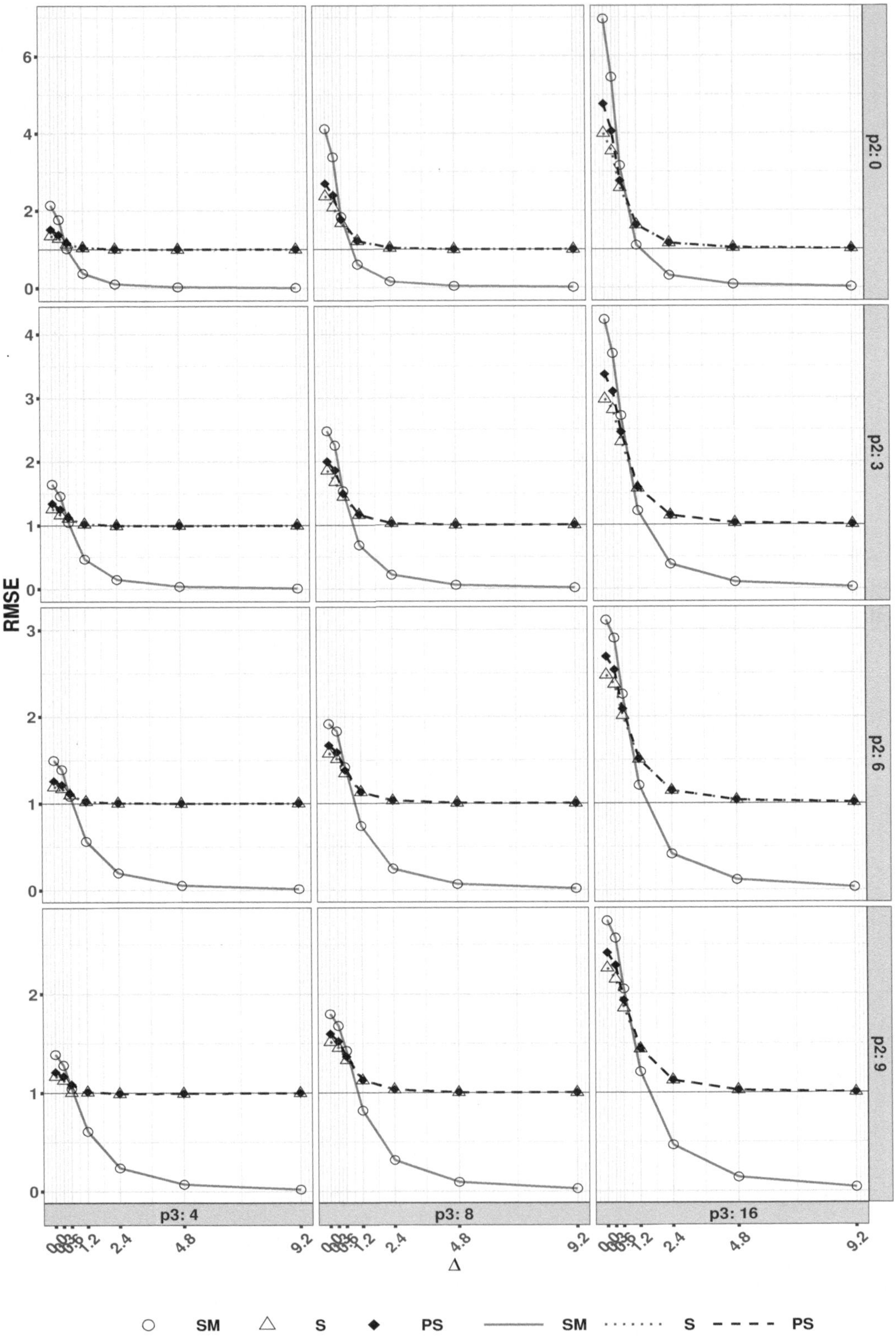

FIGURE 10.3: RMSEs of the Estimators as a Function of Δ when $n = 100$, $p_1 = 4$ and $\gamma = 0.9$.

TABLE 10.4: The RMSE of the Estimators for $n = 100$ and $p_1 = 5$.

γ	p_2	p_3	SM	S	PS	Ridge	ENET	LASSO	ALASSO	SCAD
0.3	0	5	2.26	1.47	1.68	1.07	1.11	1.44	1.83	1.92
		10	3.57	2.38	2.67	1.07	1.09	1.78	2.72	3.01
		15	5.14	3.38	3.86	0.98	0.99	2.01	3.75	4.46
	5	5	1.61	1.32	1.38	1.09	1.12	1.53	1.54	1.50
		10	2.29	1.83	1.98	0.99	1.00	1.77	2.10	2.10
		15	3.11	2.43	2.65	1.14	1.14	2.32	2.68	2.77
	10	5	1.40	1.20	1.26	1.01	1.04	1.57	1.36	1.19
		10	1.95	1.63	1.74	1.15	1.16	2.06	1.79	1.64
		15	2.44	2.04	2.18	1.08	1.08	2.06	2.19	1.91
0.6	0	5	2.39	1.53	1.71	0.88	1.19	1.37	1.75	1.83
		10	3.51	2.42	2.67	0.92	1.04	1.72	2.43	2.48
		15	5.50	3.46	3.99	1.07	1.10	2.32	3.62	3.31
	5	5	1.65	1.32	1.40	0.94	1.16	1.54	1.61	1.59
		10	2.37	1.86	2.01	1.10	1.16	2.06	2.33	2.06
		15	3.02	2.38	2.60	1.16	1.18	2.44	2.93	2.59
	10	5	1.41	1.22	1.26	1.11	1.25	1.76	1.53	1.28
		10	1.91	1.61	1.71	1.16	1.22	2.10	1.97	1.62
		15	2.36	2.01	2.13	1.32	1.33	2.76	2.51	2.05
0.9	0	5	2.22	1.52	1.65	1.14	1.34	1.35	0.81	0.76
		10	3.42	2.37	2.63	1.25	1.67	1.72	1.09	0.97
		15	5.40	3.39	3.94	1.66	2.16	2.42	1.64	1.42
	5	5	1.66	1.33	1.39	1.31	1.60	1.62	1.00	0.90
		10	2.29	1.82	1.97	1.68	2.10	2.19	1.46	1.29
		15	3.10	2.44	2.66	1.71	2.27	2.58	1.90	1.68
	10	5	1.38	1.20	1.25	1.80	2.04	2.06	1.31	1.15
		10	1.91	1.60	1.72	1.76	2.30	2.41	1.65	1.51
		15	2.49	2.09	2.21	1.86	2.34	2.66	1.83	1.55

TABLE 10.5: The RMSE of the Estimators for $n = 100$ and $p_1 = 10$.

γ	p_2	p_3	SM	S	PS	Ridge	ENET	LASSO	ALASSO	SCAD
0.3	0	5	1.62	1.32	1.38	1.06	1.12	1.21	1.36	1.47
		10	2.32	1.85	2.00	1.02	1.06	1.44	1.88	2.07
		15	3.13	2.44	2.66	1.02	1.04	1.67	2.41	2.71
	5	5	1.40	1.20	1.26	1.05	1.12	1.34	1.34	1.34
		10	1.95	1.64	1.74	1.03	1.08	1.57	1.82	1.90
		15	2.47	2.06	2.20	1.03	1.05	1.73	2.30	2.37
	10	5	1.36	1.20	1.23	1.05	1.14	1.45	1.35	1.28
		10	1.70	1.49	1.56	1.04	1.08	1.61	1.72	1.60
		15	2.10	1.83	1.94	1.14	1.15	1.96	2.04	1.89
0.6	0	5	1.66	1.33	1.40	0.88	1.16	1.18	1.22	1.41
		10	2.40	1.88	2.04	1.13	1.30	1.54	1.71	1.84
		15	3.05	2.40	2.62	1.00	1.17	1.70	2.08	2.20
	5	5	1.41	1.22	1.26	1.15	1.32	1.42	1.38	1.40
		10	1.92	1.62	1.71	1.01	1.27	1.58	1.71	1.72
		15	2.38	2.02	2.15	1.25	1.35	2.09	2.25	2.16
	10	5	1.32	1.18	1.21	1.06	1.32	1.50	1.40	1.35
		10	1.73	1.52	1.59	1.25	1.43	1.87	1.81	1.65
		15	2.24	1.93	2.04	1.27	1.42	2.12	2.16	1.97
0.9	0	5	1.67	1.33	1.40	1.25	1.18	1.18	0.62	0.61
		10	2.31	1.84	1.99	1.12	1.36	1.36	0.75	0.74
		15	3.16	2.48	2.71	1.24	1.73	1.73	1.06	0.95
	5	5	1.39	1.20	1.25	1.22	1.35	1.35	0.73	0.74
		10	1.93	1.60	1.73	1.29	1.68	1.68	0.99	0.92
		15	2.54	2.12	2.26	1.42	1.92	1.92	1.12	1.04
	10	5	1.33	1.17	1.22	1.38	1.65	1.65	0.93	0.90
		10	1.73	1.52	1.58	1.47	1.87	1.87	1.06	1.01
		15	2.12	1.82	1.93	1.78	2.16	2.17	1.20	1.00

TABLE 10.6: The RMSE of the Estimators for $n = 200$ and $p_1 = 5$.

γ	p_2	p_3	SM	S	PS	Ridge	ENET	LASSO	ALASSO	SCAD
0.3	0	5	2.31	1.57	1.69	0.82	1.00	1.33	2.00	2.11
		10	3.49	2.33	2.67	0.91	0.96	1.73	2.87	3.11
		15	4.79	3.21	3.73	0.93	0.94	2.09	3.87	4.42
	5	5	1.58	1.31	1.37	0.93	1.06	1.37	1.12	1.02
		10	2.20	1.78	1.92	0.94	0.99	1.68	1.53	1.43
		15	2.87	2.32	2.51	1.00	1.01	1.93	1.88	1.78
	10	5	1.36	1.21	1.24	0.96	1.06	1.39	0.93	0.78
		10	1.79	1.55	1.64	1.01	1.04	1.63	1.19	0.95
		15	2.16	1.86	1.99	0.99	1.00	1.87	1.40	1.12
0.6	0	5	2.36	1.52	1.70	0.65	1.24	1.40	1.91	1.94
		10	3.19	2.29	2.54	0.84	1.17	1.81	2.49	2.52
		15	4.93	3.32	3.81	0.95	1.08	2.40	3.81	3.68
	5	5	1.58	1.28	1.37	0.86	1.30	1.46	1.27	1.17
		10	2.29	1.85	1.98	0.96	1.29	1.91	1.81	1.60
		15	2.74	2.23	2.42	0.92	1.13	2.17	2.28	1.97
	10	5	1.43	1.23	1.27	1.00	1.37	1.59	1.10	0.92
		10	1.77	1.54	1.62	0.96	1.29	1.84	1.40	1.12
		15	2.22	1.89	2.02	1.00	1.21	2.02	1.69	1.34
0.9	0	5	2.34	1.52	1.70	0.59	1.40	1.40	1.28	1.62
		10	3.65	2.44	2.75	0.73	1.83	1.84	1.83	2.09
		15	5.03	3.29	3.84	0.89	2.21	2.23	2.41	2.68
	5	5	1.56	1.28	1.35	0.78	1.68	1.68	1.44	1.61
		10	2.13	1.74	1.88	0.92	2.04	2.05	1.91	2.03
		15	2.86	2.31	2.50	1.11	2.49	2.52	2.45	2.42
	10	5	1.40	1.22	1.27	0.96	1.85	1.85	1.47	1.46
		10	1.80	1.56	1.64	1.16	2.29	2.29	1.89	1.79
		15	2.25	1.95	2.06	1.00	2.31	2.35	2.17	2.15

TABLE 10.7: The RMSE of the Estimators for $n = 200$ and $p_1 = 10$.

γ	p_2	p_3	SM	S	PS	Ridge	ENET	LASSO	ALASSO	SCAD
0.3	0	5	1.59	1.31	1.37	0.89	1.12	1.20	1.40	1.49
		10	2.20	1.78	1.92	0.92	1.04	1.46	1.92	2.07
		15	2.88	2.32	2.52	0.85	0.95	1.60	2.48	2.74
	5	5	1.36	1.21	1.24	0.94	1.11	1.26	1.12	1.07
		10	1.80	1.55	1.64	0.86	1.06	1.43	1.45	1.40
		15	2.17	1.87	1.99	0.93	1.00	1.65	1.72	1.69
	10	5	1.31	1.17	1.21	0.88	1.15	1.27	0.97	0.87
		10	1.61	1.44	1.51	0.93	1.11	1.44	1.18	1.06
		15	1.94	1.73	1.81	0.99	1.05	1.63	1.40	1.27
0.6	0	5	1.58	1.28	1.37	0.67	1.18	1.18	1.23	1.43
		10	2.29	1.85	1.99	0.79	1.45	1.53	1.78	1.94
		15	2.75	2.24	2.43	0.76	1.38	1.68	2.09	2.36
	5	5	1.43	1.24	1.28	0.82	1.34	1.35	1.25	1.23
		10	1.77	1.54	1.62	0.79	1.43	1.54	1.49	1.51
		15	2.23	1.90	2.03	0.78	1.53	1.69	1.89	1.85
	10	5	1.29	1.15	1.20	0.81	1.36	1.37	1.09	1.03
		10	1.59	1.42	1.48	0.81	1.50	1.55	1.33	1.26
		15	1.96	1.74	1.82	0.77	1.53	1.71	1.57	1.46
0.9	0	5	1.57	1.28	1.36	0.66	1.17	1.17	0.77	1.09
		10	2.15	1.75	1.89	0.64	1.43	1.43	1.04	1.38
		15	2.88	2.32	2.52	0.67	1.66	1.66	1.35	1.76
	5	5	1.41	1.23	1.27	0.69	1.37	1.37	0.96	1.29
		10	1.80	1.56	1.65	0.70	1.59	1.59	1.25	1.61
		15	2.28	1.97	2.08	0.76	1.85	1.85	1.52	1.83
	10	5	1.27	1.14	1.18	0.75	1.52	1.52	1.12	1.40
		10	1.59	1.43	1.49	0.78	1.74	1.74	1.38	1.60
		15	1.91	1.71	1.78	0.80	1.92	1.92	1.60	1.86

full model estimator in the absence of a shrinkage strategy. The shrinkage strategy is simple to implement and computationally attractive when comparing penalty estimators.

10.5 Application to Air Pollution Data

We consider air pollution data originally used by McDonald and Schwing (1973). Yüzbaşı et al. (2020) illustrated generalized ridge shrinkage methods and Yüzbaşı et al. (2021) illustrated restricted bridge estimation method. This data set includes 15 explanatory variables related to air pollution, socioeconomic factors, and meteorological measurements to measure mortality rate, the dependent variable, for 60 US cities in 1960. The data is freely available from Carnegie Mellon University's StatLib [1]. The variable descriptions are given in Table 10.8.

TABLE 10.8: Lists and Descriptions of Variables.

Response Variable	
mortality	Total age-adjusted mortality from all causes (Annual deaths per 100,000 people)
Predictors	
Precip	Mean annual precipitation (inches)
Humidity	Percent relative humidity (annual average at 1:00pm)
JanTemp	Mean January temperature (degrees F)
JulyTemp	Mean July temperature (degrees F)
Over65	Percentage of the population aged 65 years or over
House	Population per household
Educ	Median number of school years completed for persons 25 years or older
Sound	Percentage of the housing that is sound with all facilities
Density	Population density (in persons per square mile of urbanized area)
NonWhite	Percentage of population that is nonwhite
WhiteCol	Percentage of employment in white collar occupations
Poor	Percentage of households with annual income under $3,000 in 1960
HC	Pollution potential of hydrocarbons
NOX	Pollution potential of oxides of nitrogen
SO2	Pollution potential of sulfur dioxide

In addition, we provide the variance inflation factors (VIF) and tolerance values for the explanatory variables in Table 10.9. The variables HC and NOX have quite large VIF values of 98.637 and 104.981, respectively. This shows that there is a collinearity problem with this data set. Moreover, the condition number of the standardized data matrix is 930.6907 indicating severe collinearity.

In real data applications, if there is no prior information regarding significant importance of covariates, one might do stepwise or variable selection techniques to select the

[1] http://lib.stat.cmu.edu/datasets/

TABLE 10.9: VIFs and Tolerance Values for the Variables.

Variables	Tolerance	VIF
Precip	0.243	4.114
Humidity	0.525	1.907
JanTemp	0.163	6.145
JulyTemp	0.252	3.968
Over65	0.134	7.471
House	0.232	4.310
Educ	0.206	4.866
Sound	0.250	3.998
Density	0.602	1.660
NonWhite	0.148	6.773
WhiteCol	0.352	2.842
Poor	0.115	8.715
HC	0.010	98.637
NOX	0.010	104.981
SO2	0.236	4.229

best subsets. In this study, we use Schwarz Bayes information criteria method via the *ols_step_best_subset* a function of the *olsrr* package in R. We find that 4 explanatory variables significantly explain the response variable and the remaining predictors may be ignored. We fit the submodel with the help of this auxiliary information and the full and submodel are given in Table 10.10.

TABLE 10.10: Fittings of Full and Submodel.

Models	**Formulas**
Full model	mortality $= \beta_0 + \beta_1$(Precip) $+\beta_2$(Humidity)$+\beta_3$(JanTemp) $+\beta_4$(JulyTemp)$+\beta_5$(Over65) $+\beta_6$(House)$+\beta_7$(Educ) $+\beta_8$(Sound)$+\beta_9$(Density)$+\beta_{10}$(NonWhite)$+\beta_{11}$(WhiteCol) $+\beta_{12}$(Poor)$+\beta_{13}$(HC)$+\beta_{14}$(NOX)$+\beta_{15}$(SO2)
Submodel	mortality $= \beta_0 + \beta_1$(Precip) $+\beta_4$(JanTemp)$+\beta_7$(NonWhite) $+\beta_{15}$(SO2)

In order to calculate the prediction error of the estimators, we randomly split the data into a training set that has 75% of the observations and a testing set that has the remaining 25%. The response is centered and the predictors are centered and scaled based on the training data set before fitting the model. Since the splitting data is a random process, we repeat it 1000 times. There is no noticeable variation using a larger number of replications thus we did not consider further values. To evaluate the performance of the suggested estimators we calculate the predictive error (PE) of an estimator. For ease of comparison, we define the relative predictive error (RPE) of $\widehat{\boldsymbol{\beta}}^*$ in terms of the full model Liu regression estimator $\widehat{\boldsymbol{\beta}}^{\text{FM}}$ and is evaluated as

$$\text{RPE}\left(\widehat{\boldsymbol{\beta}}^*\right) = \frac{\text{PE}(\widehat{\boldsymbol{\beta}}^{\text{FM}})}{\text{PE}(\widehat{\boldsymbol{\beta}}^*)},$$

where $\widehat{\boldsymbol{\beta}}^*$ can be any of the listed estimators. If the RPE is smaller than one it indicates superiority to the full model. The results are given in Table 10.11.

TABLE 10.11: The Average PE, SE of PE and RPE of Methods.

	PE	SE	RPE
FM	2484.721	155.606	1.000
SM	1340.586	21.474	**1.853**
S	1606.978	72.654	**1.546**
PS	1602.944	72.650	**1.550**
LSE	2685.776	171.681	0.925
ENET	2463.034	159.568	1.009
LASSO	2603.936	165.521	0.954
Ridge	1517.200	46.420	**1.638**
ALASSO	2610.677	167.860	0.952
SCAD	2177.077	159.864	1.141

Table 10.11 reveals the PE, SE (standard error) and RPE of the listed estimators. The submodel estimator LSM has the smallest PE since it is computed based on the assumption that the selected submodel is the true model. As expected, the Liu shrinkage and pretest estimators is better than penalty estimators LASSO, ENET, ALASSO and SCAD in terms of PE and SE. Thus, the data analysis corroborates with our simulation and theoretical findings.

10.6 R-Codes

```
> library('MASS') # It is for 'mvrnorm' function
> library('glmnet') # For Penalized Methods
> library('corpcor') # For rank.condition
> library('ncvreg') # For SCAD
> set.seed(2023)
> ####
> # The function of MSE
> MSE<-function(beta,beta_hat){mean((beta-beta_hat)^2)}
> # The function of PE
> PE.funct_beta <- function(newx, newy, opt.beta)
+ {
+   fit = newy - newx %*% opt.beta
+   return(t(fit)%*%fit / length(newy))
+ }
> #
> n  <- 100  # The number of sample
> p1 <- 5   # The number of strong signals
> p2 <- 10   # The number of weak signals
> p3 <- 20  # The number of zeros, no signal
> p <- p1+p2+p3
> beta1_true <- rep(1  ,p1)
> beta2_true <- rep(0.1,p2)
> beta3_true <- rep(0  ,p3)
> # The ture value of covariates
```

```
> beta_true <- c(beta1_true,beta2_true,beta3_true)
> # The design matrix
> gamma = 0.90 # correlation coefficient
> X = matrix(0,n,p)
> w=matrix(rnorm(n*p,mean=0,sd=1), n, p)
> for (i in 1:n){
+   for (j in 1:p){
+     X[i,j]=sqrt(1-gamma^2)*w[i,j]+(gamma)*w[i,p];
+   }
+ }
> #Eigenvalues and eigenvectors of X'X
> EIG <- eigen(t(X)%*%X)
> VEC <- EIG$vectors
> VAL <- EIG$values
> C <- t(X)%*%X
> # conditon test
> CNT <- rank.condition(t(X)%*%X)$condition
> CNT
[1] 5086.671
> ## assigning colnames of X to "X1","X2",....
> v<-NULL
> for(i in 1:p){
+   v[i] <- paste("X",i,sep="")
+   assign(v[i],X[,i])
+ }
> epsilon <- rnorm(n)# The errors
> y <- X%*%beta_true+epsilon
> # Split data into train and test set
> all.folds <- split (sample (1 : n), rep (1 : 2, length = n))
> train_ind <- all.folds$'1'
> test_ind  <- all.folds$'2'
> y_train <- y[train_ind]
> X_train <- X[train_ind,]
> ### Centering train data of y and X
> X_train_mean  <- apply(X_train,2,mean)
> X_train_scale <- scale(X_train,X_train_mean,F)
> y_train_mean  <- mean(y_train)
> y_train_scale  <- y_train - y_train_mean
> train_scale_df <- data.frame(y_train_scale,X_train_scale)
> #test data
> y_test <- y[test_ind]
> X_test <- X[test_ind,]
> ### Centering test data of y and X based on train means
> y_test_scale <- y_test - y_train_mean
> X_test_scale <- scale(X_test,X_train_mean,F)
> # Formula of the Full model
> xcount.FM <- c(0,paste("X", 1:(p1+p2+p3), sep=""))
> Formula_FM<- as.formula(paste("y_train~",
+                              paste(xcount.FM, collapse= "+")))
> # Formula of the Sub model
> xcount.SM <- c(0,paste("X", 1:p1, sep=""))
> Formula_SM<- as.formula(paste("y_train~",
+                              paste(xcount.SM, collapse= "+")))
> # Calculation of test stat
> Full_model <- lm(Formula_FM, data =train_scale_df)
```

```
> Sub_model  <- lm(Formula_SM, data =train_scale_df )
> test <- anova(Sub_model, Full_model, test='Chisq')
> tn <- test$Sum[2]
> ###### LSE based on Full and Sub Models
> beta_LSE_FM <- coef(Full_model)
> beta_LSE_SM <- c(coef(Sub_model),rep(0,p2+p3))
> # Estimation of variance
> sigma_LSE <- sigma(Full_model)
> # Estimation of Liu parameter d
> a <- VEC%*%beta_LSE_FM
> di <- (VAL*(a^2-sigma_LSE))/(sigma_LSE+VAL*a^2)
> d <- as.numeric(max(0,max(di)))
> # Estimation of Liu
> beta_LFM <- solve(C+diag(p))%*%(C+d*diag(p))%*%beta_LSE_FM
> beta_LSM <- solve(C+diag(p))%*%(C+d*diag(p))%*%beta_LSE_SM
> #shrinkage estimator based on b.enet and alasso
> beta.S <-beta_LFM-(beta_LFM-beta_LSM)*(p2+p3-2)/tn
> beta.PS<-beta_LSM+ max(0,(1-(p2+p3-2)/tn))*(beta_LFM-beta_LSM)
> ##PENALIZED Methods
> #ENET based on BIC
> alphas=seq(0.1,0.9,0.2)
> fits.enet <- list()
> for (ind in 1:length(alphas)){
+   fits.enet[[ind]] <- glmnet(X_train_scale, y_train_scale,
+                              alpha = alphas[ind])
+ }
> cvs.enet <- sapply(fits.enet, function(x) {
+   BIC = deviance(x) + log(NROW(X))*x$df
+   beta.enet = coef(x)[-1,which.min(BIC)]
+   return(
+     list(PE=PE.funct_beta(X_test_scale,
+                           y_test_scale,beta.enet),
+          beta.enet = beta.enet))})
> ind_opt  <- which.min(do.call(rbind,cvs.enet[1,]))
> beta.enet <- do.call(rbind,cvs.enet[2,])[ind_opt,]
> #LASSO based on BIC
> lasso.fit <- glmnet(X_train_scale, y_train_scale, alpha = 1)
> lasso.bic = deviance(lasso.fit) +
+   log(NROW(X_train_scale))*lasso.fit$df
> beta.lasso = coef(lasso.fit)[-1,which.min(lasso.bic)]
> #Ridge based on BIC
> ridge.fit <- glmnet(X_train_scale, y_train_scale, alpha = 0)
> ridge.bic = deviance(ridge.fit) +
+   log(NROW(X_train_scale))*ridge.fit$df
> beta.ridge = coef(ridge.fit)[-1,which.min(ridge.bic)]
> #ALASSO based on BIC
> weight <- beta.lasso
> weight <- ifelse(weight == 0, 0.00001, weight)
> alasso.fit <- glmnet(X_train_scale, y_train_scale,
+                      alpha = 1, penalty.factor=1/abs(weight))
> alasso.bic = deviance(alasso.fit) +
+   log(NROW(X_train_scale))*alasso.fit$df
> beta.alasso = coef(alasso.fit)[-1,which.min(alasso.bic)]
> #SCAD based on BIC
> scad.fit <- ncvreg(X_train_scale, y_train_scale, penalty=c("SCAD"))
```

```
> lam <- scad.fit$lambda[which.min(BIC(scad.fit))]
> beta.scad <- coef(scad.fit, lambda=lam)[-1]
> # Calculte MSE of estimators for betas
> MSE_beta<-
+   c(FM=MSE(beta_true,beta_LFM),
+     SM=MSE(beta_true,beta_LSM),
+     PS=MSE(beta_true,beta.PS),
+     LSE=MSE(beta_true,beta_LSE_FM),
+     ENET=MSE(beta_true,beta.enet),
+     LASSO=MSE(beta_true,beta.lasso),
+     ALASSO=MSE(beta_true,beta.alasso),
+     RIDGE=MSE(beta_true,beta.ridge),
+     SCAD=MSE(beta_true,beta.scad))
> cbind(MSE=MSE_beta,Best_Ranking= rank(MSE_beta))
              MSE Best_Ranking
FM     0.34143347            8
SM     0.03308041            2
PS     0.03142317            1
LSE    0.44745670            9
ENET   0.07428583            4
LASSO  0.07341646            3
ALASSO 0.08682496            5
RIDGE  0.08921905            6
SCAD   0.19079641            7
> # Calculte PE of estimators based on test data
> MSE_y<-
+   c(FM=PE.funct_beta(X_test_scale,y_test_scale,beta_LFM),
+     SM=PE.funct_beta(X_test_scale,y_test_scale,beta_LSM),
+     PS=PE.funct_beta(X_test_scale,y_test_scale,beta.PS),
+     LSE=PE.funct_beta(X_test_scale,y_test_scale,beta_LSE_FM),
+     ENET=PE.funct_beta(X_test_scale,y_test_scale,beta.enet),
+     LASSO=PE.funct_beta(X_test_scale,
+ y_test_scale,beta.lasso),
+     ALASSO=PE.funct_beta(X_test_scale,
+ y_test_scale,beta.alasso),
+     RIDGE=PE.funct_beta(X_test_scale,
+ y_test_scale,beta.ridge),
+     SCAD=PE.funct_beta(X_test_scale,y_test_scale,beta.scad))
> cbind(PE=MSE_y,Best_Ranking= rank(MSE_y))
              PE Best_Ranking
FM     2.8589604            8
SM     0.9347810            1
PS     0.9399276            2
LSE    3.0353042            9
ENET   1.2995597            3
LASSO  1.3353945            4
ALASSO 1.4540323            6
RIDGE  1.3742826            5
SCAD   2.3107331            7
```

10.7 Concluding Remarks

This chapter presents the relative performance of the Liu regression estimator, shrinkage estimators, and penalty estimators in the context of multiple linear regression models in the presence of multicollinearity in the data matrix when the model may be sparse. We provide the asymptotic theory of the Liu regression estimator and shrinkage estimators. The asymptotic analysis reveals that the shrinkage estimator is a better strategy than both full estimators. If the model sparsity assumption does not hold then shrinkage estimators are far superior than submodel estimators.

We also numerically compared the relative performance of the shrinkage estimators with five penalty estimators. The numerical results show that the shrinkage strategy has an edge over on penalized procedures in almost all cases, especially in the presence of weak coefficients. The performance of the submodel estimator reigns supreme when the sparsity assumption is true. However, the results show that shrinkage estimators outshine the submodel estimators when sparsity does not hold.

We also apply the theses strategies to real data. The data analysis provides the same conclusion. Hence, we suggest to use shrinkage estimators as alternatives to penalty estimators as advocated by Ahmed (2014).

Bibliography

Syed Ejaz Ahmed. Improved estimation in a multivariate regression model. *Computational Statistics & Data Analysis*, 17(5):537–554, 1994.

Syed Ejaz Ahmed. Asymptotic shrinkage estimation: the regression case. *Applied Statistical Science II*, 113–139, 1997a.

Syed Ejaz Ahmed. Improved r-estimation of regression coefficients. *Journal of Statistical Research*, 31(1):53–73, 1997b.

Syed Ejaz Ahmed. Improved estimation of the parameters of an autoregressive gaussian process under uncertain. *Applied Statistical Science, III: Papers in Honor of AK Md. E. Saleh*, page 97, 1998.

Syed Ejaz Ahmed. Shrinkage estimation of regression coefficients from censored data with multiple observations. In *Empirical Bayes and Likelihood Inference*, pages 103–120. Springer, 2001.

Syed Ejaz Ahmed. *Penalty, shrinkage and pretest strategies: variable selection and estimation.* Springer, 2014.

Syed Ejaz Ahmed and A. K. Basu. Least squares, preliminary test and stein-type estimation in general vector ar (p) models. *Statistica Neerlandica*, 54(1):47–66, 2000.

Syed Ejaz Ahmed and Saber Fallahpour. Shrinkage estimation strategy in quasi-likelihood models. *Statistics & Probability Letters*, 82(12):2170–2179, 2012.

Syed Ejaz Ahmed and A. K. Md. E. Saleh. Improved nonparametric estimation of location vectors in multivariate regression models. *Journal of Nonparametric Statistics*, 11(1-3): 51–78, 1999.

Syed Ejaz Ahmed and Christopher J. Nicol. An application of shrinkage estimation to the nonlinear regression model. *Computational Statistics & Data Analysis*, 56(11):3309–3321, 2012.

Syed Ejaz Ahmed and Eugene Afiamoah Opoku. Submodel selection and post-estimation of the linear mixed models. In *Proceedings of the Tenth International Conference on Management Science and Engineering Management*, pages 633–646. Springer, 2017.

Syed Ejaz Ahmed and S. M. Enayetur Raheem. Shrinkage and absolute penalty estimation in linear regression models. *Wiley Interdisciplinary Reviews: Computational Statistics*, 4 (6):541–553, 2012.

Syed Ejaz Ahmed and Bahadır Yüzbaşı. Big data analytics: integrating penalty strategies. *International Journal of Management Science and Engineering Management*, 11(2):105–115, 2016.

Syed Ejaz Ahmed and Bahadır Yüzbaşı. High dimensional data analysis: Integrating submodels. In *Big and Complex Data Analysis*, pages 285–304. Springer, 2017.

Syed Ejaz Ahmed, A. A. Hussein, and P. K. Sen. Risk comparison of some shrinkage m-estimators in linear models. *Journal of Nonparametric Statistics*, 18:401–415, 2006.

Syed Ejaz Ahmed, Kjell A. Doksum, S. Hossain, and Jinhong You. Shrinkage, pretest and absolute penalty estimators in partially linear models. *Australian & New Zealand Journal of Statistics*, 49(4):435–454, 2007.

Syed Ejaz Ahmed, Shakhawat Hossain, and Kjell A. Doksum. Lasso and shrinkage estimation in weibull censored regression models. *Journal of Statistical Planning and Inference*, 142(6):1273–1284, 2012.

Syed Ejaz Ahmed, A. A. Hussein, and M. Al-Momani. Efficient estimation for the conditional autoregressive model. *Journal of Statistical Computations and Simulation*, 85: 2569–2581, 2015.

Syed Ejaz Ahmed, Hwanwoo Kim, Gökhan Yıldırım, and Bahadır Yüzbaşı. High-dimensional regression under correlated design: An extensive simulation study. In *International Workshop on Matrices and Statistics*, pages 145–175. Springer, 2016.

Syed Ejaz Ahmed, Dursun Aydın, and Ersin Yılmaz. Kernel ridge estimator for the partially linear model under right-censored data. *Journal of the Iranian Statistical Society*, 20(1): 1–26, 2021.

John Aitchison and SD Silvey. Maximum-likelihood estimation of parameters subject to restraints. *The Annals of Mathematical Statistics*, 29(3):813–828, 1958.

Fikri Akdeniz and Hamza Erol. Mean squared error matrix comparisons of some biased estimators in linear regression. *Communications in Statistics-Theory and Methods*, 32 (12):2389–2413, 2003.

Marwan Al-Momani, Abdulkadir A. Hussein, and S. Ejaz Ahmed. Penalty and related estimation strategies in the spatial error model. *Statistica Neerlandica*, 71(1):4–30, 2017.

Saeed Aldahmani and Hongsheng Dai. Unbiased estimation for linear regression when $n < v$. *International Journal of Statistics and Probability*, 4(3), 2015.

Uri Alon, Naama Barkai, Daniel A. Notterman, Kurt Gish, Suzanne Ybarra, Daniel Mack, and Arnold J Levine. Broad patterns of gene expression revealed by clustering analysis of tumor and normal colon tissues probed by oligonucleotide arrays. *Proceedings of the National Academy of Sciences*, 96(12):6745–6750, 1999.

L. An, S. Nkurunziza, K. Y. Fung, D. Krewskia, and I. Luginaah. Shrinkage estimation in general linear models. *Computational Statistics and Data Analysis*, 53(7):2537–2549, 2009.

Mohammad Arashi, Mina Norouzirad, S. Ejaz Ahmed, and Bahadır Yüzbaşı. Rank-based liu regression. *Computational Statistics*, 33(3):1525–1561, 2018.

Mohammad Arashi, Yasin Asar, and Bahadır Yüzbaşı. Slasso: a scaled lasso for multicollinear situations. *Journal of Statistical Computation and Simulation*, 91(15):3170–3183, 2021.

Yasin Asar, S Ejaz Ahmed, and Bahadır Yüzbaşı. Efficient and improved estimation strategy in zero-inflated poisson regression models. In *International Conference on Management Science and Engineering Management*, pages 329–342. Springer, 2018.

Dursun Aydın, Bahadır Yüzbaşı, and S. Ejaz Ahmed. Modified ridge type estimator in partially linear regression models and numerical comparisons. *Journal of Computational and Theoretical Nanoscience*, 13(10):7040–7053, 2016.

R. Barro and J. W. Lee. *Dataset for a panel of 138 countries*, 1994. URL `http://www.nber.org/pub/barro.lee`. Discussion paper.

G. Bassett and R. Koenker. Asymptotic theory of least absolute error regression. *Journal of American Statistical Association*, 78:618–622, 1978.

Alexandre Belloni and Victor Chernozhukov. Least squares after model selection in high-dimensional sparse models. *Bernoulli*, 19(2):521–547, 2013.

David A. Belsley. Conditioning diagnostics. *Wiley StatsRef: Statistics Reference Online*, 2014.

Ernst R. Berndt. *The practice of econometrics: classic and contemporary*. Reading, Mass.; Don Mills, Ont.: Addison-Wesley Publishing Company, 1991.

Marianne Bertrand and Sendhil Mullainathan. Are emily and greg more employable than lakisha and jamal? a field experiment on labor market discrimination. *American economic review*, 94(4):991–1013, 2004.

Peter J. Bickel, Ya'acov Ritov, and Alexandre B. Tsybakov. Simultaneous analysis of lasso and dantzig selector. *The Annals of Statistics*, 37(4):1705–1732, 2009.

Peter Bühlmann and Sara Van De Geer. *Statistics for high-dimensional data: methods, theory and applications*. Springer Science & Business Media, 2011.

Peter Bühlmann, Markus Kalisch, and Lukas Meier. High-dimensional statistics with a view toward applications in biology. *Annual Review of Statistics and Its Application*, 1: 255–278, 2014.

A. Colin Cameron and Pravin K. Trivedi. *Microeconometrics: methods and applications*. Cambridge University Press, 2005.

M. Chaowagul, D. Phahuyuth, and N. Pooncharoen. Report on the socio-economic situation monitoring and evaluation, Kwae Noy dam under his majesty the king's initiation, budjet year 2015, 2015.

Sourav Chatterjee. High dimensional regression and matrix estimation without tuning parameters. *arXiv preprint arXiv:1510.07294*, 2015.

Scott Shaobing Chen, David L. Donoho, and Michael A. Saunders. Atomic decomposition by basis pursuit. *SIAM review*, 43(1):129–159, 2001.

Shaobing Chen and David Donoho. Basis pursuit. In *Proceedings of 1994 28th Asilomar Conference on Signals, Systems and Computers*, volume 1, pages 41–44. IEEE, 1994.

Izabela Chojnicka and Aleksander Wawer. Social language in autism spectrum disorder: A computational analysis of sentiment and linguistic abstraction. *PLoS One*, 15(3): e0229985, 2020.

Angela Unna Chukwu and Samuel Oluwafemi Oyamakin. On hyperbolic monomolecular growth model in height/diameter growth of pines. *British Journal of Mathematics & Computer Science*, 9:320–327, 2015.

Xiangqin Cui, J. T. Gene Hwang, Jing Qiu, Natalie J. Blades, and Gary A. Churchill. Improved statistical tests for differential gene expression by shrinking variance components estimates. *Biostatistics*, 6(1):59–75, 2005.

Roger R. Davidson and William E. Lever. The limiting distribution of the likelihood ratio statistic under a class of local alternatives. *Sankhyā: The Indian Journal of Statistics, Series A*, pages 209–224, 1970.

Annette J. Dobson and Adrian G. Barnett. *An introduction to generalized linear models.* Chapman and Hall/CRC, 2018.

David Donoho and Jiashun Jin. Higher criticism thresholding: Optimal feature selection when useful features are rare and weak. *Proceedings of the National Academy of Sciences*, 105(39):14790–14795, 2008.

Bradley Efron, Trevor Hastie, Iain Johnstone, and Robert Tibshirani. Least angle regression. *The Annals of Statistics*, 32(2):407–499, 2004.

Robert F. Engle, Clive W. J. Granger, John Rice, and Andrew Weiss. Semiparametric estimates of the relation between weather and electricity sales. *Journal of the American statistical Association*, 81(394):310–320, 1986.

Camila Epprecht, Dominique Guegan, Álvaro Veiga, and Joel Correa da Rosa. Variable selection and forecasting via automated methods for linear models: Lasso/adalasso and autometrics. *Communications in Statistics-Simulation and Computation*, 50(1):103–122, 2021.

R. L. Eubank. A note on smoothness priors and nonlinear regression. *Journal of the American Statistical Association*, 81(394):514–517, 1986.

Ludwig Fahrmeir and Heinz Kaufmann. Consistency and asymptotic normality of the maximum likelihood estimator in generalized linear models. *The Annals of Statistics*, 13 (1):342–368, 1985.

S. Fallahpour and Syed Ejaz Ahmed. Shrinkage estimation and variable selection in multiple regression models with random coefficient autoregressive errors. *Statistics Probability Letters*, 92:199–208, 2014.

Jianqing Fan and Runze Li. Variable selection via nonconcave penalized likelihood and its oracle properties. *Journal of the American statistical Association*, 96(456):1348–1360, 2001.

Fang Fang, Jiwei Zhao, S Ejaz Ahmed, and Annie Qu. A weak-signal-assisted procedure for variable selection and statistical inference with an informative subsample. *Biometrics*, 77 (3):996–1010, 2021.

Ronald Fisher. Linear discriminant analysis. *Ann. Eugenics*, 7:179, 1936.

LLdiko E. Frank and Jerome H. Friedman. A statistical view of some chemometrics regression tools. *Technometrics*, 35(2):109–135, 1993.

Jerome H. Friedman. Estimating functions of mixed ordinal and categorical variables using adaptive splines. Technical report, Stanford Univ CA Lab for Computational Statistics, 1991.

Wenjiang J. Fu. Penalized regressions: the bridge versus the lasso. *Journal of Computational and Graphical Statistics*, 7(3):397–416, 1998.

Xiaoli Gao and Syed Ejaz Ahmed. Efficient adaptive estimation strategies in high-dimensional partially linear regression models. *Perspectives on Big Data Analysis: Methodologies and Applications*, 622:61–73, 2014.

Xiaoli Gao, Syed Ejaz Ahmed, and Yang Feng. Post selection shrinkage estimation for high-dimensional data analysis. *Applied Stochastic Models in Business and Industry*, 33 (2):97–120, 2017a.

Xiaoli Gao, Syed Ejaz Ahmed, and Yang Feng. (rejoinder) post selection shrinkage estimation for high-dimensional data analysis. *Applied Stochastic Models in Business and Industry*, 33(2):97–120, 2017b.

Paul Geladi and Bruce R. Kowalski. Partial least-squares regression: a tutorial. *Analytica chimica acta*, 185:1–17, 1986.

Charles J. Geyer. On the asymptotics of convex stochastic optimization. *Unpublished manuscript*, 37, 1996.

Diane Galarneau Gibbons. A simulation study of some ridge estimators. *Journal of the American Statistical Association*, 76(373):131–139, 1981.

Richard Goldstein. Conditioning diagnostics: Collinearity and weak data in regression, 1993.

Peter J. Green and Bernard W. Silverman. *Nonparametric regression and generalized linear models: a roughness penalty approach.* CRC Press, 1993.

Keelin Greenlaw, Elena Szefer, Jinko Graham, Mary Lesperance, Farouk S. Nathoo, and Alzheimer's Disease Neuroimaging Initiative. A bayesian group sparse multi-task regression model for imaging genetics. *Bioinformatics*, 33(16):2513–2522, 2017.

Bruce E. Hansen. The risk of james–stein and lasso shrinkage. *Econometric Reviews*, 35 (8-10):1456–1470, 2016.

Trevor Hastie, Robert Tibshirani, Jerome H. Friedman, and Jerome H. Friedman. *The elements of statistical learning: data mining, inference, and prediction*, volume 2. Springer, 2009.

Trevor Hastie, Robert Tibshirani, and Martin Wainwright. Statistical learning with sparsity. *Monographs on Statistics and Applied Probability*, 143:143, 2015.

Ezra Hauer, Douglas W. Harwood, Forrest M. Council, and Michael S. Griffith. Estimating safety by the empirical bayes method: a tutorial. *Transportation Research Record*, 1784 (1):126–131, 2002.

Joseph M. Hilbe. *Logistic regression models.* Chapman and Hall/CRC, 2009.

Joseph M. Hilbe. *Negative binomial regression.* Cambridge University Press, 2011.

H. Ho. *Essays on housing markets.* PhD thesis, University of Toronto, 1997.

Arthur E. Hoerl and Robert W. Kennard. Ridge regression: Biased estimation for nonorthogonal problems. *Technometrics*, 12(1):55–67, 1970.

David W. Hosmer Jr, Stanley Lemeshow, and Rodney X. Sturdivant. *Applied logistic regression*, volume 398. John Wiley & Sons, 2013.

S. Hossain, Kjell A. Doksum, and S. Ejaz Ahmed. Positive shrinkage, improved pretest and absolute penalty estimators in partially linear models. *Linear Algebra and Its Applications*, 430(10):2749–2761, 2009.

S Hossain, Syed Ejaz Ahmed, Grace Y. Yi, and B. Chen. Shrinkage and pretest estimators for longitudinal data analysis under partially linear models. *Journal of Nonparametric Statistics*, 28(3):531–549, 2016.

Shakhawat Hossain and Ejaz Ahmed. Shrinkage and penalty estimators of a poisson regression model. *Australian & New Zealand Journal of Statistics*, 54(3):359–373, 2012.

Shakhawat Hossain and Syed Ejaz Ahmed. Shrinkage estimation and selection for a logistic regression model. *CRM Proceedings-Contemporary Mathematics*, 622:159–176, 2014.

Shakhawat Hossain, Syed Ejaz Ahmed, and Hatem A. Howlader. Model selection and parameter estimation of a multinomial logistic regression model. *Journal of Statistical Computation and Simulation*, 84(7):1412–1426, 2014.

Shakhawat Hossain, Syed Ejaz Ahmed, and Kjell A. Doksum. Shrinkage, pretest, and penalty estimators in generalized linear models. *Statistical Methodology*, 24:52–68, 2015.

Harold Hotelling. Analysis of a complex of statistical variables into principal components. *Journal of Educational Psychology*, 24(6):417, 1933.

Jian Huang, Shuangge Ma, and Cun-Hui Zhang. Adaptive lasso for sparse high-dimensional regression models. *Statistica Sinica*, pages 1603–1618, 2008.

C. M. Hurvish and C. L. Tsai. Model selection for least absolute deviations regression in small samples. *Statistics and Probability Letters*, 9:259–265, 1990.

Richard Arnold Johnson, Dean W. Wichern, et al. *Applied multivariate statistical analysis*, volume 6. Pearson London, UK, 2014.

George G. Judge and Mary Ellen Bock. *The statistical implications of pre-test and Stein-rule estimators in econometrics*, volume 25. North-Holland, 1978.

Azme Khamis et al. Nonlinear growth models for modeling oil palm yield growth. *Journal of Mathematics and Statistics*, 1(3):225–233, 2005.

Yongdai Kim, Hosik Choi, and Hee-Seok Oh. Smoothly clipped absolute deviation on high dimensions. *Journal of the American Statistical Association*, 103(484):1665–1673, 2008.

Keith Knight and Wenjiang Fu. Asymptotics for lasso-type estimators. *Annals of statistics*, pages 1356–1378, 2000.

Nan M. Laird and James H. Ware. Random-effects models for longitudinal data. *Biometrics*, pages 963–974, 1982.

Jerald F. Lawless. Negative binomial and mixed poisson regression. *The Canadian Journal of Statistics/La Revue Canadienne de Statistique*, pages 209–225, 1987.

J. F. Lawless and Kishore Singhal. Efficient screening of nonnormal regression models. *Biometrics*, pages 318–327, 1978.

Chenlei Leng, Yi Lin, and Grace Wahba. A note on the lasso and related procedures in model selection. *Statistica Sinica*, pages 1273–1284, 2006.

Yanming Li, Hyokyoung G. Hong, Syed Ejaz Ahmed, and Yi Li. Weak signals in high-dimensional regression: Detection, estimation and prediction. *Applied Stochastic Models in Business and Industry*, 35(2):283–298, 2019.

H. Liang and W. X. Song. Improved estimation in multiple linear regression models with measurement error and general constraint. *Journal Multivariate Analysis*, 100:726—741, 2009.

Hua Liang. Estimation in partially linear models and numerical comparisons. *Computational Statistics & Data Analysis*, 50(3):675–687, 2006.

Supranee Lisawadi, Muhammad Kashif Ali Shah, and Syed Ejaz Ahmed. Model selection and post estimation based on a pretest for logistic regression models. *Journal of Statistical Computation and Simulation*, 86(17):3495–3511, 2016.

Supranee Lisawadi, Syed Ejaz Ahmed, and Orawan Reangsephet. Post estimation and prediction strategies in negative binomial regression model. *International Journal of Modelling and Simulation*, 41(6):463–477, 2021.

Hanzhong Liu and Bin Yu. Asymptotic properties of lasso+ mls and lasso+ ridge in sparse high-dimensional linear regression. *Electronic Journal of Statistics*, 7:3124–3169, 2013.

Kejian Liu. A new class of blased estimate in linear regression. *Communications in Statistics-Theory and Methods*, 22(2):393–402, 1993.

Kejian Liu. Using liu-type estimator to combat collinearity. *Communications in Statistics-Theory and Methods*, 32(5):1009–1020, 2003.

NT Longford. Regression analysis of multilevel data with measurement error. *British Journal of Mathematical and Statistical Psychology*, 46(2):301–311, 1993.

Tao Lu, Ying Pan, Shyan-Yuan Kao, Cheng Li, Isaac Kohane, Jennifer Chan, and Bruce A. Yankner. Gene regulation and dna damage in the ageing human brain. *Nature*, 429(6994): 883–891, 2004.

J. MacQueen. Classification and analysis of multivariate observations. In *5th Berkeley Symp. Math. Statist. Probability*, pages 281–297, 1967.

Kristofer Månsson, B. M. Golam Kibria, and Ghazi Shukur. On liu estimators for the logit regression model. *Economic Modelling*, 29(4):1483–1488, 2012.

P. McCullagh and J. Nelder. Generalized linear models second edition chapman & hall, 1989.

Gary C. McDonald and Richard C. Schwing. Instabilities of regression estimates relating air pollution to mortality. *Technometrics*, 15(3):463–481, 1973.

Lukas Meier, Sara Van De Geer, and Peter Bühlmann. The group lasso for logistic regression. *Journal of the Royal Statistical Society: Series B (Statistical Methodology)*, 70(1):53–71, 2008.

Nicolai Meinshausen and Peter Bühlmann. High-dimensional graphs and variable selection with the lasso. *The Annals of Statistics*, 34(3):1436–1462, 2006.

D. C. Montgomery, E. A. Peck, and G. G. Vinig. *Introduction to Linear Regression Analysis (3nd ed)*. John Wiley & Sons, 2011.

Thomas A. Mroz. The sensitivity of an empirical model of married women's hours of work to economic and statistical assumptions. *Econometrica*, 55(4):765, Jul 1987. ISSN 0012-9682. doi: 10.2307/1911029. URL http://dx.doi.org/10.2307/1911029.

Keith E. Muller and Paul W. Stewart. *Linear model theory: univariate, multivariate, and mixed models.* John Wiley & Sons, 2006.

Raymond H. Myers et al. *Generalized linear models: with applications in engineering and the sciences.* John Wiley & Sons, 2nd edition, 2010.

Yunlong Nie, Eugene Opoku, Laila Yasmin, Yin Song, Jie Wang, Sidi Wu, Vanessa Scarapicchia, Jodie Gawryluk, Liangliang Wang, Jiguo Cao, et al. Spectral dynamic causal modelling of resting-state fmri: an exploratory study relating effective brain connectivity in the default mode network to genetics. *Statistical Applications in Genetics and Molecular Biology*, 19(3), 2020.

M Norouzirad, Mohammad Arashi, and Syed Ejaz Ahmed. Improved robust ridge m-estimation. *Journal of Statistical Computation and Simulation*, 87(18):3469–3490, 2017.

Eugene A. Opoku, Syed Ejaz Ahmed, and Farouk S. Nathoo. Sparse estimation strategies in linear mixed effect models for high-dimensional data application. *Entropy*, 23(10):1348, 2021.

Michael J. Panik. *Growth curve modeling: theory and applications.* John Wiley & Sons, 2014.

Mee Young Park and Trevor Hastie. L1-regularization path algorithm for generalized linear models. *Journal of the Royal Statistical Society: Series B (Statistical Methodology)*, 69 (4):659–677, 2007.

Trevor Park and George Casella. The bayesian lasso. *Journal of the American Statistical Association*, 103(482):681–686, 2008.

Karl Pearson. Principal components analysis. *The London, Edinburgh, and Dublin Philosophical Magazine and Journal of Science*, 6(2):559, 1901.

Siwaporn Phukongtong, Supranee Lisawadi, and S. Ejaz Ahmed. Penalty, post pretest and shrinkage strategies in a partially linear model. *Communications in Statistics-Simulation and Computation*, pages 1–22, 2020.

Janjira Piladaeng, Syed Ejaz Ahmed, and Supranee Lisawadi. Penalised, post-pretest, and post-shrinkage strategies in nonlinear growth models. *Australian & New Zealand Journal of Statistics*, 2022.

Christopher D. Powell et al. Models based on the mitscherlich equation for describing typical and atypical gas production profiles obtained from in vitro digestibility studies using equine faecal inoculum. *Animals*, 10(2):308, 2020.

A. Qu and P. Shi. Weak signal identification and inference in penalized model selection. *Annals of Statistics*, 2017, Vol. 45, No. 3, 1214–1253 DOI: 10.1214/16-AOS1482

Peter Radchenko and Gareth M. James. Improved variable selection with forward-lasso adaptive shrinkage. *The Annals of Applied Statistics*, 5(1):427–448, 2011.

SM Enayetur Raheem, Syed Ejaz Ahmed, and Kjell A. Doksum. Absolute penalty and shrinkage estimation in partially linear models. *Computational Statistics & Data Analysis*, 56(4):874–891, 2012.

C. Radhakrishna Rao. Inference on discriminant function coefficients, 1969. Inference on discriminant coefficients. In Essays in probability and statistics, Bose et al. (Eds), 587–602. North Carolina: University of North Carolina.

O. Reangsephet, S. Lisawadi, and Syed Ejaz Ahmed. Adaptive estimation strategies in gamma regression model. *Journal of Statistical Theory and Practice*, 14(1):1–27, 2020.

Orawan Reangsephet, Supranee Lisawadi, and Syed Ejaz Ahmed. Weak signals in high-dimensional poisson regression models. *Applied Modeling Techniques and Data Analysis 1: Computational Data Analysis Methods and Tools*, 7:203–219, 2021.

Mahdi Roozbeh. Robust ridge estimator in restricted semiparametric regression models. *Journal of Multivariate Analysis*, 147:127–144, 2016.

Frank Rosenblatt. The perceptron: a probabilistic model for information storage and organization in the brain. *Psychological Review*, 65(6):386, 1958.

Todd E. Scheetz, Kwang-Youn A. Kim, Ruth E. Swiderski, Alisdair R. Philp, Terry A. Braun, Kevin L. Knudtson, Anne M. Dorrance, Gerald F. DiBona, Jian Huang, Thomas L. Casavant, et al. Regulation of gene expression in the mammalian eye and its relevance to eye disease. *Proceedings of the National Academy of Sciences*, 103(39):14429–14434, 2006.

Michael G. Schimek. Estimation and inference in partially linear models with smoothing splines. *Journal of Statistical Planning and Inference*, 91(2):525–540, 2000.

Ahmed S. E. Estimation of regression coefficients in an exponential regression model with censored observations. *Journal of the Japan Statistical Society*, 29(1):55–64, 1999.

Paul Speckman. Kernel smoothing in partial linear models. *Journal of the Royal Statistical Society: Series B (Methodological)*, 50(3):413–436, 1988.

Thomas A. Stamey, John N. Kabalin, John E. McNeal, Iain M. Johnstone, Fuad Freiha, Elise A. Redwine, and Norman Yang. Prostate specific antigen in the diagnosis and treatment of adenocarcinoma of the prostate. ii. radical prostatectomy treated patients. *The Journal of Urology*, 141(5):1076–1083, 1989.

RDC Team. R programming. *R Development Core Team*, 2021.

Robert Tibshirani. Regression shrinkage and selection via the lasso. *Journal of the Royal Statistical Society: Series B (Methodological)*, 58(1):267–288, 1996.

Minh-ngoc Tran. The loss rank criterion for variable selection in linear regression analysis. *Scandinavian Journal of Statistics*, 38(3):466–479, 2011.

J. W. R. Twisk, H. C. G. Kemper, and G. J. Mellenbergh. Longitudinal development of lipoprotein levels in males and females aged 12–28 years: the amsterdam growth and health study. *International Journal of Epidemiology*, 24(1):69–77, 1995.

Fatih Uckardes. A modified mitscherlich model and its degradation kinetics equations. *Archives Animal Breeding*, 56(1):1005–1013, 2013.

H. Wang, G. Li, and G. Jiang. Robust regression shrinkage and consistent variable selection through the lad-lasso. *Journal of Business & Economic Statistics*, 25:347–355, 2007.

Hansheng Wang and Chenlei Leng. Unified lasso estimation by least squares approximation. *Journal of the American Statistical Association*, 102(479):1039–1048, 2007.

Mingqiu Wang, Lixin Song, and Xiaoguang Wang. Bridge estimation for generalized linear models with a diverging number of parameters. *Statistics & Probability Letters*, 80(21-22): 1584–1596, 2010.

Lingtao Wu and Dominique Lord. Examining the influence of link function misspecification in conventional regression models for developing crash modification factors. *Accident Analysis & Prevention*, 102:123–135, 2017.

Huiliang Xie and Jian Huang. Scad-penalized regression in high-dimensional partially linear models. *The Annals of Statistics*, 37(2):673–696, 2009.

Ersin Yilmaz, Bahadır Yüzbaşı, and Dursun Aydin. Choice of smoothing parameter for kernel type ridge estimators in semiparametric regression models. *REVSTAT-Statistical Journal*, 19(1):47–69, 2021.

Ersin Yılmaz, Dursun Aydın, and S. Ejaz Ahmed. Regression with right-censored high-dimensional data: An application with different imputation techniques: 10.48129/kjs. splml. 18961. *Kuwait Journal of Science*, 2022.

Ming Yuan and Yi Lin. Model selection and estimation in regression with grouped variables. *Journal of the Royal Statistical Society: Series B (Statistical Methodology)*, 68(1):49–67, 2006.

Bahadır Yüzbaşı and S Ejaz Ahmed. Ridge type shrinkage estimation of seemingly unrelated regressions and analytics of economic and financial data from "fragile five" countries. *Journal of Risk and Financial Management*, 13(6):131, 2020.

Bahadır Yüzbaşı and Syed Ejaz Ahmed. Shrinkage ridge regression estimators in high-dimensional linear models. In *Proceedings of the Ninth International Conference on Management Science and Engineering Management*, pages 793–807. Springer, 2015.

Bahadır Yüzbaşı and Syed Ejaz Ahmed. Shrinkage and penalized estimation in semi-parametric models with multicollinear data. *Journal of Statistical Computation and Simulation*, 86(17):3543–3561, 2016.

Bahadır Yüzbaşı, Syed Ejaz Ahmed, and Mehmet Güngör. Improved penalty strategies in linear regression models. *REVSTAT-Statistical Journal*, 15(2):251–276, 2017a.

Bahadır Yüzbaşı, Mohammad Arashi, and Syed Ejaz Ahmed. Big data analysis using shrinkage strategies. *arXiv preprint arXiv:1704.05074*, 2017b.

Bahadir Yüzbaşi, Syed Ejaz Ahmed, and Yasin Asar. L1 correlation-based penalty in high-dimensional quantile regression. In *2018 4th International Conference on Big Data and Information Analytics (BigDIA)*, pages 1–5. IEEE, 2018.

Bahadır Yüzbaşı, Yasin Aşar, Şamil Şık, and Ahmet Demiralp. Improving estimations in quantile regression model with autoregressive errors. *Thermal Science*, 22(1):S97–S107, 2018.

Bahadır Yüzbaşı, Syed Ejaz Ahmed, Mohammad Arashi, and Mina Norouzirad. Lad, lasso and related strategies in regression models. In *International Conference on Management Science and Engineering Management*, pages 429–444. Springer, 2019.

Bahadır Yüzbaşı, Syed Ejaz Ahmed, and Dursun Aydın. Ridge-type pretest and shrinkage estimations in partially linear models. *Statistical Papers*, 61(2):869–898, 2020.

Bahadır Yüzbaşı, Mohammad Arashi, and S. Ejaz Ahmed. Shrinkage estimation strategies in generalised ridge regression models: Low/high-dimension regime. *International Statistical Review*, 88(1):229–251, 2020.

Bahadır Yüzbaşı, Mohammad Arashi, and Fikri Akdeniz. Penalized regression via the restricted bridge estimator. *Soft Computing*, 25(13):8401–8416, 2021.

Bahadır Yüzbaşı, Yasin Asar, and Syed Ejaz Ahmed. Liu-type shrinkage estimations in linear models. *Statistics*, 56(2):396–420, 2022.

Hossein Zareamoghaddam, Syed E Ahmed, and Serge B Provost. Shrinkage estimation applied to a semi-nonparametric regression model. *The International Journal of Biostatistics*, 17(1):23–38, 2021.

Cun Hui Zhang. Penalized linear unbiased selection. *Department of Statistics and Bioinformatics, Rutgers University*, 3(2010):894–942, 2007.

Cun-Hui Zhang. Nearly unbiased variable selection under minimax concave penalty. *The Annals of Statistics*, 38(2):894–942, 2010.

Cun-Hui Zhang and Stephanie S. Zhang. Confidence intervals for low dimensional parameters in high dimensional linear models. *Journal of the Royal Statistical Society: Series B (Statistical Methodology)*, 76(1):217–242, 2014.

Jiwei Zhao, Yang Yang, and Yang Ning. Penalized pairwise pseudo likelihood for variable selection with nonignorable missing data. *Statistica Sinica*, 28(4):2125–2148, 2018.

Peng Zhao and Bin Yu. On model selection consistency of lasso. *The Journal of Machine Learning Research*, 7:2541–2563, 2006.

Ji Zhu and Trevor Hastie. Classification of gene microarrays by penalized logistic regression. *Biostatistics*, 5(3):427–443, 2004.

Hui Zou. The adaptive lasso and its oracle properties. *Journal of the American Statistical Association*, 101(476):1418–1429, 2006.

Hui Zou and Trevor Hastie. Regularization and variable selection via the elastic net. *Journal of the Royal Statistical Society: Series B (Statistical Methodology)*, 67(2):301–320, 2005.

Yajie Zou, Lingtao Wu, and Dominique Lord. Modeling over-dispersed crash data with a long tail: examining the accuracy of the dispersion parameter in negative binomial models. *Analytic Methods in Accident Research*, 5:1–16, 2015.

Verena Zuber and Korbinian Strimmer. High-dimensional regression and variable selection using car scores. *Statistical Applications in Genetics and Molecular Biology*, 10(1), 2011.

Index